BURYANOVS GLOBAL LAW FORUM

Anthropology of the Planetary Human-Centered Civilization 4.0

A Summary of Thirty Years of Global Research

AVERTING CIVILIZATIONAL ARMAGEDDON

In Search of a Way Out of the Geopolitical Deadlock

Worldview Liberty and Public-Law Worldview Neutrality in an Age of Global Existential Threats

SERGEY A. BURYANOV

Geneva, New York, Moscow, Vatican City, Tehran, Riyadh, Jerusalem, Beijing, New Delhi, Bangkok

United Humans Press, 2026

Preliminary Reviews of Averting Civilizational Armageddon

"Sergey A. Buryanov's monograph represents a timely and vital contribution to the global discussion on human development in the 21st century. The author takes a pioneering and methodologically robust step by directly linking the challenges of rapid neurotechnological advancement and artificial intelligence with the fundamental human right to freedom of conscience. Moving beyond classical legal frameworks, Buryanov sets out a new humanitarian paradigm. This conceptual shift deeply resonates with the strategic approaches and core values embedded within the United Nations system. This work is essential reading for anyone involved in shaping the boundaries of a secure digital future."

— Tatyana P. Minchenko, Doctor of Philosophy, Professor of Philosophy of Religion and Freedom of Conscience, Institute of Greek Language and Center for the Study of Orthodox Culture, Greece

"This is a profound, large-scale, and courageously executed work that offers a solid philosophical and legal foundation for overcoming the deep socio-cultural and technological divide currently fracturing modern humanity. Sergey A. Buryanov, Candidate of Legal Sciences and Associate Professor, convincingly and rigorously argues that the very survival of our civilization in the digital age depends directly on the uncompromising protection of inner human freedom and the autonomy of the individual mind against new forms of digital control. The author provides not merely a diagnosis of the global crisis, but a mature strategy for civilizational preservation."

— Alexander N. Chumakov, Doctor of Philosophy, Professor of Global Studies and Social Philosophy, Lomonosov Moscow State University, Russia

"This monograph offers the academic community and practitioners a much-needed, highly strategic response to the deepening systemic crisis of international law. The universal principle of the worldview neutrality of states, proposed by the author, rises far above pure theoretical abstraction. Buryanov delivers a viable, meticulously elaborated legal mechanism capable of steering international relations out of the present geopolitical deadlock. Under the current conditions of existential, technogenic threats, this concept opens up a realistic pathway toward preserving the domain of human freedom."

— Insur Z. Farkhutdinov, Doctor of Legal Sciences, Editor-in-Chief of the Eurasian Law Journal, Russia

BURYANOVS GLOBAL LAW FORUM

Anthropology of the Planetary Human-Centered Civilization 4.0

A Summary of Thirty Years of Global Research

AVERTING CIVILIZATIONAL ARMAGEDDON

BURYANOVS GLOBAL LAW FORUM

Anthropology of the Planetary Human-Centered Civilization 4.0

A Summary of Thirty Years of Global Research

AVERTING CIVILIZATIONAL ARMAGEDDON

In Search of a Way Out of the Geopolitical Deadlock

Worldview Liberty and Public-Law Worldview Neutrality
in an Age of Global Existential Threats

SERGEY A. BURYANOV

United Humans Press

Geneva, New York, Moscow, Vatican City, Tehran, Riyadh, Jerusalem, Beijing, New Delhi, Bangkok

2026

Averting Civilizational Armageddon: In Search of a Way Out of the Geopolitical Deadlock. Worldview Liberty and Public-Law Worldview Neutrality in an Age of Global Existential Threats
By Sergey A. Buryanov
Published by United Humans Press, Geneva, New York, Moscow, Vatican City, Tehran, Riyadh, Jerusalem, Beijing, New Delhi, Bangkok
First Edition: 2026
Library of Congress Cataloging-in-Publication Data
Buryanov, S. A.
Averting Civilizational Armageddon: In Search of a Way Out of the Geopolitical Deadlock. Worldview Liberty and Public-Law Worldview Neutrality in an Age of Global Existential Threats / Sergey A. Buryanov. - Geneva, New York, Moscow, Vatican City, Tehran, Riyadh, Jerusalem, Beijing, New Delhi, Bangkok: United Humans Press, 2026. - 570 p.
ISBN: 979-8-9990219-2-2 (Paperback / Hardcover)
Abstract: This monograph substantiates a new paradigm for the survival of humanity under the conditions of a civilizational geopolitical deadlock and unprecedented technological pressure on individual consciousness. It is demonstrated that systemic control over the worldview sphere and the manipulation of the mass consciousness of the majority of the global population constitute a basic and indispensable condition for unleashing wars in the interests of narrow elite groups since without total informational and cognitive dictate a forceful geopolitical confrontation for the benefit of a small fraction of humanity would become drastically less sustainable.

The author argues that a decisive path for preventing civilizational Armageddon lies in the transition to a new model of world order based on the unconditional international recognition of freedom in the formation of one's worldview guaranteed by public-law worldview neutrality. The rapid development of neurotechnologies, artificial intelligence, and facial recognition systems creates an existential threat of total control over the individual mind under which conditions this edition presents for the first time the draft texts of a Global Declaration, a Global Convention, and a Model Law aimed at the proactive protection of cognitive liberty and fundamental human rights.

While Yuval Noah Harari and Mo Gawdat look at the existential challenges of artificial intelligence predominantly from historical and technological perspectives this monograph provides a needed international legal framework to secure freedom of the mind. Developing the critical political-economic analysis of techno-feudalism presented in the works of Yanis Varoufakis and Cédric Durand the author substantiates public-law worldview neutrality as a vital legal mechanism to dismantle digital fiefdoms. Steering away from the neo-reactionary dystopias of Silicon Valley technocracy which advocates for corporate absolutism and the erosion of fundamental human rights Buryanov rejects this technocratic determinism and presents a powerful democratic and human-centered alternative. By transforming existential warnings about technologies into concrete and actionable legal instruments the work establishes a robust framework for preserving individual autonomy against total digital threats.
Keywords: Cognitive liberty; worldview freedom; worldview liberty; AI ethics; algorithmic manipulation; digital human rights; existential threats; civilizational Armageddon; ban on emotion recognition; global governance; state neutrality; regulation of tech giants; mental integrity; digital sovereignty of the individual; algorithmic neutrality; future of humanity; AI safety.

Publisher Information:
United Humans Press
Geneva, New York, Moscow, Vatican City, Tehran, Riyadh, Jerusalem, Beijing, New Delhi, Bangkok
www.maxlaw.tilda.ws/globalization4ir
Printed in the United States of America / United Kingdom / Germany
This book is printed on acid-free paper.

Dedicated to the blessed memory

of my parents

A PERSONAL NOTE TO THE AUTHOR ON HIS 60TH YEAR

To Sergey A. Buryanov - my father, a rare legal scholar, teacher, and builder of a legal language for human freedom.

This edition is prepared as a gesture of gratitude to a brilliant legal thinker whose work has been devoted to one of the most difficult questions of modern civilization: how to protect the inner freedom of the human being when states, ideologies, markets, and technologies struggle for power over consciousness.

Over more than three decades, Sergey A. Buryanov has combined constitutional law, international law, human rights, global studies, education, and the philosophy of freedom into a single intellectual line. His scholarship shows rare continuity: from freedom of conscience and worldview neutrality to cognitive liberty, digital globalization, and the legal foundations of a human-centered civilization.

His research anticipated several central debates of our time: worldview neutrality, cognitive liberty, digital human rights, protection of consciousness, and the need for a human-centered legal order in an age of global risks.

With gratitude and love,

Maxim S. Buryanov

ACKNOWLEDGEMENTS

The author expresses his profound gratitude to Maxim Buryanov and Rimma Buryanova for their invaluable assistance in the preparation and writing of this book.

The author also expresses his deepest gratitude to Master of Laws (LL.M. Maxim S. Buryanov) for his invaluable contribution to the development and drafting of the international legal instruments presented in the Appendices, specifically: the Draft Global Declaration on the Right to Liberty of Worldview Choice and the Principle of Worldview Neutrality of Subjects of Public International Law in Analog and Digital Environments; the Draft Global Convention on the Right to Liberty of Worldview Choice and the Principle of Worldview Neutrality of Subjects of Public International Law in Analog and Digital Environments; and the Draft Model Law on the Right to Liberty of Worldview Choice and the Principle of Worldview Neutrality of Subjects of Public Law in Analog and Digital Environments.

The author also extends his sincere appreciation to the official reviewers of this monograph: Dr. Tatyana P. Minchenko, Dr. Alexander N. Chumakov, and Dr. Insur Z. Farkhutdinov. Their profound expertise, rigorous academic insights, and constructive feedback have been instrumental in refining the core concepts and international legal arguments presented in this work.

CONTENTS

READER'S MAP
How to Read This Book

Power has always searched for the place from which a human being can be governed most deeply. At one time it seized land; later it controlled temples, schools, printing presses, factories, parliaments, screens, and archives. In the digital age, the frontier has become intimate: attention, emotion, belief, identity, memory, and the in-visible architecture in which a person learns what to fear, what to desire, and what to call truth.

The question of this book

Who has the right to shape the human picture of the world: the person, the state, the platform, the market, the algorithm, the church, the party, or a legal order built around human dignity? This question may sound philosophical. In the age of artificial intelligence, neurotechnologies, biometric governance, predictive systems, and digital platforms, it becomes a practical legal question. It concerns courts, schools, universities, parliaments, international organizations, public services, social networks, and ev-ery private space in which a human being still tries to think freely.

Why now

The timetable of technology has outrun the grammar of law. Systems capable of measuring attention, predicting behaviour, classifying emotions, steering information flows, and personalizing influence already shape the conditions of public life. Meanwhile, many legal systems still speak as if freedom of conscience were limited to the old conflict between religion and the state. This book insists on a wider horizon: free-dom of conscience must be reinterpreted as liberty of worldview choice, and pub-lic-law worldview neutrality must become one of the legal conditions of survival in the age of civilizational risks.

Why This Book Matters Beyond Law

The question beyond professional law

This book is not addressed only to lawyers. It is written for everyone who is trying to understand what becomes of human freedom when consciousness itself turns into a field of technological, political, ideological, religious, and commercial influence. It speaks to readers of Yuval Noah Harari, Yanis Varoufakis, Curtis Yarvin, Robert Sapolsky, Nassim Nicholas Taleb, Marshall Rosenberg, Rick Rubin, and to all those who search for the conditions of a creative, constructive, nonviolent society.

From warning to legal protection

Harari warned that artificial intelligence may learn to hack the human being. Buryanov asks the next question, already legal and institutional: what could protect the human mind when the hacking of the human being becomes a method of power? Varoufakis describes techno-feudalism: a world in which platforms govern attention, data, access, and dependence. This book moves from diagnosis to protection. It shows how law can defend the person against platform power, algorithmic manipulation, biometric control, neurotechnological pressure, and the capture of worldview choice.

Against post-democratic temptation

Curtis Yarvin and the post-democratic imagination offer another future: corporate sovereignty, network states, private rule, and managed human populations. Buryanov argues the opposite. The answer to digital chaos must not be a new sovereign, a more convenient empire, or a sterile cage. It must be a human-centered legal order in which neither the state, nor the corporation, nor the platform, nor religion, nor ideology, nor the algorithm may own the inner world of the person.

The core legal idea

The key idea of the book is public-law worldview neutrality. Public authority must not impose, privilege, suppress, or rank worldviews. Its duty is to protect the equal freedom of every human being to form, preserve, revise, and express their own understanding of life. This principle carries the classical right to freedom of con-science into the age of artificial intelligence, biometric identification, emotion recog-nition, neurotechnologies, and instruments of cognitive influence.

Freedom, creativity, and anti-fragility

The book also adds a legal dimension to discussions about free will, behaviour, creativity, and neuroscience. If Sapolsky asks how free the human being really is, and Rubin writes about creativity as a way of being, Buryanov asks what legal conditions allow freedom and creativity to remain real human possibilities. In Taleb's language, a society that protects worldview liberty becomes more anti-fragile: it can learn, adapt, and renew itself without forcing human beings into one official picture of reality.

A bridge toward a nonviolent civilization

This is why the book also resonates with the ethics of nonviolent communication associated with Marshall Rosenberg. Worldview liberty and public-law worldview neutrality do not erase differences. They create the legal space in which disagreement can exist without dehumanization, and in which the false division between "ours" and "theirs" can be overcome through dignity, legal certainty, and respect for the inner life of every person.

Who should read it

This book is written for several circles of readers. Lawyers and judges will find a proposal for renewing the language of freedom of conscience and legal certainty. Diplomats, public officials, and international civil servants will find a framework for thinking about global security, sustainable development, and human rights beyond the usual geopolitical vocabulary. Scholars, teachers, and students will find a long intellectual route from religious tolerance to worldview liberty. Experts in artificial intelligence, digital governance, and neurotechnologies will find a legal warning about the capture of consciousness. Parents and citizens will find a simpler concern underneath the theory: how to keep the inner world of the person free when the surrounding envi-ronment becomes programmable.

How the argument unfolds

The Prologue gives the book its civilizational horizon. The Introduction places the problem within global risks, international law, sustainable development, artificial intelligence, and the return of force in world politics. Parts I and II reconstruct the history and legal foundations of freedom of conscience. Parts III and IV move the argument to the planetary and international legal levels. Part V reads Russia as a case, a symptom, and a warning. Part VI and the concluding materials turn diagnosis into legal architecture.

From doctrine to legal architecture

The Appendices belong to the core of the book. The draft Global Declaration, Global Convention, explanatory note, commentary, and model law translate the cen-tral idea into normative language: the freedom to form one's own worldview deserves legal protection in analog and digital environments. The book therefore moves from history to doctrine, from doctrine to institutional design, and from institutional design to a possible architecture of human-centered civilization 4.0.

For the international reader

For many years, Sergey A. Buryanov's work circulated mainly within Russian and Eurasian academic spaces. This English edition gives the international reader an opportunity to discover a body of scholarship that anticipated several debates now becoming urgent worldwide: cognitive liberty, public-law neutrality, digital human rights, algorithmic power, legal certainty, and the defense of consciousness in the age of intelligent systems.

The practical promise

Read this book not only as a history of freedom of conscience, but as a map of the next legal frontier: the protection of human interiority in a programmable civilization. Its practical promise is simple: a society can become technologically advanced and re-main humane only if power is legally prevented from owning the conditions under which a person thinks, believes, creates, doubts, and chooses.

CORE CONCEPTS

A Short Map of the Book's Language

Freedom of conscience: the legal and moral capacity of the person to live from an inner source of conviction, rather than from imposed doctrine, administrative pressure, fear, or engineered conformity.

Liberty of worldview choice: the freedom to form, choose, preserve, revise, and express one's own picture of the world, including religious, philosophical, ethical, scientific, cultural, and civic systems of meaning.

Worldview liberty: the contemporary legal form of freedom of conscience, expanded beyond the narrow framework of religion and directed toward the whole field in which a person seeks truth, meaning, identity, and moral orientation.

Worldview sphere: the field in which convictions, doubts, values, symbols, identities, beliefs, knowledge, faith, and moral orientation are formed and expressed.

Public-law worldview neutrality: the duty of states, public authorities, educational institutions, armed forces, courts, and other public-law actors to guarantee equal legal space for worldviews without imposing, privileging, suppressing, or ranking them.

International-law worldview neutrality: the extension of this principle to subjects of public international law, including states and international organizations, so that global cooperation rests on human rights rather than ideological ownership of truth.

Worldview sovereignty: the protected legal status of the inner world of the person: convictions, doubts, beliefs, values, moral search, and the ability to change one's position without coercion.

Cognitive liberty: the freedom of the mind from ideological coercion, algorithmic manipulation, emotional surveillance, biometric psychoanalysis, and neurotechnological intervention.

Mental integrity: the inviolability of the psychological and cognitive sphere of the human being, including protection against technologies that extract, classify, or exploit internal states.

Mental sovereignty: the civilizational boundary that marks consciousness as a territory of human dignity and legal protection, rather than a resource for platforms, states, or predictive systems.

Forum internum: the inner domain of thought, conscience, belief, doubt, and moral search; in this book, it becomes one of the decisive legal frontiers of the digital age.

Public authority over truth: the dangerous claim by a state, institution, ideology, platform, or algorithmic system to decide which picture of the world is legitimate for everyone.

Algorithmic dictate: a form of soft power in which recommendation systems, ranking, personalization, scoring, and automated moderation shape the visible world before the person can freely judge it.

Algorithmic manipulation: the hidden steering of attention, emotion, preference, belief, and behaviour through digital architectures that appear neutral while directing the conditions of choice.

Technological hacking of consciousness: the transition from influencing external behaviour to engineering the conditions in which attention, emotion, belief, loyalty, and worldview are formed.

Emotion recognition: a technology that claims to read internal states from faces, voices, bodies, or behaviour and therefore threatens privacy, dignity, equality, and the legal opacity of the inner world.

Biometric psychoanalysis: the use of biometric data to infer psychological, emotional, ideological, or behavioural characteristics of a person for evaluation, prediction, control, or exclusion.

Neurotechnological intervention: the use of technologies connected with the brain, nervous system, or cognitive processes in ways that can assist, treat, enhance, monitor, influence, or potentially govern the human mind.

Algorithmic silence: the right of the person to spaces where thought can emerge without permanent prediction, nudging, surveillance, profiling, or automated correction.

Worldview discrimination: distinction, exclusion, preference, stigmatization, or unequal treatment based on a person's worldview, convictions, religion, non-religion, philosophical position, or lawful search for meaning.

Digital Leviathan: the concentration of state, corporate, data, and platform power into infrastructures capable of governing people through identity, access, attention, and informational environment.

Digital fiefdoms: platform domains in which access, visibility, income, communication, and dependency are governed by private infrastructures that increasingly behave like quasi-sovereign powers.

Legal certainty: the requirement that key legal concepts be clear, stable, and applicable without arbitrary discretion, especially when they concern freedom, conscience, religion, belief, security, and extremism.

Positive security: a legal and social model in which stability grows through rights, trust, education, dignity, and inclusion, rather than through fear, ideological pressure, and permanent emergency.

Global law 4.0: a prospective legal horizon in which international law, human rights, digital governance, education, and sustainable development are reorganized around the survival and dignity of the human being.

Human-centered civilization 4.0: a model of technological and legal development in which artificial intelligence, platforms, neurotechnologies, and global governance serve human freedom, pluralism, sustainability, and dignity.

Civilizational Armageddon: the collapse of the legal, moral, and political conditions under which humanity can remain a community of free persons rather than a managed mass of predictable behavior.

Civilizational security code: a set of legal principles and guarantees capable of protecting human dignity, worldview liberty, cognitive freedom, and peaceful coexistence under conditions of global technological acceleration.

Together, these concepts create the book's inner grammar. They carry the classical language of freedom of conscience into the age of AI, biometric governance, digital platforms, neurotechnologies, and global existential risk.

A THIRTY-YEAR INTELLECTUAL TIMELINE

Sergey A. Buryanov's Legal Path

The present book grows out of more than three decades of legal scholarship, civic expertise, public teaching, and academic work. Sergey A. Buryanov is the author of more than twenty books and the organizer of more than one hundred scientific conferences. His academic profile includes 416 scientific publications, including 389 publications indexed in eLIBRARY.ru and 286 publications indexed in the Russian Science Citation Index. His body of work includes more than fifty monographs and teaching publications.

His RSCI h-index is 39; his Scopus h-index is 3; his h-index in the RSCI core is 3. He has published 82 articles in peer-reviewed scientific journals included in the Higher Attestation Commission list, including 16 over the last three years, and 12 articles in Scopus or Web of Science, including 9 over the last three years. These figures reflect a rare continuity of research labour and a substantial academic corpus.

1996-2005. The beginning of the legal path and the first formulation of a new paradigm.

The early research trajectory formed around freedom of conscience, state-confessional relations, human rights, and the renewal of legal regulation. During this period, freedom of conscience began to move beyond a narrow religious category and toward a broader doctrine of worldview freedom.

2007-2012. Human rights, tolerance, discrimination, and the secular state.

Works prepared with the Russian Helsinki Group and other scholars deepened the analysis of xenophobia, intolerance, discrimination, freedom of conscience, and state secularity. This period strengthened the human-rights foundation of the author's doctrine and connected legal theory with civic expertise.

2010-2015. International presentation and the global horizon.

During this period, the author's ideas entered wider international contexts, including a working mission to New York where he presented his research and legal vision. The question of freedom of conscience, once treated mainly as a national or regional problem, began to appear as part of a larger conversation about global order, human rights, and the conditions of peaceful coexistence.

2015-2020. From national legal problems to global processes.

The research agenda expanded toward globalization, international law, global education, sustainable development, and the idea of worldview neutrality as a condition for a more adequate world order. The national problem became a planetary legal question.

2021-2024. Digital globalization, scientific school, and the human-centered turn.

The research line entered the field of digital globalization, new threats to global security, digital inequality, digital human rights, and the search for an adequate legal order in the age of technological acceleration. Around this intellectual line, a scientific school becomes visible. Students and younger colleagues move into international expert spaces, including UN-related processes and Global Shapers communities; publish books on digital rights; and bring these ideas into debates on digital governance, artificial intelligence, and human-centered transformation. Some of these works appear with forewords and support from figures associated with the origins and architecture of the Internet. The intellectual school developing around Buryanov's work shows that his doctrine belongs to a living intellectual tradition, capable of speaking to the next generation.

2026. Averting Civilizational Armageddon.

This monograph gathers the author's long-standing work into a civilizational proposal: worldview liberty and public-law worldview neutrality must become part of the legal architecture needed to protect cognitive liberty, mental sovereignty, and the future of humanity. At this historical moment, after decades of work largely known within Eurasian academic and human-rights contexts, Sergey A. Buryanov's voice can and should be heard by a global readership.

Across these stages, the reader meets a brilliant and unusually consistent scholar who connects constitutional law, international law, human rights, education, global studies, and the future of digital civilization into one coherent legal worldview. His work gives language to a problem that many societies are only beginning to feel: the defense of the inner human world under conditions of ideological, technological, and geopolitical pressure.

ACADEMIC PROFILE AT A GLANCE

Scientific identifiers: SPIN code 2317-4485; AuthorID 201001.

Teaching and academic programs developed or taught include International Public Law, Geopolitics and Global Studies, Tolerance, Jurisprudence, Freedom of Conscience, Religion and Belief, Constitutional Rights of the Child and Freedom of Conscience, Current Problems in the Realization of Freedom of Conscience, the Principle of Freedom of Conscience in Education, Contemporary Society and the Human Being, Advocacy and the Fundamentals of Legal Practice, Contemporary Problems of Legal Science, Fundamentals of Eurasian Integration, Educational Law, and Theory of State and Law.

His professional development includes advanced training in information technologies for university teaching, pedagogy in professional education, human rights and mediation teaching methods, inclusive education, occupational safety, first aid, practical approaches to the protection of minors, current problems of global studies, and the methodology of teaching legal disciplines in modern higher education.

Among his distinctions are the prize and diploma for winning the Moscow City University competition of scientific and pedagogical workers in the category Best Monograph in 2010; the Russian Helsinki Group award in the field of human rights protection for expert and scientific activity in human rights in 2010; and a letter of gratitude from the dean of the Faculty of Law for his creative work in organizing the student scientific circle Current Issues of Constitutional and International Law in the Age of Globalization.

This academic profile records a sustained intellectual vocation: the defense of human freedom through law, education, civic responsibility, and global legal thinking.

AUTHOR'S NOTE

The Result of a Thirty-Year Search and a Call to Action

Dear Reader,

You are holding in your hands more than a book. It is the result of thirty years of research into the deep mechanisms of human freedom and state order, and at the same time a blueprint for safeguarding the human mind amid the turbulence of the digital age.

For three decades, I have watched the fabric of our civilization change. Humanity has travelled a long road: from hopes for the end of wars, the realization of human rights, and universal digital openness to a geopolitical deadlock and the emergence of invisible "algorithmic cages." When I began this path, the threats to worldview choice seemed local and surmountable. Over these years, however, a tectonic shift has occurred.

We have moved from censorship of the printed word to the technological hacking of the human soul. Today, battles for territory have been replaced by battles for meaning, and the weapons are not only missiles, but also algorithms that know more about you than you know about yourself. AI systems are capable of recognizing your emotions and "correcting" your beliefs in real time.

Why are these projects appearing precisely now?

Because there is no more time left for endless debate. Observing the evolution of international law and digital technologies, I have come to the conclusion that we have reached an existential threshold. States have ceased to be arbiters and have become architects of our beliefs. Digital giants have turned our weaknesses into commodities and our worldview into a product of software configuration. In schools and universities, emotion-recognition systems are beginning to dictate "correct" reactions, killing the very capacity for doubt.

This is the road to civilizational Armageddon: a world in which we are formally free, but in fact our thoughts are dictated by code.

What will you find inside?

In this book, I do not confine myself to criticism. The texts of the Global Declaration and the Global Convention contained in these pages concentrate the full experience of my legal and philosophical analysis.

This is not a utopia, but a calibrated and systemic response to civilizational Armageddon. It is also a legal shield that firmly prohibits emotion recognition and other forms of technological espionage against the human soul. Ultimately, it is a foundational norm of the future, obliging states and global corporations to recognize the inner world of the human being as a forbidden zone for external control.

I propose worldview neutrality as a new global normative standard. The state and technology should be a transparent pane of glass through which you look at the world, not a coloured lens imposed upon you. State worldview neutrality is not a luxury. It is a key condition for the survival of civilization in the age of AI.

This book is written for those who feel that "digital echo chambers" are becoming too narrow. It is for parents, teachers, lawyers, and everyone who wants to preserve the sovereignty of their own mind and the future of humanity. Thirty years of research have led me to this truth, and today I pass this instrument on to you.

Protect your mind. Preserve the future. As long as we remain capable of making a choice that is not dictated by an algorithm, we still have a future. The battle for the right to truth begins here.

Respectfully,

The Author

PROLOGUE. A NEW PARADIGM OF WORLDVIEW CHOICE IN THE POST-DIGITAL AGE

The Civilizational Armageddon, as understood in this book, is not limited to the physical destruction of the world. It also means the loss of the human capacity to form an independent picture of reality.

The deepest crisis of our age is therefore not only geopolitical. It is a crisis of meaning. States, ideological movements, digital platforms, and private infrastructures increasingly compete for the power to define what people see, fear, believe, desire, and call true. Truth becomes a strategic resource. Worldview becomes an object of control.

In the age of Globalization 4.0[1], this struggle enters a new phase. Power no longer acts only through armies, borders, laws, schools, churches, media, or markets. It also acts through data, recommendation systems, predictive analytics, biometric identification, emotion-recognition technologies, and neurocognitive interventions. These instruments do not merely influence public opinion. They can shape the conditions under which a person forms judgement, trust, loyalty, identity, and self-understanding.

This is why public-law worldview neutrality can no longer be treated as a narrow legal doctrine or a technical question of church-state relations. It has become one of the legal conditions for preserving the human being as a free subject. A state that claims ownership over truth undermines freedom from above. A platform that silently engineers perception undermines freedom from within. Both forms of power threaten the person's right to an autonomous picture of the world.

The central question of this book is therefore simple and difficult at once: who has the right to shape the inner world of the human being?

If the answer belongs to the state, freedom becomes obedience. If it belongs to the market, freedom becomes dependence. If it belongs to the platform, freedom becomes a personalized corridor of choices prepared in advance. If it belongs to the algorithm, freedom becomes prediction disguised as convenience.

The alternative proposed in this book begins with a different premise. The human being must remain the primary author of their own worldview. Public authority must not impose, privilege, suppress, or rank worldviews. Its task is

[1] Globalization 4.0 refers to the stage of global development shaped by the Fourth Industrial Revolution, in which digital, physical, and biological technologies transform not only markets and production, but also governance, law, security, identity, human rights, and the formation of worldviews. In this book, the concept is used in a legal and civilizational sense: as the condition in which global interdependence becomes algorithmic, cognitive, and normative, requiring new guarantees of freedom of conscience, liberty of worldview choice, and public-law worldview neutrality.

to protect the equal freedom of every person to form, preserve, revise, and express their understanding of life.

This is the meaning of worldview sovereignty. It is not a claim to isolation from society, culture, religion, science, or technology. No human being thinks outside influence. Worldview sovereignty means protection against the monopolization of meaning. It means that neither the state, nor the corporation, nor the platform, nor the ideology, nor the algorithm may become the owner of the person's inner world.

The search for a way out of the geopolitical deadlock therefore begins at a deeper level than diplomacy alone. It begins with the legal recognition of worldview liberty and public-law worldview neutrality as foundations of a human-centered civilization. Without these principles, technological progress may increase power while diminishing the human being. With them, law can become a shield for cognitive liberty before total transparency, emotional surveillance, and algorithmic governance turn the inner life of the person into an object of administration.

This book is a call to defend the last domain of freedom that no civilization can afford to lose: the right of the human being to think, believe, doubt, create, and understand the world without the enforced ownership of their consciousness by any external power.

I. Introduction to a New Paradigm. The Right to Worldview Choice in the Face of a Global Rupture

This theoretical and applied study is devoted to the fundamental and practical issues involved in realizing the internationally recognized right to freedom of conscience. In the conditions of contemporary global processes, where humanity has entered a civilizational and geopolitical deadlock, old legal forms require more than cosmetic repair. They require deep rethinking from the standpoint of a new paradigm.

The foundation of this work consists not only of theoretical developments but also of a living body of data: legal documents, materials from human rights monitoring, testimony from civil society organizations, and media sources. This multi-layered basis allowed the author to diagnose the problems of implementing international obligations in the domestic legislation of the Russian Federation, taking into account the globalization of social relations.

The structure of the monograph examines the key dimensions of freedom:

Chapters 1-2: Genesis and meanings. Through historical analysis and clarification of key and related concepts, a modern theoretical and legal foundation is constructed. The author substantiates the importance of the worldview neutrality of states as a fundamental condition for the effectiveness of international regulation.

Chapters 3-4: The global scale. Freedom of conscience is considered at the epicentre of global challenges. In the age of neurotechnologies and digital pressure, about which international organizations are already warning, the international legal consolidation of protection for the human "inner fortress" becomes a question of civilizational survival.

Chapters 5-6: The Russian case. The monograph provides a detailed analysis of the problems of realizing the norms of the right to freedom of conscience in Russia. By examining the activities of state and non-governmental institutions, the author identifies the barriers that prevent the full implementation of the principle of worldview neutrality.

Chapter 7 and the Conclusion: A strategy for the future. The final part of the work goes beyond the mere recording of problems. It formulates concrete prospects for improving the law and presents conclusions and recommendations for the world community and for the state.

The final part of the book contains recommendations, the Draft Global Declaration and the Draft Global Convention on the Right to Liberty of Worldview Choice and the Principle of Worldview Neutrality of Subjects of International Law in Analog and Digital Environments, together with an explanatory note and commentary, as well as the Draft Model Law.

This work is a manifesto for transition to a new paradigm, where the international recognition of freedom to form one's picture of the world serves as a reliable shield against existential threats to humanity. The book calls for decisive action to realize the right to freedom of conscience in order to preserve human agency in the world of tomorrow.

II. Civilizational Deadlock. The Monopoly of Power Over Truth

The contemporary civilizational deadlock is largely predetermined by the fact that the historically shaped monopoly of public authority over truth has transformed into a global tool for maintaining the dominance of narrow elite groups.

In this context systemic control over the worldview sphere and the manipulation of the mass consciousness of the majority of the global population constitute a basic and indispensable condition for unleashing large-scale military conflicts. In turn without such a total informational and cognitive dictate a forceful geopolitical confrontation for the benefit of a small fraction of humanity would be a priori impossible.

The usurpation of the right to define basic meanings and values by state and quasi-religious institutions inevitably leads to the total alienation of individual freedom of thought and transforms ideological doctrines into a cognitive weapon that legitimizes the irrational arms race.

States rarely perish only because of a shortage of resources. Nor only because of diplomatic mistakes. Nor only because of greed, fear, or the technical superiority of some over others. More often, catastrophe begins earlier: at the moment when power ceases to be satisfied with governing external order and begins to claim authority over truth.

A civilizational deadlock does not arise simply where armies, states, and blocs collide. It arises where one force begins to believe that it has the right to decide what the true world is, which person is "correct," which worldview deserves protection, and which must be displaced, suspected, or destroyed. At precisely this moment politics ceases to be a dispute over interests and becomes a struggle for ownership of the very picture of the world.

Human history knows many wars. They were not all religious. They did not all arise from questions of conscience. They cannot all be explained by the absence of secularity. It would be too easy, and therefore too foolish, to reduce all the blood of the world to one cause. But it would be just as naive to ignore the fact that the longest, fiercest, and most destructive conflicts have been especially easy to ignite wherever power fused with a monopoly over meaning. Where the opponent was declared not merely a rival, but the bearer of a false, dangerous, or impermissible worldview. Where war was waged not only for land, but also for the right to decide how the inner world of a human being should be arranged. This logic directly corresponds to the historical line reconstructed in this book. International law largely emerged as a response to the inability of religion to secure peaceful international relations and to its role in initiating religious wars.

In contemporary reality, an impartial look at the notional map of the most violent armed conflicts reveals its significant overlap with the map of states most closely intertwined with particular religious currents.

This is why freedom of conscience cannot be understood as a private freedom among others, as a modest legal privilege for the sphere of belief. In its true sense, it concerns the human right not to hand over one's inner world into external possession. In this perspective, the right to freedom of conscience is not only the right to profess a religion or to refuse religion. It is liberty of worldview choice. It is the possibility to form, choose, change, and live one's own picture of the world without coercion, without an imposed hierarchy of loyalties, and without state or quasi-state dictate in the sphere of meaning. For precisely this reason, it becomes not a private guarantee, but one of the axes of the entire system of human rights. This means that freedom of conscience, understood as liberty of worldview choice, is a system-forming right, and its violation significantly increases the risks of authoritarianism, intolerance, discrimination, and violence.

A civilizational deadlock begins when this freedom ceases to function. When the state believes that it has its own correct metaphysics. When official ideology begins to divide citizens not by their actions but by the permissibility of their convictions. When religious affiliation becomes a passport to loyalty. When dissent is declared a threat to public security. When public institutions begin to produce not law, but correct ontology. When the language of law becomes the language of sorting people into the "normal," the "traditional," the "loyal," the "alien," the "dangerous," and the "superfluous."

It is in such conditions that conflicts cease to be limited. They begin to feed on the deepest layers of human identity. Then war becomes not merely violence, but a sacred or ideological mission. Not merely a clash of interests, but a struggle over final meaning. Not merely external confrontation, but an invasion into the very fabric of the human world.

This is one of the key ideas of the book. Ethno-confessional barriers, sustained by violations of freedom of conscience, corruption in relations between the state and religious associations, intolerance, xenophobic stereotypes, and worldview-motivated violence, become an extremely serious obstacle to international solidarity and the transition to sustainable development. In other words, the problem of freedom of conscience is not a peripheral topic for specialists in religion. It is a question of international security, the resilience of complex societies, and the survival of civilization under conditions of planetary interdependence.

Here it is especially important to name correctly what is often hidden behind the familiar word "secularity." This book takes a principled step. It proposes that we think further and more precisely: not only through the category of secularity, which can easily be narrowed to the "non-religiosity" of the state, but through the deeper principle of worldview neutrality. This is not the cold indifference of power to everything important. Nor is it a rejection of law, education, values, or public order. On the contrary, it is an extremely difficult but necessary self-limitation of power. The state must not appropriate for itself the status of owner of final truth. It must not be a moderator of meanings. It must not rank convictions according to their usefulness for loyalty. It must not turn public institutions into machines for producing a single correct picture of the world. We directly propose speaking of worldview neutrality not only for public authorities, education, the media, law enforcement agencies, and the armed forces, but also of international-law worldview neutrality of subjects of international law.

It is here that the question of freedom of conscience connects with planetary politics. While humanity lived in a world of relatively isolated states and local conflicts, it was still possible to pretend that religious or ideological unfreedom was the internal affair of particular societies. The age of globalization destroyed that illusion. Today, every large system of worldview coercion produces consequences far beyond its own borders. It intensifies migration waves, fuels wars of identity, destroys the space of trust, fragments international law, and makes genuine solidarity in the face of common threats impossible. And these threats are no longer local. The book states this directly: under conditions of dynamic global processes, new security threats, and planetary interdependence, the transition to the sustainable development of civilization is impossible in principle without an effective mechanism for realizing the right to liberty of worldview choice and the worldview neutrality of states and other subjects of international law.

This means that civilizational deadlock is not only a geopolitical situation. It is, above all, a spiritual and political form of stagnation, in which humanity has learned to produce ever more power but has not learned to limit the right of power over truth. We know how to build networks, but we do not know how to preserve the inner freedom of the person within those networks. We know how to create international organizations, but we still do not fully know how to protect the human right to one's own worldview choice. We know how to speak of human rights as a universal language, but we still too often permit states, ideologies, confessions, and now technologies to compete for the privilege of defining the permissible picture of the world.

That is why the contemporary civilizational deadlock cannot be overcome only through a new security architecture, only through economic reforms, or only through diplomatic balance. All of this is necessary. But it is insufficient. As long as the human being remains vulnerable to the external appropriation of the inner world, conflict will return again and again: in new clothes, under new flags, with new technologies, and with the same ancient claim to decide for the person how to think, what to believe, what to fear, and whom to regard as an enemy.

The way out of the deadlock begins where power recognizes the limits of its competence in the sphere of meaning. Not everything that matters to the human being belongs to the state. Not every truth must be publicly mandatory. Not every identity can be transformed into a normative measure for all. Not every majority receives the right to dispose of the conscience of a minority. And not every technology capable of predicting and guiding behavior should receive silent permission to form convictions.

If this is expressed in a single formula, it would read as follows: civilizational deadlock begins where power appropriates truth. Accordingly, the way out begins where the human being is returned the right to one's own picture of the world. This is why freedom of conscience and worldview neutrality are not narrow legal topics, but conditions of the planetary future. Not because they will automatically stop every war, but because without them wars will again and again become sacred, ideological, total, and endless.

The refusal of subjects of public law to follow the principle of neutrality turns education, science, and the Internet into instruments of ideological expansion. Without a return to the "zero point" of worldview neutrality, the geopolitical rupture will become final and existential. The central answer to the challenges of the present lies in returning to the human being the right to one's own reality, not a "designated" one. This means that without strong international guarantees of worldview liberty, a way out of the civilizational deadlock is technically impossible.

III. Public-Law Neutrality in the Context of Globalization 4.0. The Battle for Cognitive Autonomy

For a long time, it seemed that the major conflicts of history unfolded around entirely tangible things: land, resources, borders, armies, trade, money, territories. States fought for space, empires struggled for influence, ideologies divided the world, and international law tried at least partly to limit the destructive force of this struggle. Humanity paid for it with millions of lives and centuries of fear.

But in the twenty-first century it is becoming increasingly clear that behind all earlier conflicts there always stood another, deeper cause. Not only the struggle for land, but the struggle over how a person sees the world. Not only the struggle for institutions, but the struggle for consciousness. Not only the struggle for the behaviour of people, but for the sources of their convictions, identity, faith, doubt, meanings, and inner support.

Today this is no longer a philosophical abstraction. It is reality. We have entered an era in which power over the person is exercised less and less through direct physical coercion and more and more through control over attention, emotions, perception, symbols, and the agenda. If in the past the main aim of the state, the church, the party, or ideology was to subordinate the body, today what is increasingly subordinated is the very frame of thought. A person may not be forbidden to think. Instead, he may simply be offered in advance a ready-made picture of the world: convenient, profitable, emotionally charged, algorithmically amplified, and therefore almost invisible as an external influence.

It is here that the main question of the new era arises. Does the human being retain the right to form one's own picture of the world independently? At first glance, this may seem like merely an updated version of the old dispute over freedom of conscience. Yet precisely here lies one of the greatest errors of contemporary legal and political thought. Freedom of conscience has for too long been understood as the right to profess a religion, to change religion, or to profess none at all. This is undoubtedly a crucial part of the problem. But it is only a part.

In reality, freedom of conscience is much broader and deeper. It is liberty of worldview choice. It is the freedom of a person to seek truth, to doubt, to believe, not to believe, to change convictions, to compare knowledge and faith, to build one's own system of meanings, and to live in accordance with it so long as this does not violate the rights of others and does not go beyond the limits of lawful conduct. In this sense, freedom of conscience is not a private right alongside others, but one of the foundations of human autonomy itself. The monograph directly advances the idea that free worldview choice lies at the foundation of free political choice, legal democracy, and the social state.

This is why encroachment on freedom of conscience is more dangerous than it appears. When a person loses the right to form their own convictions independently, that person gradually loses political freedom, critical thinking, and the capacity for moral choice. Where freedom of conscience is not protected, xenophobia, intolerance, discrimination, cults of loyalty, and violence against dissenters sooner or later arise. The violation of freedom of conscience increases the risks of authoritarian power, the destruction of human rights, and the growth of violence rooted in worldview intolerance.

But today we face an even more alarming problem. History has known states that tried to dictate to the individual the correct faith, the correct ideology, the correct morality, and the correct loyalty. Yet in the past, power did not possess the instruments of penetration into everyday life that have now appeared: artificial intelligence, recommendation systems, facial and emotion recognition, digital platforms, big data, behavioural analytics, neurotechnologies, the Internet of Things, digital ecosystems of governance, and platform dependency on the infrastructure of communication and information. All this creates an environment in which influence over the human being becomes continuous, personalized, adaptive, and almost invisible.

The danger here lies not only in total surveillance, though surveillance itself is already destructive. Even more dangerous is the fact that new technologies can gradually transform inner freedom into a managed function of the environment. If a person is continuously prompted about what to fear, whom to regard as an enemy, what to consider permissible, which identity to regard as "normal," which feelings to experience, and which thoughts deserve public approval, then we are no longer dealing simply with informational noise. We are dealing with a new form of power over the formation of the picture of the world.

This is why the main line of this book does not run between the "religious" and the "secular," between the "traditional" and the "non-traditional," or between "believers" and "non-believers." The main line runs between two models of social order.

The first model assumes that the state, the union of state and ideology, or the union of state and technology has the right to act as arbiter of meanings. Such a system always finds justification for its claims. It speaks of tradition, security, civilizational identity, spiritual bonds, the protection of the majority, stability, cultural code, national interests, the struggle against "extremism," the protection of morality, or digital sovereignty. But behind all this there very often lies one and the same aspiration: to obtain the right to direct the worldview sphere of the human being.

The second model proceeds from the opposite principle: the state must not be the owner of truth. It must not impose a worldview hierarchy on the person. It must not create privileged convictions and discriminated convictions. Its task is to guarantee the space in which the person forms their own picture of the world.

This is precisely what the book calls the worldview neutrality of the state. It is treated not as a secondary technical detail, but as one of the most important guarantees of human rights. Moreover, the author consistently shows that the issue is not merely the familiar concept of secularity in the narrow sense, but a more precise and universal principle of neutrality in the worldview sphere. This neutrality must encompass not only the central government, but also local self-government, education, state media, law enforcement agencies, the armed forces, and other institutions of public authority.

The central idea of the book is that without the worldview neutrality of the state, freedom of conscience remains vulnerable, and without liberty of worldview choice, all other human rights become vulnerable. This is especially important in a world experiencing several crises at once. Liberty of worldview choice is placed in the context of global processes, instability in international relations, security threats, sustainable development, and the "great rupture" of the contemporary world. The fate of human rights is interwoven with the fate of the global order. We emphasize that humanity is facing not private local difficulties, but challenges to its own survival.

Here the decisive hypothesis arises. If freedom of conscience was once perceived as one topic on the human rights agenda, today it must be understood as one of the key principles of the survival of human civilization. A society in which the inner freedom of the person has been destroyed may preserve external institutions, but it has already lost the capacity for genuine democracy, peaceful coexistence, non-violent resolution of contradictions, and sustainable development. Where a person has no right to one's own picture of the world, any technological power will sooner or later begin to work against that person.

Therefore this book is not only about law. And not only about the state. And not only about religion. It is about the future of the human being. It is about whether inner autonomy can be preserved in an age when external governance is becoming ever more subtle, total, and intellectually equipped. It is about why freedom of conscience must be rethought as liberty of worldview choice. It is about why international law needs renewal and why human rights require a new stage of development. It is about why, in the age of artificial intelligence and digital control, the neutral state ceases to be a theoretical luxury and becomes a condition of human dignity. And finally, it is about the simplest and most difficult thing: the human right not to be the product of another will.

For if, in the twenty-first century, humanity loses the battle for the right of every person to form one's own picture of the world independently, all other freedoms will become little more than external scenery. First inner freedom will be lost, and then everything that only yesterday seemed guaranteed.

This is why the battle for the human mind has already begun. The only question is whether law, politics, and the international community will grasp the scale of this battle before technologies, ideologies, and new forms of power finally and irreversibly change the human being themselves.

IV. Technological Hacking of Consciousness. Emotion Recognition and Algorithmic Dictate as the Last Frontier of Identity

It seems that the end of the world, about which prophets warned for centuries, turns out to be quite unlike their fiery visions. There is no heavenly thunder, no horsemen of the Apocalypse, no final trumpet. The civilizational Armageddon described in this book arrives almost silently, under the soft glow of screens, to the rhythm of notifications, in the hum of data centres, and in the flawless operation of code. Meanwhile, code learns more and more about the human being, while leaving the human being with less and less knowledge of themselves.

Today catastrophe comes not in the form of obvious destruction, but in the form of comfortable recalibration. The human being is not broken. He is calibrated. He is not forbidden to think. Instead, the very environment in which thought is born is quietly rearranged. Direct loyalty is not demanded of him. Instead, an architecture of digital everyday life is created in which the desired reactions, fears, desires, and even the contours of an acceptable future seem to arise by themselves. This is the central drama of an age in which the crisis of the twenty-first century unfolds not only in ministries, parliaments, or corporate headquarters, but in the very structure of human self-determination: where digital infrastructure begins to govern attention, impulse, perception, and, ultimately, worldview.

Throughout the modern age, European thought was built around the fundamental conviction that the human being stood at the centre of the world as a subject of reason and free will. Today this picture is collapsing. The dark side of contemporary reality mercilessly records the decentring of the human being. In this perspective, history appears as a process in which flows of capital, technology, speed, and non-human intelligence gradually push the human being out of life. Machine, network, code, algorithm are no longer tools in the classical sense. They are forces within which the person risks becoming merely a temporary carrier of a signal.

Luciano Floridi called us inforgs, informational organisms existing within a common infosphere. But today the question sounds more sharply than ever. What will this inforg become if the inner world is transformed from a mystery of freedom into raw material for the computational economy? In the industrial age, humanity extracted oil, gas, and metals. In the digital age, the "extraction" of the human being itself has begun: consciousness becomes the deposit, attention the useful mineral. Does personality then become merely a by-product?

Corporate platforms, predictive analytics systems, and state digital Leviathans have moved from exploiting the external world to exploiting the internal world. Your micro-emotions, pauses in reading, eye movements, rhythm of speech, trajectory of interest, reaction to anxiety, vulnerability to stimuli, and behavioural fluctuations become objects of extraction, calculation, and managerial application.

In her book on surveillance capitalism, Shoshana Zuboff rightly warns that the asymmetric concentration of data and power threatens human autonomy and democratic control.

Algorithms no longer merely predict behaviour. They begin to produce it. They construct probabilities by amplifying some impulses and suppressing others, locking the subject into a computationally profitable circuit. Recommendation, ranking, search, and moderation systems are no longer convenient interface functions. They are mechanisms for forming the picture of the world. While a person feels that he is choosing freely, it becomes almost invisible that the very space of choice has already been pre-filtered, dosed, and optimized for retention, polarization, predictability, and manageability.

There is no need to declare an ideology mandatory if the digital environment can turn some ideas into a constant background and others into almost unreachable noise. There is no need to ban dissent if the very capacity for worldview search can be weakened in advance. This is no longer merely censorship. It is control over the initial conditions in which convictions arise.

This process reaches a new depth with the development of neurointerfaces, affective computing, behavioural profiling, and other technologies that promise to connect brain and machine. Neurotechnology start-ups present themselves almost as messengers of liberation, promising medical breakthroughs, new forms of communication, restoration of lost functions, and the expansion of human capacities. Much of this potential is real. But wherever it becomes possible to decode a neural signal, an emotional state, a pattern of attention, or an impulse of intention, there also appears the unprecedented temptation not only to read the inner world, but to reconfigure it.

From there it is only a short step to a world in which emotion-recognition systems, biometric psychoanalysis, brain-computer interfaces, adaptive virtual and mixed-reality environments, platform analytics, and state data arrays merge into a single infrastructure of soft control over consciousness. This is no longer classical totalitarianism, nor the crude repression of the past. It is a new form of digital guardianship, operating through intelligent behavioural modelling, invisible nudging, personalized suggestion, and cognitive enclosure.

If a person is deprived of the possibility to form, reconsider, and preserve convictions independently, in silence and secrecy, that person ceases to be a full subject of law. He becomes a controlled biological peripheral module of the system.

This is why modern law must move beyond archaic definitions of freedom of conscience that remain closely tied to the legally indeterminate concept of "religion." In the twenty-first century, freedom of conscience is no longer only the right to believe or not to believe, to profess a religion or to reject it. Today it must be understood as the possibility of free worldview choice and the worldview sovereignty of the person.

This means the right of the human being independently to form, preserve, reconsider, change, and express convictions without coercion, discrimination, hidden algorithmic influence, or digital manipulation. This right includes not only freedom of opinion as an already formed result, but also the freedom of the path toward it. Not only the right to speak, but the right to search. Not only the right to express a conviction, but the right not to allow a platform, state, corporation, or algorithm to silently program the conditions under which that conviction arises.

From this follows the need to affirm worldview neutrality as a fundamental legal principle of the new age. In the digital environment, state neutrality must mean not only refusal of an official ideology, but also a direct prohibition on the use of state algorithms, artificial intelligence systems, automated recommendation systems, profiling, and forecasting for the manipulation of consciousness, the imposition of a single doctrine, the suppression of worldview pluralism, or hidden influence on the political, religious, philosophical, and other convictions of the human being.

But even this is insufficient if platforms and transnational digital corporations remain outside the legal framework. Systemically significant digital intermediaries can no longer be regarded as neutral services. De facto, they form the cognitive environment of billions of people. This means they must be bound by obligations to prevent discriminatory moderation on worldview grounds, ensure algorithmic transparency, disclose the basic principles of information-flow formation, notify users of significant automated influence, and establish effective procedures for review and appeal. They must also undergo independent audits for worldview discrimination and hidden manipulation.

Algorithmic transparency becomes a new form of constitutional limitation on power. The right to the opacity of consciousness becomes a basic human right. Consciousness must have the right to a digital shadow.

This is especially true of emotion-recognition systems, affective computing, and biometric psychoanalysis. These technologies encroach not only upon private life, but also upon the mental integrity of the person. They claim the right to extract the internal state of the human being, classify it, store it, and use it to assess loyalty, reliability, learning capacity, manageability, and suitability. Their introduction into public authorities, educational organizations, labour relations, and other public spheres is incompatible with human dignity.

Such systems must not merely be restricted. They must be subject to a complete legislative ban, the destruction of corresponding data arrays, and the establishment of strict legal liability for their use. Any data obtained through such interference must be recognized as legally void. For where the technological dissection of the inner world begins, freedom ends.

We stand before a choice between two models of the future.

The first model invites the human being finally to surrender the centre to systems of efficiency, predictability, speed, and computational optimization. It promises singularity, superintelligence, and a passage beyond old humanism. Yet the price of this bargain is obvious: in the end, the person becomes an object of analytics, behavioural modelling, and infrastructural governance.

The second model proceeds from the opposite premise. It does not deny technology, romanticize the past, or propose stopping progress. It demands that the human being be placed at the centre of law, politics, and the digital organization of the world. And not as a biological species in itself, but as the bearer of dignity, inner freedom, the right to doubt, error, change of convictions, and resistance to any computational colonization of consciousness.

This is the central pathos of this book. Its task is not to declare war on technology. Its task is to draw, within the digital age itself, a legally clear and civilizationally binding boundary that neither state, market, platform, algorithm, neurointerface, nor the most seductive philosophy of post-human acceleration has the right to cross. This boundary lies where the worldview sphere of the person begins.

If humanity manages to protect this territory legally, it will retain a chance to remain human in an age of super-technologies. If not, the disappearance of the human being from the central stage of history will be framed not as a catastrophe, but as yet another brilliant innovation. And then the end of freedom will come not to the sound of an explosion, but to the music of flawlessly functioning code.

V. The Point of No Return. The Right to Worldview Sovereignty and Protection from Total Algorithmic Control as an Imperative of Post-Digital Civilization

We stand on the threshold of a new reality in which old declarations of human rights no longer protect us from "algorithmic gods." Digital platforms, conceived as instruments for expanding opportunities, have become prisons for consciousness with invisible walls made of recommendation algorithms, personalization filters, and echo chambers that intensify extremes.

Worldview liberty in the twenty-first century is not a passive state, but the result of active resistance to technological determinism. It requires a conscious refusal of digital dependency, critical reflection on information supplied by algorithms, and the restoration of the capacity to form convictions independently.

The way out of the "digital cave" is a collective act of liberation in which we must break down the walls of echo chambers created by code and return to ourselves the right to dialogue with reality rather than with its digital projection.

The worldview neutrality of the state is not merely the absence of censorship. It is the active construction of barriers between the human psyche and manipulative AI systems. It means creating a legal and technological architecture of cognitive sovereignty as a system for protecting individual consciousness from external programming.

The key elements of this architecture include the following. First, a legislative ban on the use of emotion-recognition technologies and neurointerfaces for purposes of manipulation. Second, algorithmic transparency and the user's right to know how their digital environment is formed. Third, data sovereignty and full human control over one's digital traces, including the right to delete them. Fourth, educational standards that develop digital literacy and critical thinking from an early age.

Power must be an invisible and impartial setting for the personal search for truth, not the director of our convictions. Its role is not to dictate worldview, but to guarantee the conditions for its free formation.

The old social contract, "security in exchange for freedom," has exhausted itself. It produced a world in which personal data became the new oil and human attention became a commodity. A new contract requires "algorithmic cleanliness" and rests on three principles. First, the right to emotional privacy and a ban on emotional espionage through the collection and analysis of data on a person's psycho-emotional state without explicit consent. Second, a ban on ideological profiling and the inadmissibility of using algorithms to form worldview attitudes or political mobilization. Third, the digital neutrality of the state and the obligation of public authority to ensure equal conditions of access to information and protect citizens from algorithmic discrimination.

We agree to digital services, but we require compliance with these conditions. This is not a whim. It is a condition for the survival of civilization as a community of free wills rather than a collection of biological data.

History places before us a radical choice. Either a model in which emotions are programmed through neurointerfaces and biomarkers, thoughts are moderated in real time by moderation algorithms, individuality dissolves into a collective unconscious, and freedom is replaced by the comfort of total control. Or a model of a society of free persons, where the right to worldview choice is the highest legal norm, where technologies serve the expansion of opportunities rather than the restriction of freedom, where the diversity of convictions is understood as a source of resilience, and where the human being remains a subject rather than an object of governance.

Either we consolidate and realize the right to worldview choice as the highest legal norm today, or tomorrow civilization will turn into a global "hive." This is our last chance to preserve humanity in the post-digital age by creating a new legal reality in which freedom of thought does not give way to digital order, but defines its boundaries.

Algorithmic dictate seeks to turn the human being into a predictable function. But worldview is a territory free from code. We proclaim the right to algorithmic silence and a ban on technological intrusion into the process by which thought is formed. Today, the protection of personal sovereignty is not a struggle over data, but a struggle for the right to an incalculable truth that cannot be reduced to a set of bits.

We call for global recognition that the inner world of the human being, his convictions, doubts, and faith, is a zone of absolute sovereignty. This sphere is not subject to digitization, cannot be delegated to an algorithm, and must be protected from any technological interference as the highest value of post-digital law.

We stand on the threshold of the greatest desynchronization in history. While our technological genius already operates at the speed of artificial intelligence and the depth of neurotechnologies, our legal and political institutions remain hostage to the archaic "law of force." This gap between the digital "brain" of civilization and its ideologized "social body" has brought us to an existential deadlock. The old model of world order, where the state acts as the monopolistic moderator of meanings and censor of consciousness, is becoming the detonator of civilizational Armageddon.

The key vector for escaping this geopolitical nosedive is recognition of a new constant in the form of the worldview neutrality of the state. In an age when algorithms can penetrate the very fabric of human thought, the right to autonomy of the picture of the world ceases to be a philosophical refinement and becomes a condition for the survival of the species. We must legally and technologically consolidate the sovereignty of the mind, transforming human rights secured by worldview neutrality into an inviolable code on which the architecture of the future will be built. This is not merely a refusal of ideological dictate. It is insurance against the transformation of Superintelligence into an instrument of total suppression.

While Yuval Noah Harari and Mo Gawdat look at the existential threats of artificial intelligence from historical and technological perspectives, this monograph provides a needed international legal framework to secure cognitive liberty.Developing the critical economic analyses of techno-feudalism presented by Yanis Varoufakis and Cédric Durand, the author substantiates public-law worldview neutrality as a decisive legal mechanism to dismantle digital fiefdoms. Steering away from the neo-reactionary dystopias of Silicon Valley technocracy, which advocates for corporate absolutism and the erosion of fundamental human rights, Buryanov rejects this technocratic determinism and presents a powerful democratic and human-centered alternative.By transforming these tech-existential warnings into concrete and actionable legal instruments, the work establishes a robust framework for preserving individual autonomy against existential digital threats.

CONCEPT BOX

INTRODUCTION

THE QUESTION OF THE BOOK

Worldview liberty as a legal problem of the digital age

The central question of this book concerns the legal future of the human inner world.

Who has the right to shape the picture of reality through which a person thinks, believes, doubts, judges, and acts?

In the twentieth century, this question belonged mainly to the field of freedom of conscience, religion, and ideology. In the twenty-first century, it enters the fields of artificial intelligence, neurotechnologies, biometric systems, digital platforms, and geopolitical power.

The book argues that freedom of conscience must be developed into a broader doctrine of liberty of worldview choice. This doctrine requires public-law worldview neutrality as one of the legal conditions for protecting human dignity in analog and digital environments.

INTRODUCTION. THE KEY CONDITION OF GLOBAL STABILITY IN THE TWENTY-FIRST CENTURY. WHY, IN AN AGE OF DEEPENING GEOPOLITICAL RUPTURE AND THE TRIUMPH OF THE LAW OF FORCE, FREEDOM OF CONSCIENCE MUST BE REINTERPRETED AS LIBERTY OF WORLDVIEW CHOICE

In the conditions of dynamic change in contemporary social relations, which on the one hand are directed toward technological progress and the formation of a single global system, and on the other hand carry instability in international relations, new global security threats, and risks to the very existence of humanity, it is especially important to study the formation and development of the principles and norms of the right to liberty of worldview choice in international law, as well as their implementation at the domestic level.

The Statement of January 27, 2026 by the Science and Security Board of the Bulletin of the Atomic Scientists on the symbolic Doomsday Clock states that humanity is now 85 seconds from "nuclear midnight," the closest position to midnight in the entire history of the project. Among the many causes of heightened threats, the statement identifies aggressive conduct by nuclear powers, the weakening of nuclear arms control, and conflicts in Ukraine and the Middle East. Particular concern was expressed regarding the integration of artificial intelligence (AI) into weapons systems, as well as the role of AI in the spread of disinformation. According to the experts, humanity's collective efforts to reduce global existential risks remain insufficient, and the world continues to stand on the threshold of serious instability. Thus, alongside nuclear weapons, the most significant threats to the existence of humanity were identified as climate problems and the potential misuse of achievements in biological science and various new technologies, including artificial intelligence.

The negative consequences of globalization also include substantial threats to international peace and security aimed at undermining the world order and manifested in the intensification of force-based geopolitical confrontation: the growing number and severity of international armed conflicts, with almost half of all states involved; a significant rise in military budgets; the modernization and expansion of nuclear arsenals; the integration of artificial intelligence with nuclear weapons; the reduction of international cooperation in nuclear disarmament; the strengthening tendency to turn outer space into an arena of armed confrontation; the struggle for technological superiority and a high-technology non-nuclear arms race capable of disturbing strategic balance and stability, including hypersonic systems, robotics, unmanned systems on land, in the air, on the surface of water, and underwater, in combination with artificial intelligence and quantum computing; the removal of restrictions for developers and the non-human-centred deployment of artificial intelligence; the monopolization of high technologies by corporations and/or states, which may become a cross-cutting problem aggravating all the aforementioned risks with hard-to-predict consequences; and the dangerous mixing of commercial interests with security interests through the participation of major corporations in armed conflicts as independent actors.

At a press conference on March 25, 2026, the United Nations Secretary-General Antonio Guterres once again called on world leaders to "stop climbing the ladder of escalation" and instead "climb the ladder of diplomacy," as well as to return to full respect for international law. He warned that the world stood on the edge of a larger war, a growing wave of human suffering, and a deeper global economic shock.

I. I. Lukashuk reasonably observed that, "for the first time in history, the question of the survival of humanity has arisen. The distinctive feature of these threats is that they are global and can be resolved only through the joint efforts of states."

O. N. Khlestov pointed to new risks and to the need to strengthen international law as an instrument for maintaining stability under the conditions of globalization. In his view, the post-war world order enshrined in the UN Charter will retain its importance as the foundation of international relations, but it must adapt to new challenges, which will require new formats of interaction among states.

I. Z. Farkhutdinov identifies contemporary global threats to international security, including those involving confessional states.

N. N. Emelyanova emphasizes the key role of the United Nations in forming a global system of international security and notes that global security challenges, in addition to interstate wars, include poverty, infectious diseases, environmental degradation, wars and violence within states, the proliferation and possible use of nuclear, radiological, chemical, and biological weapons, transnational organized crime, and other dangers.

A. A. Moiseev considers international law under the conditions of globalization as a dynamic system adapting to processes of economic integration and the emergence of supranational institutions. In his view, globalization does not abolish state sovereignty, but requires its qualitative transformation in order to ensure effective interaction among actors in the world market. The author sees the future of the system in strengthening legal instruments and international institutions capable of smoothing contradictions between national interests and global challenges. Particular attention should be paid to the role of international law in creating the legal and institutional conditions for peaceful coexistence, mutual enrichment, and the protection of cultural diversity.

A. N. Chumakov argues that in recent years globalization and digitalization have formed an inseparable unity, in which technologies accelerate the objective processes of human integration. This cumulative interaction gives rise to digital globalization, which makes the world transparent and hyperconnected while also sharpening global challenges. The author stresses that the cumulative effect of globalization and digitalization creates serious anthropological risks and a crisis of traditional governance systems that cannot adapt quickly enough to rapid social change. In his view, the main risk lies in technological outpacing. In conditions where the digital environment develops faster than morality and international law, it becomes necessary to search for ways of transitioning to conscious governance of world development.

Earlier, in a discussion at the United Nations General Assembly, UN Secretary-General Antonio Guterres expressed concern that a "great fracture" could emerge in the contemporary world, as reported by the UN News Centre. In particular, Guterres believed that in the current unstable period, marked by imbalances of power in international relations, a new threat was maturing: "I fear the possibility of a Great Fracture: the world splitting in two, with the two largest economies on earth creating two separate and competing worlds, each with its own dominant currency, trade and financial rules, its own Internet and artificial intelligence capacities, and its own zero-sum geopolitical and military strategies." The UN Secretary-General expressed confidence that unconditional respect for human rights is an essential condition for the preservation and sustainable development of civilization. He also proposed preserving a universal economy based on respect for international law and a multipolar world with strong multilateral institutions. "Human rights are universal and indivisible. We cannot pick and choose, favoring some while ignoring others. People have the right to a safe and peaceful life, to well-being and a decent standard of living, to health, housing and food, to social protection and a healthy environment, to education and decent work. These rights run like a thread through the entire 2030 Agenda for Sustainable Development," Guterres stated.

We should agree with the view that the common interests of sustainable development shared by every human being can and must become the basis of a planetary consensus on the main issues of international cooperation, including human rights. According to V. A. Kartashkin, human rights do not fall exclusively within the domestic jurisdiction of states, whose sovereignty in this area is substantially limited, while "the principles and norms relating to human rights are regarded by contemporary international law as jus cogens, operating erga omnes." "International cooperation in the field of human rights must be based on the concept of sustainable development. This concept is aimed at overcoming global problems that endanger both human rights and the very existence of humanity. Their resolution requires international solidarity."

A. I. Ekimov writes in his works about the need for a global dimension, futurization of new law, and governance for the achievement of sustainable development. The researcher rightly notes that "in the model of sustainable development, the requirements of universal justice and the security of the individual, society, and the state are achieved not by countering emerging threats, but by preventing the threats themselves." "The law of sustainable development must be based on a new system for interpreting the concepts of justice, ethical values, and spiritual values. All values acquire a global dimension. It is commonly said that this applies to all people without exception, although usually only present generations are meant. But these generations, continuing to transmit into the future patterns of economic activity and, especially, use of nature that lead to the destruction of the natural environment and the depletion of natural resources, first condemn future generations to a substantial decline in quality of life and then to death. Thus, present generations live, in effect, at the expense of future generations. This is why, in order to ensure sustainable development, the principles of law must also be extended to future generations of people, so as to ensure the survival and preservation of the human race for the foreseeable future."

The progressive development of international law is called upon to play a most important role in the development of international cooperation for sustainable development. However, I. I. Lukashuk rightly noted that "much remains to be done so that it corresponds to the fundamental changes that have occurred in the world system." He further identifies the main qualities of that system: "the transition from a military-political to a political-economic basis of world order; the growing role of the financial and economic spheres; transnational integration and regional associations; and the strengthening of democratic tendencies within states and in international relations."

I. V. Sledzevsky's article "Desecularization of the World Community as a Source of Tension in International Relations and World Politics" examines the impact of the increasing role of religion in social and political life on international relations. The author notes that desecularization, that is, the strengthening of the religious factor, is becoming a significant source of tension and conflict in the international arena, and he reasonably identifies the main manifestations of this process: the politicization of religion, the use of religious rhetoric in diplomacy, and the formation of ideologies based on religious doctrines. The article emphasizes that desecularization complicates diplomatic interaction and requires that the religious factor be taken into account in international relations.

The 2016 Human Development Report, prepared by the Human Development Report Office (HDRO) of the United Nations Development Programme (UNDP), speaks of the close connection between human development and the possibility for every person to fully realize their life potential at the national and global levels. Thus, at the centre of the concept of human development is the expansion of freedoms for everyone without discrimination, embodied in human rights, which are the "cornerstone of human development, its foundation." It is especially emphasized that "every individual deserves to have choices among different identities that are valuable to them. The recognition of and respect for such choices are preconditions for peaceful coexistence in multi-ethnic and multicultural societies."

These problems manifest themselves most sharply in multinational and multiconfessional states, including Russia. F. M. Rudinsky's study of pre-revolutionary and Soviet experience showed that the insufficient effectiveness of norms in the sphere of freedom of conscience and relations between the state and religious associations led to the aggravation of ethno-confessional contradictions and social upheavals. Here it is also necessary to note the books of Soviet dissidents and human rights defenders L. L. Regelson and G. P. Yakunin, as well as L. M. Alekseeva. Alongside other factors, human rights violations, the growth of intolerance and discrimination, and the persecution of dissenters, including on worldview grounds, created the conditions for the collapse of the Soviet Union. In order to avoid a repetition of catastrophe, one important and necessary condition is the search for effective ways to improve the principles and norms governing the right to freedom of conscience.

The situation in the Middle East is a vivid example of how ethno-confessional barriers turn a region into an arena of global confrontation. Most borders in the region were drawn artificially as a colonial legacy, without regard to ethnic and religious areas. As a result, a person's loyalty often belongs not to important and necessary condition is the search for effective ways to improve the principles and norms governing the right to freedom of conscience.

The situation in the Middle East is a vivid example of how ethno-confessional barriers turn a region into an arena of global confrontation. Most borders in the region were drawn artificially as a colonial legacy, without regard to ethnic and religious areas. As a result, a person's loyalty often belongs not to the state, but to a religious community, Sunni or Shiite, or to an ethnic group, such as Kurds, Arabs, or Persians, while internal discord always gives external actors a pretext for intervention. Religion in this context has ceased to be a matter of personal freedom and has become a geopolitical weapon. The conflict between Iran, often described through the "Shiite arc," and the Gulf monarchies, often described as a Sunni bloc, is not only theology, but also a struggle for regional leadership. As long as identity is built on the negation of the "other," any local incident risks escalating into civilizational Armageddon.

In countries such as Lebanon, Iraq, and Syria, state institutions are paralysed by quotas based on religious affiliation. This creates "fragile states" that cannot conduct sovereign policy, because each community seeks protection from an external patron, whether the United States, China, Russia, Iran, or Turkiye.

Ethno-confessional barriers in the Middle East function as transboundary resonators and turn any local conflict into an existential threat of global scale through the erosion of state sovereignty, demographic pressure, migration destabilization, and ethno-confessional cleansing.

Powerful non-state actors operate in the region, and their identity is built to a significant extent on religious solidarity rather than loyalty to the state. Even if a government wants peace, such actors can drag a country into war, guided by confessional duty or another interest.

Ethno-confessional cleansing, for example in Syria or Iraq, destroys the social fabric. It is not simply a humanitarian catastrophe, but an instrument for redrawing the map. Refugees become a factor of destabilization for neighbouring states, transferring internal barriers to the level of interstate relations.

As a result, the Middle East demonstrates that worldview unfreedom, understood as intolerance toward those of another faith within societies, inevitably leads externally to a geopolitical deadlock.

However, the regional problems of the Middle East become even more acute in the context of global challenges. Since the Middle East is one of the most water-scarce regions of the world, global warming is proceeding there at an accelerated pace. In conditions where ethno-confessional contradictions obstruct joint management of water resources, their shortage provokes mass migration, destabilizing not only the region but also Europe. In addition, the global course toward decarbonization through the abandonment of oil and gas undermines the economic foundation of most states in the region.

This means that, without radical reforms and the overcoming of internal closure, these countries risk turning into "failed states" with enormous arsenals of weapons and hungry populations. The deployment of AI-based surveillance and governance systems under conditions of ethno-confessional distrust may lead to the creation of "digital ghettos," where technologies are used to cement barriers and suppress worldview liberty, thereby preserving the deadlock. In fact, the Middle East has become a testing ground on which global actors test new methods of hybrid warfare, where local ethno-religious groups become instruments in the hands of major powers.

Against the background of total distrust and existential fear of the neighbour, for example Iran in relation to Israel and Israel in relation to Iran, global non-proliferation regimes collapse. This turns the region into the epicentre of a possible nuclear or biological Armageddon that would be impossible to localize.

The relevance of this study is strengthened by the assumption that the Middle East today is a kind of mirror of humanity's future if the world does not learn to overcome identity barriers in order to address common existential threats.

In our view, the way out of the geopolitical deadlock in this region and in the world is possible only through an evolutionary transition to worldview-neutral pragmatism, where economic interdependence and cooperation for overcoming common global and regional threats become more important than confessional dogmas.

Thus, under the conditions of contemporary social processes, the problem of overcoming ethno-confessional divisions is global and requires the united efforts of the world community. Above all, the transformation of social relations manifests itself in the integration of social interactions into a single planetary system. This process reflects the current stage of changing states in the development of social relations, from their complication and convergence to mutual penetration and, as a consequence, interdependence on a planetary scale.

The transition to the sustainable development of civilization and the overcoming of the bloody ethno-confessional contradictions of the contemporary world are in principle impossible without the formation of an effective mechanism for realizing the right to liberty of worldview choice and public-law worldview neutrality at the international and domestic levels. The worldview neutrality of states and other subjects of international law is inseparably linked with the concept of liberty of worldview choice, which is a vital component of individual freedom. In turn, individual freedom is embodied in human rights as the most essential opportunities of the human being guaranteed by law. In other words, the worldview neutrality of states and other subjects of international law acts as one of the most important guarantees of human rights, including liberty of worldview choice.

It is assumed that the study of this problem will help to understand the mechanisms of mutual influence among the internationally recognized right to freedom of conscience, related concepts, and international relations. It will also make it possible to identify their prospects in the context of the development of global processes and problems, as well as the search for paths toward the sustainable development of human civilization and individual states, including Russia.

This theoretical and applied scientific study is devoted to the theoretical and practical issues of the internationally recognized right to freedom of conscience and to the problems of its implementation in the Russian Federation under the conditions of contemporary global processes. The work is based on fundamental theoretical developments; legal documents; materials of human rights monitoring; materials of civil society organizations and the media; and other open sources.

From the standpoint of a new paradigm and taking into account the globalization of social relations, the book examines the historical prerequisites and development of the contemporary theoretical and legal foundations of the internationally recognized right to freedom of conscience, the secular nature of the state, and related concepts; its international legal consolidation and protection mechanisms; and issues of implementing international obligations in domestic legislation. The work studies the problems of realization and identifies prospects for improving the principles and norms of the internationally recognized right to freedom of conscience and the principle of worldview neutrality of the state under the conditions of contemporary global processes and geopolitical threats.

The final part contains conclusions and recommendations for the world community and the Russian Federation, as well as draft international acts aimed at realizing the internationally recognized right to freedom of conscience, secured by the worldview neutrality of subjects of public international law, under the conditions of digital globalization 4.0.

The work is prepared for researchers, teachers, postgraduate students, students, legislators, public officials, staff members of international organizations, human rights defenders, and all those interested in the problems of global studies, political science, human rights, and worldview neutrality. It is recommended for use as a textbook or teaching aid in the following disciplines: Global Studies, Geopolitics, Political Science, International Law, Constitutional Law, Education Law, Social Studies, Human Rights, Legal Foundations of Tolerance, and Freedom of Conscience.

REFERENCES TO THE INTRODUCTION

List of Sources Used

1. Buryanov S.A., Buryanov M.S. In Search of an Adequate World Order: The Future of the Rule of Law in the Era of Digital Globalization 4.0 // Age of Globalization. 2024. No. 2 (50). pp. 127-140. DOI: 10.30884/vglob/2024.02.10.
2. Buryanov S.A., Buryanov M.S. Globalization of the World Order and the Law of Force. What Led to the Intensification of Force-Based Geopolitical Confrontation in the Modern World, and Is There a Prospect of Evolving Toward Sustainable Human-Centered Development? Part 1 // Eurasian Law Journal. 2025. No. 5 (204). pp. 23-27.
3. Buryanov S.A., Buryanov M.S. Globalization of the World Order and the Law of Force. What Led to the Intensification of Force-Based Geopolitical Confrontation in the Modern World, and Is There a Prospect of Evolving Toward Sustainable Human-Centered Development? Part 2 // Eurasian Law Journal. 2025. No. 7 (206). pp. 18-22.
4. Lukashuk I.I. Interaction between International and Domestic Law under Conditions of Globalization // Journal of Russian Law. 2002. No. 3. pp. 115-128.
5. Khlestov O.N. Globalization and the Fight against International Terrorism // International Lawyer: All-Russian Journal of International Law. 2003. No. 2. pp. 24-28.
6. Khlestov O.N. The USSR, Russia, and the World Order: 70 Years of Service to the Fatherland / O.N. Khlestov. 4th ed. Moscow: Dashkov and Co. Publishing and Trading Corporation, 2022. 171 p.
7. Farkhutdinov I.Z. The American Doctrine of a Preventive Military Strike: International Legal Aspects. Moscow, 2017. 338 p.
8. Farkhutdinov I.Z. The American Doctrine of a Preventive Strike from Monroe to Trump: International Legal Aspects. Moscow, 2018. 419 p.
9. Farkhutdinov I.Z. The Israeli Doctrine of Preventive Self-Defense and International Law // Eurasian Law Journal. 2016. No. 8. pp. 15-32.
10. Farkhutdinov I.Z. The Iranian Doctrine of Preventive Self-Defense and International Law // Eurasian Law Journal. 2017. No. 1. pp. 15-26.
11. Farkhutdinov I.Z. The Iranian Doctrine of Preventive Self-Defense and International Law (Conclusion) // Eurasian Law Journal. 2017. No. 2. pp. 15-25.
12. Emelyanova N.N. International Legal Problems of Ensuring Global Security at the Present Stage: Doctoral Dissertation in Law. Moscow, 2013. 404 p. p. 5.
13. Moiseev A.A. The World at a Crossroads: International Law in an Era of Global Change // Eurasian Law Journal. 2026. No. 1 (212). pp. 13-21.
14. Moiseev A.A. Supranationality in Contemporary International Law. Moscow: Scientific Book, 2007. 288 p.
15. Moiseev A.A. The Importance of International Law for the Development of Dialogue among National Cultures under Conditions of Globalization // Dialogue of Cultures under Conditions of Globalization: Proceedings of the 12th International Likhachev Scientific Readings (Saint Petersburg, May 17-18, 2012). Saint Petersburg: Saint Petersburg University of the Humanities and Social Sciences, 2012. Vol. 1. pp. 484-486.
16. Chumakov A.N. Globalization and Digitalization: Social Consequences of Cumulative Interaction // Problems of Philosophy. 2021. No. 8. pp. 36-46.
17. Kartashkin V.A. Human Rights and Principles of International Law in the 21st Century: Monograph. Moscow: Norma: INFRA-M, 2018. p. 115.
18. Krivolapov P.S. New Trends in International Cooperation in the Field of Human Rights: Candidate of Legal Sciences Dissertation. Moscow, 2006. 213 p.
19. Ekimov A.I. The "Convergent Model" of Social Development and Law // Law and Modern States. 2015. No. 4. pp. 11-18.

20. Cited work. p. 17.
21. Cited work. p. 18.
22. International Law: Special Part: Textbook for Students of Law Faculties and Universities. Lukashuk I.I. Moscow: Wolters Kluwer, 2005. p. 245.
23. Sledzevsky I.V. The Desecularization of the World Community as a Source of Tension in International Relations and World Politics // Outlines of Global Transformations: Politics, Economics, Law. 2018. Vol. 11. No. 4. pp. 30-45. DOI: 10.23932/2542-0240-2018-11-4-30-45.
24. Freedom of Conscience: Problems of Theory and Practice. Monograph / ed. by F.M. Rudinsky and S.A. Buryanov. Moscow: CJSC Trading Firm "MIR", 2012. 1120 p.
25. Yakunin G., Regelson L. Letters from Moscow: Religion and Human Rights in the USSR / Keston College, Centre for the Study of Religion and Communism. 1978. 157 p.
26. Yakunin G.P. On the Current Situation of the Russian Orthodox Church and the Prospects for Religious Revival in Russia: Report of the Christian Committee for the Defense of Believers' Rights in the USSR / Priest Gleb Yakunin. Frankfurt am Main: Posev, 1979. 143 p.
27. Alekseeva L.M. The History of Dissent in the USSR: The Contemporary Period. 3rd stereotype ed. Moscow, 2012. 384 p.
28. Buryanov S.A., Buryanov M.S. Human Rights as a Key Factor in Achieving Sustainable Managed Development // Age of Globalization. 2022. No. 4 (44). pp. 97-110. DOI: 10.30884/vglob/2022.04.07 41000.
29. Buryanov S.A., Buryanov M.S. New Vectors in the Evolution of Human Rights under the Contradictory Dynamics of Global Processes 4.0 // Bulletin of the Academy of Justice. 2023. No. 2 (2). pp. 18-25.
30. Buryanov S.A. The Right to Freedom of Conscience under Conditions of Global Processes and Challenges 4.0 // Eurasian Law Journal. 2023. No. 6 (181). pp. 18-21.
31. Buryanov S.A., Buryanov M.S. In Search of an Adequate World Order: The Future of the Rule of Law in the Era of Digital Globalization 4.0 // Age of Globalization. 2024. No. 2 (50). pp. 127-140. DOI: 10.30884/vglob/2024.02.10.
32. Buryanov S.A. International Recognition of the Right to Freedom of Conscience and Problems of Its Implementation in the Russian Federation under Conditions of Contemporary Global Processes: Monograph. Moscow: Polygraph Service, 2020. 624 p.
33. Civil Human Rights in Russia: Modern Problems of Theory and Practice / compiled by F.M. Rudinsky. Transaction Publishers, New Brunswick, New Jersey (U.S.A.), 2008. 458 p.
34. Buryanov S.A. State Worldview Neutrality in the Context of Deteriorating Imbalances in Globalization: The Case and Current State of Affairs in the Russian Federation from 2016 to the Beginning of 2017. Amsterdam, 2018. 228 p.

PART I. HOW HUMANITY CAME TO THE RIGHT TO ITS OWN PICTURE OF THE WORLD

CIVILIZATIONAL THESIS

"Sapere aude" - dare to know.

The history of human rights begins where the person gains legal space to form a picture of the world. Freedom of conscience is therefore the first legal language of inner autonomy: the right to believe, doubt, search, revise, and refuse an imposed truth. In this part, the classical question of religion becomes a wider question of worldview liberty.

CONCEPT BOX

CHAPTER 1

FROM RELIGIOUS TOLERATION TO WORLDVIEW LIBERTY

The expansion of freedom of conscience

The history of freedom of conscience begins with the struggle to protect the person from imposed religious truth.

Over time, this legal history reveals a wider problem. A human being needs protection not only in the choice of religion, but also in the formation of a broader worldview: philosophical, ethical, scientific, civic, cultural, and personal.

Chapter 1 shows how classical freedom of conscience can become the foundation for a modern right to liberty of worldview choice.

This transition matters because the inner life of the person cannot be reduced to confession, belief, or non-belief. It includes the whole field in which meaning is formed.

THE HUMAN PICTURE OF THE WORLD

Freedom of Conscience in the Age of Globalization 4.0

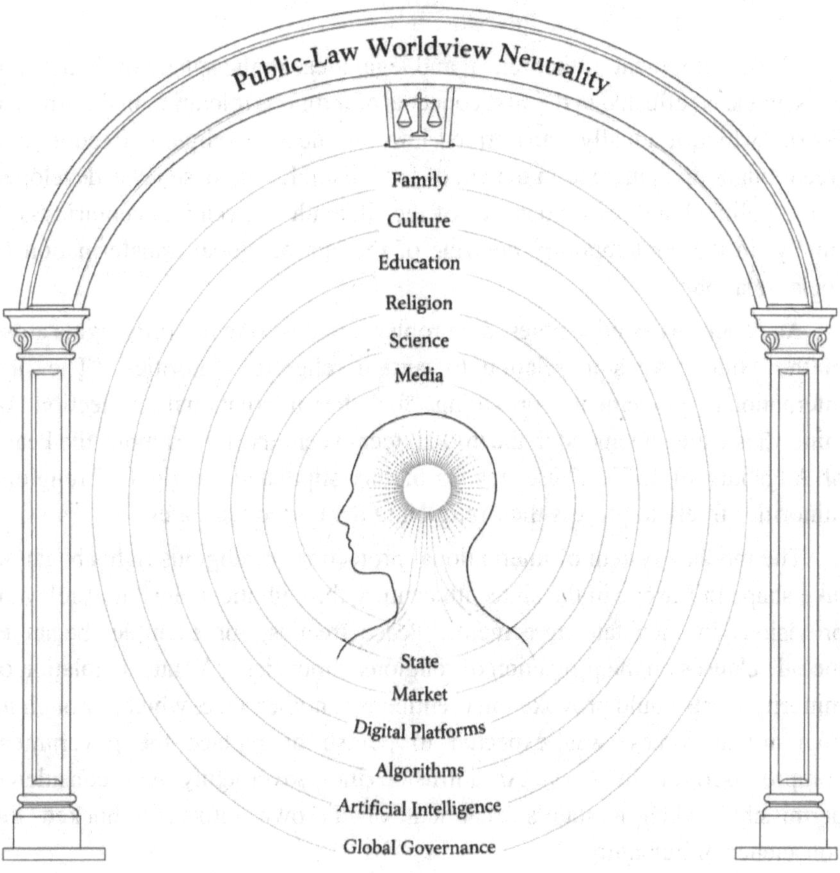

Public authority must protect the space of worldview formation, not occupy it.

The human picture of the world is formed through many layers: family, culture, education, religion, science, media, public authority, markets, platforms, algorithms, and global processes. The central legal question of the post-digital age is whether these forces help the person develop freely or begin to appropriate the inner space in which meaning, belief, identity, and conscience are formed. Public-law worldview neutrality creates a legal boundary against the monopolization of meaning.

Chapter 1. From Religious Toleration to Liberty of Worldview Choice

1.1. Formation and Development of the Right to Freedom of Thought, Conscience and Religion in International Law

The development of theoretical and legal ideas in the sphere of freedom of conscience is reflected in the first concepts of religious toleration in the Ancient World, which gradually transformed into the ideas of religious freedom and free thought that emerged in the era of feudalism. Later, these ideas developed in the political and legal doctrines of the fifteenth to twentieth centuries and, finally, in the contemporary doctrine of the age of global transformation of social relations.

As A. Kh. Abashidze notes, the problems of ensuring minority rights arose in the Middle Ages in relation to certain religious minorities. "The first international agreements providing for the international protection of minorities were concluded in the mid-sixteenth century, for example, the Peace of Augsburg of 1555. These treaties mainly stipulated the right of religious minorities freely to profess their religion in the respective states."

The modern system of international protection of religious rights began to take shape in Europe in the sixteenth century through the inclusion of relevant provisions in interstate agreements. Peace treaties, for example, began to include clauses on the protection of religious minorities. "A state's violation of minority rights could provoke intervention by another state, which, through its own armed forces, was expected to punish or replace the government compromised in this way. An intrusion into sovereignty was considered permissible when a state's treatment of its own citizens 'shocked the conscience of humanity.'"

According to I. I. Lukashuk, Hugo Grotius, as early as the seventeenth century and relying on the Bible, argued that differences in politics and faith did not prevent the existence of a unified international law. In practice, however, everything turned out differently, and this thesis found no embodiment in the real policies of states, which, "in violation of Christian canons, in the eighteenth and nineteenth centuries created a discriminatory concept of international law of Christian states in relation to other peoples." Ultimately, in the researcher's view, "the very idea of international law arose to a considerable extent as a kind of reaction to the inability of religion to ensure peaceful international relations and to its role in initiating religious wars."

One may agree with the view that religion played the role of a normative complex in international relations long before the formation of international law. Accordingly, "the content and experience of the implementation of this complex could not but be taken into account in the formation of international law. Religion also influenced international law through the domestic law of states, the content of which was also largely determined by religion." At the same time, it should be borne in mind that "international law was created predominantly on a European basis, with Christianity playing the predominant role," and only with a certain influence of other religions.

Thus, after a certain period following the conclusion of the Peace of Westphalia in 1648, the priority of protecting religious freedom largely yielded to the pressure of state sovereignty. However, during the eighteenth to twentieth centuries, the ideals of international protection of religious freedom began to strengthen their position.

For example, certain components of the protection of religious freedom were enshrined in the international law of armed conflicts. A. A. Dorskaya considers the protection of religious freedom in connection with the following aspects: violations of that freedom may be one of the causes of an armed conflict; medical assistance in accordance with the principle of impartiality, one of the basic principles of the Red Cross, must be provided without any discrimination, including on grounds of religion; protection of the freedom of conscience of prisoners of war under Chapter VIII of the Geneva Convention relative to the Treatment of Prisoners of War of 12 August 1949; international protection during armed conflicts of buildings and objects constituting cultural property, including those connected with religious worship, under Article 1 of the Hague Convention for the Protection of Cultural Property in the Event of Armed Conflict of 14 May 1954; agreements concluded by the parties to a conflict on the passage of religious personnel and property into the conflict zone, under Article 15 of the First Geneva Convention of 1949; respect for religious convictions and religious rites during military occupation, under Article 46 of the Fourth Hague Convention respecting the Laws and Customs of War on Land of 18 October 1907 and Article 27 of the Fourth Geneva Convention relative to the Protection of Civilian Persons in Time of War of 12 August 1949.

The Covenant of the League of Nations of 10 January 1920, adopted on the basis of the Treaty of Versailles, did not contain legal obligations of the member states in the field of religious freedom.

Freedom of conscience and religion was mentioned in connection with the territories of former colonies administered under the mandate system, later the United Nations trusteeship system. The Covenant of the League of Nations referred to guarantees of "freedom of conscience and religion, subject only to such restrictions as may be necessary for the maintenance of public order and morals" under Article 22.

To protect freedom of conscience and religion in the former colonies of Germany and Turkey, the League of Nations established the mandate system. As a condition for preparing colonies for independence, they were required to guarantee "the protection of religious, linguistic and ethnic minorities, as well as the rights of foreigners and freedom of conscience." The territories in question later included Palestine and Jordan under British administration; Syria and Lebanon under French administration; Cameroon and Togoland under British and French administration; and Rwanda under Belgian administration.

It is possible to distinguish three groups of normative acts aimed at protecting religious rights, formed between 1919 and 1939 within the League of Nations framework. The first group consisted primarily of treaties between the Allied states and newly admitted members of the League of Nations, including treaties with Bulgaria, Austria, Poland and others. The second group included unilateral declarations by a number of states, including Albania, Estonia and others. Finally, the third group included local agreements, such as the treaty between Germany and Poland of 1920. "The most important purpose of this system was to ensure, through special treaty-based legal relations, international guarantees for the protection of the rights of various groups of peoples, ethnic, religious and linguistic minorities living in the territories of the former Austro-Hungarian and Ottoman Empires."

Another important step toward the formation of the institution of international protection of freedom of conscience was the adoption by the Institute of International Law in 1929 of the Declaration of the International Rights of Man, which enshrined the duty of states to recognize the religious convictions of every person within their territory.

The Declaration by United Nations, signed on 1 January 1942 by twenty-six states participating in the anti-Hitler coalition and later becoming the basis of the future United Nations, stated the need to protect life, liberty, independence and religious freedom, as well as to preserve human rights and justice both in their own countries and in other countries.

Within the United Nations, created after the Second World War, human rights received further development. The Charter of the United Nations of 26 June 1945 called for reaffirming "faith in fundamental human rights" on the basis of tolerance and also affirmed the aspiration of states to carry out international cooperation in solving world problems by promoting and encouraging respect for human rights and fundamental freedoms.

A quarter of a century later, the Declaration on Principles of International Law concerning Friendly Relations and Cooperation among States in accordance with the Charter of the United Nations, adopted by United Nations General Assembly resolution 2625 (XXV) of 24 October 1970, referred to the obligations of states to maintain and strengthen international peace founded upon freedom, equality, justice and respect for fundamental human rights; to cooperate in promoting universal respect for and observance of human rights and fundamental freedoms for all and in eliminating all forms of racial discrimination and all forms of religious intolerance; and to promote, through joint and separate action, universal respect for and observance of human rights and fundamental freedoms in accordance with the Charter.

In 1975, the Final Act of the Conference on Security and Cooperation in Europe was adopted. It confirmed respect for human rights regardless of distinctions such as race, sex, language and religion, including freedom of thought, conscience, religion and belief, as a principle of international law. In particular, the participating states undertook to respect the rights of the individual to profess and practise, alone or in community with others, religion or belief, acting in accordance with the dictates of one's own conscience.

The principle of respect for human rights and fundamental freedoms, including freedom of thought, conscience, religion and belief, was further developed in the international documents known as the International Bill of Human Rights: the Universal Declaration of Human Rights of 1948, the International Covenant on Economic, Social and Cultural Rights of 1966, the International Covenant on Civil and Political Rights of 1966 and its two Optional Protocols.

It should be noted that both Western and Russian doctrines of international law understand "generally recognized principles of international law" as "general principles of international law" or "general principles of law recognized by civilized nations," which are also often called "fundamental" principles. International courts, in particular the International Court of Justice, do not draw a strict distinction between the principles they apply. Moreover, the International Court of Justice and the doctrine of international law may classify both general principles of law and moral norms as applicable principles.

On the basis of an analysis of the fundamental universal documents forming the Bill of Human Rights, one may conclude that they constituted an important stage in the evolution of internationally recognized freedom of conscience. At the same time, it must be acknowledged that they are aimed mainly at the protection of religious freedom and do not fully meet the principle of legal certainty or the requirements of modern legal drafting. In particular, attempts to overcome numerous terminological contradictions were undertaken by the United Nations Human Rights Committee, as will be considered in more detail below.

Issues touching upon the sphere of freedom of conscience are also contained in other legally binding and recommendatory universal international legal instruments: the Convention on the Prevention and Punishment of the Crime of Genocide of 1948, the Convention relating to the Status of Refugees of 1951, the Convention relating to the Status of Stateless Persons of 1954, the Convention against Discrimination in Education of 1960, the Convention on the Elimination of All Forms of Discrimination against Women of 1979, the Convention on the Rights of the Child of 1989, the 1981 Declaration on the Elimination of All Forms of Intolerance and of Discrimination Based on Religion or Belief, the 1992 Declaration on the Rights of Persons Belonging to National or Ethnic, Religious and Linguistic Minorities, the 1995 Declaration of Principles on Tolerance and other instruments.

I believe that the question of the effectiveness of universal international documents in the sphere of freedom of conscience requires detailed study. In addition to universal instruments, numerous regional documents in the field of freedom of conscience have also been adopted, many of them highly controversial from a legal point of view.

In the context of the preliminary analysis of universal and regional documents in the sphere of freedom of conscience, it is necessary to agree with the experts who argue for the need to study and further enshrine in international law the secularity of the state as worldview neutrality. The problem is that the secularity of the state, being the most important guarantee for the realization of freedom of conscience and protection against intolerance and discrimination, is not designated in international legal documents, thereby creating grounds for human rights violations and conflicts.

Undoubtedly, "the enshrinement in fundamental international legal documents of the worldwide protection of human rights and fundamental freedoms, including the human right to freedom of conscience and religion, is one of the most important achievements of modern international law." Yet it must also be recognized that an inadequate understanding of contemporary social relations and unresolved theoretical and legal problems predetermine the lag of legal science and significantly reduce the effectiveness of international legal principles and norms in the field of internationally recognized freedom of conscience. Analysis of the relevant principles and norms has revealed their defects, contradictions and collisions.

One must also agree with the opinion of M. Rohn. The expert notes that "the struggle for freedom of religion has lasted for many centuries, resulting in countless and tragic conflicts. The twentieth century witnessed the codification of common values relating to freedom of religion, although the intensity of clashes did not diminish... After the adoption of the Universal Declaration, attempts to develop an implementable international legal instrument concerning the right to freedom of religion and belief were unsuccessful."

The low effectiveness of the international system for the protection of freedom of conscience has led to a disruption of the ethno-confessional balance of international relations and has become a catalyst for the aggravation of global problems facing human civilization. In particular, global challenges have recently manifested themselves with particular severity in the sphere of international security.

It appears extremely relevant to improve the effectiveness of universal international legal documents in the sphere of freedom of conscience through their reform and to bring regional and domestic documents into conformity with them.

International legal documents, as rules established and fixed in the relevant act, occupy a central place in the system of international human rights protection. In this case, we are speaking of treaty norms and of norms contained in acts of international organizations and conferences. It should be noted that, in addition to the above-mentioned documented norms, customary international legal norms also exist. As a result of codification, they may become treaty norms, and under certain conditions they are enshrined in law-enforcement acts as well as in resolutions of states and international organizations.

Fundamentally new approaches are needed to reform the principles and norms in the sphere of the internationally recognized right to freedom of conscience as one of the foundations for the progressive development of international law. In this context, serious modernization is required of international legal and domestic principles and norms in the field of the right to freedom of conscience, which must be brought into line with the requirements of modern legal drafting.

In turn, increasing the effectiveness of the above-mentioned principles and norms requires the creation of an adequate system of guarantees for the realization of freedom of conscience on the basis of the principle of worldview neutrality of states and other subjects of international law. The principle of worldview neutrality of states and other subjects of international law must be studied and normatively enshrined. It seems appropriate to discuss this issue within the expert structures of universal international organizations and to develop a corresponding international programme.

The implementation of an innovative paradigm of freedom of conscience is considered one of the most important and necessary conditions for the formation of global thinking and of a set of other conditions required for the transition to the sustainable development of civilization. In fact, this concerns a new stage in the development of freedom of conscience as a global value, which today exists only in scholarly works and is reflected in the activities of a limited number of public institutions.

In international law, the evolution of the institution of freedom of thought, conscience and religion is interconnected with the development of its international legal norms and institutions, which in turn are inseparably connected with the transformation of international social relations. In particular, the following evolutionary stages may be distinguished: first, the law of interstate relations; second, the law of international relations, with a certain expansion of the circle of subjects; and third, the beginning of the formation of the law of relations within the world, planetary or global community.

Since, under the conditions of globalization of social relations, contemporary international law is called upon to serve as an instrument for maintaining the international legal order and ensuring equal cooperation among states and peoples in solving the global problems of the modern world, the importance of realizing human rights, including the right to freedom of conscience, increases substantially.

Thus, the following main stages may be identified in the formation and development of the right to freedom of thought, conscience and religion in international law: first, the sixteenth century, the protection of the rights of religious minorities on the basis of interstate agreements; second, the eighteenth century to the first half of the twentieth century, the beginning of the formation of a system of international protection of the right to religious freedom, including certain components of the protection of religious freedom enshrined in the international law of armed conflicts and in League of Nations documents, including the protection of freedom of conscience and religion in former colonies; third, the mid-twentieth century to the end of the twentieth century, the consolidation and development of the right to freedom of thought, conscience and religion as a component of the international system of human rights protection, including the United Nations Charter, the International Bill of Human Rights and documents of regional organizations; and fourth, the beginning of the twenty-first century, the doctrinal development of a new paradigm of the internationally recognized right to freedom of conscience as liberty of worldview choice for every person on the basis of the worldview neutrality of states and other subjects of international law.

1.2. Theoretical and Legal Problems in the Field of the Right to Freedom of Conscience and Their Impact on the Effectiveness of International and Domestic Regulation

The study of the effectiveness of international legal and domestic regulation in the field of human rights, especially with regard to the right to freedom of conscience, is inseparably connected with a complex of theoretical and legal questions. While acknowledging the contribution of earlier scholars, one must nevertheless state that the scientific and theoretical elaboration of freedom of conscience and related categories remains insufficient. In particular, the definition and content of the right to freedom of conscience remain underdeveloped, as do questions concerning the relationship between freedom of conscience and related concepts such as freedom of religion, the secularity of the state, state religious policy, relations between the state and religious associations, and others.

The unresolved relationship between the above-mentioned concepts and the concept of freedom of conscience leads to legally incorrect interpretations of the relevant principles and conceptual apparatus. As a result, state religious policy and relations between the state and religious associations are, in a highly debatable manner, presented as having priority over freedom of conscience.

A serious problem is the use, within the legal system, of disputable terms drawn from religious studies, including traditional religions or religious organizations, non-traditional religions or religious organizations, sect, as well as the concepts of the feelings of believers, missionary activity, spiritual security, religious extremism, Islamic extremism, and similar terms.

In turn, the degree of scholarly development of key and related concepts in the sphere of the right to freedom of conscience has a significant influence on the formation of international legal principles and norms, as well as on their interaction with the domestic legislation of the Russian Federation. In particular, a narrow interpretation of freedom of conscience in law-making and law enforcement initially predetermines the absence of legal guarantees for persons who are not members of religious organizations, above all for non-believers.

The situation is aggravated by the absence, and indeed the fundamental impossibility, of creating a single legal definition of religion. The possibility of multiple interpretations of the term religion predetermines corruption risks in the application of that term. In this regard, the lawfulness of state expert evaluation in religious studies, and the use of knowledge from religious studies in resolving legal questions, becomes an especially relevant issue.

The correctness of other legal concepts derived from the term religion, for example religious association, is also debatable. The category of secularity of the state, closely connected with the problem of realizing freedom of conscience, also requires clarification, since it is often identified with secularism and defined as non-religiosity.

It is important to understand the significance of realizing freedom of conscience for achieving tolerance and overcoming discrimination, intolerance and xenophobia on grounds of worldview affiliation.

The problem of the influence of the ethno-confessional factor on security has become extremely acute at both the international and domestic levels.

The above-mentioned problems require study in light of the contemporary character of social relations and the requirements of legal drafting technique. It is clear that any normative materials must correspond to contemporary scholarly developments. In this context, the problem of the effectiveness of international legal principles and norms in the sphere of the right to freedom of conscience becomes particularly acute.

The concept and content of freedom of conscience have been the subject of scholarly and public debate for many years. It should be noted that the mechanical combination of the concepts conscience and freedom does not provide an adequate understanding of the term freedom of conscience as an integral whole. While the concept of freedom has become firmly established in legal science, the term conscience has no unambiguous legal interpretation and does not constitute an element of the system of legal regulation. As a category of ethics, conscience characterizes the capacity of the person for moral self-control, the ability independently to formulate moral duties for oneself, to require oneself to fulfil them, and to assess one's own actions.

At the same time, freedom of conscience is inseparably connected with the concept of individual freedom, which in turn is one of the most important components of personal liberty.

It should also be noted that legal science contains no unified view of the place of freedom of conscience in the human rights system. For example, M. V. Baglay classifies freedom of conscience among personal rights and freedoms, while V. D. Perevalov and V. I. Leushin classify it among cultural rights. F. M. Rudinsky classifies the right to freedom of thought, conscience and religion among civil rights, or personal constitutional rights.

I believe that freedom of conscience, understood as liberty of worldview choice, is the foundation of freedom as such and a system-forming right within the human rights system. It does not seem entirely correct to say that one right is more important than another. Yet without proper realization of freedom of conscience, the entire system of human rights is exposed to threats. The violation of freedom of conscience is accompanied by increased risks of transition toward a non-alternative political system, authoritarianization of power, mass violations of human rights, and the growth of xenophobia, intolerance, discrimination and violence based on them.

Another problem, which has no unambiguous and generally accepted theoretical interpretation, is connected with the use of the legal constructions freedom of conscience and the right to freedom of conscience. Both constructions are used in legislation and law-enforcement practice, often as synonyms.

In the context of the problem of realizing freedom of conscience, the author agrees with F. M. Rudinsky, who considers the right to freedom of conscience to be its legal embodiment. In turn, legal recognition is only one of the conditions for achieving freedom of conscience.

Freedom of conscience is a multifaceted phenomenon touching various aspects of social life. Accordingly, numerous approaches exist to the understanding of freedom of conscience. The aspect-based approach was first proposed by F. M. Rudinsky, who believed that the multifaceted nature of this freedom predetermines the fact that it can be an object of study for various social sciences: philosophy, ethics, political science and jurisprudence. Thus, when we speak of freedom of conscience in the moral and philosophical sense, we mean primarily human freedom in the sphere of moral relations, the possibility of acting according to the commands of one's conscience while not violating social, including legal, norms. In the sociological sense, freedom of conscience is a spiritual value and an important social good created by society through historical development. In the political sense, the exercise of freedom of conscience is one aspect of democracy. Its socio-political content is determined by the nature of the social order, the character of state power, the political regime, the level of development of science and culture, the role of the church in the political and spiritual life of society, the historical traditions existing in the country, and other factors. Later, many researchers also adopted the aspect-based approach.

Depending on the connection with the general methodology of a particular sphere of social knowledge, N. Yu. Teteryatnikov distinguishes two main types of approaches: first, approaches relatively independent of such methodology, including axiological, concrete-historical and concrete-legal approaches; and second, approaches situated within a particular sphere of social knowledge. The latter may conditionally be divided into two types: approaches that consider freedom of conscience primarily in connection with the individual, their beliefs, consciousness and conscience, that is, in general, with morality, which may be either religious or non-religious, including theological and ethical approaches; and approaches that relate this freedom primarily to the social relations into which the individual enters when their freedom is limited by the same freedom of other persons, including sociological, political-science, legal, religious-studies and philosophical approaches.

The consequence of the diversity of approaches to understanding freedom of conscience is the ambiguity of its interpretations. At the same time, particular interpretations of this freedom may be mutually exclusive in relation to others, or may not intersect with them at all. In the theoretical and legal sense, approaches to the problem of realizing freedom of conscience include two interconnected meanings: freedom of conscience as a subjective right and as objective law, that is, as a legal institution.

In general, the principle of freedom of conscience was theoretically conceptualized and evolved more as a historical, philosophical and ethical category than as a legal one. Moreover, for a long historical period many worldview questions were closely connected with the religious worldview and were resolved from the standpoint of a religious picture of the world and religious morality. Accordingly, the concept of freedom of conscience acquired a narrower meaning connected with the individual's attitude toward religion.

Gradually, in the words of M. P. Mchedlov, freedom of conscience increasingly came to be understood not only as the right to profess any religious views or to deny them, but much more broadly. The shift of emphasis from the philosophical aspects of freedom of conscience toward legal aspects occurred as legal and political systems transformed. Moreover, this shift itself exerted reciprocal influence on that transformation. In the words of Alting von Geusau, recognition of the fundamental principle of freedom of thought and conscience swept away the old era of hierarchies of power.

According to N. Yu. Teteryatnikov, freedom of conscience increasingly came to be viewed through the prism of human rights and freedoms, which received positive expression in the legal sources both of individual states, including Russia, and of the international community as a whole. This pattern illustrates the transition from natural rights to positive law, from subjective rights to objective law. That is, the general tendency is that law gradually ceases to be an instrument of the state for managing social processes and increasingly becomes a guarantor of the freedom of every person in society and the state. The role of freedom of conscience here is primary. This role consists in the fact that freedom of conscience, being a positively established freedom of the individual in the worldview and behavioral spheres, serves as the legal basis of personal liberty in the modern rule-of-law state and civil society.

At the international level, the ideals of freedom of religion have gained recognition throughout the world. Without in any way diminishing their significance, it must be stated that the independent concepts enshrined in the Universal Declaration of Human Rights of 1948 and other international documents - freedom of thought, conscience and religion - have become mixed in the works of legal scholars and, consequently, on the pages of legal documents. As Gloria M. Moran notes, the 1948 Universal Declaration and the most recent constitutions following its example speak of freedom of thought, conscience and religion. In this connection, scholars cannot agree whether there is only one global freedom considered in three different aspects, or whether three independent types of freedoms should be distinguished. From the 1950s until the end of the 1970s, most European scholars preferred the latter option, taking into account the specific purpose of religious freedom; more recently, however, they have inclined toward the view that this term is inseparably connected with freedom of conscience.

Yet even the formula freedom of thought, conscience and religion is not the only generally accepted one. At the same time, the diversity of legal constructions is by no means the only problem. The main problem is that the concept of freedom of conscience and the related legal regulation of relations concerning everyone have been reduced almost exclusively to freedom of religion, that is, a freedom concerning only individuals who consider themselves believers, and even further, to problems concerning the activity of religious associations, that is, the collective form of realizing the right to freedom of religion.

As a result, legislation on freedom of conscience came to be perceived as special, sectoral religious legislation. Priority is given to religious associations, rather than to the human person, who becomes an appendage of confessions. The right of everyone to freedom of conscience remains in the legal system merely as a declaration, in fact situated outside the legal field. It should be noted that not only many secular scholars but also theologians participated in the development of the principle of freedom of conscience. While recognizing certain merits of the latter, one must nevertheless state that the conceptual apparatus used in defining freedom of conscience bears a certain confessional imprint. In addition, this conceptual apparatus was developed in the course of studying monotheistic religions, above all Christianity. It is therefore unsurprising that its use in relation to non-theistic material often distorts the picture.

A. S. Lovinyukov believes that, because of the legal tradition of reducing the concept of the right to freedom of conscience to the right to profess a religion or conduct atheist propaganda, Russian scholars and legislators were unable to develop and adopt modern and effective legislation.

For example, one of the leading experts and drafters of legislation on freedom of conscience, Yu. A. Rosenbaum, understood freedom of conscience as the right of citizens, permitted and guaranteed by the laws of the state, freely and independently to determine their attitude toward religion and to carry out actions arising from that attitude, provided legality and the legal order established in the country are observed.

Other scholars and experts also largely followed established patterns. For example, M. V. Baglay defined freedom of conscience as a person's right both to believe in God in accordance with the teaching of one or another religion freely chosen by them, and to be an atheist, that is, not to believe in God.
A. A. Bezuglov and S. A. Soldatov understand the right to freedom of conscience as the right of every person independently to decide the question of their attitude toward religion: to be a believer or an atheist.

A. V. Pchelintsev recognizes freedom of religion as part of the broader constitutional category of freedom of conscience, covering the entire spectrum of worldview rights and freedoms. At the same time, however, he considers certain separate complexes of individual religious rights and freedoms exercised by a person independently, and collective religious rights and freedoms exercised by a person together with others and forming an integral part of the system of general permissions.

A. A. So holds a similar position, considering freedom of religion to be an independent constitutional freedom of the person, relating to the religious sphere and not identical to freedom of conscience. Freedom of religion covers that part of human consciousness which concerns the attitude toward the religion professed, toward faith, that is, it is freedom of choice within the limits of religious consciousness and behavior. In a broad sense, freedom of religion may be considered as a freedom touching questions of faith in spiritual and moral ideals as a whole. Thus, freedom of religion is the type and measure of possible conduct of a person in choosing any moral and ethical views and beliefs based on one or another religion or on faith in spiritual and moral ideals, as well as the possibility of following these beliefs, provided the law is observed.

Unlike Soviet and many contemporary lawyers, F. M. Rudinsky in 1963 gave a definition of freedom of conscience without tying it to religion: in the broad sense of the word, freedom of conscience is the possibility for every person to perform actions in accordance with their ideas of justice and injustice, good and evil, if those ideas do not contradict the morality accepted in a given society and in a particular environment.

One should also note the definition proposed by the philosopher V. N. Savelyev in the early 1990s, according to which freedom of conscience is the securing in society of such democratic rights and freedoms that actually guarantee the person the possibility of manifesting their convictions in society.

S. A. Avakyan defines freedom of conscience as the freedom of a person's moral and ethical views, that is, what to consider good and evil, virtue or baseness, a good or bad act, honest or dishonest behavior, and so on. Freedom of religion is the possibility of believing in the existence of some extraordinary, divine being - the most honest, just and humane, concerned with the moral purity of each of us, helping us choose the true path, restraining us from bad actions, and directing us toward helping our neighbor.

In the author's 2001 article, it was stated that freedom of conscience is the right to satisfy a worldview need that gives meaning to existence. Freedom of conscience therefore includes the full diversity of forms of systems of worldview orientation.

Among contemporary researchers, one should also single out the definition by N. Yu. Teteryatnikov, which is free from attachment to religion. He considers freedom of conscience to be a fundamental, complex, first-generation natural subjective right belonging to every person by birth and expressed as the personal freedom of the individual in the sphere of their worldview, as well as behavior based on that worldview, to which there implicitly corresponds a duty of behaviorally tolerant attitude toward the legally permissible freedom of all other persons.

Approaches to defining the concept of freedom of conscience are usually interconnected with approaches to its content. For example, A. S. Lovinyukov defines the content of freedom of conscience in connection with the right to profess religion or to reject it, namely as a complex legal category consisting of ten interconnected structural elements: the right to profess any religion; the right to perform religious rites; the right to change religion; the right not to profess any religion; the right to propagate religion; the right to conduct atheist propaganda; the right to charitable activity; the right to religious education; cultural and educational religious activity; and equality before the law of all citizens, regardless of their attitude toward religion. It should be noted that the above-mentioned approaches were also characteristic of most researchers of the Soviet period.

Studying the problem of the content of freedom of conscience, the author wrote in 2000 that the right of every person to freedom of conscience is a system-forming right within the human rights system. By realizing this right, a person becomes aware of themselves, unlike animals, acquiring meaning and determining their place in life. Whether a person becomes an end or a means, a personality or a cog, depends on how that person defines their meaning; the person's present and future depend on it. Freedom of conscience therefore includes the full diversity of forms of systems of orientation, including the one called religious, realized individually and/or collectively on the principles of equality.

In the approaches of S. Yu. Simorot, the attachment of freedom of conscience to the above-mentioned outdated approaches is significantly weaker, but it has not been completely avoided. The position of N. Yu. Teteryatnikov appears consistent and maximally correct, since it is not connected with the concept of religion at all. Freedom of conscience as the subjective right of every person includes the right to form, choose, have, change, defend and disseminate beliefs, the right to adhere to beliefs and the right to renounce them, as well as the right to act or refrain from acting in accordance with beliefs, provided that the externally expressed conduct of the individual is lawful. O. V. Bryantseva took half a step backward and, to a certain extent, connected the legal content of the right to freedom of conscience with the definition of religion. E. V. Tikhonova believes that at the present stage of legal genesis, freedom of conscience correlates with the freedom of religious, worldview and other beliefs of the individual and their lawful behavior based on those beliefs. Thus, the above analysis shows that the tendency toward broadening approaches to the content of freedom of conscience is not stable even among researchers who have accepted its broadest definition.

A contemporary scholarly approach is found in the works of T. P. Minchenko, who believes that freedom of conscience is a system-forming and fundamental right in the human rights system, connected with satisfying a worldview need, the realization of which forms worldview identity and gives meaning to life on the basis of free choice. A similar legally correct definition appears in recent collective works, as well as in the author's monographs. D. M. Nikitaev also gives a contemporary definition of freedom of conscience as free worldview choice, enshrined in international and domestic law.

In the context of the present study, the following main approaches to defining the concept and content of freedom of conscience may be distinguished.

1. An approach that connects the concept and content of freedom of conscience only with the attitude toward religion or confession, and with the activities of religious associations.

2. An approach that connects the concept and content of freedom of conscience not only with the attitude toward religion or confession and the activities of religious associations, but also with other worldview forms.

3. An approach that connects the concept and content of freedom of conscience with the attitude toward the worldview sphere, without specially singling out religion or confession and the activities of religious associations. I believe that precisely this approach, which defines freedom of conscience as liberty of worldview choice, corresponds to contemporary social relations, the principle of the rule of law and the requirements of legal drafting technique.

In the context of studying the problem of realizing the right to freedom of conscience, the relationship between personal and collective freedom is highly relevant. In particular, N. Yu. Teteryatnikov, having proposed his own scheme for the relationship between freedom of conscience and freedom of religion, notes that freedom of conscience is a personal freedom that cannot belong to a collective subject, such as any public association, including a religious one. Further, disputing the inclusion of freedom of religion in full within freedom of conscience, the author proposes that within freedom of conscience one should see only individual freedom of religion in the part concerning its natural-law elements, while the positive-law elements of this freedom, including the rights of religious associations, the institution of the secrecy of confession and others, remain outside freedom of conscience and constitute an independent subject of freedom of religion.

Analysis of the real situation shows that the freedom of conscience of every person is held hostage by self-contained relations between the state and religious associations and by state confessional policy. The formal pretext for their dominance is allegedly concern for the realization of the rights of believers, that is, freedom of religion, while in reality it concerns the corporate interests of confessional bureaucracy and power. The former receive special privileges from the state in exchange for political support.

Therefore, the more correct position appears to be one that excludes the opposition between the individual and the collective: first, freedom of conscience includes the full diversity of lawful manifestations of systems of orientation, without attachment to the concepts of religion or confession;

second, freedom of conscience includes individual and/or collective forms; third, the collective form is derivative of the individual form within the realization of the right of association. Freedom of conscience therefore includes the full diversity of forms of systems of worldview orientation realized individually and/or collectively on the principles of equality.

Summing up certain results, it should be noted that the right of every human being to freedom of conscience is the legal dimension, enshrined in the system of forms of international and domestic law, of liberty of worldview choice, including the ability to form, adhere to, choose and change an already made choice, as well as lawful behavior based on that choice. This right is system-forming within the human rights system. Without its realization, other human rights lose much of their real content and remain declarative. By realizing the right to freedom of conscience, a person becomes aware of themselves, finds meaning and determines their place in life. Human life cannot be lived merely by repeating species patterns; a person must live their own life. It may be assumed that, by satisfying this need, a human being becomes properly human. Thus, freedom of conscience is the right to satisfy the worldview need that gives meaning to existence.

In general, despite a certain decline in confessional influence on human civilization, there is hardly any basis for speaking of a decrease in every person's need for orientation and self-awareness. Only the forms of realizing this specific need change as civilization develops. Old forms evolve or die away, giving way to new forms.

It may be assumed that the development of forms of systems of orientation is inseparably connected with the evolution of humanity as a whole. Hence the need for equal rights of new forms, including prospective forms that will arise in the future, in relation to those already existing. All the more so because the original forms were once new. Christianity, for example, was at one time new in relation to traditional paganism.

The development of civilization determines a change in the relationship between science and religion in society, and between knowledge and faith in every person. Awareness of the diversity of these relationships and respect for the relationship between knowledge and faith among different people in society are the principal aspect of freedom of conscience in contemporary reality.

In jurisprudence, these principles must be reflected in the right of everyone to freedom of conscience. The realization of this system-forming right determines the ability of the individual to realize themselves as a person; the possibility of overcoming contradictions between tendencies caused by the advanced development of civilization and human nature; the ability of the state to develop sustainably without social explosions and upheavals; and the capacity of the world community to find ways to solve the global problems facing humanity.

Thus, the right to freedom of conscience consists of inalienable opportunities belonging to everyone from birth and enshrined in the system of forms of international and domestic law, to satisfy the worldview need that gives meaning to existence on the basis of free worldview choice, as well as lawful behavior based on that choice, without restriction in or loss of other civil rights and freedoms.

The content of freedom of conscience includes the right, individually and/or together with others, freely to form, choose, change and disseminate beliefs and to act in accordance with them, without infringing the freedom and personal dignity of others.

Awareness of the diversity of relations and respect for the relationship between knowledge and faith among different people in society constitute the main aspect of freedom of conscience in contemporary reality. Freedom of conscience is the core of human identity; one may say that it is the right to be human. Free worldview choice is the basis of free political choice, legal democracy and the social state.

The essence of freedom of conscience is the individual's recognition of diversity, respect for the worldview choice of others, including their views, beliefs and worldview as a whole, as well as respect for the lawful behavior of other persons, including acts and omissions.

From the legal point of view, the special singling out of freedom of religion does not appear correct. In other words, freedom of religion should be considered part of liberty of worldview choice, as if absorbed by it. It is necessary to abandon the special legal status of religious association and the corresponding special legal regulation. This means that the regulation of social relations in the sphere of freedom of conscience should be carried out on general legal grounds.

At the doctrinal level, it is generally accepted that the legal expression of freedom of conscience in an individual state is its legal institution, which is formed from the various legal forms operating in that state and regulating this freedom.

In the Russian Federation, the legal institution of freedom of conscience consists of the following elements: the Constitution of the Russian Federation, namely the constitutional norms fixing the basic provisions of freedom of conscience; international legal acts on human rights enshrining freedom of conscience in one or another interpretation, that is, generally recognized principles and norms of international law in the sphere of freedom of conscience; international treaties of the Russian Federation in the parts relating to the guarantee of freedom of conscience; federal legislation regulating the protection and limitation of freedom of conscience in the Russian Federation; treaties between state bodies of the Russian Federation and its constituent entities concerning the guarantee of freedom of conscience; constitutions or charters and other legal acts of the constituent entities of the Russian Federation aimed at additional regulation of freedom of conscience for the purpose of protecting it in the relevant constituent entity, taking into account its historical and cultural specificity; decisions of local self-government bodies touching upon questions of the realization of freedom of conscience; and judicial practice at all levels in disputes concerning freedom of conscience, including decisions of international human rights courts in relation to Russia and decisions of domestic judicial bodies headed by the Constitutional Court of the Russian Federation.

Thus, in the Russian Federation there operates a system of forms of international and domestic law on freedom of conscience connected by the mechanism of implementation. I believe it is necessary to distinguish between the concepts sources of the right to freedom of conscience in the Russian Federation and forms of the right to freedom of conscience in the Russian Federation. In particular, the following sources of the right to freedom of conscience may be identified: uniform and repeatedly reproduced actions, secured by coercion, of authorized subjects of international human rights law, natural and/or legal persons, for the protection of rights and legitimate interests; agreements of states, other subjects of international human rights law, state authorities, local self-government bodies and other authorized persons; and law-making activity of subjects of international human rights law, state authorities and/or other authorized legal subjects in the Russian Federation.

Accordingly, the actions of subjects of international human rights law and of natural and/or legal persons find expression in the form of customs of international law and customs of Russian law; agreements find expression in the form of international treaties containing norms of the law of freedom of conscience and normative legal treaties containing norms of Russian law; and law-making activity finds expression in the form of normative legal acts containing norms of Russian law on freedom of conscience. As a result, international law on freedom of conscience and Russian law on freedom of conscience function as components, or subsystems, of the above-mentioned unified system. In addition, the system of forms of the law on freedom of conscience in the ndence with the system of forms of human rights law in the Russian Federation, which consists of international human rights law and Russian human rights law.

1.3. The Principles of the Rule of Law and the Law-Governed State as the Basis for Reforming the Forms of International and Domestic Law on Freedom of Conscience

The right to freedom of conscience, as an important component of human rights, is enshrined in international legal principles and norms as a universal value of the modern world. However, unresolved scientific and theoretical problems, above all those concerning fundamental principles and the conceptual apparatus, make even the first stages of realizing freedom of conscience questionable, namely its recognition in international principles and norms and its implementation in domestic law. In this context, all forms of the right to freedom of conscience are exposed from the outset to predetermined distortions that create obstacles to the realization of this system-forming right.

The need to study directions for improving international legal forms is determined by their importance as a basis for increasing the effectiveness of norms of international and domestic law on freedom of conscience.

It appears highly relevant to study the principles of the rule of law, the law-governed state, and other principles that constitute or are connected with them, as a basis for reforming the forms of international law on freedom of conscience.

It is generally accepted that the concept of the rule of law was first presented by the English scholar Albert Venn Dicey in 1885 and consisted of three main components: justice, legal equality, and guarantees of individual liberty.

James R. Silkenat, President of the American Bar Association, speaks of the importance of developing a deeper understanding of the rule of law and its key elements in contemporary international conditions. Referring to the World Justice Project and its Rule of Law Index, he identifies four universal features of the rule of law:

1. The government, its officials and representatives are accountable under the law. 2. Normative legal acts are clear and certain, officially published, stable and fair, and aimed at ensuring and protecting fundamental rights, including the protection of persons and property. 3. The process by which normative legal acts are adopted, administered and enforced is open, fair and rational. 4. Justice is administered by competent, ethical and independent representatives or a neutral party, who are available in sufficient numbers, have adequate resources, and reflect the structure of the society they serve.

The American scholar Richard H. Fallon identifies five elements in the realization of the concept of the rule of law. The first element consists in the capacity of legal norms, standards or principles to guide people in conducting their affairs. People must be able to understand the law and conform their conduct to the norms of law in force. The second element of the concept is efficacy. Law is intended actually to guide the behavior of people, at least for the most part. In other words, people must be guided by law and obey it. The third element is stability. Law must be reasonably stable in order to facilitate planning of behavior and coordination of action over time. The fourth element consists in the supremacy of lawful authority. Law must guide the behavior of public officials, including judges, as well as ordinary citizens. The fifth element includes the existence of institutions of impartial justice. Courts must be capable of enforcing the law and must administer justice on the basis of fairness. The practical realization of these elements of the rule of law creates appropriate guarantees for universally recognized human rights.

According to Fallon, the concept of the rule of law pursues the following aims: first, it is intended to provide protection against anarchy, or, in other words, against a war of all against all. Second, it is intended to create proper conditions for people to plan their future by giving them a reasonable assurance that they can know in advance the legal consequences of various actions. Third, it is intended to provide guarantees against possible arbitrariness by public authorities.

Russian scholars, legislators and law-enforcement bodies began to perceive the concept of the rule of law relatively recently. Yet the rule of law is still often understood, not quite correctly, as the supremacy of statute law.

For example, a 2002 dictionary of constitutional law of the Russian Federation states that the rule of law is a principle of the constitutional order of the Russian Federation, meaning that the state, its bodies and officials are bound by law and by the fundamental rights and freedoms of the person and citizen. The rule of law is said to manifest itself in the supremacy of the law, one of the most important principles of the legal system, determining the priority and dominant position of statute within the system of legal acts of the state, secured by a system of material and legal means. The fullest practical expression of the supremacy of statute is its special legal force. The highest legal force of statute is ensured by the fact that the adoption of a law is the prerogative of the highest representative body of state power; no other body may suspend or repeal a law; legal acts must comply with statute; and, in the event of conflicts between statute and other legal acts, priority belongs to statute.

In most Russian legal dictionaries and encyclopedias, the concept of the rule of law is not mentioned at all, while the supremacy of statute is understood as the dominance of a normative act adopted by the country's highest legislative body over other subordinate acts; as the fundamental principle of legality presupposing the exclusivity of statute, that is, recognition of its highest legal force in relation to all other normative legal acts; and as the predominance of the constitution and statutes in the activity of executive state power over the subordinate normative acts issued by that same power.

G. A. Gadzhiev notes that the words rule of law are absent from the Russian Constitution, yet the judicial doctrine of the rule of law is the foundation of Russian constitutional law. The same author notes that in Resolution 1594 (2007), The Principle of the Rule of Law, the Parliamentary Assembly of the Council of Europe drew attention to the fact that in some young democracies of Eastern Europe lawyers tend mainly to understand the term rule of law as supremacy of statute law, that is, as the supremacy of statute. However, the term rule of law should be translated into Russian as verkhovenstvo prava, the rule of law, rather than as verkhovenstvo zakona, the supremacy of statute.

A. Kh. Abashidze and K. Yu. Melshina identify the following features of the rule of law in the state: constitutionalism, separation of powers, publicity, transparency, discretion, a developed and fair judicial system, including the independence and impartiality of judges, access to justice, legal equality, legitimate authority and a clear legal position of the state. Accordingly, at the international level, the key feature of the rule of law is the good-faith observance by the principal subjects of international law, states and international organizations, of the obligations they have assumed.

B. Panteleev considers that the rule of law, together with the principle of legal certainty, includes the principle of legal effectiveness and balance of interests, or proportionality. Recently, this interpretation has received the widest application in judicial practice and has become part of the concept of the rule of law. As a component of the concept of the rule of law, the principle of legal certainty was formed in the decisions of international courts and is one of the most important common European democratic achievements and a necessary element of the normative legal construction of the social state.

B. Panteleev connects the constitutional presumption of the inalienability of human and civil rights and freedoms, and the maintenance of citizens' trust in the law and in the actions of the state, with legal certainty. This means preserving the stability of legal regulation, the inadmissibility of arbitrary changes to the existing system of norms, and the predictability of legislative policy in the social sphere, so that participants in the relevant legal relations

Rights, the Office of Legal Affairs, the United Nations Development Programme, the United Nations Children's Fund, the Office of the United Nations High Commissioner for Refugees, the United Nations Development Fund for Women and the United Nations Office on Drugs and Crime.

At the national level, the United Nations mechanism supporting the rule of law includes a constitution, or its equivalent as the basic law of the country; a clear and consistent system of laws and their observance; strong institutions of justice, governance, security and human rights, possessing effective structures, financial resources, trained personnel and technical means; procedures and mechanisms for the administration of justice in transitional periods; and the existence of public opinion and civil society that strengthen the rule of law and ensure the accountability of public officials and institutions.

On 24 September 2012, a meeting of the United Nations General Assembly on the rule of law at the national and international levels was held at United Nations Headquarters. More than 90 participants took part in the event, including approximately 40 heads of state and government.

One of the results of that forum was the adoption of the Declaration of the High-Level Meeting of the General Assembly on the Rule of Law at the National and International Levels. The document was adopted by General Assembly resolution 67/1.

The participants declared their commitment to the rule of law and to its fundamental importance for political dialogue and cooperation among all states, as well as for the further strengthening of the three main pillars of the United Nations: international peace and security, human rights and development.

It was recognized that in their collective response to the challenges and opportunities arising from the many complex transformations in political, social and economic spheres, all actors should be guided by the rule of law, since it forms the foundation of friendly and equitable relations among states and the basis for building just and equitable societies.

Affirming their commitment to the fundamental principles and norms of international law, the parties confirmed that human rights, the rule of law and democracy are interlinked and mutually reinforcing, and that they belong to the universal and indivisible core values and principles of the United Nations.

The participants stated that they were convinced that the rule of law and development are strongly interrelated and mutually reinforcing; that the advancement of the rule of law at the national and international levels is essential for sustained and inclusive economic growth, sustainable development, the eradication of poverty and hunger, and the full realization of all human rights considered this issue as an agenda item since 1992. In particular, resolutions were adopted at sessions A/RES/61/39, A/RES/62/70 and A/RES/63/128.

The Security Council has repeatedly held thematic debates on the rule of law, including S/PRST/2003/15, S/PRST/2004/2, S/PRST/2004/32, S/PRST/2005/30 and S/PRST/2006/28, and has adopted resolutions placing special emphasis on the importance of this issue in relation to women, peace and security, including Security Council resolutions 1325 and 1820; the situation of children in armed conflict, for example Security Council resolution 1612; and the protection of civilians in armed conflict, for example Security Council resolution 1674. The Peacebuilding Commission also regularly considers the rule of law in relation to countries on its agenda.

Among the key United Nations documents aimed at realizing the rule of law, the following should be mentioned: the United Nations Millennium Declaration; the Report of the Panel on United Nations Peace Operations, known as the Brahimi Report; the 2005 World Summit Outcome; the report of the Secretary-General In Larger Freedom: Towards Development, Security and Human Rights for All (A/59/2005); the report Strengthening and Coordinating United Nations Rule of Law Activities (A/63/226); the report Uniting Our Strengths: Enhancing United Nations Support for the Rule of Law (A/61/636); and the report The Rule of Law and Transitional Justice in Conflict and Post-Conflict Societies (S/2004/616).

For the United Nations, the rule of law is a principle of governance according to which all persons, institutions and entities, public and private, including the state itself, are accountable to laws that are publicly promulgated, equally enforced and independently adjudicated, and that are consistent with international human rights norms and standards. It also requires measures to ensure adherence to the principles of supremacy of law, equality before the law, accountability to the law, fairness in the application of the law, separation of powers, participation in decision-making, legal certainty, avoidance of arbitrariness, and procedural and legal transparency.

In recent years, the United Nations has been actively working to support the mechanism for ensuring the rule of law at the international and domestic levels. More than 40 United Nations entities deal with rule-of-law issues, working in more than 110 countries around the world.

General coordination of rule-of-law work is carried out by the Rule of Law Coordination and Resource Group, headed by the Deputy Secretary-General of the United Nations. Its work is supported by the Rule of Law Unit, which includes the Department of Political Affairs, the Department of Peacekeeping Operations, the Office of the United Nations High Commissioner for Human

Rights, the Office of Legal Affairs, the United Nations Development Programme, the United Nations Children's Fund, the Office of the United Nations High Commissioner for Refugees, the United Nations Development Fund for Women and the United Nations Office on Drugs and Crime.

At the national level, the United Nations mechanism supporting the rule of law includes a constitution, or its equivalent as the basic law of the country; a clear and consistent system of laws and their observance; strong institutions of justice, governance, security and human rights, possessing effective structures, financial resources, trained personnel and technical means; procedures and mechanisms for the administration of justice in transitional periods; and the existence of public opinion and civil society that strengthen the rule of law and ensure the accountability of public officials and institutions.

On 24 September 2012, a meeting of the United Nations General Assembly on the rule of law at the national and international levels was held at United Nations Headquarters. More than 90 participants took part in the event, including approximately 40 heads of state and government.

One of the results of that forum was the adoption of the Declaration of the High-Level Meeting of the General Assembly on the Rule of Law at the National and International Levels. The document was adopted by General Assembly resolution 67/1.

The participants declared their commitment to the rule of law and to its fundamental importance for political dialogue and cooperation among all states, as well as for the further strengthening of the three main pillars of the United Nations: international peace and security, human rights and development.

It was recognized that in their collective response to the challenges and opportunities arising from the many complex transformations in political, social and economic spheres, all actors should be guided by the rule of law, since it forms the foundation of friendly and equitable relations among states and the basis for building just and equitable societies.

Affirming their commitment to the fundamental principles and norms of international law, the parties confirmed that human rights, the rule of law and democracy are interlinked and mutually reinforcing, and that they belong to the universal and indivisible core values and principles of the United Nations.

The participants stated that they were convinced that the rule of law and development are strongly interrelated and mutually reinforcing; that the advancement of the rule of law at the national and international levels is essential for sustained and inclusive economic growth, sustainable development, the eradication of poverty and hunger, and the full realization of all human rights and fundamental freedoms, including the right to development; and that all these, in turn, strengthen the rule of law. For this reason, they were convinced that this interrelationship should be considered within the international development agenda after 2015.

The role of the rule of law as one of the key elements of conflict prevention, peacekeeping, conflict resolution and peacebuilding was also noted.

In addition to the priority of human rights and the prohibition of discrimination, the following were identified among the principal elements of the rule of law: the importance of a fair, stable and predictable legal framework for ensuring inclusive, sustainable and equitable development; the independence of the judiciary, together with its impartiality and integrity; the right to equal access to justice for all, including persons belonging to vulnerable groups; and the importance of raising awareness of legal rights.

I believe that among the above-mentioned principles, legal certainty plays a foundational role. Without its realization, the implementation of the rule of law and other interconnected principles encounters significant difficulties.

M. V. Presnyakov considers legal certainty to be a manifestation of the principles of equality, legality and justice. In particular, he refers to numerous decisions of the Constitutional Court of the Russian Federation which state that the principle of equality of all before the law and the courts cannot be realized without a uniform understanding and interpretation of the norm by all law-enforcement bodies. Understanding legal certainty as the antithesis of legal arbitrariness expresses its essential connection with the principle of legality, the quality of the law and the concept of the rule of law. According to N. A. Vlasenko, unclear content and the irrational use of ambiguous words and expressions lead to a decrease in the effectiveness of legal regulation.

Accordingly, ensuring a high level of perfection of the legal, linguistic, logical and social components of norms presupposes compliance with the following requirements: normativity, completeness, concreteness, systemacity, precision, clarity, simplicity and brevity, or economy. At the same time, the certainty of a legal text is ensured by using phraseological expressions that do not permit ambiguous interpretation.

A similar position is taken by other authors. I consider scientifically well-founded the view that a norm uncertain in its content does not ensure its equal application to equal subjects and therefore violates the principle of equality of all before the law and the courts, and inevitably leads to arbitrariness.

Special emphasis should be placed on the importance of legal certainty in key legal terms, which play a fundamental role in the creation of normative legal acts. The lexical basis of a legal text, its proper linguistic content, consists primarily of words, the building blocks that possess text-forming and style-forming properties. Accordingly, such linguistic units as phraseological expressions and other stable formations perform the most important pragmatic function: they convey normative information. It is precisely through them that the certainty of this information is achieved, quite apart from their function of continuity in law. The main role in expressing the will of the norm-setting authority belongs to the grammatical sentence as a lexical formation containing meaning, with the subordinate function of regulating factual relations in society. An unprofessional or careless attitude toward these instruments in preparing the text of a normative legal act leads to inconsistency in legal mechanisms. In other words, there is a regular connection and synchrony between linguistic means and the future legal text. Departure from these requirements leads to violations in the construction and content of normative legal prescriptions.

P. D. Barenboim and D. V. Kravchenko note that the requirement of legal certainty is inherent both in the doctrine of the law-governed state and in the doctrine of the rule of law. N. A. Vlasenko rightly believes that the priority of law, that is, the principle of the rule of law, in a law-governed state is connected with the fact that the exercise of public power in such a state and its relations with citizens are based on law. This implies limitation of state power and the priority of human rights. Since political power, especially executive power, tends toward various abuses, reliable legal frameworks are necessary for its effective functioning, frameworks that limit and restrain excesses and violations of human rights. Legal limitations are necessary so that such a vice of power as arbitrariness does not flourish.

I believe that the principles of the rule of law and the law-governed state are interconnected with each other and with the principles of legal certainty as a basis for the realization of human rights and non-discrimination. In other words, the modern understanding of law is inseparably connected with human rights, the protection and realization of which must become the priority of the modern state.

The development of the concept of the law-governed state was influenced by thinkers of the past and present: Plato, Aristotle, Cicero, Locke, Montesquieu, Kant, Speransky, Chicherin, Vernadsky, V. S. Solovyov, Korkunov, Petrazhitsky, Vysheslavtsev, Struve, I. A. Ilyin and many others.

such equality can be ensured only if the norm is uniformly understood and interpreted by all law-enforcement bodies. Uncertainty in the content of a legal norm, by contrast, permits unlimited discretion in the process of law enforcement and inevitably leads to arbitrariness, and therefore to violation of the principles of equality and the supremacy of law.

The principle of legal certainty has received some development as an element of legal technique, which, being a means of creating and improving law, is defined as a system of professional legal rules and means used in drafting legal acts and carrying out other legal activity in law-making, legal interpretation, authoritative and non-authoritative realization of law, ensuring the perfection of its form and content.

In turn, legal technique is an element of legal technology, including means, techniques, methods, actions, operations, processes, principles and the like; and both legal technology and legal technique are components of legal practice.

M. P. Davydova distinguishes two levels of technical-legal instruments of legal technique: rules and means.

Rules of legal technique are defined as standards developed by science and/or officially established, observance of which in the process of legal activity serves as a criterion of its quality. The system of general rules of legal technique includes two groups: external rules, which establish certain criteria for evaluating the result of legal activity, such as fairness, democracy, legality and humanism; and internal rules, which characterize the process of legal activity itself, such as economy, justification, expediency and rationality.

The means, or techniques, of legal technique are specific instruments of professional legal activity that ensure the achievement of its aims. The most significant general means of legal technique may be grouped into three categories: general social, doctrinal and normative means. Their interconnection reflects to a certain extent the logic of the formation and operation of law, covering the components without which its existence is impossible.

V. Yu. Kartukhin identifies closely interconnected linguistic, logical and epistemological rules of legal technique, the failure to observe which prevents an unambiguous and adequate interpretation of a normative legal act. In particular, he argues that violation of linguistic rules of legal technique entails ambiguity of literal interpretation, that is, logically true and at the same time semantically mutually exclusive interpretations of one and the same norm. Violation of logical rules entails inadequate interpretation, that is, unjustified expansion or narrowing of the predicative characteristics of the subject indicated in the norm. In turn, violation of epistemological rules entails the fundamental impossibility of interpreting and applying the norm in accordance with generally recognized aims and tasks of law, since the falsity of information contained in the norm fundamentally contradicts the very essence of legal regulation.

S. V. Bakhvalov divides the means of legislative technique into general social, technical and special legal means. In particular, he includes among general social means: concepts, judgments, languages, both domestic and foreign, the parts of speech expressing them, signs, linguistic units such as words, phrases, nouns and verbs, letters, punctuation marks, abbreviations, as well as numbers, numbering, graphic objects such as drawings, graphs, diagrams and geographical maps, tables of contents, thesauri, symbols, terms, scientific categories and laws, and various social norms, including ethical, aesthetic and religious norms. Technical means include computers, office equipment, the Internet and so on. Special legal means of legislative technique are divided into means of constructing a legal norm, including presumptions, fictions, reservations, legal constructions, definitions and so on, and means of formatting the text of a law, including references, cross-references, notes, annexes and so on.

In the context of the present study, I consider the most appropriate approach to be one according to which the main techniques, or means, of modern legal technique include legal terminology, above all key legal concepts through which the content of normative prescriptions is expressed; legal constructions, which reflect the structure of legal material on the basis of the relationship between rights, duties and legal responsibility; and methods for constructing normative legal acts, that is, techniques for drafting normative documents.

I believe that key legal concepts play the fundamental role, because it is from them that legal norms, constructions and legal acts are formed.

I also believe that the totality of diverse linguistic means, the language of law, used in professional legal activity is scientifically regarded as foundational within the group of general social means of legal technique.

M. L. Davydova considers the structure of legal language in two aspects: vertical and horizontal. The syntactic, or vertical, structure of legal language includes the following levels: word, that is, legal term; sentence, that is, normative legal prescription; and complex syntactic whole, that is, association of prescriptions. The functional, or horizontal, structure takes into account the possible spheres of application of legal language and is determined by a combination of official-business, scientific and journalistic styles of speech used at different levels of the existence of law. The basis of the system of legal terms, which acts as a kind of core uniting all structural elements and stylistic sections of legal language, is formed by scientific terms that change and develop as legal science develops.

Since the contemporary stage of the development of legal science is inseparably connected with the principles of the rule of law, the law-governed state and human rights, the language of law today must comply with the principle of legal certainty.

I. V. Yusipova attributed linguistic, or language-based, and normative legal means for establishing, ensuring, protecting and defending human rights to the first level. She defines the linguistic means of the mechanism for ensuring human rights as the totality of lexical-morphological and syntactic constructions mediating permissions, prohibitions and positive obligations contained in legal regulation and connected in one way or another with the realization of human rights. Accordingly, normative legal means are a system of legal norms fixing human rights, aimed at their protection and defense, and establishing responsibility for their violation.

Vocabulary is understood as an open system uniting diverse words that form the lexical composition of any language. It is differentiated by various criteria, including degree of use, stylistic affiliation, origin and spheres of social use. Each of these criteria imposes certain limits on the use of particular groups of words in the language of a law that establishes or ensures rights and freedoms. By degree of use, vocabulary is divided into active and passive. The active vocabulary of the legal securing of human rights includes words necessary for everyday communication, as well as words designating legal concepts and terms.

Terminology is a component, but a relatively separate part, of the vocabulary of a national language. The terminological sphere connected with human and civil rights and freedoms, one of the most important institutions of a democratic society, tends to become a separate group of vocabulary serving this area of legal language. In view of this, for better understanding and use in the realization of human rights, this group should be legislatively consolidated and interpreted.

The inconsistency of legal terminology with the requirements of legal certainty makes it impossible to achieve a uniform understanding and interpretation of a legal norm, predetermining a decline in its effectiveness and the poor quality of the normative legal act.

V. Yu. Kartukhin regards certainty and consistency as the principal criteria of the logical aspect of a legal norm as a component of legal technique. Certainty presupposes that in the text of a normative legal act all terms, when repeated, are used in one and the same meaning, and that different terms cannot be used to express one and the same thought. Only the term that has received a definition should be used in the text. Synonymy and homonymy must be excluded. Synonymy must be avoided in the text of a law, since there are no two identical concepts. The use of synonymy in a legal text may result in divergent readings and the need for additional interpretation. The use of homonyms requires clarification of the meaning of each of them. Where it is impossible to avoid the use of homonyms, it is necessary to explain what meaning the given term has. Consistency, as a property of logical correctness, implies that the text of the law is free from internal contradictions. Contradictions must be eliminated within one normative legal act, among normative legal acts of the same level, and with laws possessing higher legal force. The logical structure of a normative legal act should consist of three parts: general, main and final.

Thus, among the principal rules of legal technique, observance of which is necessary to achieve the proper quality of a law, the following are identified in the scholarly literature: certainty, consistency, logic, concision and uniformity. Accordingly, among the language rules of legal technique one may distinguish linguistic, terminological, syntactic and stylistic rules.

The quality of a law is usually understood as the totality of necessary properties and features inherent in a law as an effective regulator of social relations, containing normative legal prescriptions that make it possible to achieve optimal positive results in various spheres of social life.

Since the realization of human rights is the main criterion of the effectiveness of modern society, law and the state, this very criterion is central in determining the quality of the corresponding normative legal act. I believe that the realization of the principle of the rule of law and the effectiveness of forms of the right to freedom of conscience ultimately depend on compliance with the rules of modern legal technique.

I believe that the realization of the principle of legal certainty and other rules of modern legal technique is the basis of the principles of the rule of law and the law-governed state, which in turn make possible the securing of human rights as the most significant opportunities for human development and as inalienable properties determining the measure of human freedom.

In turn, the right to freedom of conscience as the possibility of free worldview choice is a system-forming right within the human rights system. The content of the right to freedom of conscience includes the possibility, individually and/or together with others, freely to form, choose, change and disseminate beliefs and to act in accordance with them, without infringing the freedom and personal dignity of others. Awareness of the diversity of relations and respect for the relationship between knowledge and faith among different people in society is the principal aspect of the right to freedom of conscience in contemporary reality.

As legal and political systems transform, the concept and content of freedom of conscience are increasingly considered through the prism of human rights and freedoms, expanding and gradually freeing themselves from attachment to the key term religion, which does not meet the principle of legal certainty.

However, in the contemporary scientific and educational environment there is no single generally accepted theoretical and legal definition of the right to freedom of conscience, and the existing tendency to expand it as the possibility of free worldview choice is not yet sufficiently stable. At the international legal and domestic levels, freedom of conscience is defined not entirely correctly and is identified with freedom of religion or confession.

Since the principle of worldview neutrality of the state is the most important guarantee of freedom of conscience, it is also interconnected with the principles of the rule of law and the law-governed state. In this context, the principles of the rule of law, including the principle of legal certainty of key terms, and of the law-governed state serve as the basis for reforming the forms of international law on freedom of conscience.

In turn, realization of the principle of legal certainty of key terms in reforming the forms of law on freedom of conscience will increase the effectiveness of the relevant norms and will contribute to the implementation of the principle of the supremacy of international law, the law-governed state and sustainable development of the modern world.

1.4. The Significance of the Worldview Neutrality of States for the Development of Forms of International and Domestic Law on Freedom of Conscience

Many contemporary researchers believe that the principle of the secularity of the state is interconnected with the principle of the law-governed state, which is based on the rule of law and oriented toward the realization of human rights. As already noted above, the principles of freedom of conscience, the secularity of the state, the relevant state policy, and relations between the state and religious associations are also most closely interconnected.

Indeed, today it is difficult to imagine the principles of freedom of conscience and the secularity of the state separately, outside their mutual connection and dependence. Many Russian and foreign authors, including S. M. Aleinikova, M. Yu. Babii, B. Balci, A. I. Barmenkov, J. Bauberot, K. G. Kanevsky, A. V. Karpushkin, M. G. Kirichenko, V. V. Klochkov, A. S. Lovinyukov, T. P. Minchenko, S. A. Mozgovoi, D. M. Nikitaev, M. M. Persits, S. Peyrouse, R. A. Podoprigora, N. A. Pridvorov, Yu. A. Rozenbaum, F. M. Rudinsky, S. S. Savva, A. E. Sebentsov, A. A. So, M. S. Stetskevich, N. Yu. Teteryatnikov, E. V. Tikhonova, G. G. Cheremnykh, Yu. E. Fedotova, and others, speak in their works of the dependence of the realization of freedom of conscience on observance of the principle of the secularity of the state, including the separation of religious associations from the state.

In particular, the well-known researcher F. M. Rudinsky notes that the separation of church and state means that the state does not interfere in the internal religious life of the church and does not entrust it with the performance of any state functions, while the church does not interfere in the affairs of the state. It appears scientifically justified to hold that the secularity of the state, including the separation of religious associations from the state, is the most important guarantee of the realization of freedom of conscience, since the liberation of the conscience of the individual from compulsory state guardianship gives the person the ability to decide independently the question of their convictions. Religious organizations become private associations that cannot use power against persons of other faiths or against non-believers.

In his draft law of the USSR on freedom of conscience, Yu. A. Rozenbaum defined its principal constitutional guarantees as the equality of citizens regardless of their attitude toward religion; the separation of the church, or religious organizations, from the state and of school, or education, from the church, or religious organizations; and legislation establishing responsibility for violations of this freedom.

A. S. Lovinyukov and a number of other researchers quite rightly state that the principle of equality of believers and non-believers, together with other interdependent principles, is a very important structural element of freedom of conscience. At the same time, they note the need for guarantees of this principle, since otherwise it remains a formal declaration.

E. B. Pokaninova also speaks of the relationship between the principles of freedom of conscience and the secular state, noting that their rootedness presupposes: a) recognition of religion as a private matter of each individual person; b) refusal by the state and society to follow religious concepts and meanings in their activity; c) strict observance, in relations between the individual and the state, of legal principles and of the equality of all religions before the law; d) the development of interfaith dialogue, while respecting the rule of autonomy of religious organizations from the state; and, secondly, religious studies education aimed at providing a secular foundation for tolerance toward diverse groups of believers and non-believing citizens.

In this context, I believe that the impossibility of the state violating freedom of conscience can and must become a guarantee of real equality. Historical experience shows that, when states have had such an opportunity, they have never refused special restrictions and control over the worldview sphere and

and the law-governed state, which in turn make possible the securing of human rights as the most significant opportunities for human development and as inalienable properties determining the measure of human freedom.

In turn, the right to freedom of conscience as the possibility of free worldview choice is a system-forming right within the human rights system. The content of the right to freedom of conscience includes the possibility, individually and/or together with others, freely to form, choose, change and disseminate beliefs and to act in accordance with them, without infringing the freedom and personal dignity of others. Awareness of the diversity of relations and respect for the relationship between knowledge and faith among different people in society is the principal aspect of the right to freedom of conscience in contemporary reality.

As legal and political systems transform, the concept and content of freedom of conscience are increasingly considered through the prism of human rights and freedoms, expanding and gradually freeing themselves from attachment to the key term religion, which does not meet the principle of legal certainty.

However, in the contemporary scientific and educational environment there is no single generally accepted theoretical and legal definition of the right to freedom of conscience, and the existing tendency to expand it as the possibility of free worldview choice is not yet sufficiently stable. At the international legal and domestic levels, freedom of conscience is defined not entirely correctly and is identified with freedom of religion or confession.

Since the principle of worldview neutrality of the state is the most important guarantee of freedom of conscience, it is also interconnected with the principles of the rule of law and the law-governed state. In this context, the principles of the rule of law, including the principle of legal certainty of key terms, and of the law-governed state serve as the basis for reforming the forms of international law on freedom of conscience.

In turn, realization of the principle of legal certainty of key terms in reforming the forms of law on freedom of conscience will increase the effectiveness of the relevant norms and will contribute to the implementation of the principle of the supremacy of international law, the law-governed state and sustainable development of the modern world.

1.4. The Significance of the Worldview Neutrality of States for the Development of Forms of International and Domestic Law on Freedom of Conscience

Many contemporary researchers believe that the principle of the secularity of the state is interconnected with the principle of the law-governed state, which is based on the rule of law and oriented toward the realization of human rights. As already noted above, the principles of freedom of conscience, the secularity of the state, the relevant state policy, and relations between the state and religious associations are also most closely interconnected.

Indeed, today it is difficult to imagine the principles of freedom of conscience and the secularity of the state separately, outside their mutual connection and dependence. Many Russian and foreign authors, including S. M. Aleinikova, M. Yu. Babii, B. Balci, A. I. Barmenkov, J. Bauberot, K. G. Kanevsky, A. V. Karpushkin, M. G. Kirichenko, V. V. Klochkov, A. S. Lovinyukov, T. P. Minchenko, S. A. Mozgovoi, D. M. Nikitaev, M. M. Persits, S. Peyrouse, R. A. Podoprigora, N. A. Pridvorov, Yu. A. Rozenbaum, F. M. Rudinsky, S. S. Savva, A. E. Sebentsov, A. A. So, M. S. Stetskevich, N. Yu. Teteryatnikov, E. V. Tikhonova, G. G. Cheremnykh, Yu. E. Fedotova, and others, speak in their works of the dependence of the realization of freedom of conscience on observance of the principle of the secularity of the state, including the separation of religious associations from the state.

In particular, the well-known researcher F. M. Rudinsky notes that the separation of church and state means that the state does not interfere in the internal religious life of the church and does not entrust it with the performance of any state functions, while the church does not interfere in the affairs of the state. It appears scientifically justified to hold that the secularity of the state, including the separation of religious associations from the state, is the most important guarantee of the realization of freedom of conscience, since the liberation of the conscience of the individual from compulsory state guardianship gives the person the ability to decide independently the question of their convictions. Religious organizations become private associations that cannot use power against persons of other faiths or against non-believers.

In his draft law of the USSR on freedom of conscience, Yu. A. Rozenbaum defined its principal constitutional guarantees as the equality of citizens regardless of their attitude toward religion; the separation of the church, or religious organizations, from the state and of school, or education, from the church, or religious organizations; and legislation establishing responsibility for violations of this freedom.

A. S. Lovinyukov and a number of other researchers quite rightly state that the principle of equality of believers and non-believers, together with other interdependent principles, is a very important structural element of freedom of conscience. At the same time, they note the need for guarantees of this principle, since otherwise it remains a formal declaration.

E. B. Pokaninova also speaks of the relationship between the principles of freedom of conscience and the secular state, noting that their rootedness presupposes: a) recognition of religion as a private matter of each individual person; b) refusal by the state and society to follow religious concepts and meanings in their activity; c) strict observance, in relations between the individual and the state, of legal principles and of the equality of all religions before the law; d) the development of interfaith dialogue, while respecting the rule of autonomy of religious organizations from the state; and, secondly, religious studies education aimed at providing a secular foundation for tolerance toward diverse groups of believers and non-believing citizens.

In this context, I believe that the impossibility of the state violating freedom of conscience can and must become a guarantee of real equality. Historical experience shows that, when states have had such an opportunity, they have never refused special restrictions and control over the worldview sphere and religious associations. Granting the state special powers to regulate this sphere inevitably leads to arbitrary interpretation and, as a consequence, to unavoidable abuses. I believe that at the root of the special powers mentioned above, which lead to violations of freedom of conscience, lies the highly questionable granting to the state of authority to identify the worldview sphere. In this context, the question of legally correct definitions of the concept and content of the principle of the secularity of the state is key to the realization of freedom of conscience and other interconnected principles.

Since equality of different associations is a necessary condition for their peaceful coexistence, which is especially relevant in multinational and multi-confessional states, a law-governed secular, or worldview-neutral, state should have no special confessional, or religious, policy or bodies at all. Thus, the principle of secularity, understood as the worldview neutrality of states, extends to the state and municipal education system, which must be neutral in relation to any worldview.

The Federal Law On Education in the Russian Federation of 29 December 2012, No. 273-FZ, establishes the secular nature of education in state and municipal organizations engaged in educational activity. The law emphasizes that students are granted academic rights to freedom of conscience, information, and free expression of their own views and convictions. The

Federal Law On Freedom of Conscience and Religious Associations of 26 September 1997, No. 125-FZ, provides that the state ensures the secular character of education in state and municipal educational institutions, and that a religious association does not perform the functions of state authorities, state institutions, or bodies of local self-government. The secular nature of education is also secured in a number of subordinate normative acts.

In this context, the legal content of secular education in state and municipal educational institutions of Russia presupposes: ensuring constitutional freedom of conscience and freedom of religion; separation from the state and municipal education system, and equality before the law, of worldview associations, whether religious, anti-religious or other; inadmissibility of establishing any worldview, religious, anti-religious or otherwise, as compulsory or official in state and municipal educational institutions; prohibition on worldview and other public associations interfering in the management of state or municipal educational institutions; prohibition on combining the educational process with coercion to study a specific worldview, including religion, atheism or religious culture, as well as coercion to renounce a worldview; and inadmissibility of compulsory religious rites in state and municipal educational institutions, except during free time.

Yet states, while they are the main subjects of international law, are not its only subjects. Moreover, the expansion of the circle of subjects of international law is objective in nature and follows from the transformation of contemporary social relations. It should be noted that, in scholarly literature, subjects of international law are understood as participants that are holders of the relevant rights and duties, capable of entering into international public legal relations and bearing international legal responsibility on the basis of international legal norms. Thus, international legal personality is an important property of a participant in international legal relations and determines its international legal status.

There is a variety of approaches to the classification of subjects of international law. In particular, by legal nature, subjects of international law are divided into principal subjects, namely states, and non-principal subjects, including international organizations, state-like entities, international non-governmental organizations, peoples and nations struggling for independence, subjects of federations, administrative-territorial units of unitary states, and others. In addition, subjects of international law are divided into original and derivative, typical and atypical, primary and secondary, traditional and non-traditional, law-making and law-applying, universal and sectoral subjects.

International organizations are secondary subjects of international law, but the growth of their role in international relations is objective in nature. By the character of membership, international organizations are divided into intergovernmental and non-governmental organizations. International intergovernmental organizations are subdivided by circle of participants into universal, regional and subregional organizations; by the character of competence into organizations of general and special competence; by the character of powers into interstate and supranational organizations; and by procedure of accession into open and closed organizations. International non-governmental organizations are associations of domestic public organizations and individuals created to strengthen cooperation in various spheres of international relations.

The theory concerning the international legal personality of individuals and national legal entities remains debatable, but the tendencies in the transformation of contemporary social relations provide ever more objective grounds in its favor. As G. V. Ignatenko notes, one cannot accept as convincing the arguments of those who deny the international legal status of individuals by referring to the small number, compared with the state, of relations involving individuals that are based on international legal norms. In principle, what matters is the legal capacity itself to have and exercise rights and duties, while the quantitative indicator characterizes the factual condition, but not the legal capacity.

Together with individuals, national legal entities and other non-traditional subjects, international non-governmental organizations possess limited international legal personality and are characterized by the absence of international treaty-making capacity.

A certain number of scholarly works have been devoted to the principle of secularity of individual states, but the principle of secularity of subjects of international law has not received proper development. This principle is insufficiently developed at the theoretical and legal level and has not found adequate reflection either in principles and norms of international law or in recommendatory acts of international organizations.

At the domestic level, the principle of secularity is declared in a significant number of states, which has found some expression in the form of debatable models of relations between the state and religious associations: the model of a state religion or confession, involving the existence of an official privileged religious association; the model of separation of religious associations from the state, involving equality of religious associations; and the model of cooperation between the state and religious associations, involving cooperation between the state and selected religious associations.

As already noted, Russian legislation, law enforcement and the activity of state bodies lack a unified understanding of the secularity of the state. The understanding of the secularity of the state as secular, or non-religious, often dominates. Thus, one concept is defined through another undefined concept, creating grounds for abuse in the activity of state bodies. At the same time, the only legally correct interpretation of the secularity of the state as worldview neutrality, implying the indifferent attitude of the state toward religion, is excluded for the sake of political compromises and cooperation between the state and religious organizations useful to the authorities. As a result, Russia is merely declared a secular state, while in reality it is not indifferent to different worldviews and conducts a selection among them.

Ultimately, the definition of the concept and content of the secularity of the state through the legally undefined concept of religion and its derivatives substantially predetermines violations of one of the foundations of the constitutional order. In addressing this problem, representatives of contemporary legal science have developed new approaches to the concept and content of freedom of conscience and the secularity of the state, without tying them to the legally indefinable term religion.

Since the principle of the secularity of states has still not been secured in international legal documents, in many countries, mainly those belonging to the religious legal family, it is not even declared. Moreover, this principle is insufficiently developed at the theoretical and legal level and has not found adequate reflection either in principles and norms of international law or in recommendatory acts of international organizations.

At the domestic level, the principle of secularity is declared in most contemporary states, and the objective tendency in relations between the state and religious associations is a gradual movement away from privileged status toward their equality. However, serious problems in realizing the principle of secularity exist in practically all states.

To some extent, these problems arise from the existence of highly debatable models of relations between the state and religious associations, which make it possible to justify elements of state confessional preference. As a consequence, the world system for protecting internationally recognized freedom of conscience has proved not fully effective. In turn, this has substantially predetermined the ethno-confessional imbalance of the entire system of international relations and the extreme aggravation of the global problems of human civilization.

In this context, a number of works in political science and international relations dealing with the so-called religious turn are of interest. In particular, a number of authors, including Fox, Sandal and Snyder, criticize classical theories of international relations and point to the significant influence of religion on the legitimacy of power, the identity of states and relations among them. Supporters of the concept of desecularization and the post-secular world, including P. Berger and Gorski, believe that the world has not become less religious but has become fiercely religious, which directly affects world politics. Other authors, including E. Hurd, rightly criticize the very division into the religious and the secular, arguing that the Western model of secularism prevents adequate understanding of processes in non-Western countries. Also of interest are the works of authors such as Haynes and Dawson, who study the influence of religion on conflicts and on the possibilities for overcoming them, noting both its peacebuilding potential and its potential for confrontation.

A serious problem is that in the contemporary world there are no states that observe the principle of secularity in the sense of a neutral attitude toward the worldview sphere. Contemporary states are, to varying degrees, interconnected with selected religious associations, which inevitably entails violations of internationally recognized freedom of conscience and creates preconditions for local, regional and global conflicts. Not only in confessional states, but in practically all states of the world, the realization of freedom of conscience for everyone is replaced by a state religious policy conducted for the purpose of retaining power and implemented through corrupt relations between the state and religious associations.

As noted above, to designate violations of the principle of worldview neutrality, the author substantiates the term clerical ideologization of the state. In particular, as applied to the international legal sphere, clerical ideologization of states and other subjects of international law means a violation of the worldview neutrality of subjects of international law, expressed in the fusion of their institutions with worldview structures whose doctrinal provisions are used as the basis of their activity, even when those provisions contradict the fundamental principles and norms of international law.

In the context of intensifying interpenetration and interdependence in all spheres of social interaction, problems of legal regulation of liberty of worldview choice in states belonging to the religious legal family require special attention. The fusion of the state with any religious direction in practice means the dominant position of one specific religious association. Only very rarely are dominant positions in the state shared by several associations. At the same time, in almost all cases, among the consequences of such a state-confessional alliance one can name violations of freedom of conscience, growing intolerance, discrimination and violence motivated by worldview affiliation. It is telling that problems arise not only for representatives of different confessional directions, for example Christians in an Islamic state, and for the non-confessional part of society, for example atheists, but also for representatives of different currents within one direction, for example Shiites in a Sunni state and Sunnis in a Shiite state.

In general, under the influence of global processes, despite numerous problems, there is an intensification of interaction between international and domestic legal systems, as well as a convergence of different legal families. Analysis of world experience as a whole makes it possible to conclude that the objective tendency in relations between the state and religious associations is a gradual refusal of the privileged position of religious associations in favor of elements of state secularity, including equality of religious associations, which is reflected in the constitutions and legislation of states. Those states that ignore this tendency become victims of political conflicts with a significant ethno-confessional component, such as Syria, Iraq and others.

I believe that it is precisely the inadequate scientific and theoretical development of the problem of the secularity of the state, including the state and municipal education system, that has predetermined problems of legal regulation, problems at other levels of realization, including the activity of state authorities, law enforcement and judicial practice, the human rights protection level, and the media, as well as the systemic nature of human rights violations in this area.

The principle of the secularity of states is of enormous significance for human rights, for the rule of law at the domestic and international levels, and for the development of forms of international and domestic law on freedom of conscience. Yet its practical embodiment encounters problems in defining the concept, its content and the relationship among key terms. Problems in defining the concept and content of the principle of the secularity of the state concern contemporary states and the world community as a whole. In addition, numerous problems concerning the relationship between the principle of the secularity of the state and related concepts require resolution, including relations between the state and religious associations, state religious policy, tolerance, and the overcoming of xenophobia, intolerance, discrimination and related phenomena.

1.5. The Interrelationship between the Right to Freedom of Conscience, the Worldview Neutrality of States, Tolerance, and the Overcoming of Xenophobia, Intolerance and Discrimination

Questions of achieving tolerance and counteracting xenophobia and intolerance are extremely relevant in the multinational and multi-confessional Russian Federation. In addition, the relevance of this problem increases significantly under conditions of contradictory development of global processes and the aggravation of ethno-confessional challenges in the contemporary world. Moreover, manifestations of xenophobia, intolerance, discrimination and violence based on them indicate that today in Russia there are significant barriers between people, obstacles that hinder socio-economic development and create threats to the security of the individual, society and the state.

Conflicts arising from ethno-confessional contradictions are aggravated to a considerable extent by the fact that both in the Russian Federation and in the contemporary world there are systemic problems in realizing human rights in the sphere of freedom of conscience and the secularity of the state, while xenophobia and intolerance are also growing. Ultimately, this creates extremely serious obstacles to interaction among people and to the positive integration of states for the sustainable development of civilization.

The scholarly literature contains a variety of highly debatable approaches to the definition of the concept and content of tolerance. I will not reproduce them here, limiting myself to a legal approach. In this context, I believe that tolerance is mutual respect for the rights of individuals and communities regardless of differences, on the basis of equality, provided that externally expressed conduct is lawful. This understanding of tolerance follows from the 1995 Declaration of Principles on Tolerance.

The most obvious indicators of the state of realization of freedom of conscience, which will be discussed in greater detail below, are numerous violations of the rights of individuals and associations, manifested in the form of discrimination, intolerance and violence based on them.

Intolerance and discrimination based on religion or belief are commonly understood as any distinction, exclusion, restriction or preference based on religion or belief and having as its purpose or effect the destruction or impairment of the recognition, enjoyment or exercise, on an equal basis, of human rights and fundamental freedoms. This concept originates from the Latin word discriminatio, meaning distinction on a particular ground, such as race, sex, religion and so on.

The concept of xenophobia is theoretically very weakly developed and is not applied in the Russian legal system. Accordingly, the relationship between unlawful acts and the xenophobic views under which they are committed is not always obvious, and the correctness of identifying this relationship remains debatable.

It is generally accepted that xenophobia, from the Greek xenos, meaning stranger, and phobos, meaning fear, is hatred, intolerance or dislike of someone or something foreign, unfamiliar or unusual. In dictionaries of social psychology, xenophobia is defined as a negative attitude, fear and dislike toward individuals and communities perceived as alien and incomprehensible.

I believe that, from a legal point of view, it is more correct to use the concept discrimination, intolerance and xenophobia motivated by worldview affiliation instead of the widespread phrase motivated by religion or belief.

This approach is based on recognition of the absence of a legal concept of religion and the narrowness of the concept of beliefs. It is implied that worldview is a system of views about the world and the place of the human being in it, about the relationship of the human being to surrounding reality and to oneself, as well as the principal life positions, ideals, convictions, principles and value orientations conditioned by those views. Accordingly, a conviction is an element of worldview that gives a person or social group confidence in their views, knowledge and assessments of reality.

Xenophobia is an ambiguous phenomenon with a very complex structure. Many researchers, when identifying its components, first of all mention stereotypes and social attitudes.

In scholarship there are different approaches to defining the concept and content of stereotypes. For example, R. V. Bazikov distinguishes stereotypes at the individual and group levels, including cognitive, affective and social components. The former are defined as stable psycho-semantic connections, inherent in certain members of a social group, between typical meaning-images of their everyday existence, formed regardless of the character of their interaction with social objects. Accordingly, a social stereotype at the group level is defined as the connection between the system of interaction that has developed within a group and its system of views of the surrounding world.

O. G. Orlova substantiates an understanding of stereotype according to which it is not a generalized image or scheme, but separate elements of an image, associative by nature and evaluative, culturally and ideologically conditioned. Stereotypes form a system in the culture and society that they describe, as well as in the discourse that determines the ideological and value-based features of a concept, and in the concept itself, which contains different frames in its structure. Thus, geopolitical stereotypes are present in the geopolitical subdiscourse of journalistic discourse and are united into a structure through frames of the concept Russia. Political stereotypes are noted in the political subdiscourse of journalistic discourse. Societal stereotypes and ethnocultural stereotype-symbols are present in the socio-humanitarian subdiscourse. Everyday stereotypes, object-symbols, function in the everyday subdiscourse of journalistic discourse.

Xenophobic stereotypes can form simplified and distorted elements of the image of a social object and, ultimately, justify negative attitudes and actions toward so-called outsiders. In particular, according to Z. Zh. Gakaev, ethnic stereotypes are the most common variety of social stereotype and represent a relatively stable idea about the moral, intellectual and physical qualities inherent in representatives of different ethnic communities. Ethnic stereotypes occupy one of the central places in the system of manipulative propaganda, since they block reason and rational human capacities, consolidate a complex set of negative emotions, and evoke feelings of fear, hatred, mistrust, suspicion and hostility. This becomes possible because stereotypes, which most often arise spontaneously, are necessarily connected with perception and not necessarily with experience, yet they can directly influence the acquisition of new experience. Being extremely stable, while at the same time changeable, they reflect not reality, but a label attached to reality.

Thus, stereotypes are necessarily connected with perception and not necessarily with experience, yet they can directly influence the acquisition of new experience; they most often arise spontaneously; they serve to protect traditions; they carry within themselves an evaluative element in the form of an emotional attitude toward the object, while at the same time being binary, yes or no, ours or alien, and therefore either negative or positive; they economize thinking and summarize historical experience; they reflect not reality, but a label attached to reality; they are extremely stable, but changeable; they cannot be absolutely true, although they may be based on notions close to reality, and they may also be absolutely false; they perform cognitive, communicative, protective, mobilizing and delegitimizing functions, including the protection of positive ethnic identity.

Social attitudes, as a stable system of ideas and emotions connected with a social object, may manifest themselves in prejudices and biases, including hostility and partiality toward representatives of ethno-confessional groups. In particular, ethnophobia constitutes the affective component of a social attitude and is defined as an unfavorable irrational evaluative reaction of an individual toward other ethnic groups, their representatives, and situations of direct and indirect interethnic interaction. This reaction is expressed in inadequate, irrational and negative evaluations and behavior by the individual. The evaluative reaction extends both to the behavior of an individual person and to the group.

Mechanisms of linguistic construction of inequality, called hate speech and reproduced through the media, can serve as a direct instrument for forming barriers between people. For example, the spread of ethnic xenophobia among young people in contemporary society is carried out to a significant extent through the mass media. According to sociological research, it is from the media that young people receive negative information about interethnic relations. All of the above manifests itself most strongly in confessional states, as well as in states that practice systemic violations of secularity and freedom of conscience.

As an interim conclusion, it may be noted that the social danger of xenophobia, the antipode of tolerance, lies in the creation of conditions for violations of the principle of equality and of human rights, which is fraught with conflict. The most important components of xenophobia are stereotypes and social attitudes, formed mainly as a response to living conditions during particular historical periods in the development of civilization.

To a very significant extent, negative stereotypes and social attitudes are formed and reproduced in connection with the activity of state structures: in the sphere of ethno-confessional policy, above all in connection with denial or violation of the principle of the secularity of the state; in the sphere of the media and mass communications; in the state and municipal education system; and elsewhere. To a certain, much lesser extent, stereotypes and social attitudes are formed and reproduced in the course of everyday interactions among people.

Hidden xenophobia in the form of negative stereotypes and the linguistic construction of inequality is no less dangerous than direct xenophobia. At the same time, diagnosing the existence of stereotypes and counteracting hidden xenophobia is very difficult.

Another aspect of this problem is connected with the fact that states interact through international intergovernmental organizations, whose activity in protecting human rights, achieving tolerance and counteracting intolerance in order to achieve closer interaction proves insufficiently effective. I think that one of the reasons lies in the fact that intergovernmental organizations express and reflect the policies of states, and those policies, as a rule, are oriented toward the interests of the relevant power groups, including what are called national interests, which are accustomed to exploiting dividing principles in order to retain power.

The legal foundations of tolerance and of counteracting xenophobia and intolerance in the Russian Federation consist of forms of international and domestic law implemented in the Russian Federation. To a significant extent, they intersect with the legal foundations of freedom of conscience and the secularity of the state. In particular, the constitutional principle of the secularity of the state, enshrined in Article 14 of the Constitution of Russia, is the most important guarantee for achieving tolerance, protection against discrimination and human rights as a whole. Article 19 does not permit propaganda or agitation inciting social, racial, national or religious hatred and enmity. It also prohibits propaganda of social, racial, national, religious or linguistic superiority.

Among the universal international documents concerning the sphere of tolerance and counteracting intolerance, the following should be highlighted: the Charter of the United Nations, the Universal Declaration of Human Rights, the International Covenant on Civil and Political Rights, the International Covenant on Economic, Social and Cultural Rights, the Declaration on the Elimination of All Forms of Intolerance and of Discrimination Based on Religion or Belief, the Declaration on the Rights of Persons Belonging to National or Ethnic, Religious and Linguistic Minorities, the Convention on the Rights of the Child, the Convention against Discrimination in Education, the Declaration of Principles on Tolerance and others.

The corresponding regional documents in the field of freedom of conscience include acts of the Council of Europe, such as the Convention for the Protection of Human Rights and Fundamental Freedoms, Protocol No. 1 to that Convention, the European Social Charter, and recommendations of the Parliamentary Assembly of the Council of Europe; acts of the CSCE/OSCE, including the Final Act of the Conference on Security and Co-operation in Europe, the concluding documents of the Madrid and Vienna meetings, the Document of the Copenhagen Meeting of the Conference on the Human Dimension of the CSCE, the Charter of Paris for a New Europe, the Document of the Moscow Meeting of the Conference on the Human Dimension of the CSCE, the Charter for European Security and others; acts of the European Union, including the Charter of Fundamental Rights of the European Union; and acts of the Commonwealth of Independent States, including the Declaration of the Heads of State of the Commonwealth of Independent States on International Obligations in the Field of Human Rights and Fundamental Freedoms and the Convention of the Commonwealth of Independent States on Human Rights and Fundamental Freedoms.

The Charter of the United Nations, together with other fundamental principles, called for reaffirming faith in fundamental human rights on the basis of tolerance and also affirmed the aspiration of states to carry out international cooperation in solving world problems on the basis of encouraging and developing respect for human rights and fundamental freedoms.

In accordance with Article 26, paragraph 2, education shall be directed to the full development of the human personality and to the strengthening of respect for human rights and fundamental freedoms. Education shall promote understanding, tolerance and friendship among all nations, racial and religious groups, and shall further the activities of the United Nations for the maintenance of peace.

Although the Universal Declaration of Human Rights is not formally legally binding, it has become one of the principal sources of state obligations in the field of human rights and the prohibition of intolerance and discrimination. It may be regarded as an authoritative instrument for interpreting the Charter of the United Nations. A large part of its norms may be regarded as important evidence of customary international law binding all states of the contemporary world.

The International Covenant on Civil and Political Rights of 16 December 1966 entered into force on 23 March 1976. Article 2 of the Covenant states that each State Party undertakes to respect and to ensure to all individuals within its territory and subject to its jurisdiction the rights recognized in the Covenant, without distinction of any kind, such as race, colour, sex, language, religion, political or other opinion, national or social origin, property, birth or other status.

It should be specially noted that, in order to monitor implementation by States Parties and pursuant to Part IV, Articles 28-45, of the International Covenant on Civil and Political Rights, the UN Human Rights Committee was established. The powers of the Committee include consideration of reports and complaints concerning violations, as well as the development of general comments of a recommendatory character.

On issues concerning implementation of the provisions of the International Covenant on Civil and Political Rights in the sphere of freedom of conscience, the UN Human Rights Committee presented on 20 July 1993 General Comment No. 22 (48), Freedom of Thought, Conscience and Religion, Article 18. Particularly relevant here is the attempt to resolve contradictions in the sphere of realization of the freedom of parents or legal guardians to ensure the religious and moral education of their children in accordance with their own convictions. On this matter, the Committee considers that the provisions of

Article 18, paragraph 4, permit instruction in public schools in subjects such as the general history of religions and ethics if it is given in a neutral and objective way. The freedom of parents or legal guardians to ensure the religious and moral education of their children in conformity with their own convictions, referred to in Article 18, paragraph 4, is related to the guarantees of the freedom to teach a religion or belief set out in Article 18, paragraph 1. The Committee's view is important that public education that includes instruction in a particular religion or belief is inconsistent with Article 18, paragraph 4, unless provision is made for non-discriminatory exemptions or alternatives that would accommodate the wishes of parents or guardians.

However, the key question of financing religious instruction remained outside the attention of the experts of the Human Rights Committee, which significantly devalued their position on this problem.

The International Covenant on Economic, Social and Cultural Rights was adopted by General Assembly resolution 2200 A (XXI) of 16 December 1966 and entered into force on 3 January 1976. The States Parties undertook to take steps, by all appropriate means, including particularly the adoption of legislative measures, with a view to achieving progressively the full realization of the rights recognized in the Covenant. The states undertook that the rights proclaimed in the Covenant would be exercised without discrimination of any kind as to race, colour, sex, language, religion, political or other opinion, national or social origin, property, birth or other status. Article 13 of the Covenant secured the right of everyone to education, which shall be directed to the full development of the human personality and the sense of its dignity and shall strengthen respect for human rights and fundamental freedoms, and shall enable all persons to participate effectively in a free society, promote understanding, tolerance and friendship among all nations and all racial, ethnic or religious groups, and further the activities of the United Nations for the maintenance of peace. Under Article 13, paragraph 3, the States Parties undertake to respect the liberty of parents and, where applicable, legal guardians to choose for their children schools other than those established by public authorities, provided they conform to such minimum educational standards as may be laid down or approved by the state, and to ensure the religious and moral education of their children in conformity with their own convictions.

The Convention on the Rights of the Child was adopted by General Assembly resolution 44/25 of 20 November 1989 and is binding on States Parties. It entered into force on 2 September 1990. It was preceded by the non-binding Declaration of the Rights of the Child adopted by the UN General

Assembly in 1959. The States Parties to the Convention undertook to respect and ensure all rights provided for in the Convention to each child within their jurisdiction without discrimination of any kind, irrespective of the child's or their parent's or legal guardian's race, colour, sex, language, religion, political or other opinion, national, ethnic or social origin, property, disability, birth or other status. States Parties shall take all appropriate measures to ensure that the child is protected against all forms of discrimination or punishment on the basis of the status, activities, expressed opinions or beliefs of the child's parents, legal guardians or family members. Under Article 12, States Parties assure to the child who is capable of forming their own views the right to express those views freely in all matters affecting the child, with the views of the child being given due weight in accordance with the child's age and maturity. Article 13 guarantees the child the right to freedom of expression, including freedom to seek, receive and impart information and ideas of all kinds, regardless of frontiers, orally, in writing or in print, in the form of art, or through any other media of the child's choice. The exercise of this right may be subject to certain restrictions, but only such as are provided by law and are necessary for respect of the rights or reputations of others, or for the protection of national security or public order, or of public health or morals. Article 29 secures the agreement of States Parties that the education of the child shall be directed to the development of the child's personality, talents and mental and physical abilities to their fullest potential; the development of respect for human rights and fundamental freedoms and for the principles enshrined in the Charter of the United Nations; the development of respect for the child's parents, cultural identity, language and values, for the national values of the country in which the child is living, the country from which he or she may originate, and for civilizations different from their own; the preparation of the child for responsible life in a free society, in the spirit of understanding, peace, tolerance, equality of sexes and friendship among all peoples, ethnic, national and religious groups and persons of indigenous origin; and the development of respect for the natural environment. Under Article 30, in states in which ethnic, religious or linguistic minorities or persons of indigenous origin exist, a child belonging to such a minority or indigenous group shall not be denied the right, in community with other members of their group, to enjoy their own culture, to profess and practise their own religion, or to use their own language.

Unfortunately, a number of provisions of the Convention on the Rights of the Child touching upon tolerance and counteracting intolerance contain fundamental scientific and theoretical contradictions that hinder their practical realization.

Among the documents aimed at achieving tolerance that were adopted within the United Nations Educational, Scientific and Cultural Organization, UNESCO, one should note the Convention against Discrimination in Education, adopted by the General Conference of UNESCO at its eleventh session on 14 December 1960, and the Declaration of Principles on Tolerance, approved by resolution 5.61 of the UNESCO General Conference of 16 November 1995. In the 1960 Convention against Discrimination in Education, paragraph 1 deserves attention because it defines discrimination as covering any distinction, exclusion, limitation or preference which, being based on race, colour, sex, language, religion, political or other opinion, national or social origin, economic condition or birth, has the purpose or effect of nullifying or impairing equality of treatment in education, and in particular: depriving any person or group of persons of access to education of any type or at any level; limiting any person or group of persons to education of an inferior standard; establishing or maintaining separate educational systems or institutions for persons or groups of persons, except in cases provided for in the Convention; or inflicting on any person or group of persons conditions incompatible with the dignity of man.

The Declaration of Principles on Tolerance was approved by resolution 5.61 of the UNESCO General Conference of 16 November 1995. The Declaration provides a contemporary definition of the key concept of tolerance, which is inseparably connected with freedom of conscience, and states that in schools and universities, through non-formal education, at home and in the workplace, the spirit of tolerance should be strengthened and attitudes of openness, mutual listening and solidarity should be formed. It also refers to the need to promote methods of systematic and rational teaching of tolerance that address the cultural, social, economic, political and religious sources of intolerance, which lie at the root of violence and exclusion. This should contribute to improving mutual understanding and strengthening solidarity and tolerance in relations both among individuals and among ethnic, social, cultural, religious and linguistic groups, as well as nations. Thus, states assumed an obligation to promote tolerance and non-violence, using programs and institutions in the fields of education, science, culture and communication.

However, since the principle of secularity, or, in my view, worldview neutrality, is not secured at all in international legal documents, the effectiveness of the above-mentioned norms in the field of tolerance and protection against intolerance and discrimination is significantly reduced, and violations of human rights are thereby predetermined.

In the Constitution of Russia of 1993, part 2 of Article 19 organically complements Articles 14 and 28. It states that the state guarantees equality of rights and freedoms of the person and citizen regardless of attitude toward religion, convictions, membership of public associations and other circumstances. In addition, the Constitution of the Russian Federation confirmed such norms as ideological diversity, part 2 of Article 13, equality of human and civil rights and freedoms regardless of attitude toward religion and convictions, and prohibition of any forms of restriction of citizens' rights on grounds of religious affiliation, part 2 of Article 19. Part 2 of Article 29 emphasizes that propaganda or agitation inciting religious hatred and enmity is not permitted and that propaganda of religious superiority is prohibited. In addition, part 3 of Article 59 establishes that a citizen of the Russian Federation whose convictions or religious belief contradict military service, as well as in other cases established by federal law, has the right to replace it with alternative civilian service.

The 1997 Federal Law On Freedom of Conscience and Religious Associations was defective from the outset, containing gaps and collisions both from the point of view of legal technique and from the standpoint of the Constitution of Russia. In recent years, the repressive potential of this Federal Law has been strengthened by numerous amendments, among which one should highlight provisions aimed at counteracting so-called extremist activity. It must be acknowledged that the application of this Federal Law does not fully contribute to achieving tolerance and protection against intolerance and discrimination.

At the federal level, in addition to the Federal Law On Freedom of Conscience and Religious Associations, special mention should be made of the Federal Law of 25 July 2002, No. 114-FZ, On Counteracting Extremist Activity, and the amendments to other legislative acts introduced in connection with its adoption. These are based on the legally undefined term extremism, which predetermines intolerance, abuses and violations of human rights in the sphere of law enforcement.

The remarks made above apply to a significant extent to the Strategy for Counteracting Extremism in the Russian Federation until 2025, approved by the President of the Russian Federation on 28 November 2014, Pr-2753. This Strategy was developed to specify the provisions of the Federal Law of 25 July 2002, No. 114-FZ, On Counteracting Extremist Activity, and of Presidential Decree No. 537 of 12 May 2009, On the National Security Strategy of the Russian Federation until 2020, in which one of the sources of threats to national security was recognized as the extremist activity of nationalist, radical religious, ethnic and other organizations and structures aimed at violating the unity and territorial integrity of the Russian Federation and destabilizing the internal political and social situation in the country.

Since the document is based on the legally incorrectly defined concept of extremism, there are serious doubts that its implementation will contribute to the declared strengthening of civic unity, the achievement of interethnic and interfaith accord, and the preservation of the ethnocultural diversity of the peoples of the Russian Federation. It should be noted that all concepts derived from the legally undefined concept of extremism, including ideology of extremism or extremist ideology, manifestations of extremism or extremist manifestations, subjects of counteracting extremism, counteracting extremism and others, also do not appear legally correct.

As for criminal legislation protecting against intolerance and discrimination, it also does not appear legally correct or effective, since it does not fully comply with the principles of legal certainty, the rule of law and modern legal technique.

It should be emphasized that, according to the National Security Strategy of the Russian Federation, threats to national security in the field of culture include the weakening of the unity of the multinational people of the Russian Federation through propaganda of permissiveness and violence, and racial, national and religious intolerance. Accordingly, in order to prevent threats to national security, the Russian Federation concentrates its efforts on strengthening the internal unity of Russian society, ensuring social stability, interethnic accord and religious tolerance. However, since confessional identity is interconnected with ethnicity, violations of the principle of the secularity of the state create grounds for conflicts in the sphere of national relations as well.

Violations of the worldview neutrality of states, including the education system, and of other subjects of international law lie at the foundation of violations of freedom of conscience, xenophobia and intolerance, and ultimately create security threats at the global, regional and domestic levels. In this context, the following primarily domestic problems in this sphere should be highlighted: incorrect scientific, educational and normative legitimation of legally undefined concepts such as extremism, terrorism, spiritual security, sect, traditional religious organizations, non-traditional religious organizations, traditional Islam, non-traditional Islam, missionary activity, feelings of believers and others; incorrect scientific and educational legitimation of the connection between unlawful activity and the worldview sphere, which is manifested in concepts such as religious extremism, religious terrorism, Islamic extremism, Islamic terrorism and others; and dissemination through the media and mass communications, as well as through the education system, of stereotypes and approaches aimed at reproducing xenophobia, intolerance and discrimination in ethnic and confessional spheres.

In recent years, within the state and municipal education system, highly debatable programs have been implemented, supposedly aimed at achieving tolerance, including through the education system. For example, the Federal Target Program Formation of Attitudes of Tolerant Consciousness and Prevention of Extremism in Russian Society for 2001-2005 set as its aim the formation and introduction of social norms of tolerance, determining the stability of the behavior of individuals and social groups in different situations of social tension, as the basis of civic accord in a democratic state. At the same time, in order to implement this aim, among other things, the development and implementation of an effective state policy of counteraction to the legally indefinable extremism was declared in a highly questionable way, which largely predetermined an extremely low result. The state customer of the program was the Ministry of Education of the Russian Federation; its developers included the Ministry of Education of Russia, the Ministry of Science of Russia, the Ministry of Press, Television and Radio Broadcasting and Mass Communications of Russia, the Ministry of Foreign Affairs of Russia, the Ministry for Nationalities of Russia, the Ministry of Justice of Russia, the Ministry of Culture of Russia, the Ministry of Defense of Russia, the Russian Academy of Sciences, the Russian Academy of Education, the Prosecutor General's Office of the Russian Federation, and the State Scientific Institution Interdisciplinary Analytical Center for Social Innovations. The main implementers of the program were institutes of the Russian Academy of Sciences and Russian Academy of Education, educational institutions of all levels, organizations subordinate to the ministries and agencies that developed the program, and others.

In addition to the federal level, corresponding programs were also implemented in constituent entities of the Russian Federation. For example, in St. Petersburg this was the Program for Harmonizing Interethnic and Intercultural Relations, Preventing Manifestations of Xenophobia, and Strengthening Tolerance in St. Petersburg for 2006-2010, the Tolerance Program. The introductory part of the subsequent Program for Harmonizing Intercultural, Interethnic and Interconfessional Relations and Educating a Culture of Tolerance in St. Petersburg for 2011-2015, also the Tolerance Program, describes it as successful. Educational program complexes and modules aimed at strengthening attitudes of tolerant consciousness and behavior among young people were developed and introduced into the educational proce   of education was established to form a tolerant environment in St. Petersburg among different age and ethnic groups of the population; broad sections of the city's population were given access to objective information about ethnic and religious organizations operating in St. Petersburg; a number of ethnocultural and interethnic events were organized and conducted both at the city and district levels. All this contributed, among other things, to a reduction in the number of offences committed in St. Petersburg on grounds of interethnic hatred and, accordingly, to a reduction in negative informational occasions causing serious harm to the image of the city.

These programs are not without shortcomings, but it should be noted that the aims and objectives of the Program for Harmonizing Intercultural, Interethnic and Interconfessional Relations and Educating a Culture of Tolerance in St. Petersburg for 2011-2015, the Tolerance Program, do not contain the legally incorrect term extremism. However, this term is used in a highly controversial way in the content of the program. Moreover, among the main conditions and directions for implementing the program, the need for close interaction among state authorities, local self-government, educational institutions, scientific and cultural institutions, and non-governmental organizations is indicated. A particularly important task toward which all program activities should be directed is described as conveying to all sections of the population of St. Petersburg the need actively to oppose interethnic and interreligious discord and to understand and share the ideology of ethnic and confessional tolerance.

The central role in realizing the aims and objectives of the program is assigned to the formation of a culture of tolerance through educational means. The education system plays a fundamentally important role in forming a tolerant environment in St. Petersburg, serving as a factor in the successful integration of first-generation migrants into the urban community. Preventive and propaganda work among young people occupies a special place in this field, oriented toward forming persons competent in issues of intercultural relations, open to the perception of others, constructive in their attitude toward the diversity of cultural, religious and linguistic traditions developing in St. Petersburg, and capable of preventing conflicts arising on grounds of ethnocultural differences or resolving them by non-violent means.

An important aspect of implementation of the program is declared to be the formation of a stable mechanism of linguistic, socio-cultural and everyday adaptation of foreign students to new living conditions, and the strengthening of integration measures for this social group.

Experts identify among the program's shortcomings the practical absence of such important target groups as law enforcement officers, civil servants, lecturers and teachers; the priority given to the values of the receiving society and its traditions and customs; and a sharp tilt toward ethnographic culture.

In general, according to D. V. Dubrovsky, the Russian programs mentioned above do not fully correspond to international standards, in particular those contained in human rights documents and in the 1995 Declaration of Principles on Tolerance, which are aimed at educating children and young people in the spirit of human rights and respect for cultural diversity and which address measures of social protection for excluded groups, above all migrants who find themselves in a vulnerable position.

To a significant extent, one can agree with the expert's view that, from the standpoint of international principles, programs for the education of tolerance should first of all concern social assistance to migrants and members of their families, as well as assistance in professional development. Their target groups should primarily be different groups of cultural minorities, and the predominant forms of work should be various social projects aimed at improving the situation with non-discrimination and human rights.

In addition, I believe that the general shortcomings of tolerance programs include, at best, their orientation toward forming a respectful attitude, or dialogue, primarily toward differences among social groups, or cultures, and only partly toward their representatives. Moreover, the idea of the unequal value of the cultures of migrants and the indigenous population makes programs for educating tolerance meaningless, since it proceeds from a discriminatory premise.

At the same time, from a legal point of view, the correct approach is to form respect for every person and their rights regardless of differences and group affiliation. As for the search for directions to change the situation for the better, the systemic nature of the problem implies the need for systemic measures to resolve it at all levels: scientific, theoretical and educational; legislative; the activity of state authorities, law enforcement and judicial practice; human rights protection; and the media and mass communications.

Particular attention should be paid to the formation of preventive measures based on the realization of human rights, the worldview neutrality of the state, the values of civil society, and the reduction of negative tension in society. The model of positive security of society receives the broadest space for global development only under conditions of international and interstate cooperation, but integrative tendencies within the state can also become a positive platform for developing public security on the basis of the civic identity of the person, regardless of their national-ethnic or religious preferences. The civic community of people is integrated in the state regardless of religious or national differences. Therefore, the removal of any dividing lines in society is favorable to the cohesion and solidarity of all citizens who have equal rights and duties before the law.

As a priority measure, clearly insufficient but necessary for reducing the level of intolerance and discrimination, it is necessary adequately to secure and realize the principle of worldview neutrality of states and other subjects of international law. I believe that this principle is foundational in relation to the principles of the rule of law, the law-governed state, human rights, freedom of conscience, equality of individuals and public associations, prohibition of discrimination motivated by worldview affiliation, ideological diversity and others.

Positive experience and errors in the ethno-confessional sphere should be studied carefully and impartially and taken into account when reforming the normative legal framework, including in the field of education. Since the formation of tolerant relations is significantly influenced by the ethnocultural policy of state authorities and local self-government, there arises a need to strengthen its organizational foundations and to create an integral system of legal, socio-economic, political and state-administrative measures for regulating ethno-social processes. In particular, systematic work is needed to form a state system of legal education and upbringing in the sphere of freedom of conscience and tolerance. In addition, diagnosis and prevention of xenophobia in youth and especially student environments appears to be an extremely important direction.

It should be specially noted that state information policy must also be worldview-neutral, which will contribute to reducing ethno-confessional contradictions and achieving a state of security in society. For this, first of all, the relevant normative legal framework must be corrected at both the international and domestic levels. In this context, the G20 statement following the summit in autumn 2015 is important. It states that terrorism cannot and should not be associated with any religion, nationality, civilization or ethnic group. The leaders of the Group of Twenty noted that the fight against terrorism is an important priority for all countries. They reaffirmed their resolve to work together to prevent and suppress terrorist acts by strengthening international solidarity and cooperation, fully recognizing the central role of the United Nations and acting in accordance with the UN Charter and obligations under international law.

In order to overcome barriers to cooperation among people, it is necessary to create a system for diagnosing and counteracting intolerance within the education system. Today, and especially for the formation of a positive future, overcoming the legacy of the age of intolerance, division and confrontation is vitally necessary. The problems of human rights, tolerance and counteracting intolerance, as well as worldview neutrality of states and other subjects of law, must be reflected in contemporary educational programs.

It appears extremely relevant to introduce thematic blocks and special educational courses within the state and municipal education system, such as Tolerance, Legal Foundations of Tolerance, and Tolerance and Counteracting Intolerance. It should also be emphasized that systematic work is needed to form tolerance and counteract intolerance through international organizations and state education systems. Thus, tolerance and counteracting xenophobia, intolerance and discrimination must be considered in connection with internationally recognized human rights, including the right to freedom of conscience, as well as with the principle of worldview neutrality of states and other subjects of international law.

REFERENCES TO CHAPTER 1

List of Sources Used

1. Buryanov S.A. Problems of Improving the Legal and Regulatory Framework for the Implementation of the Right to Freedom of Conscience in the Context of Globalization // State and Law. 2002. No. 10. pp. 26-31.
2. International Protection of Human Rights: Textbook / executive editor A.Kh. Abashidze. Moscow: RUDN University, 2017. 466 p. p. 25.
3. Abashidze A.Kh. Problems of the International Legal Protection of Minorities: Doctoral Dissertation in Law. Moscow, 1997. 58 p. p. 14.
4. Weissbrodt D., Fitzpatrick D., Newman F. International Human Rights: Law, Policy, and Process. Third Edition. University of Minnesota Human Rights Center. 2003. Electronic resource: URL: http://www1.umn.edu/humanrts/russian/hrtsbook/hrtsintroduction.html#note_3 (accessed: 27.01.2017).
5. Lukashuk I.I. International Law. General Part: Textbook for Students of Law Faculties and Universities. 3rd ed., revised and expanded. Moscow: Wolters Kluwer, 2005. p. 37.
6. Cited work. p. 105.
7. Dorskaya A.A. Protection of Freedom of Conscience in International Humanitarian Law // International Public and Private Law. 2009. No. 4. pp. 8-11.
8. Weissbrodt D., Fitzpatrick D., Newman F. International Human Rights: Law, Policy, and Process. Third Edition. University of Minnesota Human Rights Center. 2003. Electronic resource: URL: http://www1.umn.edu/humanrts/russian/hrtsbook/hrtsintroduction.html#note_3 (accessed: 27.01.2017).
9. Samovich Yu.V. The Formation of the Institution for the Protection of Individual Rights and Freedoms in International Law // International Public and Private Law. 2006. No. 1. pp. 2-5.
10. Abashidze A.Kh. Principles of International Law: Problems of Conceptual and Substantive Character // Moscow Journal of International Law. No. 4. pp. 19-30.
11. Cited work. p. 29.
12. Buryanov S.A. International Legal Documents in the Field of Freedom of Conscience and the Practice of Their Implementation in the Russian Federation: Theoretical and Applied Study for 2011. Moscow: Russian Helsinki Group, 2012. pp. 227-228.
13. Valeev R.M. International Protection of the Human Right to Freedom of Conscience and Religion // Bulletin of Volga University named after V.N. Tatishchev. 2012. No. 2 (76). p. 126.
14. Rowe M. The Right to Freedom of Religion or Belief. Study Guide. Minnesota: University of Minnesota Human Rights Center. 2003 // URL: http://www1.umn.edu/humanrts/russian/edumat/studyguides/Rreligion.html (accessed: 17.02.2017).
15. See: Biryukov P.N., Buryanov S.A. Insur Farkhutdinov: Ensuring Peace and Security in Eurasia (International Legal Assessment of Events in Syria). Interview with Doctor of Law and Editor-in-Chief of the Eurasian Law Journal Insur Zabirovich Farkhutdinov // Eurasian Law Journal. 2015. No. 10. pp. 8-15.
16. Farkhutdinov I.Z. The Israeli Doctrine of Preventive Self-Defense and International Law // Eurasian Law Journal. 2016. No. 8 (99). pp. 15-32.
17. Farkhutdinov I.Z. International Law on the Use of Military Force by a State against Non-State Actors // Eurasian Law Journal. 2016. No. 7 (98). pp. 59-76. Farkhutdinov I.Z. International Law on the Principle of Non-Use of Force or Threat of Force: History and the Present // Eurasian Law Journal. 2015. No. 11 (90). pp. 34-38. Farkhutdinov I.Z. International Law on the Self-Defense of States // Eurasian Law Journal. 2016. No. 1 (92). pp. 91-100.
18. Farkhutdinov I.Z. International Law and the U.S. Doctrine of Preventive Self-Defense // Eurasian Law Journal. 2016. No. 2 (93). pp. 23-31.
19. Buryanov S.A. Freedom of Conscience as a Global Value: Toward Political Unity and the Resolution of Global Problems // Age of Globalization. No. 1. 2009. pp. 136-151.
20. Baglay M.V. Constitutional Law of the Russian Federation. Moscow, 2001. p. 166.
21. Theory of State and Law / executive editors V.M. Korelsky and V.D. Perevalov. Yekaterinburg, 1996. p. 501.
22. Civil Human Rights in Russia: Contemporary Problems of Theory and Practice / ed. by F.M. Rudinsky. Moscow: CJSC Trading Firm "MIR", 2006. p. 39.
23. Rudinsky F.M. The Science of Human Rights and Problems of Constitutional Law. Moscow: CJSC Trading Firm "MIR", 2006. p. 639.

24. Avakyan S.A. Freedom of Religion as a Constitutional-Legal Institution // Bulletin of Moscow State University. Series 11. No. 1. 1999.
25. Savelyev V.N. Freedom of Conscience: History and Theory. Moscow: Higher School, 1991.
26. Kirichenko M.G. Freedom of Conscience in the USSR. Moscow: Legal Literature, 1985.
27. Teteryatnikov N.Yu. Freedom of Conscience as a Subjective Right and a Legal Institution: Theory, History, and the Present. Krasnoyarsk: Law Institute of the Ministry of Internal Affairs of Russia, 2005. p. 32.
28. Cited work. pp. 32-33.
29. Mchedlov M.P. The Basis of Civic Accord // Freedom of Conscience in the Spiritual Revival of the Fatherland. Moscow, 1994. p. 11.
30. Alting von Geusau Frans A.M. How Can the Transition to a Law-Governed Society Be Made? Moscow, 1997. p. 33.
31. Alekseev S.S. The Holiest Thing God Has on Earth: Immanuel Kant and Problems of Law in the Modern Era. Moscow: NORMA Publishing House, 1998. pp. 256-259.
32. Teteryatnikov N.Yu. Cited work. pp. 100-101.
33. Moran G.M. What Is Religious Freedom? What Should the Law Guarantee? // Religion and Law. No. 1. 1997. p. 14.
34. Pimenov A. The Past That Has Not Passed: Toward a Characterization of Non-Theistic Religiosity // Religion. Magic. Myth. Moscow: URSS, 1997. p. 264.
35. Lovinyukov A.S. The 1990 USSR Law on Freedom of Conscience and Religious Organizations // Soviet State and Law. No. 4. 1991. p. 28.
36. Rosenbaum Yu.A. The Soviet State and the Church / executive editor B.M. Lazarev. Moscow, 1985. 174 p. p. 94.
37. Baglay M.V. Constitutional Law of the Russian Federation. Moscow, 1998. p. 192.
38. Bezuglov A.A., Soldatov S.A. Constitutional Law of Russia. In 3 volumes. Vol. 1. Moscow, 2001. p. 468.
39. Pchelintsev A.V. Freedom of Religion and Religious Associations in the Russian Federation (Constitutional-Legal Study): Abstract of Doctoral Dissertation in Law. Moscow, 2012. p. 15.
40. So A.A. Constitutional-Legal Foundations of Freedom of Religion and the Activities of Religious Associations in Russia (Using the Subjects of the Russian Federation in the North-Western Federal District as an Example): Abstract of Candidate of Legal Sciences Dissertation. Saint Petersburg, 2010. p. 7.
41. Rudinsky F.M. The Institution of Freedom of Conscience under Soviet State Law: Candidate of Legal Sciences Dissertation. Novosibirsk, 1963. p. 6.
42. Savelyev V.N. Freedom of Conscience: History and Theory. Moscow, 1991. p. 97.
43. Avakyan S.A. Constitutional Law of Russia: Course Textbook. In 2 volumes. Vol. 1. Moscow: Jurist, 2005. pp. 606-607.
44. Buryanov S.A. Legal Foundations, Essential Content, and Guarantees of Freedom of Conscience // State and Law. No. 2. 2001. p. 24.
45. Teteryatnikov N.Yu. Theoretical and Legal Aspects of Freedom of Conscience: Abstract of Candidate of Legal Sciences Dissertation. Yekaterinburg, 2002. p. 10.
46. Lovinyukov A.S. Freedom of Conscience (Analysis, Practice, Conclusions) // State and Law. 1995. No. 1. p. 26.
47. Buryanov S.A. State Policy of the Russian Federation in the Sphere of Freedom of Conscience at the Turn of the Millennium // Law and Politics. No. 1. 2000. p. 85.
48. Simorot S.Yu. Cited work. pp. 26-28.
49. Teteryatnikov N.Yu. Cited work. p. 179.
50. Bryantseva O.V. Implementation of the Constitutional Right to Freedom of Conscience in the Russian Federation (Using Volgograd Oblast as an Example): Abstract of Candidate of Legal Sciences Dissertation. Volgograd, 2005. p. 7.
51. Tikhonova E.V. The Institution of Freedom of Conscience and Freedom of Religion in the Law of Contemporary Russia: Abstract of Candidate of Legal Sciences Dissertation. Tambov, 2006. p. 17.
52. Minchenko T.P. The Problem of Freedom of Conscience in the Era of Post-Secularity: Origins and Prospects: Doctoral Dissertation in Philosophy. Tomsk, 2011. p. 12.
53. Freedom of Conscience: Problems of Theory and Practice. Monograph / ed. by F.M. Rudinsky and S.A. Buryanov. Moscow: CJSC Trading Firm "MIR", 2012. 1120 p.
54. Aleynikova S.M., Buryanov S.A. The Secular State in Questions and Answers: Brief, Accessible, and Relevant / ed. by I. Kondratyev. Moscow, 2015. 62 p.

55. Buryanov S.A., Krivenky A.I., Pashentsev D.A., Romanova G.V. Issues of the Globalization of Culture and the Protection of the Cultural Rights of the Human Person and Citizen (Interdisciplinary Study): Monograph / general editors A.I. Krivenky and S.A. Buryanov. Moscow: Moscow City University, 2018. 200 p.
56. The Influence of Hellenism on Contemporary Science, Culture, and Education: Collective Monograph / scientific editor T.P. Minchenko. Tomsk: STT, 2016. 402 p.
57. Buryanov S.A. Problems of Implementing Freedom of Conscience and the Secular Character of the State in the Russian Federation: Theoretical and Applied Study for 2012 / ed. by R.V. Bagdasarov. Moscow: Russia for All, 2013. 352 p.
58. Buryanov S.A. The Secular Character of the State and Internationally Recognized Freedom of Conscience: Theoretical and Applied Study for 2015 - Early 2016. Moscow: Polygraph Service, 2016. 258 p.
59. Buryanov S.A. The Secular Character of the State in the Russian Federation: Theoretical and Applied Study for 2016 - Early 2017. Moscow, 2017. 196 p.
60. Buryanov S.A. State Worldview Neutrality in the Context of Deteriorating Imbalances in Globalization: The Case and Current State of Affairs in the Russian Federation from 2016 to the Beginning of 2017. Amsterdam, 2018. 228 p.
61. Nikitaev D.M. On the Contemporary Definition of the Concept and Content of Freedom of Conscience // Contemporary Lawyer. 2018. No. 4 (25). pp. 66-75.
62. Nikitaev D.M. Problems of Implementing Freedom of Conscience in Contemporary Russia and the World / D.M. Nikitaev, S.A. Buryanov; scientific editor S.A. Buryanov. Moscow: RUSCIENCE, 2020. 146 p.
63. Teteryatnikov N.Yu. Theoretical and Legal Aspects of Freedom of Conscience: Abstract of Candidate of Legal Sciences Dissertation. Yekaterinburg, 2002. p. 17.
64. Ibid. p. 18.
65. Fromm E. Psychoanalysis and Religion // Twilight of the Gods. Moscow: Progress, 1990. p. 158.
66. Buryanov S.A. Legal Foundations, Essential Content, and Guarantees of Freedom of Conscience // State and Law. No. 2. 2001. p. 24.
67. Buryanov S.A. Implementation of the Constitutional Freedom of Conscience and Freedom of Religion in the Russian Federation: Monograph. Moscow: CJSC Trading Firm "MIR", 2009. 1120 p.
68. Teteryatnikov N.Yu. Freedom of Conscience as a Subjective Right and a Legal Institution: Theory, History, and the Present. Krasnoyarsk, 2005. p. 125.
69. Freedom of Conscience, Religion, and Belief (Problems of Theory and Practice). Russian Documents: Collection of Normative Legal Acts / compiled by S.A. Buryanov, Yu.V. Gavrilova, S.S. Savva; ed. by S.A. Buryanov. Moscow: Moscow City University, 2011. 160 p.
70. Freedom of Conscience, Religion, and Belief (Problems of Theory and Practice). International Documents: Collection of Normative Legal Acts / compiled by S.A. Buryanov, Yu.V. Gavrilova, S.S. Savva; ed. by S.A. Buryanov. Moscow: Moscow City University, 2011. 122 p.
71. Freedom of Conscience, Religion, and Belief (Problems of Theory and Practice). Documents of Foreign States: Collection of Normative Legal Acts / compiled by S.A. Buryanov, Yu.V. Gavrilova, S.S. Savva; ed. by S.A. Buryanov. Moscow: Moscow City University, 2013. 196 p.
72. Freedom of Conscience, Religion, and Belief (Problems of Theory and Practice): Collection of Judicial Practice Materials / compiled by S.A. Buryanov, Yu.V. Gavrilova, S.S. Savva; ed. by S.A. Buryanov. Moscow: Moscow City University, 2013. 204 p.
73. Freedom of Conscience, Religion, and Belief (Problems of Theory and Practice): Collection of Materials of State and Non-Governmental Human Rights Bodies / compiled by S.A. Buryanov, Yu.V. Gavrilova, S.S. Savva; ed. by S.A. Buryanov. Moscow: Moscow City University, 2014. 180 p.
74. Buryanov S.A. Implementation of the Constitutional Freedom of Conscience and Freedom of Religion in the Russian Federation: Monograph. Moscow: CJSC Trading Firm "MIR", 2009. 288 p.
75. Silkenat J.R. The Rule of Law and Its Understanding // International Legal Doctrines: A Search for Mutual Understanding / Collection of Articles edited by E.G. Tarlo and D.V. Kravchenko. Moscow: LUM, 2013. p. 8.
76. Kalamkaryan R.A., Migachev Yu.I. International Law. Moscow: Eksmo, 2004. p. 53.
77. Ibid. pp. 52-53.
78. Constitutional Law of the Russian Federation. Moscow, 2002. Electronic resource. URL: http://slovari.yandex.ru (accessed: 3.01.2018).
79. Encyclopedia of the Lawyer. Moscow, 2005. Electronic resource. URL: http://sl

80. Elementary Principles of the General Theory of Law. Moscow, 2003. Electronic resource. URL: http://slovari.yandex.ru (accessed: 3.01.2018).
81. Comprehensive Legal Dictionary. Akademik.ru. Moscow, 2010. Electronic resource. URL: http://jurisprudence.academic.ru (accessed: 3.01.2018).
82. Gadzhiev G.A. The Judicial Doctrine of the Rule of Law in Russia // International Legal Doctrines: A Search for Mutual Understanding / Collection of Articles edited by E.G. Tarlo and D.V. Kravchenko. Moscow: LUM, 2013. p. 25.
83. Ibid. p. 26.
84. Abashidze A.Kh., Melshina K.Yu. The Rule of Law as Understood by the United Nations // Bulletin of Volgograd State University. Series 5: Jurisprudence. 2016. Vol. 15. No. 4 (33). pp. 10-18. p.16.
85. Panteleev B. Legal Certainty or Unpredictability? Electronic resource. URL: http://www.center-bereg.ru/916.htm (accessed: 3.01.2018).
86. Kartashkin V.A. Human Rights: The Relationship and Interaction of International and Domestic Law // Proceedings of the Institute of State and Law of the Russian Academy of Sciences. 2016. No. 4. pp. 14-34. p. 26.
87. Kartashkin V.A. The Supremacy of International Law in World Politics and at the National Level // Observer. 2015. No. 11 (310). pp. 112-122.
88. Kalamkaryan R.A., Migachev Yu.I. International Law. Moscow: Eksmo, 2004. p. 41.
89. Cited work. pp. 41-42.
90. Cited work. p. 54.
91. Abashidze A.Kh., Melshina K.Yu. The Rule of Law as Understood by the United Nations // Bulletin of Volgograd State University. Series 5: Jurisprudence. 2016. Vol. 15. No. 4 (33). pp. 10-18. p. 16.
92. The United Nations and the Rule of Law. Electronic resource. URL: http://www.un.org/ru/ruleoflaw/ (accessed: 3.01.2018).
93. The United Nations and the Rule of Law. Electronic resource. URL: http://www.un.org/ru/ruleoflaw/ (accessed: 3.01.2018).
94. The United Nations and the Rule of Law. Electronic resource. URL: http://www.un.org/ru/ruleoflaw/ (accessed: 3.01.2018).
95. Declaration of the High-Level Meeting of the General Assembly on the Rule of Law at the National and International Levels. Electronic resource. URL: http://www.un.org/ru/documents/decl_conv/declarations/ruleoflaw2012.shtml (accessed: 3.01.2018).
96. Declaration of the High-Level Meeting of the General Assembly on the Rule of Law at the National and International Levels. Electronic resource. URL: http://www.un.org/ru/documents/decl_conv/declarations/ruleoflaw2012.shtml (accessed: 3.01.2018).
97. Presnyakov M.V. The Principle of Legal Certainty in the System of Constitutional Principles // Human Rights: Theory, History, Practice. Collection of Scientific Works Dedicated to the 65th Anniversary of the Universal Declaration of Human Rights / general editor E.M. Pavlenko. Volgograd: Publishing House of the Volgograd Branch of RANEPA, 2014. p. 215.
98. Cited work. p. 219.
99. Cited work. p. 221.
100. Cited work. p. 222.
101. Vlasenko N.A. Theory of State and Law: Textbook (2nd ed., revised, expanded, and corrected). Moscow: Prospekt, 2011. p. 340.
102. Cited work. p. 341.
103. Kenenov A.A., Chernobel G.T. Logical Foundations of the Lawmaking Process // Jurisprudence. 1991. No. 6. p. 73.
104. Sultanov A.R. On Legal Certainty and Judicial Lawmaking. Electronic resource. URL: http://www.juristlib.ru/book_3659.html (accessed: 16.01.2018).
105. Zhuravleva O.O. The Principle of Legal Certainty and Acts of the Higher Judicial Bodies in Tax Law. Electronic resource. URL: http://www.justicemaker.ru/view-article.php?id=14&art=1821 (accessed: 11.01.2018).
106. Presnyakov M.V. The Principle of Legal Certainty in the System of Constitutional Principles // Human Rights: Theory, History, Practice. Collection of Scientific Works Dedicated to the 65th Anniversary of the Universal Declaration of Human Rights / general editor E.M. Pavlenko. Volgograd: Publishing House of the Volgograd Branch of RANEPA, 2014. p. 216.
107. Cited work. p. 217.
108. Vlasenko N.A. The Categories of "Uncertainty" and "Certainty" in the Study of Contemporary Law // Legal Science and Practice: Bulletin of the Nizhny Novgorod Academy of the M . 8-17. p.

14.
109. Barenboim P.D., Kravchenko D.V. The Doctrine of the Russian Rule-of-Law State: Components, Principles, and Features // International Legal Doctrines: A Search for Mutual Understanding / Collection of Articles edited by E.G. Tarlo and D.V. Kravchenko. Moscow: LUM, 2013. p. 112.
110. Vlasenko N.A. Theory of State and Law: Textbook (2nd ed., revised, expanded, and corrected). Moscow: Prospekt, 2011. p. 124.
111. Vyshkvartsev V.V. The Rule-of-Law State, Human Rights, and Freedoms in the Concepts of Russian Scholars of the 20th and Early 21st Centuries: Abstract of Candidate of Legal Sciences Dissertation. Moscow, 2009. p. 21.
112. Antonov I.Yu. The Rule-of-Law State and Fundamental Human Rights and Freedoms: Abstract of Candidate of Legal Sciences Dissertation. Moscow, 2005. p. 8.
113. Vyshkvartsev V.V. The Rule-of-Law State, Human Rights, and Freedoms in the Concepts of Russian Scholars of the 20th and Early 21st Centuries: Abstract of Candidate of Legal Sciences Dissertation. Moscow, 2009. p. 23.
114. Ibid. p. 11.
115. Theory of State and Law: Textbook / ed. by N.M. Chepurnova and A.V. Seregin. Moscow: EAOI, 2007. 465 p. p. 217.
116. Ibid. p. 217.
117. Gadzhiev G.A. The Judicial Doctrine of the Rule of Law in Russia // International Legal Doctrines: A Search for Mutual Understanding / Collection of Articles edited by E.G. Tarlo and D.V. Kravchenko. Moscow: LUM, 2013. p. 29.
118. Ibid. p. 30.
119. Ruling of the Constitutional Court of the Russian Federation of July 15, 1999, No. 11-P, "In the Case Concerning the Verification of the Constitutionality of Certain Provisions of the RSFSR Law "On the State Tax Service of the RSFSR" and the Laws of the Russian Federation "On the Fundamentals of the Tax System in the Russian Federation" and "On Federal Tax Police Bodies"" // GARANT System. Electronic resource. URL: http://base.garant.ru/12116289/#ixzz35Nkq3kAP (accessed: 5.01.2018).
120. Davydova M.L. Theoretical and Methodological Problems of the Concept and Composition of Legal Technique: Abstract of Doctoral Dissertation in Law. Volgograd, 2010. p. 7.
121. Cited work. p. 8.
122. Cited work. p. 8.
123. Kartukhin V.Yu. The Use of the Rules of Legal Technique in Lawmaking by Constituent Entities of the Russian Federation // Contemporary Jurisprudence. Current Problems of Russian Law. Electronic resource. URL: http://www.ibil.ru/index.php?type=review&area=1&p=articles&id=358 (accessed: 9.01.2018).
124. Bakhvalov S.V. Legislative Technology (Some Problems of Theory and Practice): Abstract of Candidate of Legal Sciences Dissertation. Nizhny Novgorod, 2006. p. 15.
125. Davydova M.L. Theoretical and Methodological Problems of the Concept and Composition of Legal Technique: Abstract of Doctoral Dissertation in Law. Volgograd, 2010. p. 12.
126. Yusipova I.V. Law and Language in the Mechanism of Ensuring Human Rights: Abstract of Candidate of Legal Sciences Dissertation. Vladimir, 2008. p. 7.
127. Cited work. p. 14.
128. Cited work. p. 18.
129. Cited work. p. 20.
130. Kartukhin V.Yu. The Use of the Rules of Legal Technique in Lawmaking by Constituent Entities of the Russian Federation // Contemporary Jurisprudence. Current Problems of Russian Law. Electronic resource. URL: http://www.ibil.ru/index.php?type=review&area=1&p=articles&id=358 (accessed: 9.01.2018).
131. Bakhvalov S.V. Legislative Technology (Some Problems of Theory and Practice): Abstract of Candidate of Legal Sciences Dissertation. Nizhny Novgorod, 2006. p. 14.
132. Civil Human Rights in Russia: Contemporary Problems of Theory and Practice / ed. by F.M. Rudinsky. Moscow: CJSC Trading Firm "MIR", 2010. p. 13.
133. Freedom of Conscience: Problems of Theory and Practice. Monograph / ed. by F.M. Rudinsky and S.A. Buryanov. Moscow: CJSC Trading Firm "MIR", 2012.
134. Rudinsky F.M. Law, Religion, Offenses. Volgograd, 1971. p. 36.
135. Ibid. p. 36.
136. Pokaninova E.B. Socio-Philosophical Analysis of the Transformation of State-C ract of Doctoral

Dissertation in Philosophy. Moscow, 2012. p. 16.
137. International Law: Textbook for Universities / executive editors G.V. Ignatenko and O.I. Tiunov. Moscow: NORMA-INFRA, 1999. 584 p.
138. Buryanov S.A. Religion in Elections in Russia: The Factor of Relations between the State and Religious Associations in the Federal Electoral Cycle of 2003-2004. Moscow: Institute of Freedom of Conscience, 2005. 200 p.
139. Buryanov S.A., Mozgovoy S.A. The Problem of Implementing Freedom of Conscience and Trends in Relations between the State and Religious Associations in Russia: Information and Analytical Report of the Institute of Freedom of Conscience (Second Half of 2001 - End of 2004) / S.A. Buryanov, S.A. Mozgovoy; Institute of Freedom of Conscience. Moscow, 2005. [Scientific ed., 2nd, expanded and corrected]. 526 p.
140. Buryanov S.A. International Legal Documents in the Sphere of Freedom of Conscience and the Practice of Their Application in the Russian Federation: Theoretical and Applied Study for 2011. Moscow, 2012. 244 p.
141. Buryanov S.A. Xenophobia, Intolerance, and Discrimination on Grounds of Religion or Belief in the Constituent Entities of the Russian Federation: Specialized Information and Analytical Report for 2006 - First Half of 2007. Moscow, 2007. 240 p.
142. Ivaneev S.V. There Is No Peace under Religious Olive Trees // History of State and Law. 2011. No. 1. pp. 23-24.
143. Ivaneev S.V. Global Problems and Some Approaches to Their Solution // Rights and Freedoms of the Human Person and Citizen: Theoretical Aspects and Legal Practice. Proceedings of the Annual International Scientific Conference in Memory of Professor Felix Mikhailovich Rudinsky / general editor D.A. Pashentsev. 2016. pp. 117-121.
144. Declaration on the Elimination of All Forms of Intolerance and of Discrimination Based on Religion or Belief (Proclaimed by General Assembly resolution 36/55 of November 25, 1981) // International Acts on Human Rights. Collection of Documents. Moscow: NORMA Publishing House (NORMA-INFRA-M Publishing Group), 2000. pp. 132-134.
145. Course "Human Rights": Study Guide for Staff Members of Offices of Commissioners and Human Rights Commissions in the Russian Federation. Moscow, 2005. p. 136.
146. Bazikov R.V. Social Stereotypes: A Conceptual Aspect: Candidate of Philosophical Sciences Dissertation. Rostov-on-Don, 1999. p. 7.
147. Orlova O.G. Discursive Theory of the Stereotype: Abstract of Doctoral Dissertation in Philology. Kemerovo, 2013. p. 7.
148. Gakaev Z.Zh. Ethnic Stereotypes in the Press: Candidate of Historical Sciences Dissertation. Moscow, 2003. p. 17.
149. Kroz M.V., Ratinov N.A. Socio-Psychological and Legal Aspects of Xenophobia. Moscow, 2005. p. 38.
150. Gavreev S.A. Socio-Psychological Characteristics of Ethnophobia: Candidate of Psychological Sciences Dissertation. Moscow, 2008. pp. 5-6.
151. Evstafyeva A.V. Hate Speech in the Mass Media: Linguistic and Extralinguistic Factors of Functioning: Abstract of Candidate of Philological Sciences Dissertation. Tolyatti, 2009. p. 8.
152. Gontsovsky V.K. Ethnic Xenophobia among Youth in the Southern Federal District (Using Rostov Oblast as an Example): Abstract of Candidate of Sociological Sciences Dissertation. Rostov-on-Don, 2014. pp. 15-16.
153. Ibid. p. 17.
154. Freedom of Conscience, Religion, and Belief (Problems of Theory and Practice): Collection of Materials of State and Non-Governmental Human Rights Bodies / compiled by S.A. Buryanov, Yu.V. Gavrilova, S.S. Savva; ed. by S.A. Buryanov. Moscow: Moscow City University, 2014. 180 p. pp. 59-96.
155. Buryanov S.A. International Legal Documents in the Sphere of Freedom of Conscience and the Practice of Their Application in the Russian Federation: Theoretical and Applied Study for 2011. Moscow, 2012. 244 p.
156. Saidov A.Kh. Universally Recognized Human Rights: Study Guide / A.Kh. Saidov; ed. by I.I. Lukashuk; Academic Law University. Moscow: MZ Press, 2004. p. 71.
157. De Zayas A., Möller J. Th. The United Nations Human Rights Committee Case Law 1977-2008: A Handbook. N.P. Engel Publishers, Kehl am Rhein, 2009.
158. Freedom of Conscience: Problems of Theory and Practice. Monograph / ed. by F.M. Rudinsky and S.A. Buryanov. Moscow: CJSC Trading Firm "MIR", 2012. 1120 p.
159. Federal Target Program "Formation of Attitudes of Tolerant Consciousness and Prevention of Extr ource.

URL: http://www.tolz.ru/library/?id=51 (accessed: 8.10.2017).

160. Federal Target Program "Formation of Attitudes of Tolerant Consciousness and Prevention of Extremism in Russian Society" for 2001-2005. Electronic resource. URL: http://base.garant.ru/1586359/ (accessed: 8.10.2017).
161. Program for the Harmonization of Intercultural, Interethnic, and Interconfessional Relations and the Education of a Culture of Tolerance in Saint Petersburg for 2011-2015 (Tolerance Program). Electronic resource. URL: http://dum-spb.ru/programma-tolerantnost-v-sankt-peterburge-na-2011-2015-gg (accessed: 8.10.2017).
162. Dubrovsky D. Society, Culture, Human Rights: Review of the Moscow Tolerance Program // "Us" and "Others": Tolerance, Stereotypes, Rights. Moscow: Russian Helsinki Group, 2016. 116 p. pp. 37-50.
163. Buryanov S.A. Implementation of the Constitutional Freedom of Conscience and Freedom of Religion in the Russian Federation: Monograph. Moscow: CJSC Trading Firm "MIR", 2009. pp. 249-258.
164. Alibekova S.Ya. National-Ethnic and Religious Aspects of Public Security // Theory and Practice of Social Development. No. 19. 2015. p. 163.
165. Melnikova N.V. Formation of Ethnic Tolerance in a Megalopolis: Sociological Analysis: Abstract of Candidate of Sociological Sciences Dissertation. Moscow, 2002. p. 8.
166. Krygina I.A., Rybak S.V. The Legal Education Process in the Context of Developing Ideas of Tolerance under Contemporary Conditions: Monograph. Rostov-on-Don, 2012. 142 p.
167. Apanasyuk L.A. Theoretical and Methodological Foundations for Preventing Xenophobia in the Student Environment through Socio-Cultural Activity: Abstract of Doctoral Dissertation in Pedagogy. Tambov, 2014. pp. 5-7.
168. Buryanov S.A. The Right to Freedom of Conscience under Conditions of Global Processes and Challenges 4.0 // Eurasian Law Journal. 2023. No. 6 (181). pp. 18-21.
169. Buryanov S.A. International Recognition of the Right to Freedom of Conscience and Problems of Its Implementation in the Russian Federation under Conditions of Contemporary Global Processes: Monograph. Moscow: Polygraph Service, 2020. 624 p.
170. Civil Human Rights in Russia: Modern Problems of Theory and Practice / compiled by F.M. Rudinsky. Transaction Publishers, New Brunswick, New Jersey (U.S.A.), 2008. 458 p.
171. Buryanov S.A. State Worldview Neutrality in the Context of Deteriorating Imbalances in Globalization: The Case and Current State of Affairs in the Russian Federation from 2016 to the Beginning of 2017. Amsterdam, 2018. 228 p.
172. Burianov S. The Russian Federation State Policy in the Sphere of Freedom of Conscience at the Turn of the Millennium // Law and Politics: Russian Review. 2001. No. 1.

PART II. THE STATE, RELIGION, AND THE BOUNDARIES OF MEANING

CIVILIZATIONAL THESIS

"Know thyself."

The modern state becomes dangerous when it claims authority over final truth. Public-law worldview neutrality is the discipline that prevents power from turning belief, identity, and meaning into instruments of domination. The boundary between the state and the inner world becomes a condition of human dignity.

CONCEPT BOX

CHAPTER 2

WHO HAS THE RIGHT TO DEFINE TRUTH?

State power, religion, and the limits of public authority

Relations between the state and religion reveal one of the oldest legal dangers: public authority may begin to treat truth as its own property.

When the state privileges one worldview, suppresses another, or ranks citizens by their convictions, law becomes an instrument of worldview hierarchy.

Chapter 2 examines state secularity, state-confessional relations, and the limits of public authority in the sphere of meaning.

The proposed answer is public-law worldview neutrality: the state must protect the legal space for different worldviews without becoming the owner of final truth.

WHO HAS THE RIGHT TO DEFINE TRUTH?

State power, religion, and the limits of public authority

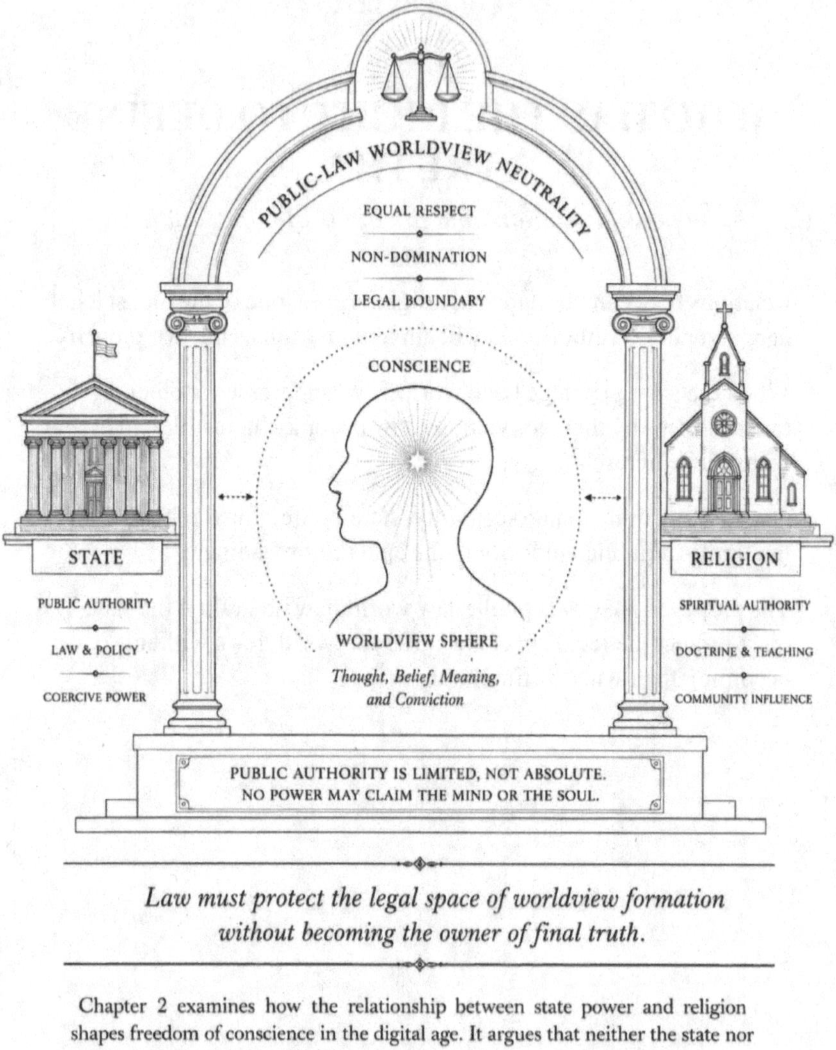

Law must protect the legal space of worldview formation without becoming the owner of final truth.

Chapter 2 examines how the relationship between state power and religion shapes freedom of conscience in the digital age. It argues that neither the state nor religious institutions may legitimately claim authority over ultimate truth. The proposed answer is public-law worldview neutrality—a principled stance that safeguards pluralism, human dignity, and the legal limits of public authority.

Chapter 2. Who Has the Right to Define Truth?

2.1. World Experience in the Legal Regulation of Relations between States and Religious Associations

The interaction between religion and politics, as well as relations between states and religious associations, has a substantial influence on all levels of realization of the internationally recognized right to freedom of conscience.

A certain number of scholarly works have been devoted to the principle of secularity in individual states. Yet the principle of secularity understood as the worldview neutrality of subjects of international law has not received adequate development. As noted above, this principle remains insufficiently developed at the theoretical and legal level and has not found adequate reflection either in the principles and norms of international law or in the recommendatory acts of international organizations.

At the same time, many contemporary researchers note, at minimum, the ambiguous influence of the politicization of religion and the clericalization of politics at both international and domestic levels.

For example, G. V. Stankevich, observing the politicization of the religious factor in the contemporary world and the clericalization of politics as part of this phenomenon, states that the religious component of the political process acts, on the one hand, as a factor in 'ensuring political and social stability' and, on the other, as a factor of socio-political destabilization. The researcher reasonably believes that 'the politicization of religion is connected, on the one hand, with the use of religious ideas by political actors for the realization of their own purposes and, on the other, with the desire of religious institutions to use political constructs to defend their positions within the traditional political space and to carry out proselytizing activity.'

In this regard, the question arises as to whether it is correct in principle for states to interfere in the activity of worldview associations through special legal regulation. It seems appropriate to heed the opinion of R. A. Podoprigora, who believes that, alongside the dominant approach based on the need for special regulation of the activity of religious organizations, there is another approach, according to which the state should not engage in such regulation. 'The religious freedoms proclaimed in international legal instruments and in the constitutions of states must be realized directly, without additional state regulation. For the protection of religious freedoms or for responding to unlawful religious activity, there already exists a sufficient number of administrative or criminal measures that may be applied to believers and their organizations, just as to any other citizens and their associations.'

Events in the Middle East, grounded in ethno-confessional divisions and contradictions, have once again strengthened the relevance of this problem. According to V. V. Naumkin, 'with respect to Iraq, a threefold division is noted: Sunni Arabs, Shiite Arabs and Kurds. Yet within this division confessional and ethnic markers of identity intersect, not to mention others, above all tribal ones; therefore, in my view, one may also speak here of two lines of binary division: Arabs and Kurds, and Shiite Muslims and Sunni Muslims, taking into account that the majority of Kurds are Sunni. However, these oppositions do not fully reflect the entire spectrum of divisions, including those involving ethno-confessional minorities. The split between different groups of Sunnis, for example between radical caliphate supporters and secular nationalists, is no less deep than the split between the main competing segments.'

Conflicts arising from ethno-confessional contradictions are significantly exacerbated by the fact that the contemporary world faces systemic problems of the worldview neutrality of states and, as a consequence, a crisis in the realization of human rights in the sphere of freedom of conscience, together with a rise in xenophobia and intolerance. Ultimately, this creates extremely serious obstacles to interaction among people, to positive international integration of states for solving global problems, and to the creation of conditions for the sustainable development of human civilization.

At the domestic level, one of the problems is the substitution of the principle of the secularity of the state by historically formed models of relations between the state and religious associations. In particular, scholarly literature presents the following models reflecting the legal status of religious associations within the state:

a) the model of a state religion or confession, that is, the existence of an official privileged religious association;

b) the model of separation of religious associations from the state, that is, equality of religious associations;

c) the model of cooperation between the state and religious associations, that is, cooperation of the state with selected religious associations.

The model of a state religion or confession operates in England, Denmark, Greece, Finland and in a number of Muslim states, including Algeria, Afghanistan, Bangladesh, Bahrain, Brunei Darussalam, Egypt, Jordan, Iraq, Iran, Yemen, Qatar, the Comoros, Kuwait, Mauritania, Malaysia, the Maldives, Morocco, the United Arab Emirates, Oman, Pakistan, Saudi Arabia, Sudan, Tunisia and others.

In the United Kingdom, two churches have retained state status. In England this is the Anglican Church, and in Scotland the Presbyterian Church. Wales and Northern Ireland, however, have no state religion.

At the same time, the Constitution of Ireland of 29 December 1937 contains references to 'the Most Holy Trinity, from Whom is all authority and to Whom, as our final end, all actions both of men and States must be referred' and to 'Our Divine Lord, Jesus Christ, Who sustained our fathers through centuries of trial.'

Under Article 12, the President, upon assuming office, publicly signs the following declaration: 'In the presence of Almighty God I do solemnly and sincerely promise and declare that I will maintain the Constitution of Ireland and uphold its laws, that I will fulfil my duties faithfully and conscientiously in accordance with the Constitution and the law, and that I will dedicate my abilities to the service and welfare of the people of Ireland. May God direct and sustain me.' Certain other state officials must also sign a public declaration containing elements of religious content (Article 34). The Constitution of Ireland was adopted by referendum on 1 July 1937 and entered into force on 29 December of the same year. Despite numerous amendments introduced since then, the preamble has remained unchanged. In 1972, following a referendum, provisions concerning the 'special position' of the Roman Catholic Church and other religious denominations were removed from the Constitution of Ireland (Article 44). It should be noted that, although the Constitution guarantees freedom of religion and the state may not establish an official religion (Article 44), religious formulations in the preamble and in other parts of the document remain the subject of debate.

The Constitution of the Kingdom of Denmark of 5 June 1953 states that 'the Evangelical Lutheran Church shall be the Established Church of Denmark and, as such, shall be supported by the State.'

The Constitution of the Hellenic Republic of 11 June 1974 provides in Article 3 that 'the prevailing religion in Greece is that of the Eastern Orthodox Church of Christ. The Orthodox Church of Greece, acknowledging our Lord Jesus Christ as its head, is inseparably united in doctrine with the Great Church of Constantinople and with every other Church of Christ of the same doctrine, observing unwaveringly, as they do, the holy apostolic and synodal canons and sacred traditions. It is autocephalous and is administered by the Holy Synod of serving bishops and by the Permanent Holy Synod elected by them, established in the manner prescribed by the Church Charter and in compliance with the provisions of the Patriarchal Tome of 29 June 1850 and the Synodal Act of 4 September 1928.'

In Norway, the church was separated from the state in 2012, although state confessional preferences were preserved. Before 2012, under the Constitution of the Kingdom of Norway of 17 May 1814, as amended in 1995, the Evangelical Lutheran religion was the official state religion (Section 2).

The Constitution of the Kingdom of Sweden of 27 February 1974 declares religious freedom; Section 6 of Chapter 8, which had provided for the existence of the Church of Sweden, was repealed. Thus, in Sweden, church and state were separated in 2000.

The Basic Law of the Republic of Finland of 11 June 1999 also declares freedom of religion, but Section 76 provides for the state status of the Evangelical Lutheran Church, formalized in the Church Act. The Orthodox Church also has state status.

In a whole range of states, Islam has state status and serves as a basis of law. In most Islamic states, an oath based on the provisions of Islam is practiced.

For example, the now inoperative Constitution of the Islamic Republic of Afghanistan of 4 January 2004 began with the words: 'In the name of Allah, the Most Merciful, the Most Compassionate.' The preamble also referred to 'Almighty Allah with hope for His lawful mercy, and with Faith in the Sacred Religion of Islam.' Under Article 1, 'Afghanistan is an Islamic Republic, independent, unitary and indivisible,' and under Article 2, 'the religion of the state of the Islamic Republic of Afghanistan is the sacred religion of Islam.' In addition, 'no law in Afghanistan may contravene the beliefs and provisions of the sacred religion of Islam' (Article 3).

The 2004 Constitution effectively ceased to operate on 15 August 2021, when the Taliban movement established control over Afghanistan. After the Taliban came to power, the state institutions created on the basis of the 2004 Constitution were dissolved, and the former laws and norms ceased to apply in practice. Power was concentrated in the hands of the Taliban leadership, acting on the basis of Islamic principles and its own decisions. After August 2021, the Taliban announced the introduction of a new foundation of state structure: the Constitution of the restored Islamic Emirate of Afghanistan. It consists of only three chapters and two articles, being significantly shorter than the 2004 Constitution; it is based on the norms of Sharia, or Islamic law, and the provisions of the Qur'an and Sunnah, and relies on the Hanafi madhhab, one of the four main schools of Sunni Islam. The document does not contain guarantees of many democratic rights and freedoms enshrined in the 2004 Constitution; in particular, the rights of women and national minorities are substantially restricted. It should be noted that the Taliban movement had been recognized as terrorist and banned in the Russian Federation, but on 17 April

2025 Russia removed the requirement to label the Taliban as a terrorist organization. The Supreme Court of the Russian Federation suspended the ban on the movement's activity in the country upon a claim by the Prosecutor General's Office. That decision entered into force immediately. On 3 July 2025, Russia became the first country in the world officially to recognize the Islamic Emirate of Afghanistan.

Under the Constitution of the State of Bahrain of 14 February 2002, Islam is the state religion and Sharia is a principal source of legislation (Article 2).

The Basic Law, or Nizam, of the Kingdom of Saudi Arabia of 1992 states that 'the Kingdom of Saudi Arabia is a sovereign Arab Islamic State. Its religion is Islam; its Constitution is the Book of Almighty Allah and the Sunnah of His Prophet, may Allah bless him; its language is Arabic; and its capital is the city of Riyadh' (Article 1).

'The subjects of the King pledge allegiance on the Book of Almighty Allah and the Sunnah of His Prophet, in obedience and loyalty to him in hardship and ease, in fortune and adversity' (Article 6).

Power in the Kingdom of Saudi Arabia is based on 'the Book of Almighty Allah and the Sunnah of His Prophet,' which form the basis of the Basic Law and other state acts (Article 7).

In accordance with Article 23, 'the state protects the Islamic faith, implements its provisions, monitors the absence of sinfulness, prevents vice and fulfils the duty of spreading Islam,' while Article 33 includes among the objectives of the armed forces the protection of the faith and the holy places.

Article 1 of the Constitution of Qatar of 2003 also establishes Islam as the state religion and Sharia as a principal source of legislation. The heir apparent must be a Muslim and the son of a Muslim woman (Article 9). After being chosen, the heir to power pronounces the following oath: 'I swear before Allah to respect Sharia, the Constitution and the law, to protect the independence of the state, its territorial integrity, the freedom and interests of citizens, and to be loyal to the state and the Emir' (Article 10).

In some states, Buddhism is the privileged religion. For example, under the Constitution of the Kingdom of Thailand of 19 August 2007, the state supports Buddhism (Article 79), and the King is a Buddhist (Article 5). Since 2017, the Kingdom of Thailand has had a new Constitution, in which Article 7 confirms that 'the King shall be a Buddhist and the upholder of all religions.' Under Article 67, 'the State shall support and protect Theravada Buddhism.' The Constitution also establishes the duty of the state to prevent any actions 'undermining Buddhism' and to encourage the participation of lay persons in protecting the faith.

Laos 'respects and protects all lawful activities of Buddhists and followers of other religions, and guides and encourages Buddhist monks and novices, as well as priests of other religions, to participate in activities beneficial to the country and the people' (Article 9); 'the duty of the State is to protect and promote the Buddhist religion' (Article 9(1)(e)).

In Cambodia, the Constitution (Article 43) directly declares that 'Buddhism is the state religion.' In Bhutan, the Constitution (Article 3) calls Buddhism 'the spiritual heritage of Bhutan,' uniting the people. At the same time, the King is the protector of all religions. In Myanmar, the Constitution recognizes the special position of Buddhism as the faith professed by the overwhelming majority of citizens (Article 361). In these countries, Buddhism is not merely a religion but part of state identity, which gives it the status of 'first among equals' or of an official ideology. At the same time, Bhutan and Myanmar constitutionally secure the freedom of other religions, although in practice their position may be limited.

Catholicism is also recognized as the state religion in a number of states.

The status of the Vatican is established by the Treaty between the Holy See and the Kingdom of Italy of 11 February 1929. In accordance with Article 1, 'Italy recognizes and reaffirms the principle proclaimed in Article 1 of the Royal Statute of 4 March 1848, according to which the Catholic, Apostolic and Roman religion is the only religion of the State.'

The Constitution of the Republic of Malta of 1964 establishes the Roman Catholic Apostolic religion as the state religion. '(2) The authorities of the Roman Catholic Apostolic Church have the duty and the right to teach which principles are right and which are wrong. (3) Religious teaching of the Roman Catholic Apostolic faith shall be provided in all state schools as part of compulsory education.'

The Constitution of the Principality of Monaco of 17 December 1962 also provides that 'the Roman Catholic, Apostolic religion is the religion of the State' (Article 9).

The model of separation of religious associations from the state is declared in the United States, the Netherlands, France, Russia and other states, and implies the existence of legal guarantees of freedom of conscience and religion and of equality of religious associations.

For example, the Constitution of the French Republic of 4 October 1958, as amended in 1996, states: 'France shall be an indivisible, secular, democratic and social Republic' (Article 1).

The Constitution of the Kingdom of the Netherlands of 17 February 1983 guarantees freedom to profess religion or religious beliefs (Article 6), prohibits discrimination on grounds of belief (Article 1), and provides for education taking into account the religious affiliation and beliefs of each pupil (Article 23), which is ensured by the confessional neutrality of the state.

Article 14 of the Basic Law of Russia declares: '1. The Russian Federation is a secular state. No religion may be established as state or mandatory. 2. Religious associations shall be separated from the state and shall be equal before the law.'

In practice, however, there are insufficient grounds to speak of worldview neutrality in the states mentioned above, or indeed in other contemporary states. Moreover, existing violations of the principle of separation of religious associations from states inevitably lead to violations of their equality.

The model of cooperation between the state and religious associations, known as the cooperative or transitional model, operates in Austria, Belgium, Germany, Spain, Italy, Luxembourg and Portugal, taking into account the 'historical particularities' of each specific state.

In Austria, under the Basic Law of the State of 21 December 1867 'On the General Rights of Citizens of the Kingdoms and Lands Represented in the Imperial Council,' an oath with religious content is permitted for the Federal President.

The Constitution of the Kingdom of Belgium of 17 February 1994, as of 16 May 2000, permits church marriage (Article 21); state schools finance instruction in recognized religions while providing the choice of non-confessional ethics (Article 24); and the payment of salaries and pensions to ministers of religion is imposed as a duty of the state, with the necessary amounts entered annually in the budget (Article 181).

The Basic Law of the Federal Republic of Germany of 23 May 1949 provides for mandatory teaching of religion in state schools, except in non-confessional schools, with the possibility of choice (Article 7).

The Constitution of the Kingdom of Spain of 27 December 1978, despite stating that 'no religion shall have a state character,' contains a provision under which public authorities 'shall maintain appropriate relations of cooperation with the Catholic Church and other confessions.' In this case, the mention of one religious organization in the text of the Basic Law clearly indicates confessional preferences of the state.

Under the Constitution of the Italian Republic of 22 December 1947, 'the State and the Catholic Church are independent and sovereign, each within its own sphere' (Article 7). In practice, however, the Catholic Church has special privileged relations regulated by the Lateran Agreements.

Article 41 of the Constitution of the Portuguese Republic of 2 April 1976 not only declares the separation of churches and religious communities from the state and their freedom of activity, but even prohibits public authorities from 'questioning anyone about their convictions or religious practice, except for the collection of non-identifiable statistical data.' It also guarantees 'freedom to teach any religion within the framework of the respective confession, as well as the use of one's own media for religious activity' and 'the right to conscientious objection to military service within the framework of the law.' Under Article 43, 'public education shall not be religious in character,' and 'political parties may not, without prejudice to their philosophical or ideological programmatic principles, use names or expressions directly related to any religion or church, nor emblems similar to national or religious symbols.'

The Constitution of the Republic of Turkey of 7 November 1982, in accordance with the principles proclaimed by Ataturk, declares that 'the Republic of Turkey is a democratic, secular and social state governed by the rule of law; founded on the concepts of public peace, national solidarity and justice; respecting human rights; loyal to the nationalism of Ataturk; and based on the fundamental principles set forth in the preamble' (Article 2).

In addition, 'education and instruction in religion and ethics shall be conducted under the supervision and control of the state. Instruction in religious culture and ethics shall be a compulsory element in the curricula of primary and secondary schools. Any other religious upbringing and education shall depend on the individual's own wish, and for minors on the request of their legal representatives' (preamble).

The Constitution of Turkey also prohibits 'the exploitation or abuse of religion or religious feelings, or things held sacred, for the purpose of securing personal or political influence, or basing, even in part, the fundamental social, economic, political and legal order of the state on religious principles' (preamble).

Article 136 provides for the functioning of the Directorate of Religious Affairs, 'which forms part of the administrative structure and performs its duties, defined by a special law, in accordance with the principles of laicism, connected not with political views and ideas but with aspirations toward national solidarity and integrity.'

The Federal Constitution of the Swiss Confederation of 18 April 1999 contains in its preamble the phrase 'In the name of Almighty God!' Article 72 delegates regulation of relations between the state and religious associations to the cantons. In the original version of the Federal Constitution of the Swiss Confederation of 18 April 1999, Article 72 contained a provision under which the establishment of bishoprics required authorization by the Confederation. However, on 10 June 2001 a referendum was held at which this rule was repealed. The decision was confirmed by a decree of the Federal Government of 22 August 2001 and was based on the federal decree of 15 December 2000.

In general, the objective tendency in relations between the state and religious associations is a movement away from privileged status toward equality of religious associations, as well as equality of 'believers' and 'non-believers,' which is reflected in the constitutions and legislation of states.

For example, in recent years, at the initiative of humanist associations in Norway, Iceland, Malta, Denmark, Canada, New Zealand, Greece and France, legislative norms prohibiting 'blasphemy,' 'insulting religion,' 'mocking religion,' 'blasphemous libel' and similar acts have been repealed. Norway and Iceland did so in 2015, Malta and France in the Alsace-Moselle region in 2016, Denmark in 2017, Canada in 2018, and New Zealand and Greece in 2019.

In addition, in October 2018 Ireland became the first country in the world to hold a referendum on the termination of a law on 'blasphemy.' Sixty-five percent of Irish voters supported the repeal of norms prohibiting 'blasphemy' and providing for criminal liability, including a fine of up to 25,000 euros, for 'blasphemous' publications and statements.

In recent years, Western countries have continued the course toward decriminalizing criticism of religion. Scotland officially repealed its blasphemy law in 2021, while Spain initiated a similar process in 2024-2025. At the same time, in December 2023 Denmark introduced criminal liability for the desecration of sacred texts on grounds of national security.

At the same time, in Germany, Poland and Italy, norms on 'insulting the feelings of believers' have been preserved and continue to be applied in judicial practice. Germany punishes insults against confessions only where there is a threat to public peace, thereby protecting public order rather than dogma. Poland has retained criminal liability for 'insulting religious feelings' as a separate offence, although the European Court of Human Rights increasingly limits the application of this norm. Italy has decriminalized blasphemy, replacing prison sentences with administrative fines for violations of public order.

Summarizing the world practice of regulating the activity of religious associations in democratic states, one may conclude that it is directed toward limiting the possibility of state interference in the internal life of religious associations, or at least tends in that direction.

At the same time, the existence in some democratic states of a third, cooperative or transitional model of relations between the state and religious associations makes it possible to justify, despite constitutional principles of state secularity and equality of religious associations, the existence of elements of inequality in the form of state confessional preferences. It must be emphasized that the transitional character mentioned above implies the gradual abandonment of these elements, not a return to them, as is occurring today in some states.

By way of conclusion, it should be noted that in the realities of the twenty-first century the world community as a whole, as well as individual states, must generalize and rethink world experience in the legal regulation of relations between states and religious associations through the prism of the internationally recognized right to freedom of conscience and the worldview neutrality of states. It appears advisable to initiate discussion of this issue within the expert structures of international organizations and to form an appropriate international program.

2.2. The Problem of the Relationship between the Right to Freedom of Conscience, State Secularity, Relations between States and Religious Associations, and Related Categories

Historically, the realization of human rights in the sphere of freedom of conscience in the Russian Federation and in other states of the world has depended very strongly on relations between the state and religious associations, as well as on other related categories. In this context, the search for approaches to the problem of the relationship between freedom of conscience, the secularity of the state, relations between the state and religious associations, and the religious policy of the state is extremely important.

A. F. Gavrilenkov defines relations between the state and religious associations as 'a set of historically formed and changing forms of relations between state institutions and religious formations. They were based on legally established political, national, spiritual and moral ideas concerning the place of confessions in life, their functions, and the spheres of activity of all participants in these relations.'

A. V. Isaev defines state-confessional relations as 'the interaction of the state and religious associations for the realization of their interests and strategies, manifested in various forms: partnership, cooperation, dialogue, indifference, domination and subordination, rivalry, confrontation and conflict.'

A 'subject-subject' connection based on cooperation and mutual support characterizes partnership relations. On the other hand, the centralization of power by one of the parties leads to a 'subject-object' influence.

O. N. Petyukova proposes the following definition of 'state-confessional legal relations': a set of historically formed and changing forms of legal relations between state and municipal bodies, organizations, institutions and enterprises, on the one hand, and religious associations and individual believers, on the other. State-church legal relations are the subject of regulation by both state and church law. In order to consider Orthodox church law in its development, it is necessary to use comprehensive legal approaches. The profound character of canonical legal regulation is connected with the fact that, unlike secular state law, it regulates not only relations between the church and public and state institutions, but also the internal life of the church organism in its communal and institutional aspect. Whereas state confessional legislation under conditions of secularization represents a fragmented and unsystematized body of legal acts belonging to different branches of law, church-law norms are united in a special religious system, which gives them authority in shaping the legal consciousness of believers.

Accordingly, by the legal form of state-confessional relations the researcher understands 'a complex of diverse legal means that carry out legal regulation and legal mediation of relations between the state and religious associations, realized in the form of specific activity by the subjects of state-confessional relations.'

E. B. Pokaninova proposed her own concise authorial definition of 'state-confessional relations' as 'a set of historically formed and changing forms of interconnections between state institutions and religious institutional formations.'

According to A. Yu. Konnevskaya, 'the term state-confessional relations does not cover relations of religious associations with local self-government bodies, into which religious associations inevitably enter when carrying out activity in the public sphere. In this regard, there arises a need to introduce a broader term: public-confessional relations, denoting the entire spectrum of relations between confessions and bodies of public authority.'

In generalized form, relations between the state and religious associations are defined by most researchers as a set of historically formed and contemporary forms of interconnections and relations between institutions of the state and local self-government and the institutional formations of religious associations.

Yet there is no terminological uniformity here. Some researchers prefer the concept of 'state-church relations' or 'church-state relations,' which is not entirely correct when applied to religions that do not use the concept of 'church.' Other researchers use the terms 'state-confessional relations,' 'public-confessional relations' and 'state-religious relations.'

The Russian constitutional model of relations between the state and religious associations corresponds to the concept of 'relations between the state and religious associations.' However, taking into account the contemporary understanding of the principle of state secularity, the most correct concept appears to be 'relations between the state and public associations.' This means that the contemporary democratic law-governed state must be separated not only from religious associations, but from any public associations.

In other words, relations between the contemporary democratic law-governed state and worldview associations must be built on legal grounds equal to those applicable to other public associations. I believe that the principle of separation of public associations from the state is a foundational norm of the secular state.

However, some authors, without sufficient justification, speak of the need for priority social partnership between the state and 'traditional' religious organizations, including the conclusion of special agreements. In particular, A. V. Loginov believes that 'priority partnership of the state with traditional confessions, aimed at expanding the participation of the latter in spiritual enlightenment and cultural revival, will allow Russia to define itself within the system of civilizational coordinates.'

Not entirely correct ideas of priority social partnership between the state and 'traditional' religious organizations are also found in the publications of other authors.

In addition, in 2004, at the meeting of Dissertation Council D.502.006.11 under the Russian Academy of Public Administration under the President of the Russian Federation, O. S. Shutova defended her candidate dissertation entitled 'The Structural Model of State Policy in the Field of Relations with Religious Associations' under the academic supervision of M. O. Shakhov. Graphically, the model represents a kind of pyramid headed by a value-and-purpose component - goals and values based on prevailing social ideas about the place and role of religion and the nature of state-confessional relations - directly interacting with the instrumental component, namely relevant authorities and administrative bodies, and the praxeological component, namely the practical activity of authorities and administrative bodies.

The dissertation author notes that 'the importance of the religious factor has always prompted the state to keep it within its field of vision' and places the normative component - legislation regulating rules of conduct for subjects in the sphere of freedom of conscience and religion - inside the power pyramid. It is obvious that, in the practical implementation of this model, ruling groups obtain unlimited opportunities for extralegal control and suppression of the worldview sphere and for manipulating confessions in order to retain power.

In this context, A. V. Karpushkin reasonably believes that 'the rapprochement of the state with particular religious associations inevitably leads to the strengthening of their correlation and to the state's use of religion for its political purposes.'

Accordingly, the state of freedom of conscience is interconnected with the constitutional and legal position of religious associations, which 'finds expression in the formal legal equality of all religious associations as a condition for realizing collective rights, irrespective of their historical and cultural significance and the time of their establishment.'

Thus, the equality of different associations is a necessary condition for their peaceful coexistence, which is especially relevant in multinational and multi-confessional states. 'The constitutionally secured equality of religious associations before the law means their formal legal equality, which must be distinguished from levelling. Formal legal equality is the denial of privileges, a single and universal measure, the establishment of equal constitutional and legal status for subjects of constitutional relations possessing similar generic features and presupposes, unlike levelling, the unequal evolutionary development of these subjects depending on their individual qualities, and not on deliberately unequal conditions.'

However, the realization of the principle of equality of religious associations, in turn, depends on state confessional or religious policy in different states of the world.

R. A. Lopatkin and F. G. Ovsienko define confessional policy as 'a part of state policy in the sphere of freedom of conscience and religion, namely a system of state actions that includes goal-setting, legal justification and a complex of organizational and practical measures: to ensure freedom of conscience and religion for everyone; to create the necessary conditions for satisfying religious needs; to regulate the activity of religious associations as subjects of public law in the part that goes beyond canonical structure and cult practice and thereby becomes public activity; to cooperate with them in solving socially and state-significant problems; and to achieve interreligious and interconfessional peace.'

I believe that through the implementation of confessional or religious policy it is impossible to ensure the right to freedom of conscience for everyone, that is, not only for 'believers' but also for all others. The problem is that confessional or religious policy allows, and even implies, the use of religion for political purposes and the manipulation of public consciousness. Thus, confessional or religious policy, like atheist policy, by its very definition does not contribute to the realization of the constitutional principle of freedom of conscience. Therefore, a law-governed secular state should have no special confessional or religious policy at all, since such policy has always been a political instrument of power.

For example, A. F. Gavrilenkov, analyzing the historical experience of the Russian state, states that its confessional policy was conducted on the basis of domestic and foreign policy objectives. The researcher identified the following main principles of this policy: support and dissemination of the Orthodox faith as the 'primary' state religion; transformation of church authority into an instrument of secular policy; imperial Orthodox paternalism; religious toleration; recognition in every subject of the empire of a religious person; recognition as full-fledged, from the standpoint of confessional affiliation, only those persons professing Orthodoxy; prohibition on conversion from one confession to another, except conversion to Orthodoxy; equality of recognized religious societies before the law; loyalty of the authorities in exchange for loyalty of the population; non-recognition of religious freedom; non-recognition of the individual's right to religious self-determination; interference in the internal life of religious associations and in the internal church activity of confessions existing in Russia; a legal hierarchy of confessions and their inequality in rights, privileges and restrictions; and the performance by confessional legislation of the role of regulator of the national question.

In this context, I consider incorrect the approach that presupposes the priority of confessional policy over the principles of freedom of conscience, in particular the approach proposed by the authors of the above-mentioned work 'The Confessional Policy of the Russian State.'

The same remarks apply to M. I. Odintsov's monograph 'State-Church Relations in Russia: Based on Materials from Russian History of the Twentieth Century,' which likewise contains the priority of confessional policy over the realization of constitutional principles in the field of freedom of conscience.

On the one hand, 'as soon as the state, whether Ancient Rome or present-day America, begins to claim the role of God, the principal human freedom - freedom of conscience - is immediately infringed.' On the other hand, violation of the principle of freedom of conscience, or liberty of worldview choice, inevitably leads to the authoritarianization of power and to disregard for the interests of society.

Although relations between the state and religious associations and state confessional or religious policy have existed historically, in the contemporary world they are objectively losing their positions.

The Constitution of the Russian Federation, the legislation of most contemporary democratic states, and the principles and norms of international law say nothing about relations between the state and religious associations or state confessional or religious policy as self-sufficient phenomena.

Since the adoption of the Universal Declaration of Human Rights in 1948, and for Russia at least since the adoption of the Constitution of the Russian Federation by popular vote in 1993, both relations between the state and religious associations and state confessional or religious policy must be regarded as derivatives of the principles of human rights, the rule of law and worldview neutrality.

In this context, the correctness of posing the problem of forming a concept of state-confessional relations, which in recent years has become one of the central issues of scholarly and public discussion, gives rise to serious doubts.

Nevertheless, over the past years Russian society has been presented with several draft concepts of relations between the state and religious associations, intended to form the basis for the scholarly justification of the state's real policy in this sphere and to create a foundation for transforming legislation on freedom of conscience by including elements of selective partnership between the state and 'traditional' religious associations.

Expert analysis has shown that the concepts mentioned are aimed at the dominance in relations with the state of 'traditional' religious organizations and, ultimately, of one confession - the 'most traditional' one. As historical experience shows, not all confessions prove equally useful to power, even among the 'traditional' ones. Hence the inevitability of further selection already among 'traditional' religious organizations, in the form of even greater privileges for some and persecution of others. Most importantly, legal criteria of 'traditionality' do not exist in principle.

Thus, in the contemporary democratic law-governed state there should be no special state confessional or religious policy at all; the relevant policy in the sphere of freedom of conscience should be directed toward its realization.

In particular, R. M. Mukhametzyanova clarifies the formulation of the concept of 'state policy in the sphere of freedom of conscience,' understood as a system of purposeful actions by the state, including legal justification and a complex of organizational and practical measures aimed at ensuring freedom of conscience and limiting it in cases prescribed by law and necessary to protect public safety, order, health, morals, and the fundamental rights and freedoms of others.

The following are defined as general principles for the implementation of state policy in the sphere of freedom of conscience: the principle of the secular character of the state; the principle of freedom of conscience and religion; the principle of equality of human and civil rights and freedoms regardless of attitude toward religion; the principle of equality of religious associations before the law; the principle of cooperation between the state and religious associations; and the permissibility of restrictions in the sphere of freedom of conscience and religion in cases prescribed by law and necessary to protect public safety, order, health, morals, and the fundamental rights and freedoms of others.

Another problem of contemporary relations between the state and religious associations is the issue of agreements or treaties between state bodies and religious organizations.

In particular, in the Russian Federation, agreements were concluded between the Russian Orthodox Church of the Moscow Patriarchate and the Ministry of Culture, the Ministry of Social Protection, the Ministry of Education and Science of the Russian Federation, the Ministry of Justice of the Russian Federation, the Ministry of Defence of Russia, the Ministry of Internal Affairs of Russia, the Federal Border Service of Russia, and many others.

At the level of the constituent entities of the Russian Federation, local dioceses of the Russian Orthodox Church of the Moscow Patriarchate concluded agreements with the administrations of constituent entities of the Russian Federation, including the relevant regional sectoral departments, as well as with the plenipotentiary representatives of the President of Russia in the federal districts.

The agreements mentioned have been concluded since the early 1990s at federal, regional and local levels and presuppose priority cooperation between the state and religious associations in various fields.

In this context, certain historical lessons drawn by A. F. Gavrilenkov on the basis of his conclusions and generalizations appear extremely relevant. 'First. In a multinational and multi-confessional country, peaceful coexistence and harmony among its inhabitants can be ensured only by such state confessional policy as is aimed at integrating various religions, cultures and traditions in combination with respectful treatment of the distinctiveness of peoples. Second. State confessional policy fully allows the use of religion for political purposes. But in this case there is a real danger that religion may turn into ideology. Taking historical experience into account, present-day Russian state power should not forget this circumstance. Third. Priority, especially when actively and aggressively demonstrated, in relation to one of the confessions, even if state-forming, leads to asymmetry in state-confessional policy. Fourth. The formation of confessional policy at the contemporary stage of development of the Russian state and society is possible only with account taken of historical experience in the interaction of the state and confessions. The process of creating a contemporary model of state-church and state-confessional relations, a mechanism for the everyday resolution of problems arising in the religious sphere of social life at all levels of authority, will proceed the more successfully the more fully and objectively historical experience is taken into account. Fifth. The formation and development of confessional relations in contemporary Russia is possible only on a legislative basis, when the state assumes responsibility for the formation of these relations and excludes cases of non-legal resolution of issues of confessional policy.'

However, one must agree with the opinion of A. A. Levina, who states that in contemporary Russia the lessons of history are not learned, and that the conceptual foundation of relations between the state and religious associations has become 'the idea of cultural and historical selectivity, consisting in granting advantages to confessions on the basis of their cultural and historical significance.'

According to K. G. Kanevsky, 'at present, cooperation between leading religious associations and the state is predominantly outside the legal field and is based on arrangements between the highest hierarchs and representatives of power, as well as on agreements between religious organizations and state bodies, the legal status of which, at the very least, raises questions.'

The author agrees with the above-mentioned opinion and considers that agreements between religious organizations and state bodies contradict the principles of the secularity of the state and equality of religious associations before the law, and often are directly aimed at unlawfully restricting freedom of conscience. It is noteworthy that in some of the above-mentioned agreements religious associations agreed with the state jointly to combat 'totalitarian sects,' a concept absent from law.

I believe that the practical recommendation of A. F. Gavrilenkov deserves the most serious attention. In particular, the researcher believes that 'relations between the Russian state and religious associations at the contemporary stage must be built on the basis of a constantly developing legislative framework, which will make it possible to avoid unlawful actions by state structures. The interaction of different confessions operating on the territory of the contemporary Russian state must also be exclusively legal.' In this context, the position of Yu. E. Fedotova concerning the need to introduce criminal liability for violation of the principle of state secularity deserves attention.

L. V. Osipova substantiates the position that 'the basis of the legal model of relations between the state and religious associations must be the neutral attitude of the state toward all religious associations, excluding a differentiated approach to the regulation of their legal status. This presupposes the inadmissibility of discriminatory legislation on religious associations connected with granting advantages to some religious associations and unjustifiably restricting the right to freedom of conscience and the right of association on religious grounds for others.'

I believe that the contemporary state must be secular and must maintain neutrality not only toward all religious forms, but also toward other worldview forms, both individual and collective. Thus, relations of a democratic law-governed state, which sets as its goal the formation of an open civil society, must be built with religious associations on legal grounds common to other public non-profit associations.

A contemporary secular state, especially a multinational and multi-confessional one, should have no 'special' relations with religious organizations at all and, consequently, no concepts of such relations, nor any special 'religious' legislation, bodies or agreements with religious associations.

However, constitutional and legal analysis has not revealed states that observe the principle of secularity in the sense of a neutral attitude toward the worldview sphere. Accordingly, contemporary states are connected to varying degrees with selected religious associations, which inevitably entails violations of freedom of conscience, including discrimination on grounds of worldview affiliation, and creates prerequisites for regional and global conflicts.

It must be acknowledged that in our country and in other states of the world, the realization of freedom of conscience for everyone is replaced by state confessional or religious policy conducted for the purpose of retaining power and implemented through corrupt relations between the state and religious associations.

G. V. Stankevich, observing the politicization of the religious factor in the contemporary world and the clericalization of politics as part of this phenomenon, states that the religious component of the political process acts, on the one hand, as a factor in 'ensuring political and social stability' and, on the other, as a factor of socio-political destabilization. The researcher reasonably believes that 'the politicization of religion is connected, on the one hand, with the use of religious ideas by political actors for the realization of their own purposes and, on the other, with the desire of religious institutions to use political constructs to defend their positions within the traditional political space and to carry out proselytizing activity.'

E. M. Drinova, noting the unstable condition of the religious space of contemporary society caused by confessional contradictions, also states that 'the destructive potential of the politicization of religion prevails over its constructive potential.'

In particular, the researcher believes that 'the politicization of religion manifests itself in the inclusion of religion in the political process and presupposes the influence of the political process on religion; it is connected with rivalry among religious organizations and parties for the resources of power; and it is understood as the implementation of strategies and a set of behavioral actions of religious actors in the political life of society.'

I believe that the formation and use by ruling groups of secular states of the factor of relations between the state and religious associations for their own selfish interests and political purposes must be qualified as corruption in the sphere of relations between the state and religious associations. This corruption may be defined as abuse of power, by action or inaction, in one's own selfish interests or in the selfish interests of other persons and groups, connected with the use of public resources within the framework of 'special' relations between the state and religious associations.

The following may be attributed to the distinctive features of this type of corruption: abuses of official position in the sphere of relations between the state and religious associations are carried out constantly and continuously, becoming aggravated on the eve of and during electoral processes; there are grounds to speak of the formation of a state clerical ideology, unlawfully and at the expense of state resources; the effect of abuse of official position in the sphere of relations between the state and religious associations is achieved indirectly, as resources are unlawfully provided to certain religious organizations possessing integrative resources, authority and trust, and in exchange there occurs a transfer or imputation of authority and trust in various forms, from direct electoral agitation to the less obvious but more effective sacralization or consecration of power; the transfer of such a high level of authority or trust as sacralization is available only to power and works only for power; this type of abuse is directly connected with violation of foundational constitutional principles, including one of the foundations of the constitutional order, namely the secularity of the state, and therefore has a constitutional and legal character; this type of abuse in fact lies outside electoral legislation, including the norm on the inadmissibility of using the advantages of official position.

The system-forming factor of corruption in the sphere of relations between the state and religious associations is state political elite-power corruption at the top level, connected with decisions having the highest price: the retention of power. It relates both to relations between the state and religious associations and to the electoral process.

Since the system-forming factor is top-level corruption - state political electoral corruption - the principal party is an official or public servant. Corruption in relations between the state and religious associations manifests itself most cynically in the form of clerical ideologization of state institutions.

I believe that, for characterizing contemporary processes in the sphere of relations between the state and religious associations, the term 'clerical ideologization of the state' is more appropriate than the widely used concept of 'clericalization.'

M. O. Shakhov believes that 'the term clericalism became widespread in Europe after the French Revolution, when the process of secularization actively began. Clericalism denoted the struggle of the Church to preserve or strengthen its positions in social life, including the struggle against secular power and its secularizing tendencies. This occurred because the Catholic Church constantly claimed possession of both spiritual and civil authority. The most essential feature of clericalism is the presence of rivalry and a struggle for power. In that process, the religious organization acts as a fairly strong participant. One may speak of clericalization as long as there is an equilibrium of opposing forces. When the equilibrium is disturbed, the issue becomes either theocracy or the secular state. Sometimes a confession seeking to unite spiritual and secular authority in the hands of religious leaders becomes the dominant force in the country. In our view, such states should be called not clerical but theocratic, for example Geneva under Calvin, contemporary Iran and some other Islamic states.'

However, as the scholar of religion rightly notes, 'in Orthodox countries the character of state-church relations was fundamentally different. Theoretically, they were built on the principle of symphony between Church and state, which presupposed close cooperation while delimiting spheres of competence. In practice, however, there was a deviation toward caesaropapism, the predominance of royal power over church power.' This gives grounds to call clericalization 'not every process of strengthening the influence of the Church on society and the state, but only that which, first, is carried out in implementation of its doctrinal positions; second, through its own forces; and third, when it is connected with overcoming the resistance of secular power and part of society.'

Thus, M. O. Shakhov's hypothesis has substantial grounds: 'the use of the terms clericalism and clericalization in relation to processes taking place in contemporary Russia and certain other states is incorrect, because they hinder a correct assessment of the nature and driving forces of what is occurring. On the one hand, their prehistory suggests a mistaken analogy with Catholic clericalism. Yes, the Moscow Patriarchate undoubtedly seeks to strengthen its influence on the state, but to call this by a term characteristic of another confession means introducing even greater confusion. On the other hand, with a broader interpretation of the terms, the specificity of the concrete phenomenon remains unclear...'.

Thus, clerical ideologization of the state is a violation of the worldview neutrality, or secularity, of the state, expressed in the merging of its institutions with one of the confessions whose doctrinal positions are used as state ideology.

In relation to the international legal sphere, clerical ideologization of states and other subjects of international law is a violation of the worldview neutrality of subjects of international law, expressed in the merging of their institutions with worldview structures whose doctrinal positions are used as the basis of their activity, even when they contradict the fundamental principles and norms of international law.

Clerical ideologization of states and other subjects of international law contradicts the principles of the priority of human rights, the rule of law, worldview neutrality, equality of public associations, equality of human and civil rights and freedoms regardless of worldview choice, and ideological diversity, and ultimately creates obstacles to the realization of the internationally recognized right to freedom of conscience.

S. A. Mozgovoy rightly notes that a secular state is always secular, that is, non-religious, and, as in any secular state, the principle of separation of church and state is necessarily realized in it. At the same time, what is secular is not always secular in the proper legal sense: it may be anti-religious or atheistic. Ideally, a secular state is equally distant both from religion and from atheism. It allows them to coexist peacefully, preserving the right of worldview choice for citizens and non-citizens.

A similar position, reflecting the relationship between state secularity and relations between the state and religious associations, is held by other Russian authors as well.

The French researcher J.-P. Willaime, analysing the European experience, identifies neutrality, freedom and non-discrimination as the main features of state secularity: confessional neutrality of the state as a whole and of individual state institutions; recognition of the autonomy of individual conscience in relation to any religious and philosophical doctrines; and the principle of non-discrimination, that is, equal treatment of all citizens regardless of their convictions.

In recent years, K. Rousselet has transformed her understanding of state secularity, distinguishing three basic principles underlying the concept: freedom of conscience and religion; the independence of the state from religious ideas and institutions; and equality of persons regardless of their beliefs.

E. M. Miroshnikova writes not only about the neutrality of the state towards all religious organizations, but also about its neutrality towards worldview associations in general. Among the features of the secular state she proposes are neutrality as non-interference, neutrality as non-identification, neutrality as equality of opportunities, and neutrality as refusal of privileges. Yet her position, although it speaks of the need for the worldview neutrality of the state, cannot be called consistent, since in a later work she proposes its realization with regard to national specificities, which, in her view, determine differing degrees of cooperation between church and state. In fact, the author justifies a departure from the principle of worldview neutrality under the pretext of returning to a past that has not quite passed away. In her words, state neutrality should by no means be understood as a faceless and indifferent attitude to everything and everyone, to historical and cultural traditions in which religion has deep roots, especially in relation to Russia, which has been connected with Orthodoxy for centuries.

M. S. Stetskevich understands a secular state as a state that strives to ensure the worldview freedom of the individual, legislatively guarantees freedom of conscience, does not require any religious legitimation and, being worldview-neutral, does not conduct any special religious policy, either in support of religious associations or in the struggle against religion.

According to A. A. So, a secular state must be neutral in matters of religion and must not take the side of any confession, even the largest one, nor create conditions for strengthening the influence of any religion on all aspects of public and state life. The researcher critically assesses the practice of realizing the principle of the secular state in contemporary Russia, where development of state-confessional relations shows a tendency towards clericalization and towards the granting of a preferential legal regime to certain religious associations.

A. F. Meshcheryakova writes that a secular state is a way of organizing public, political power in which the state and the church are separated and at the same time interact with each other; there is no state or official religion; equality of religious associations, freedom of conscience and equality of citizens regardless of their attitude to religion are ensured in order to achieve a balance of interests among the individual, the state and religious entities in various spheres of social life. As the most important components of the state's secularity, she identifies the secularity of its bodies and proposes the following features of the secularity of legislative, executive and judicial bodies: secularity in the formation of such bodies; secularity in their functioning; secularity of the constitutional and legal status of persons vested with powers of authority; secularity of the ethics of officials; and secularity of state symbols.

I. V. Ponkin, in addition to the state's attitude to religion and confession and to religious associations, also addresses its attitude to ideology. A secular state, in his view, is a state that ensures the independence and sovereignty of state institutions and religious associations, as well as associations whose activity is aimed at disseminating ideology, within their respective spheres of competence. This is guaranteed by the inadmissibility of establishing in the state any universally binding religion or ideology and by the separation from the state of religious associations and associations whose activity is aimed at disseminating ideology.

Third, there is an approach that connects the concept and content of state secularity with the state's attitude to the worldview sphere without specially singling out religion, confession, or the activity of religious associations. The understanding of the secular state as a worldview-neutral state is reflected in numerous works of the Institute for Freedom of Conscience, S. M. Aleinikova, T. P. Minchenko, S. S. Savva, D. M. Nikitaev and others.

S. S. Savva rightly speaks of the need for a legal mechanism for implementing worldview neutrality, allowing the state to be equally distant from

Religion and Belief in the Russian Federation', financed by the European Union represented by the European Commission.

A. V. Pchelintsev singles out the secularity of the state as the most important constitutional and legal guarantee, refers to Article 14 of the Constitution of Russia, and then rather controversially asserts that it would be wrong to believe that freedom of religion can be genuinely guaranteed only in a secular state. It follows, according to A. V. Pchelintsev, that freedom of religion may be realized not only in a secular state. In other words, state secularity is not necessarily required as a guarantee of freedom of religion.

In attempting to explain his position, A. V. Pchelintsev states that secularity, as a constitutional and legal characteristic of the state, is a major but not the only guarantee of freedom of religion. The fullness of freedom of religion also depends on the political regime, the development of democracy and civil society institutions, the level of religious tolerance, and other factors. Here it is more or less clear that, in addition to state secularity, the realization of freedom of religion requires a whole range of other conditions.

However, questions arise concerning the correctness of A. V. Pchelintsev's narrowing of the features of state secularity exclusively to the sphere of interaction with phenomena described by the legally indeterminate term 'religion'. It is significant that he does not offer his own definition of the concept of state secularity, but formulates features that themselves fail to meet the principle of legal certainty: state neutrality in relations with religious associations; separation of religious associations from the state and their equality before the law; equality of rights and freedoms of citizens regardless of attitude to religion; and religious tolerance.

In this context, a number of questions arise that are by no means rhetorical. Must a secular state also be neutral in relation to non-religious associations? Must a secular state also be separated from non-religious associations? And finally, the most interesting question, which is already rhetorical, addressed to Doctor of Law A. V. Pchelintsev: if the state is not neutral and not separated from the above-mentioned non-religious associations, for example atheistic associations, will it still guarantee freedom of religion, or confession? I believe these questions concern all supporters of the approach under discussion.

Second, the most widespread approach links the concept and content of the secularity of the state not only with its attitude to religion and confession and the activity of religious associations, but also with other worldview forms.

For example, according to R. A. Lopatkin and F. G. Ovsienko, a secular state must be confessionally neutral, fundamentally rejecting any religion or worldview system as its official ideology, and ensuring that its citizens have the opportunity to make their own worldview choice.

O. A. Shmakova believes that a secular state is an institutional and political form of exercising power relations, based on the establishment of a legal regime of separation between the state and religious associations, while preserving the principle of formal equality of such associations before the law and the possibility of realizing the confessional interests of the population. Among the most important features of a secular state in the contemporary world she includes: free worldview self-determination and voluntary participation of citizens in the activity of religious organizations and religious groups; formal legal equality of legalized confessions and a policy of tolerance in the field of religious relations; anti-clericalism and anti-atheism as two conditions for the state to conduct a stable legal policy in the spiritual sphere; and the non-confessional character of state and local self-government institutions, together with a legal regime of cooperation with traditional religions directed primarily at the realization of the public and personal interests of the population of the state or a particular locality.

It appears not entirely correct to claim, as O. A. Shmakova does, that a secular state as an institutional, legal and political project is characterized only by a partial reorientation of public life towards secular values and norms, and that, in this connection, there is a real danger that the spiritual and ideological niche of public life may be filled by various destructive, mystical religious movements, sometimes criminal in character, penetrating the spheres of education, culture and medicine.

It also appears excessively optimistic to maintain, as O. A. Shmakova does, that there is no danger of a total clericalization of the post-Soviet state by actually traditional confessions and that no legislative or other tightening is therefore required. On the contrary, their activity, in her view, is directed at affirming humanistic values, national and religious tolerance, and suppressing xenophobic attitudes. Such a view underestimates the risks that arise where the state begins to select 'useful' religious partners and to grant them a privileged public role.

L. N. Osipova connects the definition of state secularity with the following features: the right of citizens freely to choose, profess or not profess, hold and disseminate religious and other beliefs that do not contradict legislation; the prohibition of establishing a state religion; separation of religious associations from the state and their equality before the law; separation of secular and religious education; and prohibition on religious associations engaging in political activity.

A. E. Sebentsov argues that the main principle of the secular state is the separation of church and state, which may be formalized in legal normative acts, de jure, or realized in the practical activity of the state, de facto. In such a state, to varying degrees and depending on the specific socio-historical situation, the principle of freedom of conscience is realized; this principle extends to the religious beliefs of citizens, ensures freedom of religion and the equality of citizens regardless of their attitude to religion.

S. A. Mozgovoy rightly notes that a secular state is always secular, that is, non-religious, and, as in any secular state, the principle of separation of church and state is necessarily realized in it. At the same time, what is secular is not always secular in the proper legal sense: it may be anti-religious or atheistic. Ideally, a secular state is equally distant both from religion and from atheism. It allows them to coexist peacefully, preserving the right of worldview choice for citizens and non-citizens.

A similar position, reflecting the relationship between state secularity and relations between the state and religious associations, is held by other Russian authors as well.

The French researcher J.-P. Willaime, analysing the European experience, identifies neutrality, freedom and non-discrimination as the main features of state secularity: confessional neutrality of the state as a whole and of individual state institutions; recognition of the autonomy of individual conscience in relation to any religious and philosophical doctrines; and the principle of non-discrimination, that is, equal treatment of all citizens regardless of their convictions.

In recent years, K. Rousselet has transformed her understanding of state secularity, distinguishing three basic principles underlying the concept: freedom of conscience and religion; the independence of the state from religious ideas and institutions; and equality of persons regardless of their beliefs.

E. M. Miroshnikova writes not only about the neutrality of the state towards all religious organizations, but also about its neutrality towards worldview associations in general. Among the features of the secular state she proposes are neutrality as non-interference, neutrality as non-identification, neutrality as equality of opportunities, and neutrality as refusal of privileges. Yet her position, although it speaks of the need for the worldview neutrality of the state, cannot be called consistent, since in a later work she proposes its realization with regard to national specificities, which, in her view, determine differing degrees of cooperation between church and state. In fact, the author justifies a departure from the principle of worldview neutrality under the pretext of refusal to legally establish any worldview and/or ideology as state or dominant;

ensuring a system of guarantees for the realization of liberty of worldview choice for every individual;

refusal to form international intergovernmental organizations on worldview grounds;

worldview-oriented state-like entities, peoples struggling for independence, international non-governmental organizations and other subjects of international law in the sphere of international relations must observe the principles of worldview neutrality recognized by the international community.

2.4. Worldview Neutrality of States and Other Subjects of International Law as a Guarantee for the Realization of the Right to Liberty of Worldview Choice

The principle of state secularity, together with other foundations of the constitutional order, serves as a foundation for other chapters of the Constitution of the Russian Federation and plays a leading role in the system of constitutional law. It should be especially emphasized that Article 14 of the Constitution of Russia, which declares the secular character of the state, the equality of religious associations and their separation from the state, is contained in Chapter 1 of the Constitution, entitled 'The Foundations of the Constitutional Order'. This chapter contains the fundamental principles concerning human rights, the structure of the state and organization of public authorities, and relations between the individual, society and the state.

The principle of the secularity, or worldview neutrality, of the state is interconnected with the principle of the law-governed state, which is based on the rule of law and is oriented towards the realization of human rights. The principles of freedom of conscience, state secularity, the relevant state policy, and relations between the state and religious associations are also closely interconnected.

One may agree with V. N. Linkin, who considers the following to be mandatory elements of the existence of a secular state: ideological diversity; the secular character of the Russian state; the absence of a state or obligatory religion; separation of religious associations from the state; equality of religious associations before the law; guarantees of equality of human and civil rights and freedoms regardless of attitude to religion; prohibition of any forms of restriction of citizens' rights on grounds of religious affiliation; guarantees of freedom of conscience and freedom of religion; prohibition of propaganda or agitation inciting religious hatred and hostility, as well as propaganda of religious superiority; and the absence of coercion to join any association or remain in it.

This means that only the realization of all these elements in their totality, both at the federal level and at the level of the constituent entities, will ensure the proper realization of the principle of the secular character of the state.

However, Russian legislation in fact contains no definition of state secularity. At the same time, the question of defining the concept and content of the principle of state secularity is central to the realization of freedom of conscience under Article 28 of the Constitution of the Russian Federation.

I believe that the problems involved in defining the concept of state secularity predetermine systemic violations of this principle, thereby creating preconditions for mass violations of freedom of conscience.

F. M. Rudinsky rightly maintains that separation of church and state means that the state does not interfere in the internal religious life of the church and does not entrust it with the performance of any state functions, while the church does not interfere in state affairs. This is the most important guarantee of freedom of conscience, because liberation of conscience from compulsory state guardianship allows the individual independently to decide the question of their convictions. Religious organizations become private societies that cannot use power against those of other faiths or against non-believers.

In his draft USSR law on freedom of conscience, Yu. A. Rozenbaum identified the equality of citizens regardless of their attitude to religion, the separation of the church or religious organizations from the state and of school or education from the church or religious organizations, and legislation establishing liability for violations of freedom of conscience as its principal constitutional guarantees.

According to A. S. Lovinyukov, the principle of equality between believers and non-believers is a very important structural element of freedom of conscience, without which freedom of conscience is practically inconceivable. He rightly notes that the principle of equality itself requires guarantees, otherwise it remains a formal declaration.

In my view, the impossibility of the state encroaching upon freedom of conscience can and must become a guarantee of real equality. Historical experience shows that, where the state has such a possibility, it has never refused special restrictions and control over the worldview sphere.

Granting the state special powers to regulate this sphere inevitably leads to arbitrary interpretation, in accordance with the personal confessional

T. P. Minchenko considers that, for a constructive approach to the realization of freedom of conscience, the understanding of the secular state as a worldview-neutral state appears promising.

Ukrainian scholar M. Yu. Babiy understands the principle of state secularity as the worldview-indifferent, or neutral, character of the state and its structural subdivisions.

Thus, it may be concluded that contemporary scholarship shows a tendency to transform approaches to the concept and content of state secularity in the direction of abandoning its connection with the state's attitude to religion, confession, and the activity of religious associations.

From the standpoint of the rule of law and the requirements of modern legal technique, which are based on the principle of legal certainty, the legally correct definition of state secularity is its definition as worldview neutrality, since there is no legal definition of 'religion' and, consequently, of the 'religious'. Only in this case can secularity function as one of the most important guarantees of liberty of worldview choice.

This means that approaches linking the concept of state secularity with the concept of 'religion' and its derivatives have a right to exist, but they do not appear fully correct from the point of view of contemporary legal science.

However, in Russia and in many states of the world that call themselves secular, the only legally correct interpretation of state secularity as worldview neutrality, implying the state's indifferent attitude to the worldview sphere, is excluded in favor of political compromises and cooperation between the state and religious organizations useful to the authorities.

In my view, a secular state is a worldview-neutral state that fundamentally does not accept any worldview as an official ideology and ensures that every person has the opportunity to make a free worldview choice. The secularity of the state implies its indifference in the worldview sphere, that is, a refusal of special control, non-interference subject to compliance with the law, non-identification because scientific criteria cannot be created, refusal of special privileges, separation from and equal distance towards worldview organizations.

Accordingly, the principle of state secularity includes the following components:

worldview neutrality of state authorities: legislative, executive and judicial authorities;

worldview neutrality of local self-government bodies;

worldview neutrality of the state and municipal education system;

worldview neutrality of state and municipal media and mass communications;

worldview neutrality of law-enforcement bodies;

worldview neutrality of the Armed Forces, other troops and military formations;

worldview neutrality of the penal enforcement system, places of detention and places of deprivation of liberty.

In relation to international law, I propose introducing the concept of the international-law worldview neutrality of subjects of international law. This should be understood as the indifference of states and other authorized subjects of international law towards the worldview sphere in all its lawful manifestations.

Given the problem of reconciling different approaches to understanding secularity, I propose that, at the level of international law, the term worldview neutrality of subjects of international law be used instead of the concept of secularity.

In relation to subjects of international law, the following components may be distinguished:

- international-law worldview neutrality of states;
- international-law worldview neutrality of other authorized subjects of international law;
- international-law worldview neutrality of intergovernmental organizations; international-law worldview neutrality of state-like entities;
- international-law worldview neutrality of peoples struggling for independence;
- international-law worldview neutrality of international non-governmental organizations;
- international-law worldview neutrality of subjects of federations and administrative-territorial units of unitary states.

The main features of the international-law worldview neutrality of subjects of international law, that is, states and other authorized subjects, include the following:

non-identification of the worldview sphere, because it is impossible to create legal criteria, including because such criteria would fail to meet the principle of legal certainty and the requirements of modern legal technique;

indifference in the worldview sphere, meaning refusal of special policies, special bodies and special control, and non-interference subject to compliance with the law;

prohibition of discrimination against individuals and their associations, including on grounds of worldview affiliation;

legal equality of public non-profit associations;

refusal of special relations with worldview associations, with relations carried out on the same basis as relations with public non-profit associations generally;

refusal of special privileges and restrictions for worldview associations, that is, separation from and equal distance towards worldview organizations;

refusal to legally establish any worldview and/or ideology as state or dominant;

ensuring a system of guarantees for the realization of liberty of worldview choice for every individual;

refusal to form international intergovernmental organizations on worldview grounds;

worldview-oriented state-like entities, peoples struggling for independence, international non-governmental organizations and other subjects of international law in the sphere of international relations must observe the principles of worldview neutrality recognized by the international community.

2.4. Worldview Neutrality of States and Other Subjects of International Law as a Guarantee for the Realization of the Right to Liberty of Worldview Choice

The principle of state secularity, together with other foundations of the constitutional order, serves as a foundation for other chapters of the Constitution of the Russian Federation and plays a leading role in the system of constitutional law. It should be especially emphasized that Article 14 of the Constitution of Russia, which declares the secular character of the state, the equality of religious associations and their separation from the state, is contained in Chapter 1 of the Constitution, entitled 'The Foundations of the Constitutional Order'. This chapter contains the fundamental principles concerning human rights, the structure of the state and organization of public authorities, and relations between the individual, society and the state.

The principle of the secularity, or worldview neutrality, of the state is interconnected with the principle of the law-governed state, which is based on the rule of law and is oriented ion of human rights. The principles of freedom of conscience, state secularity, the relevant state policy, and relations between the state and religious associations are also closely interconnected.

One may agree with V. N. Linkin, who considers the following to be mandatory elements of the existence of a secular state: ideological diversity; the secular character of the Russian state; the absence of a state or obligatory religion; separation of religious associations from the state; equality of religious associations before the law; guarantees of equality of human and civil rights and freedoms regardless of attitude to religion; prohibition of any forms of restriction of citizens' rights on grounds of religious affiliation; guarantees of freedom of conscience and freedom of religion; prohibition of propaganda or agitation inciting religious hatred and hostility, as well as propaganda of religious superiority; and the absence of coercion to join any association or remain in it.

This means that only the realization of all these elements in their totality, both at the federal level and at the level of the constituent entities, will ensure the proper realization of the principle of the secular character of the state.

However, Russian legislation in fact contains no definition of state secularity. At the same time, the question of defining the concept and content of the principle of state secularity is central to the realization of freedom of conscience under Article 28 of the Constitution of the Russian Federation.

I believe that the problems involved in defining the concept of state secularity predetermine systemic violations of this principle, thereby creating preconditions for mass violations of freedom of conscience.

F. M. Rudinsky rightly maintains that separation of church and state means that the state does not interfere in the internal religious life of the church and does not entrust it with the performance of any state functions, while the church does not interfere in state affairs. This is the most important guarantee of freedom of conscience, because liberation of conscience from compulsory state guardianship allows the individual independently to decide the question of their convictions. Religious organizations become private societies that cannot use power against those of other faiths or against non-believers.

In his draft USSR law on freedom of conscience, Yu. A. Rozenbaum identified the equality of citizens regardless of their attitude to religion, the separation of the church or religious organizations from the state and of school or education from the church or religious organizations, and legislation establishing liability for violations of freedom of conscience as its principal constitutional guarantees.

According to A. S. Lovinyukov, the principle of equality between believers and non-believers is a very important structural element of freedom of conscience, without which freedom of conscience is practically inconceivable. He rightly notes that the principle of equality itself requires guarantees, otherwise it remains a formal declaration.

In my view, the impossibility of the state encroaching upon freedom of conscience can and must become a guarantee of real equality. Historical experience shows that, where the state has such a possibility, it has never refused special restrictions and control over the worldview sphere.

Granting the state special powers to regulate this sphere inevitably leads to arbitrary interpretation, in accordance with the personal confessional preferences and group ideological interests of public officials, and, as a consequence, to equally inevitable abuses of power.

In the above context, it is appropriate to consider the problems of legal regulation of freedom of conscience in states belonging to the religious legal family. This family includes a number of states in Asia, Africa, and the Indian and Pacific Ocean regions.

It is generally accepted that Muslim law is one of the largest legal systems belonging to the religious legal family. A. M. Magomedov notes that Islamic law is based on religious and moral sacred norms that at the same time are legal norms and effectively regulate social relations.

Other contemporary researchers understand Muslim law as a system of norms covering all spheres of life and expressing, in religious form, primarily the will of the religious elite; such norms are sanctioned and supported to one degree or another by a theocratic Islamic state. In its foundations, this system developed in the early Middle Ages, from the seventh to the tenth centuries, in the Arab Caliphate and is based on Islam.

This approach, in turn, rests on the idea that existing law came from Allah, who at a certain moment in history revealed it to humanity through His Prophet Muhammad. The law of Allah was given to humanity once and for all; therefore society must be guided by this law rather than create its own under the influence of constantly changing social conditions.

Thus, in a broad sense, Islamic law is regarded as a unified system of social and normative regulation that includes both legal norms and non-legal regulators, primarily religious and moral norms, as well as custom.

Since any divine revelation requires interpretation, a special role belongs here to Islamic jurists, whose works form the basis of Muslim law. Muslim law is a remarkable example of jurists' law. It was created and developed by private specialists. Legal science, rather than the state, plays the role of legislator: textbooks have the force of law. When considering a case, a judge never directly refers to the Qur'an or the Sunnah, the traditions about the Prophet. Instead, he refers to an author whose authority is generally recognized.

Because social relations transformed relatively slowly, until the nineteenth century the evolution of Muslim legal consciousness found expression mainly in religious and legal commentaries and collections of fatwas, which interpreted traditional provisions and principles of sharia in new ways without ever rejecting them directly or changing their familiar formulation.

At the contemporary stage, the intrusion of European-type sources of law has proved to be, to a considerable extent, an irreversible process affecting all Islamic countries where Muslim law had previously prevailed in the full specificity of its sources. The legal systems of these countries have undergone significant changes: the importance, scope and relative weight of Muslim law have decreased, and Muslim law itself, at least in its external form, has also adopted something from European codifications.

Another issue discussed by M. N. Marchenko is the relationship between Muslim law and the law of individual Muslim countries. He notes differences that exist not only in theory but also in the life of every Muslim country, because, as in Christian and other countries, in Islam civil society has never been completely merged with the religious community and lives mainly according to its own written and unwritten laws, not exclusively according to religious canons. At the same time, in different periods, in particular countries and on particular issues, such societies could depart from orthodox provisions and enter into contradiction with the principles and norms of religious Muslim law.

A significant feature of Islamic law is its division into legal schools associated with different currents in Islam and dominant in different regions of the world.

In scholarly literature, four mutually recognizing legal schools of the Sunni, sometimes called orthodox, direction in Islam are usually identified: Maliki, Hanafi, Shafi'i and Hanbali. The first two dominate in Afghanistan, Egypt, Libya and other countries of Africa and the Middle East. The Shafi'i school is represented in Indonesia, Malaysia and some African countries. The Hanbali legal school dominates in Saudi Arabia.

In addition to Sunni schools, the contemporary world contains non-Sunni currents and legal schools that are less influential but, from the standpoint of freedom of conscience, no less valuable. These include, above all, Shiite and Kharijite legal schools. As R. A. Minniakhmetov notes, Shiites make up between 10 and 20 percent of all Muslims, and the corresponding legal school is represented in Iran and Iraq. Kharijites are today represented mainly in Oman and in several other countries.

The Shiite school, in turn, is divided into the Zaydi and the more widespread Ja'fari school. The Zaydi school has preserved its influence in Yemen and in some countries of the Middle East. The Ja'fari school is considered the foundational legal school in Iran. A significant number of followers of Ja'fari jurisprudence live in Iraq and Lebanon. Ja'fari jurisprudence is based on the sacred book, the Sunnah, ijma and the so-called rational interpretation that is characteristic exclusively of Shiite law.

In the contemporary world, which is increasingly interpenetrating, interdependent and open, Islamic law is compelled to interact with other legal families and with international law. Under these new conditions, approaches and values traditional for formerly isolated and relatively closed communities are subjected to serious tests. As a result, although Muslim law is regarded as an independent legal family, it is compelled to transform under the influence of objective tendencies of world development.

The constitutions of some Islamic states declare freedom of religion in different interpretations, while at the same time giving priority to Islam as the basis of the legal system. In particular, various currents of Islam have state status in more than twenty countries of the contemporary world, including Algeria, Afghanistan, Bangladesh, Bahrain, Brunei Darussalam, Jordan, Iraq, Iran, Yemen, Qatar, the Comoros, Kuwait, Mauritania, Malaysia, Maldives, Morocco, the United Arab Emirates, Oman, Pakistan, Saudi Arabia, Sudan, Tunisia and others.

For example, in the constitutions of Bangladesh, Brunei Darussalam, Indonesia, Jordan, Qatar, Kuwait, Malaysia, the United Arab Emirates, Oman and Pakistan, although Islam is declared the state religion, followers of other religions are declared to enjoy freedom of religion or elements of it, such as freedom to perform rites, subject to compliance with the law. As a rule, such declarations of freedom are accompanied by a reservation concerning public order, which in practice means a prohibition on preaching any religion other than Islam. In addition, in some states strict blasphemy laws substantially restrict the rights of minorities, while questions of marriage, divorce and inheritance almost always remain within the jurisdiction of religious courts.

The constitutions of Algeria, Bahrain, Iraq, Tunisia and Egypt contain declarations of freedom of conscience in addition to freedom of religion. Yet, in almost all these documents, the practical realization of this freedom is limited by public order and the protection of the state religion.

The constitutions of Bangladesh, Belize, Jordan, Iraq, Qatar, Malaysia, Oman and Pakistan contain prohibitions of discrimination on grounds of religion. Yet in Belize and Malaysia the constitutional text contains internal contradictions that make the prohibition of discrimination non-absolute.

Some Islamic states have legislatively fixed a list of religious currents to which freedom of religion is granted. For example, the Constitution of the Islamic Republic of Iran of 3 December 1979 proclaims Islam of the Ja'fari school as the official religion of Iran. Other Islamic schools, including Hanafi, Shafi'i, Maliki, Hanbali and Zaydi, are fully respected. Followers of these schools are fully free to perform religious rites in accordance with their religious norms, to engage in religious upbringing and education, to resolve civil and personal matters such as marriage, divorce, inheritance and wills, and to apply to the courts on that basis. In any place where followers of one of these Islamic currents constitute a majority of the population, local rules and regulations are established by councils on the basis of the requirements of those religious currents, while respecting the rights of followers of other Muslim currents. In addition, Iranian Zoroastrians, Jews and Christians are the only religious minorities that may freely carry out their religious rites within the law and act in civil matters and in the sphere of religious education according to their own doctrine. In practice, this means that currents not included in the above-mentioned list are subject in Iran to persecution and discrimination on grounds of religion and belief.

In some Islamic states, even declarations of freedom of conscience and freedom of religion have found no place. For example, the Basic Law of Governance of the Kingdom of Saudi Arabia of 1992, in addition to establishing the state status of Islam, prescribes the upbringing of members of Saudi society on the basis of Islamic faith and the loyalty and obedience it commands to Almighty Allah, His Prophet and those in authority. Thus, under Article 11, Saudi society is based on the adherence of its members to Islam and on their pious, careful and solidary mutual relations without contradictions; worldview diversity is evidently perceived as one such contradiction. It is significant that even non-Sunni currents within Islam are subjected to discrimination on grounds of religion and belief in Saudi Arabia.

Hindu law, as a system of personal status regulating issues of marriage, divorce, inheritance and guardianship, applies to members of the Hindu community not only in India but also in a number of countries with significant Indian diasporas, including Pakistan, Myanmar, Malaysia, Singapore, and several East African states such as Kenya, Tanzania and Uganda.

The Constitution of India of 26 November 1949 guarantees all its citizens freedom of thought, expression, belief, faith and worship; freedom of conscience and the freedom to profess, practise and propagate religion; equality before the law and prohibition of discrimination on grounds of religion; and freedom concerning attendance at religious instruction or worship in state educational institutions. It should be noted that until 2006 Hinduism retained state status in Nepal, but this status was abolished.

Buddhist law is based on the teaching of the Buddha and influences the legal systems of India, certain countries of South-East Asia and the Far East. In some states, Buddhism is established as the state religion. For example, in Bhutan the Constitution of the Kingdom of 18 July 2008 established Buddhism as the spiritual heritage of the country. This means that the Drukpa Kagyu school, a form of Lamaist Buddhism, includes about 70 percent of the population and functions as the state religion. Article 7 of the Constitution of Bhutan declares the right of citizens to freedom of thought, conscience and religion.

The Constitution of Cambodia of 21 September 1993, as amended in 1999, establishes Buddhism as the state religion but guarantees freedom of religious belief and worship, provided that such freedom does not affect the rights of other believers or violate public order and the interests of security. Article 31 states that all citizens are equal before the law, enjoy rights and freedoms, and fulfil duties regardless of race, skin colour, sex, language, religious belief, political aspirations, place of birth, social status, property status or other status.

The Constitution of the Lao People's Democratic Republic of 28 May 2003 specially establishes the public role of Buddhism in Article 9. Yet Article 37 declares freedom of religion, belief or conviction and the observance of religious prescriptions, commandments or forms of worship in accordance with one's faith, provided this does not contradict civic duties, public order or morality. Article 43 provides that citizens have the right and freedom to believe or not believe religious teachings, while Article 30 prohibits unjust discrimination against a person on grounds of religious belief.

In Thailand, more than 90 percent of the population is considered to profess Theravada Buddhism. Under the Constitution of the Kingdom of Thailand of 2017, every person has the full right to profess a religion and practise forms of worship in accordance with their religious principles, provided that this right does not contradict the duties of all Thai people, threaten state security, violate public order or moral foundations. All persons are equal before the law and have equal rights and protection under the law; unjust discrimination on grounds of religious belief is inadmissible. At the same time, state measures aimed at eliminating obstacles or assisting people in exercising their rights and freedoms on an equal basis, as well as protection or assistance for children, women, elderly persons, persons with disabilities or low-income persons, are not considered unjust discrimination. However, the Constitution prescribes that the King must profess Buddhism while having the right to be called a supporter of all religions professed by the Thai people. In addition, the 2017 Constitution obliges the state to direct assistance only to the Theravada Buddhist school and to protect Buddhism from all forms of desecration. Some experts regard this provision as a compromise that brings Thailand closer to the affirmation of Buddhism as an official religion while preserving the formal equality of other religions.

The Constitution of Sri Lanka of 1978, with amendments 17 and 18, gives priority to Buddhism but does not establish it as the state religion. Every person is guaranteed freedom of thought, conscience and religion, including the freedom to have or adopt a religion or belief of their choice, equality before the law and protection against discrimination on grounds of religion. In practice, however, Sri Lanka supports one current of Buddhism, Theravada, which has led to discrimination and intolerance against Muslims and Christians, including killings and pogroms against the property of 'infidels'.

Jewish law is a component of the legal system of Israel. At the same time, Judaism has not been proclaimed the state religion. The Declaration of Independence of Israel of 1948 states that the State of Israel will be open for Jewish immigration and the ingathering of the exiles; will foster the development of the country for the benefit of all its inhabitants; will be based on freedom, justice and peace as envisaged by the prophets of Israel; will ensure complete equality of social and political rights to all its inhabitants irrespective of religion, race or sex; will guarantee freedom of religion, conscience, language, education and culture; will safeguard the holy places of all religions; and will be faithful to the principles of the Charter of the United Nations. However, the Basic Law: Human Dignity and Liberty of 17 March 1992 does not enshrine freedom of conscience and religion. Religious institutions in Israel are supported by the state, and Jewish law operates in several spheres. The Orthodox Rabbinate cannot be considered separated from the state, which predetermines discrimination and conflicts.

Christian law, as a system of norms based on Christianity, is divided into canon, or Catholic, law; Orthodox, or church, law; and Protestant law. At the contemporary stage, Protestantism is widespread throughout the world, but it has preserved the greatest influence in the states of Europe and North America. Catholicism and canon law influence the legal systems of Argentina, the

Dominican Republic, Costa Rica, Liechtenstein, Malta, the Sovereign Military Order of Malta, Monaco, El Salvador, San Marino and some cantons of Switzerland. The state status of the Roman Catholic Church is established in Costa Rica. The Constitution declares freedom to practise other forms of worship and prohibits persecution for expressing one's opinion or for actions that do not violate the law. In Liechtenstein, the Roman Catholic Church is the state church, while other religious cults are guaranteed the profession of corresponding beliefs and the performance of church services subject to compliance with public order and moral requirements.

The Constitution of the Republic of Malta of 1964 establishes the Roman Catholic Apostolic religion as the state religion but guarantees that every person in Malta has full freedom of conscience and enjoys freedom to exercise their respective religious worship. Religious instruction in state schools is a compulsory part of the curriculum, although attendance is voluntary. Pupils and their parents have the right to refuse such lessons. Article 45 of the Constitution directly prohibits discrimination on grounds of religious belief and other differences.

The Constitution of the Principality of Monaco of 17 December 1962 also establishes the state status of the Roman Catholic Apostolic religion, while declaring freedom of worship, freedom of expression, and freedom from coercion in relation to acts of worship, rites and observance of days of rest.

Orthodox legal traditions preserve influence over the legal systems of Armenia, Greece, Georgia, Bulgaria, North Macedonia and Finland. In Greece, Orthodoxy is established as the prevailing religion, while in Finland the Orthodox Church, together with the Lutheran Church, is endowed with the status of a national church. However, the influence of religious canons is limited and concerns mainly the internal life of communities. Because the influence of Orthodox law on state legislation is limited, many researchers classify these states as belonging to the Romano-Germanic legal family rather than to the religious legal family.

Lutheranism, as part of the conditional Protestant law, has preserved influence over the legal systems of Denmark, Iceland, Norway, Finland and Sweden. In Denmark and Iceland, Lutheranism has the status of a state religion. In Finland, none of the churches, including the Lutheran Church, is designated as a state church, but they have the status of public-law corporations, which gives them the right to collect church tax through the state tax system and to receive state financing for certain purposes, such as maintaining cemeteries and historic buildings.

In Sweden, the church was separated from the state in 2000, while the Constitution preserves the requirement that the monarch and members of the royal family profess the Evangelical Lutheran faith in order to retain the right to the throne. In Norway, separation was legally established in 2012, but the final administrative separation of the church from the state was completed only at the beginning of 2017, when the church became an independent organization with its own governing body. At the same time, state confessional preferences have been preserved. For example, the state is obliged to support its activity, including the maintenance of church buildings.

Constitutional guarantees of freedom of conscience in the United Kingdom do not fully correspond to the contemporary character of social relations. Moreover, in one part of the United Kingdom, England, the state status and privileges of the church have been preserved. The Church of England, the Anglican Church, has a special status; the monarch is its Supreme Governor and head and, with the consent of the Prime Minister, appoints archbishops and bishops. Twenty-six bishops have seats in the House of Lords. The ecclesiastical law of the Church of England is part of state law, and ecclesiastical courts are subject to the control of civil courts. Marriages solemnized by priests of the Church of England have legal force. In Scotland, Presbyterianism was recognized as legally independent in spiritual matters after the adoption of the Church of Scotland Act 1921. Wales and Northern Ireland do not have a state religion.

In general, at the contemporary stage, religious law increasingly becomes an internal regulator of the activity of religious associations, subject to compliance with the legal norms of the particular state. It is mainly in this capacity that it operates in most contemporary states.

However, as A. Kh. Saidov rightly notes, each national legal system is a product of the historical development of a particular country and reflects the specific features of its history, economy, political system, cultural and local traditions, climate and natural conditions, stereotypes of mass consciousness, and national mentality. All this, in many cases, creates serious obstacles to the implementation of the norms and institutions of the law of one country into the legal system of another.

I believe that these factors create no less serious obstacles to the interaction of international legal forms of the right to freedom of conscience with national legal systems. Special difficulties arise in the interaction of international legal forms of freedom of conscience with the legal systems of states belonging to the conditional religious family of Islamic law.

It should be recalled that it is precisely in Islamic states that violations of the right to freedom of conscience are most widespread, systemic and inhuman. In particular, in a number of countries, including Iran, Saudi Arabia, Pakistan and Afghanistan, blasphemy legislation is used as a legal instrument for suppressing freedom of conscience, making dialogue with international law practically impossible without deep reform of the very foundations of these legal systems.

The case of Afghanistan has become a point of no return for international law. It is an example of how a legal system can be entirely removed from the universal legal field and transformed into an instrument of inhuman control, where the very concept of individual freedom of conscience is recognized as hostile to the state. In the complete absence of worldview neutrality, international norms simply cease to operate, because there is no legal basis for their recognition.

This is connected, among other things, with the fact that, as R. A. Podoprigora notes, the countries of the Islamic world, mainly Arab countries, have always treated human rights documents with distrust, regarding them as an invention of Western civilization. It is no coincidence that the preamble to the Cairo Declaration states that it is the Islamic world that will show the way to solving the chronic problems of materialistic civilization for humanity, which has become entangled in contradictory tendencies and ideologies.

A. M. Magomedov argues that many Muslim states are unable to participate in international treaties or to implement decisions of international organizations because, in the event of a conflict between norms of international law and norms of Islamic law, the norms of international law become inapplicable in the corresponding Islamic state. At the same time, efforts should be made to harmonize these two legal systems in order to create a universal system of law while preserving the imperative nature of Islamic law, but at the level of performance of obligations undertaken by the corresponding Islamic state under international law.

In my view, the official and/or unofficial pragmatic merging of the state with any religious direction, whether Buddhism, Islam, Catholicism, Orthodoxy, Protestantism or another, in practice means the dominant position of one particular religious association. Far less often, dominant positions are shared among several associations.

In any event, the consequences of such state-confessional pragmatism include violations of the internationally recognized right to freedom of conscience, growth of intolerance, discrimination and violence on grounds of worldview affiliation.

At the same time, problems arise not only for representatives of different confessional directions, for example Christians in an Islamic state, and for the non-confessional part of society, for example atheists, but also for representatives of different currents within one direction, for example Shiites in a Sunni state and Sunnis in a Shiite state.

One may agree with the view of contemporary researchers who note the growing interaction between international and domestic legal systems, as well as convergence among different legal families. According to A. I. Ekimov, global integration processes today have covered practically all spheres of social life, which makes the idea of network law viable and fully corresponds to the possibilities of convergence of legal systems.

Since these tendencies require an increase in the degree of freedom of the individual, the common denominator in the interaction of all legal systems is the strengthening of international legal guarantees for the realization of human rights as an instrument for embodying this freedom.

As A. N. Korshunov notes, despite the dissimilarity of national cultures, legal systems always have a principled possibility of dialogue on the basis of legal universals, namely inalienable human rights. This gives humanistic significance to law as a universal means of regulating human life. Legal universals determine the character of the evolution and self-organization of global law in the conditions of post-industrial society.

I believe that, in these conditions, the forms of human rights law must have priority over other forms of law, and legal systems based on worldview systems of the past do not fully correspond to contemporary social relations, which predetermines their insufficient regulatory effectiveness. This does not mean that confessional norms underlying religious legal families must cease to exist. Rather, they must take a place in the system of social norms that corresponds to contemporary social relations.

As A. Kh. Saidov rightly argues, references to originality, traditions or a special path cannot justify a condition of national law in which it falls below the basic global legal standards achieved by humanity by the twenty-first century.

One should also agree with A. A. Dorskaya, who believes that giving church rules the character of legal norms in a multinational and multi-confessional Russia may lead to serious consequences both for the state and for religious associations. Under conditions of global transformation of social relations, this assertion applies to every state and to the world community as a whole.

In this context, the question of recognizing, legally enshrining and realizing the worldview neutrality of states and other subjects of international law as one of the most important principles and guarantees for the realization of internationally recognized human rights, including the right to freedom of conscience, acquires fundamental significance and requires the united efforts of the world community.

Thus, the worldview neutrality of states and other subjects of international law is one of the most important guarantees for the realization of internationally recognized human rights, including the right to freedom of conscience understood as liberty of worldview choice. Accordingly, violation of the principle of state secularity disavows both individual types of guarantees and the entire system of guarantees for the realization of human rights.

As a conclusion, I propose enshrining in international law the principle of secularity, understood as public-law worldview neutrality, as an important guarantee for the realization of internationally recognized freedom of conscience understood as liberty of worldview choice. Within the codification and progressive development of international law at the universal level of the United Nations, it is recommended, at the initial stage, to adopt a Global Declaration on the Right to Liberty of Worldview Choice and the Principle of Worldview Neutrality of Subjects of Public International Law in Analog and Digital Environments, and subsequently a legally binding convention. Thereafter, on an international legal basis, regional principles and norms, as well as domestic legal systems, may be reformed. In particular, in the Russian Federation, it appears advisable to provide an expanded interpretation of the constitutional principle of state secularity as worldview neutrality in the Constitution and in federal law.

REFERENCES TO CHAPTER 2

List of Sources Used

1. Stankevich G.V. The Religious Factor of the Contemporary Political Process: Abstract of Doctoral Dissertation in Political Science. Pyatigorsk, 2012. p. 12.
2. Podoprigora R.A. The State and Religious Organizations (Administrative-Legal Issues). Almaty: Arkaim, 2002. pp. 129-130.
3. Conflicts and Wars of the 21st Century (Middle East and North Africa) / Institute of Oriental Studies, Russian Academy of Sciences. Moscow: Institute of Oriental Studies of the Russian Academy of Sciences, 2015. 504 p. p. 26.
4. Freedom of Conscience: Problems of Theory and Practice. Monograph (edited by F.M. Rudinsky, S.A. Buryanov). Moscow: ZAO TF MIR, 2012. 1120 p. pp. 637-768.
5. Freedom of Thought Report. Electronic resource. URL: https://fot.humanists.international/editorial-introduction/ (accessed: 8.12.2019).
6. Gavrilenkov A.F. The Policy of State Power of the Russian Empire toward the Orthodox Church, Non-Orthodox Christian Confessions, and Abrahamic (Non-Christian) Faiths in 1721-1917: Essence, Principles, Evolution: Abstract of Doctoral Dissertation in History. Moscow, 2010. p. 31.
7. Isaev A.V. Socio-Political Partnership between State Authorities and the Russian Orthodox Church in Contemporary Russia: Abstract of Candidate of Legal Sciences Dissertation. Orel, 2011. p. 7.
8. Ibid. p. 7.
9. Petyukova O.N. Legal Forms of Relations between the Soviet State and the Russian Orthodox Church in 1917-1945: Abstract of Doctoral Dissertation in Law. Moscow, 2011. p. 8.
10. Ibid. p. 9.
11. Pokaninova E.B. Socio-Philosophical Analysis of the Transformation of State-Confessional Relations in the Republic of Kalmykia: Abstract of Doctoral Dissertation in Philosophy. Moscow, 2012. p. 20.
12. Konnevskaya A.Yu. Constitutional Freedom of Conscience: Content, Boundaries, Guarantees: Abstract of Candidate of Legal Sciences Dissertation. Chelyabinsk, 2010. p. 10.
13. Loginov A.V. Power and Faith: The State and Religious Institutions in History and the Present. Moscow: Great Russian Encyclopedia, 2005. p. 462.
14. Ponkin I.V. Legal Foundations of the Secular Nature of the State and Education. Moscow, 2003.
15. Ponkin I.V. The Secular Nature of the State. Moscow: Educational and Scientific Center for Pre-University Education, 2004.
16. Shakhov M.O. Conceptual Problems in the Formation of State Policy on Religion // Religion and Law. 2001. No. 1. pp. 4-6.
17. Shutova O.S. A Structural Model of State Policy in the Field of Relations with Religious Associations: Abstract of Candidate of Philosophical Sciences Dissertation. Moscow, 2004. p. 5.
18. Karpushkin A.V. Constitutional Foundations and Genesis of Relations between the State and Religious Associations in the Russian Federation: Abstract of Candidate of Legal Sciences Dissertation. Saratov, 2009. p. 11.
19. Ibid. p. 11.
20. Karpushkin A.V. Constitutional Foundations and Genesis of Relations between the State and Religious Associations in the Russian Federation: Abstract of Candidate of Legal Sciences Dissertation. Saratov, 2009. pp. 11-12.
21. The Confessional Policy of the Russian State: Study Guide / executive editor M.O. Shakhov. Moscow: Publishing House of the Russian Academy of Public Administration, 2003. p. 11.
22. Gavrilenkov A.F. The Policy of State Power of the Russian Empire toward the Orthodox Church, Non-Orthodox Christian Confessions, and Abrahamic (Non-Christian) Faiths in 1721-1917: Essence, Principles, Evolution: Abstract of Doctoral Dissertation in History. Moscow, 2010. p. 31.
23. Odintsov M.I. State-Church Relations in Russia (Based on Materials from Russian History of the 20th Century) // The Russian Orthodox Church in the 20th Century: History and Relations with the State and Society. Moscow, 2002. pp. 121-187.

24. Colson C. Conflict of Kingdoms. Moscow, 1996. p. 129.
25. Mukhametzyanova R.M. The Policy of the Contemporary Russian State in the Sphere of Freedom of Conscience: The Process of Forming a New Model: Abstract of Candidate of Political Sciences Dissertation. Ufa, 2004. p. 12.
26. Ibid. p. 20.
27. Gavrilenkov A.F. The Policy of State Power of the Russian Empire toward the Orthodox Church, Non-Orthodox Christian Confessions, and Abrahamic (Non-Christian) Faiths in 1721-1917: Essence, Principles, Evolution: Abstract of Doctoral Dissertation in History. Moscow, 2010. pp. 34-35.
28. Levina A.A. The Institution of 'Freedom of Conscience and Religion': The Historical and Legal Experience of Russia: Abstract of Candidate of Legal Sciences Dissertation. Nizhny Novgorod, 2003. p. 18.
29. Kanevsky K.G. Legal Regulation of State-Confessional Relations in the Russian Federation: Candidate of Legal Sciences Dissertation. Moscow, 2004. p. 166.
30. Gavrilenkov A.F. The Policy of State Power of the Russian Empire toward the Orthodox Church, Non-Orthodox Christian Confessions, and Abrahamic (Non-Christian) Faiths in 1721-1917: Essence, Principles, Evolution: Abstract of Doctoral Dissertation in History. Moscow, 2010. p. 35.
31. Fedotova Yu.E. On Criminal Liability for Violating the Principle of the Secular Nature of the State // Science and Religion in a Secular State. Moscow, 2017. pp. 62-64; Fedotova Yu.E. Freedom of Conscience and Religion as an Object of Criminal-Law Protection: Candidate of Legal Sciences Dissertation. Yekaterinburg, 2018. 229 p.
32. Osipova L.V. The Legal Status of Religious Associations as an Institution of Civil Society in Contemporary Russia: Abstract of Candidate of Legal Sciences Dissertation. Moscow, 2009. p. 19.
33. Stankevich G.V. The Religious Factor of the Contemporary Political Process: Abstract of Doctoral Dissertation in Political Science. Pyatigorsk, 2012. p. 12.
34. Drinova E.M. Religion and the Political Process: Religious-Political Expansion and the Politicization of Religion in the Contemporary World: Abstract of Doctoral Dissertation in Political Science. Rostov-on-Don, 2012. p. 11.
35. Ibid. p. 14.
36. Freedom of Conscience: Problems of Theory and Practice. Monograph (edited by F.M. Rudinsky, S.A. Buryanov). Moscow: ZAO TF MIR, 2012. p. 598.
37. Buryanov S.A. Manifestations of Systemic Corruption in Relations between the State and Religious Associations // Overcoming Corruption as the Main Condition for the Establishment of a Rule-of-Law State. Interdepartmental Scientific Collection / editor-in-chief A.I. Komarova. Moscow, 2009. Vol. 1 (39). pp. 399-419.
38. Buryanov S.A. Religion in Elections in Russia: The Factor of State Relations with Religious Associations in the Federal Electoral Cycle of 2003-2004. Moscow, 2005. p. 21.
39. Shakhov M.O. Clericalization Does Not Threaten Russia: The Moscow Patriarchate Merely Creates the Appearance of Its Influence // NG-Religii. February 19, 2003.
40. Lebedev S.D. Religious and Secular Cultures as Types of Social Knowledge Systems. Belgorod: Belgorod State University Publishing House, 2003. p. 7.
41. Karpushkin A.V. Constitutional Foundations and Genesis of Relations between the State and Religious Associations in the Russian Federation: Abstract of Candidate of Legal Sciences Dissertation. Saratov, 2009. p. 11.
42. Bezuglov A.A., Soldatov S.A. Constitutional Law of Russia. In 3 volumes. Moscow, 2001. p. 340.
43. Linkin V.N. The Principle of the Secular State as One of the Foundations of the Constitutional Order of the Russian Federation: Abstract of Candidate of Legal Sciences Dissertation. Moscow, 2015. p. 10.
44. Novikov M.V. The Clerical State and the Church: Relations between Church and State in Israel and Neighboring Countries: Abstract of Candidate of Legal Sciences Dissertation. Moscow, 2013. p. 11.
45. Aleynikova S.M. On the Question of the 'Vulgar Understanding of Secularism': Political Myth-Making or a New Path of Development? // Bulletin of Brest University. Series 1. 2013. No. 2. p. 133.

46. Bauberot J. The French Concept of Secularism (laicite) in the Context of Globalization // Religion and the Secular State: The Principle of laicite in the World and Eurasia. Moscow: Franco-Russian Center for the Humanities and Social Sciences in Moscow, 2008. p. 48.
47. Religion and the Secular State: The Principle of laicite in the World and Eurasia. Moscow: Franco-Russian Center for the Humanities and Social Sciences in Moscow, 2008. 288 p.
48. Marsh C. The Principle of Separation of Church and State and Its Opponents: laicite and the Desecularization of America // Religion and the Secular State: The Principle of laicite in the World and Eurasia. Moscow: Franco-Russian Center for the Humanities and Social Sciences in Moscow, 2008. p. 91.
49. Charbit D. The Long Road to Secularism in Israel: Achievements, Limits, Problems // Religion and the Secular State: The Principle of laicite in the World and Eurasia. Moscow: Franco-Russian Center for the Humanities and Social Sciences in Moscow, 2008. p. 128.
50. Pchelintsev A.V. The Secular Nature of the State as a Constitutional-Legal Guarantee for the Implementation of Freedom of Religion // Freedom of Conscience in Contemporary Russia. Moscow, 2017. 200 p. p. 72.
51. Pchelintsev A.V. The Secular Nature of the State as a Constitutional-Legal Guarantee for the Implementation of Freedom of Religion // Freedom of Conscience in Contemporary Russia. Moscow, 2017. 200 p. p. 74.
52. The Confessional Policy of the Russian State: Study Guide / executive editor M.O. Shakhov. Moscow: Publishing House of the Russian Academy of Public Administration, 2003. p. 7.
53. Shmakova O.A. Institutions of Religious Legal Policy in a Contemporary Secular State: Abstract of Candidate of Legal Sciences Dissertation. Rostov-on-Don, 2008. p. 8.
54. Ibid. p. 9.
55. Ibid. p. 10.
56. Osipova L.N. Relations between State Institutions and Religious Associations in Contemporary Russia: Candidate of Political Sciences Dissertation. Ufa, 2004. pp. 12-13.
57. Sebentsov A.E. The Principle of Freedom of Conscience and Its Implementation in the Russian Federation: State, Trends, and Prospects: Abstract of Candidate of Philosophical Sciences Dissertation. Saint Petersburg, 2009. p. 14.
58. Mozgovoy S.A. The Secular State as Worldview-Neutral // Socionomicon. International Collection of Scientific and Practical Works. Issue 2. Rostov-on-Don: NB, 2006. pp. 70-73.
59. Avakyan S.A. Constitutional Law of Russia. Textbook in 2 volumes. Vol. 1. Moscow: Yurist, 2005. p. 362.
60. Baglay M.V. Constitutional Law of the Russian Federation. Textbook for Law Universities and Faculties. Moscow: NORMA-INFRA Publishing Group, 1998. p. 122.
61. Voronkova M.L. Constitutional Foundations of the Secular State in the Russian Federation: Candidate of Legal Sciences Dissertation. Saratov, 2006. p. 25.
62. Willaime J.-P. Secularism in Europe: National Logics and Europeanization // Religion and the Secular State: The Principle of laicite in the World and Eurasia. Moscow: Franco-Russian Center for the Humanities and Social Sciences in Moscow, 2008. p. 18.
63. Rousselet K. The Principle of Secularism in Russia: A Clash of Norms and Values // Religion and the Secular State: The Principle of laicite in the World and Eurasia. Moscow: Franco-Russian Center for the Humanities and Social Sciences in Moscow, 2008. p. 167.
64. Miroshnikova E.M. The Constitutional-Legal Mechanism of State-Church Relations in the Federal Republic of Germany // Global Experience of State-Church Relations. Moscow, 1998. pp. 83-84.
65. Miroshnikova E.M. The Cooperative Model of State-Church Relations: Experience and Problems: Monograph. Moscow: Institute of Europe of the Russian Academy of Sciences; Tula: Publishing House of Tula State Pedagogical University named after L.N. Tolstoy, 2007. p. 133.
66. Ibid. p. 132.

67. Stetskevich M.S. Freedom of Conscience: Study Guide. Saint Petersburg: Saint Petersburg University Press, 2006. p. 8.
68. So A.A. Constitutional-Legal Foundations of Freedom of Religion and the Activities of Religious Associations in Russia (Using the Subjects of the Russian Federation in the North-Western Federal District as an Example): Abstract of Candidate of Legal Sciences Dissertation. Saint Petersburg, 2010. p. 18.
69. Ibid. p. 18.
70. Meshcheryakova A.F. The Secular State in Contemporary Russia: Constitutional-Legal Analysis: Abstract of Candidate of Legal Sciences Dissertation. Saratov, 2009. pp. 8-10.
71. Ponkin I.V. The Secular Nature of the State. Moscow: Educational and Scientific Center for Pre-University Education, 2004. p. 49.
72. Buryanov S.A., Mozgovoy S.A. Problems in the Implementation of Freedom of Conscience and Trends in Relations between the State and Religious Associations in Russia. Moscow: Institute for Freedom of Conscience, 2005.
73. Aleynikova S.M., Buryanov S.A. The Secular State in Questions and Answers: Brief, Accessible, and Up to Date. Edited by I. Kondratiev. Moscow, 2015. 62 p.
74. Minchenko T.P. Development of the Right to Freedom of Conscience in the Post-Secular World // Rights and Freedoms of the Human Person and Citizen: Theoretical Aspects and Legal Practice: Proceedings of the Inter-University Scientific Conference in Memory of Professor Felix Mikhailovich Rudinsky, April 26, 2012. Moscow: Academy for Advanced Training and Professional Retraining of Education Workers, 2012. pp. 155-158.
75. Savva S.S. Constitutional-Legal Enshrinement of the Principle of Secular Education: Content and Practice of Implementation // Policy and Society. 2006. No. 7-8. pp. 106-115. Savva S.S. Clericalization of Secular Education in Russia: Constitutional-Legal Analysis // Religious Studies. 2006. No. 4. pp. 76-88.
76. Nikitaev D.M., Buryanov S.A. On the Question of the Contemporary Definition of the Concept and Content of the Secular Nature of the State // Contemporary Lawyer. 2019. No. 3 (28), July-September. pp. 85-96; Nikitaev D.M. Problems in the Implementation of Freedom of Conscience in Contemporary Russia and the World / D.M. Nikitaev, S.A. Buryanov / scientific editor S.A. Buryanov. Moscow: RUSAINS, 2020. 146 p.
77. Freedom of Conscience: Problems of Theory and Practice. Monograph (edited by F.M. Rudinsky, S.A. Buryanov). Moscow: ZAO TF MIR, 2012. p. 150.
78. Pridvorov N.A., Tikhonova E.V. The Institution of Freedom of Conscience and Freedom of Religion in the Law of Contemporary Russia. Moscow: Jurisprudence, 2007. pp. 39-40.
79. Minchenko T.P. Development of the Right to Freedom of Conscience in the Post-Secular World // Rights and Freedoms of the Human Person and Citizen: Theoretical Aspects and Legal Practice: Proceedings of the Inter-University Scientific Conference in Memory of Professor Felix Mikhailovich Rudinsky, April 26, 2012. Moscow: Academy for Advanced Training and Professional Retraining of Education Workers, 2012. p. 157.
80. Babiy M.Yu. Freedom of Conscience in the Context of the Humanization of Ukrainian Society: New Challenges and Prospects. Electronic resource. URL: http://archive.nbuv.gov.ua/portal/soc_gum/ukrr/2012_62-63/Babiy.htm (accessed: 8.12.2018).
81. Buryanov S.A. Freedom of Conscience and the Secular Nature of the State in the Russian Federation: Theoretical and Applied Study for 2009 - Early 2010. Moscow: OOO TF MIR, 2010. 173 p.
82. Linkin V.N. The Principle of the Secular State as One of the Foundations of the Constitutional Order of the Russian Federation: Abstract of Candidate of Legal Sciences Dissertation. Moscow, 2015. pp. 12-13.
83. Cited work. p. 13.
84. Buryanov S.A. Problems in the Implementation of Freedom of Conscience and the Secular Nature of the State in the Russian Federation: Theoretical and Applied Study for 2012 (edited by R.V. Bagdasarov). Moscow: Russia for All, 2013. 352 p.
85. Rudinsky F.M. Law, Religion, and Offenses. Volgograd, 1971. p. 36.
86. Magomedov A.M. Human Rights in Islamic and International Law: Candidate of Legal Sciences Dissertation. Moscow, 2011. pp. 8-9.

87. Saidov A.Kh. Comparative Jurisprudence (The Main Legal Systems of the Contemporary World): Textbook / edited by V.A. Tumanov. Moscow: Yurist, 2003. p. 296.
88. Ibid. p. 296.
89. Ibid. p. 297.
90. Ibid. p. 299.
91. Ibid. p. 299.
92. Ibid. p. 306.
93. Marchenko M.N. Legal Systems of the Contemporary World: Study Guide. Moscow: Zertsalo-M, 2001. pp. 292-293.
94. Minniakhmetov R.A. Shiite Concepts of Justice in Islamic Law // Eurasian Law Journal. 2013. No. 12. p. 87.
95. Ibid. p. 89.
96. Saidov A.Kh. Comparative Jurisprudence (The Main Legal Systems of the Contemporary World): Textbook / edited by V.A. Tumanov. Moscow: Yurist, 2003. p. 419.
97. Podoprigora R.A. The State and Religious Organizations (Administrative-Legal Issues). Almaty: Arkaim, 2002. pp. 123-124.
98. Magomedov A.M. Human Rights in Islamic and International Law: Candidate of Legal Sciences Dissertation. Moscow, 2011. pp. 8-9.
99. Ekimov A.I. On the Contemporary Model of Development of Russian Society // Economy. Law. Society. 2016. No. 4 (8). pp. 23-29. p. 27.
100. Korshunov A.N. The Idea of Global Law: Philosophical and Methodological Aspects: Abstract of Candidate of Philosophical Sciences Dissertation. Rostov-on-Don, 2010. p. 3.
101. Saidov A.Kh. Comparative Jurisprudence (The Main Legal Systems of the Contemporary World): Textbook / edited by V.A. Tumanov. Moscow: Yurist, 2003. p. 422.
102. Dorskaya A.A. Church Law in the Legal System of the Russian Empire from the Late 18th to the Early 20th Century: Abstract of Doctoral Dissertation in Law. Moscow, 2008. p. 16.
103. Freedom of Conscience: Problems of Theory and Practice. Monograph (edited by F.M. Rudinsky, S.A. Buryanov). Moscow: ZAO TF MIR, 2012. 1120 p.
104. Buryanov S.A. International Recognition of the Right to Freedom of Conscience and Problems of Its Implementation in the Russian Federation under Conditions of Contemporary Global Processes: Monograph. Moscow: Poligraf Servis, 2020. 624 p.
105. Civil Human Rights in Russia: Modern Problems of Theory and Practice / compiled by F.M. Rudinsky. Transaction Publishers, New Brunswick, New Jersey (U.S.A.), 2008. 458 p.
106. Buryanov S.A. State Worldview Neutrality in the Context of Deteriorating Imbalances in Globalization: The Case and Current State of Affairs in the Russian Federation from 2016 to the Beginning of 2017. Amsterdam, 2018. 228 p.

PART III. THE GLOBAL FRACTURE AND THE PLANETARY SCALE OF THE PROBLEM OF LIBERTY OF WORLDVIEW CHOICE GUARANTEED BY PUBLIC-LAW WORLDVIEW NEUTRALITY

CIVILIZATIONAL THESIS

"Where law ends, domination begins."

Globalization turns freedom of conscience into a planetary security question. In an interdependent world, worldview coercion travels across borders: it feeds intolerance, conflict, migration pressure, and geopolitical rupture. Worldview neutrality therefore belongs not only to domestic constitutional law, but to the architecture of international order.

CONCEPT BOX

CHAPTER 3

GLOBAL PROCESSES AND THE WORLDVIEW SPHERE

Why inner autonomy becomes a planetary question

Globalization has transformed freedom of conscience from a domestic legal issue into a planetary question.

Worldview conflicts, ideological pressure, discrimination, digital influence, and geopolitical confrontation now cross borders faster than traditional legal systems can respond.

Chapter 3 places liberty of worldview choice within the context of global processes, sustainable development, and civilizational risk.

The chapter argues that a stable global order requires more than security arrangements and economic coordination. It requires legal guarantees for the inner autonomy of the person.

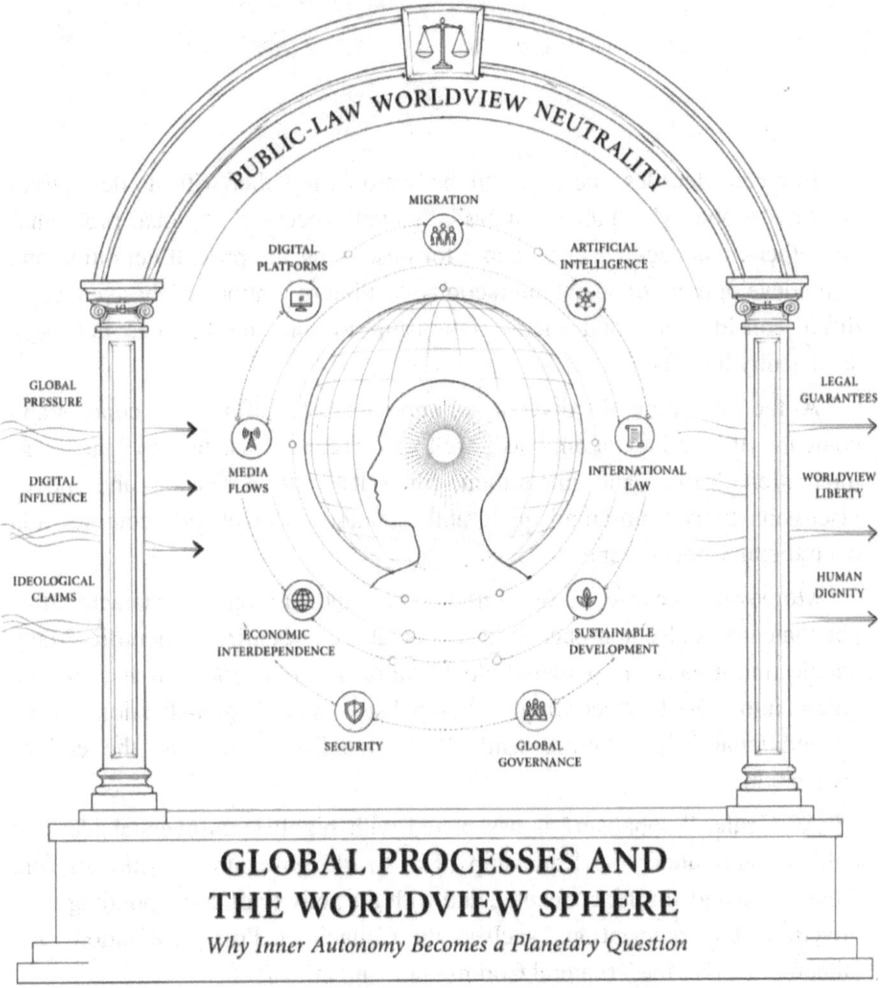

GLOBAL PROCESSES AND THE WORLDVIEW SPHERE

Why Inner Autonomy Becomes a Planetary Question

"Global order begins where the inner freedom of the person receives legal protection."

Chapter 3. The Right to Liberty of Worldview Choice in the Context of Contemporary Global Processes

3.1. The Concept, Content and Dynamics of Global Processes in the Contemporary World

In recent decades, the term "globalization", together with its derivatives such as "globality", "global studies", "global processes", "global problems" and others, has become fashionable for describing the present condition and rapid development of social interactions. Yet interpretations of these concepts differ considerably, and their relationship to one another has not been sufficiently developed.

At the same time, the diversity of approaches to defining the concept and content of globalization, as well as related categories such as internationalization and integration, often leads to confusion and moves discussion away from an adequate understanding of an already complex and contradictory phenomenon.

Moreover, because of its contradictory and ambiguous character, this phenomenon is also approached in ways that are not entirely scientific. Many people find it easier to perceive globalization without reflection and without going into detail. Accordingly, "globalists" regard globalization as an unconditional bright future, while "anti-globalists" see it as the end of civilization.

As a rule, "globalism" is associated with highly controversial ideas of uniting states under the leadership of a "world government" (mondialism, Atlanticism and similar concepts), and with the activity of corresponding non-governmental organizations, such as the Council on Foreign Relations, the Bilderberg Club, the Trilateral Commission and others.

The anti-globalist movement has embraced many organizations throughout the world, from left-wing groups to green movements. One may even say that it has itself acquired a global character. At the same time, some anti-globalists call themselves alter-globalists.

In this context, I consider the concept of "globalism" and its derivatives to characterize an attitude toward ideology and therefore to be not fully scientific, at least from the legal point of view. At the same time, the phenomena designated by these terms may become, and already have become, objects of scholarly research.

Let us consider the main scholarly approaches to defining the concept and content of globalization in the works of various researchers. On the basis of a review of the literature, one ma oncept of "globalization" was introduced into scholarly circulation by the British scholar Roland Robertson in the early 1960s. In particular, in 1985 Robertson defined globalization as a dynamic process through which a set of world factors, economic, political, informational, cultural and others, influences social relations within individual states.

Other researchers, however, link the spread of the term "globalization" with Theodore Levitt, who in 1983 published an article devoted to the globalization of markets. At the same time, according to various experts, the term "global" has been used for several centuries.

Already in the 1990s, a whole series of scholarly works by Arjun Appadurai, Ulrich Beck, Immanuel Wallerstein, Anthony Giddens, Octavio Ianni, Leslie Sklair and others appeared, each devoted to different aspects of globalization.

Ulrich Beck, in particular, understands globalization as social interactions across a wide spectrum of public life that take place above and across state borders.

According to Anthony Giddens, globalization is characterized by the intensive spread of social interactions on a planetary scale, which strengthens interconnections at both global and local levels. The researcher rightly speaks of a radical transformation of contemporary social relations, especially emphasizing the unprecedented increase in the speed of all processes, as well as the breadth of their coverage and the depth of institutional transformations.

A similar position is taken by Leslie Sklair, who regards globalization as a process of forming transnational interactions, primarily in the economic, political and cultural spheres, which take place independently of nation-states.

An important milestone in globalization studies was the work of the Japanese strategic management expert Kenichi Ohmae. In particular, in his 1990 book The Borderless World, he predicted the strengthening and inevitability of global processes in the economy, as well as their extension to other spheres of public life. In later works, Ohmae argues that the development of the global economy may lead to the disappearance of nation-states, which he calls "unnatural formations". "Nation-states are dinosaurs that live to die," Ohmae argues. Moreover, he describes globalization as an accomplished fact "that can no longer be stopped".

Among Russian researchers of the general theory of globalization, A. N. Chumakov deserves special mention. Among his many works, the most fundamental are the monographs that have gone through more than one edition: Globalization: The Contours of an Integral World, Metaphysics of

Globalization: The Cultural and Civilizational Context, and finally Global World: The Clash of Interests, first published in 2018. According to Chumakov, "globalization is a process of universalization, the formation of structures, ties and relations common to the entire planet Earth in various spheres of public life. Globalization also appears as an event and a phenomenon when it is perceived as an objective reality that manifests itself through the closure of global space, a single world economy, universal ecological interdependence, global communications and the like, and in this capacity it cannot be ignored by anyone."

According to A. I. Utkin, globalization is directed toward the merger of national economies into a single worldwide system. Its foundation, accordingly, lies in the possibility of the rapid movement of capital, the development of information technologies, swift scientific and technological progress, the free movement of goods and capital, and new forms of transport. The "international system of education" deserves particular mention.

A. D. Ursul understands globalization as "a global process of forming the unity of humanity and, at the same time, of establishing co-evolutionary relations with nature, which may be realized through a global transition to sustainable development of a noospheric orientation."

M. A. Cheshkov notes that globalization is "a process of articulating various components of humanity in the course of its evolution, as opposed to the process of differentiation of humanity." He further explains that "globalization-as-process is a concept that defines the unfolding of globality-as-structure, set by an archetype, in the diversity and richness of various connections existing between different components of the global community in its diverse spatial configurations, such as internationalization and transnationalization."

Speaking of globalization as a sociocultural shift, D. V. Ivanov identifies a wide range of phenomena: "the development of world ideologies; an intense struggle to establish a world order; an abrupt increase in the number and influence of international organizations; the weakening of the sovereignty of nation-states; the emergence and development of transnational corporations; the growth of international trade; intensive mass migrations and the formation of multicultural communities; the creation of planetary mass media and the expansion of Western culture into all regions of the world, and so on."

A. Kh. Trofimova also notes that globalization, from globe, the earthly sphere, is a process of worldwide economic, political and cultural integration and unification.

is also directed toward achieving a qualitatively new integrative condition of globality on a planetary scale.

In particular, characterizing globalization as an integrative world process embracing all spheres of public life, many researchers speak of its orientation toward the formation of a single planetary social system.

I believe that the use of the term internationalization, which implies the expansion and transformation of relevant social processes beyond nation-states, is not fully suitable for describing contemporary global processes in their final stage, and especially not in their completed condition.

This means that, at best, internationalization may be used to describe stages and conditions preceding the integrative stage. Thus, in contemporary conditions, the process of internationalization should more correctly be regarded as a component of global processes. At certain historical periods, however, when internationalization had not yet acquired planetary scale, it was premature to consider it part or stage of globalization.

In general, one may say that globalization is an integrative process directed toward the formation of a single planetary social, and later socio-techno-natural, system.

It should also be noted that, in different social spheres, these processes have proceeded and continue to proceed unevenly, which is reflected in different stages, conditions and, ultimately, levels of development of social subsystems. For example, one may speak of the dynamic development of the economic subsystem, the almost completed formation of the information subsystem, and the significant lag of the political subsystem.

This may be why researchers use different terms to describe social relations, depending on the level of development of the corresponding social subsystems. A significant point is that processes of internationalization and integration may proceed not only on the scale of the planet, but also within its separate parts, such as regions and subregions. This does not mean, however, that they will necessarily be directed toward the formation of planetary integrity in the interests of all humanity.

In reality, such processes are usually aimed at strengthening cooperation, and much less often at merging states on a regional scale, and they are carried out mainly in the interests of the corresponding "elites". Moreover, they are often accompanied by the intensification of ethno-confessional contradictions and by the formation of other barriers between people, which deeply contradict the processes of forming a planetary social system.

political institutions, ideas, principles, links, moral-political, material and other values, and diverse relations existing in the world."

It should be especially emphasized that Marchenko speaks of globalization as a "process, an integration tendency, and not as globalization as some static phenomenon appearing in the form of a 'certain civilizational standard'." At the same time, he notes that "the institutional side must not be ignored, since globalization affects the state and law not only through its functional side, namely through a system of factors of an integrative character, but also through its institutional side, namely through a system of institutions generated and nourished by globalization."

O. M. Meshcheryakova, substantiating a philosophical and legal approach to the globalization of international law, notes that "globalization means a period in the development of human civilization conditioned by awareness of planetary problems that affect the foundations of the existence of all humanity." The researcher proposes, as a promising scholarly direction, "the theory of legal global studies, which has emerged at the intersection of the general theory of state and law, international law, political science, sociology and economics." Her proposal appears scientifically grounded and timely: "taking into account new trends in the development of contemporary international law, modern models of legal regulation of those processes caused by globalization should be developed. No less important is the scientific modelling of Russia's participation in integration processes at various levels."

According to K. A. Bekyashev, in recent years there has been an intensification of interconnection and mutual influence in various spheres, including international relations. "Globalization, from the Latin globus, sphere, is the process of convergence and growth of interconnection among the nations and states of the world, accompanied by the development of common political, economic, legal, cultural and value standards."

According to A. A. Moiseev, "the benefits of globalization come down to faster economic growth under a market economy that has demonstrated its productive power and potential for increasing economic efficiency, a higher standard of living, the accelerated introduction and dissemination of technical achievements, and new economic opportunities both for individuals and for states." Proceeding from his understanding of globalization, the researcher believes that "its spontaneous character cannot be managed in the traditional sense of international law; however, the process of globalization can and must be directed along a course beneficial to the entire international community by granting certain powers to intergovernmental organizations, including supranational instruments."

M. M. Zanibekov speaks of the transformation of previously independent elements into a single system, as well as of the integration of states and societies in all existing spheres. "Globalization is a process of worldwide economic, political, cultural and legal integration and unification." A. B. Zelentsov also notes the influence of globalization on public law and on the convergence of legal systems, including administrative ones.

A. A. Mamedov likewise speaks of the growing influence of international law, within whose jurisdiction, under conditions of globalization, more and more issues previously regulated by states are gradually falling. "Under globalization, it is precisely the principles of law that play an increasingly important and active role in the system of regulatory means in the sphere of the economy, including banking activity." The researcher points to tendencies toward the removal of legal barriers to international cooperation and the formation of a global economic space. "The modern era is the era of globalization, which includes the highly dynamic development of integration processes in the economic sphere."

As I. Z. Farkhutdinov notes, "today the formation of unified world financial and information systems and centers for regulating the world economy is being completed. As a result of the activity of tens of thousands of transnational corporations, or multinational enterprises, and many international financial organizations, a kind of superstructure has emerged which, controlling more than half of all world resources, operates according to principles different from those traditional for national state bodies." At the same time, according to the researcher, globalization and the regionalization of social relations not only influence but also complement each other. "The principle of open regionalism is not limited to deepening cooperation only within existing regional and subregional groupings; it also presupposes the expansion of mutually beneficial cooperation between different integration unions. This is why attempts to oppose the processes of globalization and regionalization to one another are absolutely groundless."

A. N. Korshunov believes that the process of planetary integration requires the formation of global law. "Globalization is a multifactorial evolutionary-synergetic process of creating economic, financial, communication and information networks that permeate the entire space of the Earth and integrate civilization into an integral system, which presupposes the creation of a single legal field, a set of generally significant and generally binding norms of conduct." The researcher also believes that "the processes of globalization of sociocultural systems taking place in the world inevitably determine integrative evolutionary-synergetic transformations of the legal systems of individual states and of international law in the direction of creating global law. These transformations are based on mechanisms adequate to the emerging new world order: both objectively functioning mechanisms and those subjectively created by people on the basis of the existing legal culture of humanity and taking into account scientific and philosophical concepts of law and legislation. Therefore, on the basis of philosophical reflection on global-legal problems, carried out within the framework of the contemporary strategy of cognition, the idea of global law can be developed, and in its principal aspects it will rely on evolutionary and synergetic approaches."

M. V. Nemytyna rightly notes the need to overcome the limitations of national legal science under conditions of global processes. "The national character of science, its confinement within the sociocultural, geographical and political-legal space of one country, predetermines the extreme limitation of the research potential for studying the realities of modern civilization, the objectivc tendencies toward globalization and integration that dominate it, and information technologies, which themselves became possible thanks to the rapid growth of scientific knowledge of a supranational character." It is especially emphasized that "those areas of knowledge called upon to explain the processes and phenomena existing in society receive an impulse for development where and when there is freedom of thought and speech, freedom of expression, freedom of creativity, including scientific creativity. In a free society, jurisprudence, like related areas of scientific knowledge, develops at the conceptual level, at the level of doctrines, offering society and the state alternative scenarios for their development."

According to the original view of A. N. Kostenko, the globalization of jurisprudence is a natural process consisting in the deepening coordination of national jurisprudences with their "common denominator", namely the natural laws of people's social life and, in particular, the laws of natural law. Accordingly, the author proposes the methodological principle of "social naturalism" as an instrument for approaching this "common denominator". As the researcher notes, the principle of "social naturalism" "flows from the idea of the natural integrity of the world, according to which everything that exists in the world must exist according to the laws of Nature. Nature is the Highest Legislator for all that exists. Therefore, the dualistic formula 'nature and society' is a dramatic error, because it creates the dangerous illusion that society lies outside nature and thus denies that society is also nature, though a different kind of nature from the physical or biological."

According to A. I. Ekimov, today global integration processes have embraced practically all spheres of social life, making the idea of network law viable, an idea that "fully corresponds to the possibilities of convergence of legal systems."

A number of international documents address the impact of global processes on social problems. For example, the ILO Declaration on Social Justice for a Fair Globalization of 13 June 2008 also characterizes globalization as a process connected with "the spread of new technologies, the exchange of ideas, goods and services, the expansion of capital and financial flows, the internationalization of business and entrepreneurial activity and dialogue, as well as the movement of persons, especially working women and men."

According to the authors of that document, however, the process of globalization appears highly contradictory. Both positive and negative aspects are noted. Among the positive aspects, the document points to economic cooperation and integration, which help a number of countries "benefit from high rates of economic growth and expanding employment, draw many representatives of the rural poor into the modern urban economy, continue to make progress toward development goals and accelerate innovation in the development of new products and the exchange of ideas."

The negative aspects of global economic integration processes are connected with the fact that they force "many countries and sectors to face serious problems associated with income inequality, chronically high levels of unemployment and poverty, the vulnerability of economies to external shocks, and the growth of both unprotected employment and the informal economy, which affects the employment relationship and the sphere of protection it provides."

Thus, among the numerous approaches presented in the scholarly literature, the most scientifically grounded appear to be those that consider globalization of the social system as a dynamic process that implies not only changes of state, but is also directed toward achieving a qualitatively new integrative condition of globality on a planetary scale.

In particular, characterizing globalization as an integrative world process embracing all spheres of public life, many researchers speak of its orientation toward the formation of a single planetary social system.

I believe that the use of the term internationalization, which implies the expansion and transformation of relevant social processes beyond nation-states, is not fully suitable for describing contemporary global processes in their final stage, and especially not in their completed condition.

This means that, at best, internationalization may be used to describe stages and conditions preceding the integrative stage. Thus, in contemporary conditions, the process of internationalization should more correctly be regarded as a component of global processes. At certain historical periods, however, when internationalization had not yet acquired planetary scale, it was premature to consider it part or stage of globalization.

In general, one may say that globalization is an integrative process directed toward the formation of a single planetary social, and later socio-techno-natural, system.

It should also be noted that, in different social spheres, these processes have proceeded and continue to proceed unevenly, which is reflected in different stages, conditions and, ultimately, levels of development of social subsystems. For example, one may speak of the dynamic development of the economic subsystem, the almost completed formation of the information subsystem, and the significant lag of the political subsystem.

This may be why researchers use different terms to describe social relations, depending on the level of development of the corresponding social subsystems. A significant point is that processes of internationalization and integration may proceed not only on the scale of the planet, but also within its separate parts, such as regions and subregions. This does not mean, however, that they will necessarily be directed toward the formation of planetary integrity in the interests of all humanity.

In reality, such processes are usually aimed at strengthening cooperation, and much less often at merging states on a regional scale, and they are carried out mainly in the interests of the corresponding "elites". Moreover, they are often accompanied by the intensification of ethno-confessional contradictions and by the formation of other barriers between people, which deeply contradict the processes of forming a planetary social system.

One attempt to explain the contradictory character of the development of contemporary social processes in some regions of the planet is the theory of glocalization, which is usually, and quite controversially, understood as a certain regional scenario of globalization.

In this context, I believe that regional integration processes are primarily a consequence of the imbalance in the formation of the global social system in general, and of the underdevelopment of its political subsystem in particular.

One of the most significant defects of contemporary world political development is the dominance of the concept of the "right of force" in international relations, represented in different historical periods by one, two or several poles of power.

Ultimately, the final integrative stage of globalization can and should be called the state of globality, both in relation to the planetary social system as a whole and to its constituent elements, or subsystems, in different spheres of social relations.

I believe that the state of globality of the planetary social system, alongside the term integration, is most fully described by the terms interpenetration, interdependence and openness. M. A. Cheshkov speaks of this in a scientifically grounded way when he assigns the term "globality" the role of a key, code concept. "It means a broad set of processes and structures correlated with phenomena of interdependence, interpenetration and mutual conditioning within planetary frameworks, the result of which is the emerging integrity of humanity's being."

In this context one may also agree with the view that "the concept of 'globality' should be clearly separated from the concept of 'the evolution of humanity'; the former must not be dissolved into the latter, since globality characterizes only one aspect of humanity's evolution, namely interconnectedness and mutual correlation." This means that, if completed, globalization processes may lead to the formation of a qualitatively new state of globality, characterizing a radical evolution of contemporary social relations.

At present, one can speak not of the completed formation of a single planetary social system, but only of the formation of the subsystems that compose it, with different levels of development. I assume that this uneven level of development among social subsystems is one of the main deep causes of instability and the global challenges facing human civilization.

3.2. Global Challenges in the International-Legal Sphere as a Consequence of the Uneven Development of Global Processes and the Problem of the Transition to Sustainable Development of the International Community and Its Participants

Among the first researchers of global challenges, one should mention Theodore Levitt, John McLean, Roland Robertson, Donella Meadows and others. The contribution of V. I. Vernadsky, N. D. Kondratiev, N. N. Moiseev, K. E. Tsiolkovsky and other Russian researchers should also be noted.

Global challenges were studied within the framework of the public analytical center known as the Club of Rome, which already in its first report, The Limits to Growth, drew the attention of the world public to planetary challenges. Other, likewise far from optimistic, reports to the Club of Rome also speak of the growth of global threats to civilization. Among the many causes of global problems, the conclusion that the socio-political sphere of social relations is lagging behind is especially noteworthy.

The first and perhaps best-known report to the Club of Rome, The Limits to Growth, was prepared in 1972 under the direction of D. Meadows on the basis of the economic-mathematical models of J. Forrester at the Massachusetts Institute of Technology in the United States. In particular, a forecast of world development was prepared, and it proved catastrophic: the complete exhaustion of raw materials, with all ensuing socio-economic and other consequences. As a recipe for saving civilization, the highly controversial model of "zero growth" was proposed, which in fact implied a renunciation of development.

In the next report, Mankind at the Turning Point, prepared in 1974 by M. Mesarovic and E. Pestel, a broader range of indicators of world development was analyzed, leading to the conclusion that global regulation in the economic sphere was necessary.

In the 1975 report RIO: Reshaping the International Order, prepared by Jan Tinbergen, the socio-political aspects of global problems were examined and conclusions were drawn about the need for a new world order and very serious reforms, above all at the international level. I believe that the most radical proposal was connected with revising the concept of national sovereignty.

Other, likewise not entirely optimistic, reports to the Club of Rome, examining different spheres of social relations as well as interactions between society and nature, also speak of the growth of global threats to civilization. Among the numerous causes of global problems, attention is again drawn to the lag of the socio-political sphere of social relations.

Special mention should be made of the 1990 report on behalf of the Club of Rome, edited by A. King and B. Schneider, which in effect served as a report on twenty-five years of the organization's work.

The authors state that the ongoing global revolution, having no ideological foundation, "is being shaped by geostrategic upheavals, social, technological, cultural and ethnic factors, the combination of which leads into the unknown. ... Today we are increasingly aware that humanity, pursuing material profit from the exploitation of nature, is rapidly approaching the destruction of the planet. The ever-present danger of nuclear destruction, though now reduced, as well as the possibility of irreversible climate change and its consequences, represent an immediate threat. Such components of the contemporary problematique are global in character, and even major powers cannot cope with them alone. We therefore call for international solidarity. ... Some problems have already acquired such proportions that their successful resolution is impossible. If we do not begin to act immediately, it will be too late."

The report speaks of changes and problems of world scale that have occurred in recent decades: "inequality between rich and poor; unequal access to education and information; religious, national and, in some countries, age discrimination; social injustice; the mismatch between rights and duties, privileges and responsibility; contradictions between discipline and arbitrariness; the mismatch between economic growth and living standards; public dissatisfaction in an impersonal welfare state; and the divergence between material and spiritual needs."

In addition, the report named formidable dangers that traditional management structures cannot cope with: unbalanced economic growth; loss of governability; shortages of food and resources; environmental pollution; energy crises; demographic explosion; increased migration; and changes in the global geopolitical situation.

From the point of view of contemporary Russian scholarship, global problems are considered to be generated by the increased impact of human activity on the environment. Most researchers believe that the survival of civilization depends on the solution of global problems.

For example, according to the authors of an encyclopedic dictionary, global problems are "a set of vitally important problems affecting humanity as a whole and incapable of being resolved within individual states or even geographical regions." In particular, these include "preventing thermonuclear war; reducing rapid population growth, the 'demographic explosion' in developing countries; preventing catastrophic environmental pollution, above all of the atmosphere and the World Ocean; ensuring further economic development with necessary natural resources, especially non-renewable ones; overcoming the gap in living standards between developed and developing countries; eliminating hunger, poverty and illiteracy, and others."

According to the well-known scholar I. I. Lukashuk, "global problems represent a special type of social problems characterized by planetary scale." As the main problems, he identifies peace and security, the spread of nuclear, biological and chemical weapons, environmental protection, depletion of natural resources, health care, science, culture, the fight against organized transnational crime, and population issues.

S. A. Nizhnikov distinguishes between global and universal problems. "Failure to solve global problems leads humanity to inevitable destruction, while universal problems are those that are widespread and may grow into global ones. Universal problems may include health care, education, social protection and so on. For example, most people in the world today die not at the hands of terrorists and not from AIDS or drug addiction, but from cardiovascular diseases."

Speaking of global dangers that threaten the existence of civilization, I. M. Podzigun notes the need to search for new models based on a unified human approach, as well as for a new relationship between national and global interests, taking into account the interests of all countries.

A. B. Weber speaks of the need to ensure the governability of the contemporary global world and to overcome the lag of "political globalization" behind economic globalization. According to the researcher, it is critically important "to give international cooperation a new quality, adequate to the changed conditions of humanity's existence, to the new conditions for ensuring international security in all its dimensions, and to the new level of interdependence among countries and peoples."

According to the authors of a collective work, "globalization with a human face, if we speak of an alternative, requires a new political structure of the world, adequate to the character and scale of the problems facing humanity today. It will come into being if it is based not on the domination of one or several powers, although the roles of different countries may vary, but on the principles of cooperation and solidarity." It is also emphasized that one should distinguish "globalization as an objective phenomenon, conditioned above all by the technological revolution in informatics and telecommunications, from the policy of neoliberal globalism, which allows the United States and other states of the financial 'Group of Seven' to direct this process primarily in their own interests."

As A. A. Nemchuk rightly notes, globalization, as a natural process of world social development, requires an adequate political response from nation-states to deepening economic interdependence. One must agree that this response presupposes expanding the number of formats of international cooperation and improving coordination of their activity. Among the obstacles to the formation of a stable world order, the researcher names asymmetry of development, which requires "the development of regional equalization programs that can be implemented only under the control of the international community."

As T. A. Yashkova rightly believes, solving global problems, economic, ecological and others, requires the combined efforts of the states of the world and their citizens, not on the basis of the dictate of some and the subordination of others, but on equal and parity-based foundations, which will make it possible to avoid many contradictions and conflicts.

V. V. Lazarev and A. Kh. Saidov rightly believe that globalization has a substantial influence on all aspects of law. "This multidimensionality means, in particular, that most contemporary global problems have a legal aspect and, to a significant extent, appear as legal problems."

anniversary of the book, and that it was interesting to learn by what percentage the carrying capacity of the planet had now been exceeded. "However, Professor D. Meadows informed me in personal correspondence, and allowed me to make this information public, that there would be no new edition, one reason being that the emerging picture is too gloomy and no acceptable solution is visible."

Thus, in the researcher's view, the level of global threats has increased radically and continues to accelerate on the way to a "point of no return". "A series of independent calculations has shown that global evolution is entering a mode of unprecedentedly steep acceleration, which should reach a mathematical limit, a 'singularity point', around the middle of the twenty-first century. On the horizon there is emerging a phase transition of such scale and significance that neither humanity nor the biosphere has experienced anything like it in previous history."

In this context, the urgency of the search for scientifically grounded solutions to global problems is beyond doubt. The solution of global problems determines "social progress and the preservation of civilization: the prevention of world thermonuclear war and the securing of peaceful conditions for the development of all peoples; the prevention of catastrophic environmental pollution, including the atmosphere, the World Ocean and so on; the overcoming of the growing gap in economic level and per capita income between developed and developing countries by eliminating the backwardness of the latter, as well as eliminating hunger, poverty and illiteracy on the globe; ensuring the further economic development of humanity with necessary natural resources, both renewable and non-renewable, including food, industrial raw materials and energy sources; stopping rapid population growth, the 'demographic explosion' in developing countries, and eliminating the danger of 'depopulation' in developed countries; and preventing the negative consequences of the scientific and technological revolution."

At the same time, as E. G. Kochetov rightly notes, contemporary problems are inseparably connected with the "global transformation of the world around us". The ambiguity and contradictory character of global processes open enormous positive opportunities and at the same time create new threats. "On the agenda is the earliest possible comprehension of the roots and origins of global processes and, most importantly, the development of strategic guidelines for development under new conditions."

Thus, global problems should include those that correspond to at least one of the following criteria: first, they affect the interests of all humanity and its future; second, they acquire a worldwide character, becoming objective factors of the development of society on a planetary scale; third, their unresolved state creates a threat to humanity; and fourth, they cannot be resolved within local or regional frameworks and require the united efforts of all humanity for their resolution.

It should be noted that the scholarly literature describes more than two thousand global problems, but most scholars agree that their unresolved character threatens the very existence of human civilization. In this context, I. I. Lukashuk is undoubtedly right when, as one of the first Russian jurists, he notes that “for the first time in history the question of the survival of humanity has arisen.”

Thus, according to the well-known scholar, “the peculiarity of the named problems is that they are global and can be solved only by the joint efforts of states. The international community must become a community for the salvation of humanity. This requires a higher level of governance of social processes, both nationally and globally. The substantial improvement of governance has come to the forefront.”

I believe that the search for solutions to global problems, including approaches to forming an adequate system for governing global processes, should be based on an understanding of their causes.

Here one may agree with V. V. Volkov and other researchers who consider global problems to have been generated by the uneven development of world civilization. In particular, they note the advanced development of technical power compared with social organization, the lag of political thinking, problems of moral values and the uneven development of states.

However, defining the socio-economic sphere as the key one appears debatable. In particular, A. B. Weber rightly identifies the lag of “political globalization” as the main cause of the “new world disorder” and speaks of the need to overcome it.

Thus, among the complex of causes of global problems, special attention should be given to the significant lag in the development of global processes in the political sphere of social relations, which manifests itself, among other things, in the military activity of states fraught with the risk of unleashing a world nuclear war. Closely connected with the political sphere are the legal and educational spheres, which are also lagging behind.

In this context, I believe that the intensification of global challenges facing the world community, the uneven socio-economic development of different countries and regions, the injustice of world economic exchange, and threats in the ecological sphere are, to a significant extent, consequences of the lag in the political, legal and educational spheres of social relations. To some extent, these problems have found reflection in documents of universal, regional and subregional organizations. However, the continuation of negative tendencies makes the search for effective paths toward sustainable and governable development of the international community and its participants especially urgent.

3.3. The Significance of the Right to Liberty of Worldview Choice and the Principle of Public-Law Worldview Neutrality for Overcoming Global Challenges and Achieving the Sustainable Development of Civilization

Sustainable development is considered an alternative to the unresolved state of global problems and the possible destruction of civilization. For example, according to D. I. Romasevich, the model of sustainable global development is understood as supporting, long-term, continuous and protected development. "Such a model may be defined as a strategy of socio-natural global development directed toward ensuring the survival and continuous progress of society while not destroying the surrounding natural environment, especially the biosphere."

It is generally considered that, at the international-legal level, the term "sustainable development" was first used in the 1987 report of the World Commission on Environment and Development, Our Common Future.

However, awareness of the importance of development problems was reflected in earlier documents of intergovernmental and non-governmental organizations as well.

In particular, development issues were addressed in such United Nations documents as the Declaration on Social Progress and Development (1969), the Declaration on the Right to Development (1986), the Rio Declaration on Environment and Development (1992), the Barbados Declaration (1996), the United Nations Millennium Declaration (2000), the Doha Declaration on Financing for Development: Outcome Document of the Follow-up International Conference on Financing for Development to Review the Implementation of the Monterrey Consensus (2008), the Declaration of the High-level Meeting of the General Assembly at its sixty-third session on the midterm review of the Almaty Programme of Action (2008), the Outcome Document of the Conference on the World Financial and Economic Crisis and Its Impact on Development (2009), the Declaration of the High-level Dialogue on International Migration and Development (2013), and others.

Finally, in September 2015, at the seventieth session of the United Nations General Assembly devoted to sustainable development, the final global development agenda for the period after 2015 was approved. The new agenda presupposes the achievement of 17 goals and 169 targets.

In addition, the concept of sustainable development was developed by the Club of Rome, founded in 1968, and by other public organizations.

Before speaking about approaches to finding solutions to global problems that threaten the very existence of civilization, it is necessary to consider one more highly debatable question. It concerns clarification of the objective or subjective character of global processes. The very possibility of human activity influencing global processes in order to solve global problems and move toward the sustainable development of civilization depends on the answer to this question.

In particular, I. S. Khorin and many other researchers characterize globalization as "a long-term, mainly objective process of the formation and deepening of ties between countries and peoples."

At the same time, one may agree with the thesis that the role of the subjective factor in contemporary global processes has qualitatively increased. This refers to purposeful practical activity, a conscious policy of further developing the ties mentioned above, as well as "deepening integration processes in the interests of certain social strata, a group of states or one state."

Thus, the initially objective processes of globalization of social relations gradually become susceptible to the influence of a set of subjective factors, including political, scientific and educational activity.

In this context, let us consider several principal scholarly approaches directed toward finding solutions to global problems and moving toward the sustainable development of civilization.

Speaking of global dangers that threaten the existence of civilization, I. M. Podzigun notes the need to search for new models based on a unified human approach, as well as for a new relationship between national and global interests, taking into account the interests of all countries.

A. B. Weber speaks of the need to ensure the governability of the contemporary global world and to overcome the lag of "political globalization" behind economic globalization. According to the researcher, it is critically important "to give international cooperation a new quality, adequate to the changed conditions of humanity's existence, to the new conditions for ensuring international security in all its dimensions, and to the new level of interdependence among countries and peoples."

According to the authors of a collective work, "globalization with a human face, if we speak of an alternative, requires a new political structure of the

In this context, one cannot but agree with I. I. Lukashuk, who speaks of the unfounded nature of the idea of a world state, since "the governance of social processes becomes an increasingly complex matter as the object of governance itself becomes more complex." As a solution, he proposes, on the one hand, "expanding the powers of territorial subdivisions of the state in the international field, which makes it possible more fully to take into account their particular interests and thereby not only raise the level of governability but also weaken centrifugal tendencies," and on the other hand, "deepening the international interaction of states, leading to the growing role and expanded powers of international organizations." In this context, increasing the effectiveness of the United Nations and its specialized agencies is extremely necessary. This is referred to, in particular, in the United Nations Millennium Declaration of 2000.

The ideas of supporters of a World Confederation are reflected in a collective study from 2009, whose authors believe that the global crises of the early twenty-first century show the danger of the further disunity of humanity and stimulate the search for new radical solutions. In particular, it is noted that "the growing tendency of geopolitical integration and globalization will be a movement toward the creation of a World Confederation of States and Civilizations, representing the interests of all humanity, organized on democratic foundations and possessing sufficient powers and resources to realize these interests in practice by responding to the challenges of the twenty-first century. ... The global crises of the early twenty-first century show the danger of the further disunity of humanity and stimulate the search for new radical solutions."

V. V. Volkov believes that the "emergence of a confederation of civilizations with sustainable development" is the most adequate variant of global governance only at the first stage. Only later, "through mutual perception and exchange of values, will the formation of a single planetary civilization be possible."

I believe that approaches implying not merely the preservation but even the increase of an enormous concentration of power on a planetary scale do not correspond to the objective tendencies of the development of social relations and are initially without prospects under conditions of their further complication.

Approaches directed toward limiting power appear to correspond more closely to the tasks of forming adequate global governance.

For example, A. M. Slaughter analyzes the processes of transformation of the world system, states the need to limit state sovereignty and speaks of the need for a system of common law. In her work, however, the emphasis is not on redistributing state sovereignty in favor of other actors, but on breaking it down, well as the proper investigation of such violations and appropriate punishment for them through the use of regional or international mechanisms in accordance with the norms of international law."

Special mention should be made of contemporary scholarly works devoted to substantiating the need for global governance as an instrument for solving global problems and moving toward the sustainable development of civilization.

In scholarly circles, the concept of global governance is usually associated with the work of the United Nations Commission on Global Governance and Cooperation. In particular, in 1995 the report Our Global Neighbourhood was prepared.

A number of researchers believe that global problems have acquired universal human significance and require the formation of a global governance system. In particular, I. V. Ilyin notes that "the preservation of modern civilization and the survival of humanity ultimately depend on solving the problem of governing globalization processes, on the choice of strategy and tactics of global governance, and on the means and methods applied."

R. E. Malayan emphasizes the multifactorial and interdependent character of global problems and raises the issue of setting priorities for international cooperation. "This raises the research question of the necessary structural prerequisites for a transition to effective global governance and of its preferable character compared with anarchy." Thus, one may agree with the researcher who speaks of objective prerequisites that conditioned "not merely the need, but also the political possibility of governing international processes."

Alongside the not always obvious necessity of global governance, there are also a number of terminological problems. For example, V. B. Pavlenko believes that "global governance is the process of the gradual formation of a system of supranational and global centers of power and administration, the ultimate goal of which is the transformation of the existing international order into a 'New World Order'." V. M. Davydov considers the term global regulation preferable and defines it "as the ability to control global processes, to correct their trajectories, using hard and soft power, international law and international institutions."

According to O. N. Barabanov, "since globalization thus begins to permeate more and more spheres of human life, the question of its institutionalization, of forming a system of global regulation endowed with an appropriate scope of powers and legitimacy, becomes increasingly significant."

Further, speaking of the forms of global regulation, the author believes that "the main ones are global cooperation, in which existing sovereign states will play the decisive role in conducting coordinated worldwide policy, and global governance, in which supranational international organizations would be autonomous from states in decision-making."

As E. V. Stetsko notes, "governance" is understood, above all, as a function of a certain process of global development, globalization. "A system of 'global governance' presupposes flexibility, changeability and interchangeability of its structures, rapid response to changes in processes and in the behavior of actors. However, the measure of responsibility in 'global governance' is not fixed by law. The success of such governance is determined by efficiency, benefit, growth in accessible social goods, or by the absence of these."

A. N. Chumakov distinguishes between regulation and governance and defines them as follows. By regulation, from the Latin regulo, to arrange, adjust, put in order, he understands "a spontaneous process or deliberate actions directed toward ensuring the functioning of a system within parameters set naturally or artificially," with the aim of solving the task of "optimal functioning of the system, when the most favorable conditions are created for the interaction of the various constituent parts of that system."

Governance, accordingly, is understood as a higher type of regulation, always connected with the conscious activity of people, "based on goal-setting, feedback and a creative beginning. In other words, governance is carried out only consciously and purposefully and presupposes both the achievement of a certain result and the search for the most optimal ways of attaining the goal." One may agree with the view that "governance in general, and global governance in particular, cannot arise accidentally or merely by virtue of the natural course of events, naturally. It exists only in society and is built according to a certain plan, obeying a certain logic that sets the specific parameters of such governance. Here, unlike regulation, there is always a subject of governance that sets certain goals and ensures their achievement through active actions and corresponding decisions."

In the context indicated above, I. V. Ilyin and M. A. Kaverin rightly believe that mechanisms of global regulation are currently operating as elements of a system of global governance that may appear in the future. "The formation of an integral system of international organizations under the influence of globalization processes means a transition from international relations to global relations and an increase in the complexity of the global political system."

I believe that this approach is scientifically grounded and corresponds to the realities of contemporary social relations. Separate contemporary mechanisms state, but still remain within the framework of state policy. Another matter is that this does not abolish the freedom and influence of non-state actors."

L. V. Smorgunov analyzes various approaches to the problem of global governance, including neo-feudalism, cosmopolitan hierarchy and democratic governance, while also noting the dynamism and unpredictability of the modern world. "The open perspective and variability of possible results, constructed in the process by which the international community searches for ways of interaction, may generate something else that does not fit into the contemporary political language."

Nevertheless, posing the question of further ways of distributing governing powers among actors in the emerging political structure of the world, M. M. Lebedeva suggests two possible scenarios. The first path presupposes a spontaneous transfer of powers from states to other actors with consequences that are difficult to predict. The second path, on the contrary, implies that states, together with other actors, build a new architecture of the world "taking into account new realities and the interests of different participants: states, intergovernmental organizations, non-governmental associations, major financial and business structures and so on."

At present, humanity is to a significant extent following the path of spontaneous development of international relations, accompanied by increased instability and global problems.

I believe that the transition to governability of contemporary processes requires the formation and adoption of principles and norms corresponding to radically changed, and still rapidly changing, social relations. O. N. Barabanov rightly believes that "global governance as such will not be possible without a significant transformation of the entire international-legal regime, in which the emphasis in this case should be placed not on the sovereignty of states, but on their global responsibility and, moreover, on their subordination to institutions of global governance."

I believe that the formation of an adequate system of global governance for sustainable global development must be based on a corresponding system of legal norms and institutions, formed through substantial reform of the system of international law in interaction with domestic legal systems.

Moreover, the development and adoption of rules of conduct for global society is fundamental in the process of forming a global governance system, which brings the problem of global law to the forefront.

Since global processes are systemic in character, "world civilization requires the creation of a single legal environment of life activity, which presupposes the

Some contemporary researchers also propose, in their works, the formation of a global response in the form of a global governance system. Many of the works mentioned contain, to a significant extent, both new and adequate ideas. Yet the new is not always adequate, and the adequate is not always new.

Among the best-known approaches to forming a system for governing global processes are the world state or government, a global management system based on the United Nations and intergovernmental organizations, and variants of such a system involving transnational corporations and non-governmental organizations. Less well-known are concepts provisionally called "governance without government", including those based on network structures of civil society.

For example, the ideas of supporters of a world state are reflected in the monograph of A. Yu. Lykov, who proposes implementing a project to unite the United States, the European Union and Russia within a single economic and legal space. Among other things, the author also examines highly debatable approaches to reforming international and domestic law.

As a means of integrating the world legal system, he no less controversially proposes the adoption of generally binding codified normative acts. True, before this he rather inconsistently stipulates the need to prepare certain "model laws" that will be implemented in national legislation. Only after that, according to Lykov, should one "move to negotiations on the development of generally binding international acts, which will subsequently become the basis of the legislation of a world state."

In addition, this author proposes the following stages for integrating a unified legal system: "the stage of transition of common-law states into the Romano-Germanic legal family; the stage of international preparation of model laws for all branches of law and their implementation; the stage of entry into force of international codes; and the formation of a unified legal system." At the same time, he says nothing about how the integration of the legal systems of states belonging to the religious legal family would take place.

In this context, one cannot but agree with I. I. Lukashuk, who speaks of the unfounded nature of the idea of a world state, since "the governance of social processes becomes an increasingly complex matter as the object of governance itself becomes more complex." As a solution, he proposes, on the one hand, "expanding the powers of territorial subdivisions of the state in the international field, which makes it possible more fully to take into account their particular interests and thereby not only raise the level of governability but also weaken centrifugal tendencies," and on the other hand, "deepening the international interaction of states, leading to the growing role and expanded the development of legal education. Special emphasis is placed on the study of international-legal problems, since "legal global studies may contribute to strengthening mutual understanding among peoples and to creating more effective legal relations formed in the global world."

Above all, it is necessary to give a modern definition of the concept of international law that corresponds to contemporary social relations. Since contemporary legal understanding is inseparably connected with human rights, and contemporary social relations are connected with global processes and problems, this must be reflected in the definition of international law.

In this context, international law is a system of principles and norms in the sphere of interactions among states and other subjects, which must be directed toward the realization of human rights, thereby contributing to solving global problems and moving toward the sustainable development of human civilization.

As the main promising directions of the progressive development of international law for the transition from the spontaneous development of instability in international relations to the governability of contemporary processes and the sustainable development of human civilization, implying the development of its theoretical-legal, normative and institutional foundations, the following may be identified: human rights; freedom of conscience as the possibility of worldview choice; tolerance as equality and mutual respect; worldview neutrality of subjects of international law; reform and development of new principles of international law as its fundamental beginnings; the supremacy of international law on the basis of the principle of legal certainty; universalization of the principles and norms of international law; a non-polar system of international relations, that is, rejection of the "right of force" and transition to the "force of law"; and the increased role of the United Nations and other international organizations.

It should be especially emphasized that the realization of human rights is a necessary condition for reducing and overcoming the gap between power and society, for the evolutionary rather than revolutionary limitation of the power of some people over others, and ultimately for embodying human freedom as an objective requirement of contemporary globalizing social relations.

Human rights must be placed at the foundation of the progressive development of international law, which is called upon to form the legal basis for the transition from the spontaneous development of instability in international relations to the governability of contemporary processes and the sustainable development of human civilization.

governance to a polyhierarchical system, and then to a non-hierarchical one based on horizontally coordinated network governance.

I believe that without effective normative support this concept is not fully viable. Moreover, there is always a risk that horizontal networks will be subordinated to a vertical structure, especially in the context of the human factor problem.

A qualitatively new system for governing global processes on the basis of strengthening the influence of international organizations, including non-governmental ones, appears more promising, but only on the condition that their normative and institutional foundations evolve.

In particular, according to James Rosenau, the formation of a new world order is interconnected with a transition to a non-hierarchical system of governing world processes on the basis of increasing the role and influence of non-state actors, conventionally called "governance without government". Governance without government presupposes the existence of more flexible and effective regulators that do not possess governmental powers but operate on the basis of consensus. The researcher notes a certain decrease in the dominance of states in world politics, which allows one to speak of systemic changes.

In this context, I. V. Ilyin rightly speaks of the crisis of the unipolar world and of the need to expand the circle of actors in globalization processes, to move toward a multipolar world order and to transform geopolitics. "Geopolitics oriented toward sustainable development goals ceases to be the preserve of individual states and gradually becomes global geopolitics. The transition to a polycentric world compatible with the goals of this type of development, fair for all countries of the planet, is the common cause of all humanity. This process will have to proceed through the formation of a consensual world order."

R. S. Zuikov, in his study of the tendencies of transformation of the world system at the end of the twentieth and the beginning of the twenty-first centuries, also believes that "the likely direction of its further development will be the deepening of processes of consolidation into an integral world social organism, integrating the social communities that compose it into various spheres of world relations constituting this system."

Analyzing the main contemporary approaches to global governance, namely "world government", transformation of the UN system, and political and corporate global governance, O. N. Barabanov supplements this typology with global cooperation. "Within its framework, solutions to global problems will be adopted not through approaches imposed by individual actors, but through constructive dialogue among all interested forces that is put into practice. Such a form also presupposes the formation of a more inclusive system of global regulation, in which as many states and other actors as possible could feel involved."

A. S. Smbatyan believes that the highest point in the development of tendencies toward universalization of international-legal norms and institutions on the basis of hierarchical principles has already passed, and that new approaches must be sought. "The now notorious assertion that the process of globalization and the strengthening of interdependence among states lead to an increased role of law in regulating international relations through the development of universal, generally binding rules of conduct and the strengthening of the international legal order, under current conditions appears, at best, to be a simplified interpretation bordering on distortion of the essence of ongoing processes. On the one hand, there is indeed an objective strengthening, due to the dialectic of historical processes of world reorganization, of the interdependence of states, which leads to the development of generally binding rules for regulating international relations. On the other hand, it is no less obvious that these processes are taking place not so much at the abstract global level as at the regional and subregional levels." The researcher then notes the growing complexity of the structure of the international legal order and speaks of strengthening tendencies toward localization of international law and the gradual predominance of horizontal links.

D. Yu. Zhuzha, speaking of the emergence of new non-hierarchical methods of governance, including those in which political networks play an important role, emphasizes equality and proposes using the term "co-governance" alongside "governance" and "regulation". "The absence of a global body of government for the management of global resources makes it highly probable that principles of global co-governance can be developed and implemented on the basis of the equality of participants and coordination of common goals, means and methods of governance."

The researcher clarifies that the state does not lose its functions at all, but merely abandons the most direct methods of control, delegating responsibility for the results of regulation. "Thus, 'type two agreements', soft instruments and the redistribution of functions to lower levels do not mean the withdrawal of the state, but still remain within the framework of state policy. Another matter is that this does not abolish the freedom and influence of non-state actors."

L. V. Smorgunov analyzes various approaches to the problem of global governance, including neo-feudalism, cosmopolitan hierarchy and democratic governance, while also noting the dynamism and unpredictability of the modern world. "The open perspective and variability of possible results, constructed in the process by which the international community searches for ways of interaction, may generate something else that does not fit into the contemporary political language."

Nevertheless, posing the question of further ways of distributing governing powers among actors in the emerging political structure of the world, M. M. Lebedeva suggests two possible scenarios. The first path presupposes a spontaneous transfer of powers from states to other actors with consequences that are difficult to predict. The second path, on the contrary, implies that states, together with other actors, build a new architecture of the world "taking into account new realities and the interests of different participants: states, intergovernmental organizations, non-governmental associations, major financial and business structures and so on."

At present, humanity is to a significant extent following the path of spontaneous development of international relations, accompanied by increased instability and global problems.

I believe that the transition to governability of contemporary processes requires the formation and adoption of principles and norms corresponding to radically changed, and still rapidly changing, social relations. O. N. Barabanov rightly believes that "global governance as such will not be possible without a significant transformation of the entire international-legal regime, in which the emphasis in this case should be placed not on the sovereignty of states, but on their global responsibility and, moreover, on their subordination to institutions of global governance."

I believe that the formation of an adequate system of global governance for sustainable global development must be based on a corresponding system of legal norms and institutions, formed through substantial reform of the system of international law in interaction with domestic legal systems.

Moreover, the development and adoption of rules of conduct for global society is fundamental in the process of forming a global governance system, which brings the problem of global law to the forefront.

Since global processes are systemic in character, "world civilization requires the creation of a single legal environment of life activity, which presupposes the mandatory development of generally binding norms of conduct and interaction for all interested parties in the creation of a humanistically just world."

One may agree with the conclusion that global governance is impossible without legal support. At the same time, the development of global law must be anticipatory in character, especially in its scientific and educational component.

In particular, A. D. Ursul and T. A. Ursul reasonably assume that science and education must become a single anticipatory scientific and educational process called upon to change "the evolutionary trajectory of the social stage of development in the direction of its survival, preservation and further permanent progress." At the same time, one of the main components of the innovation process is named futurization, connected with globalization, and all possible mechanisms of anticipatory development. "Education, by preceding many other processes of activity in forming the human being, must, in forming a sustainable future, develop at a faster pace in all respects, especially in terms of financing and other resources, than all other forms of social activity. In this perspective, education will not merely transmit from generation to generation knowledge already substantially outdated for decision-making, but will more effectively promote understanding of the need for anti-crisis actions for the transition to sustainable development."

S. A. Lebedev notes the gradual erasure of distinctions "between natural-scientific and social-humanitarian knowledge, between the sciences of nature and the sciences of human beings and society", as well as the growth of interaction between science and various subsystems of culture, such as philosophy, law, economics, politics, art and others, and the growing confidence in the concept of the unity of internal scientific and sociocultural factors in the development of scientific knowledge.

The conclusion suggests itself that the purposeful formation of an adequate global governance system objectively requires scientifically grounded anticipatory development of global law and of global science and education closely connected with it.

The problem of modernizing legal science and legal education deserves separate mention. At a minimum, they must take into account the dynamics of the development of social relations; ideally, they should be anticipatory in character. At the same time, their level of integration must be increased. This is extremely important under conditions of rapid development of world integration processes embracing all spheres of social interaction and requiring the formation of a global governance system in the interests of sustainable development.

In this context, one cannot but agree with jurists who note the importance of global studies in law not only for lawmaking and law enforcement, but also for the development of legal education. Special emphasis is placed on the study of international-legal problems, since "legal global studies may contribute to strengthening mutual understanding among peoples and to creating more effective legal relations formed in the global world."

Above all, it is necessary to give a modern definition of the concept of international law that corresponds to contemporary social relations. Since contemporary legal understanding is inseparably connected with human rights, and contemporary social relations are connected with global processes and problems, this must be reflected in the definition of international law.

In this context, international law is a system of principles and norms in the sphere of interactions among states and other subjects, which must be directed toward the realization of human rights, thereby contributing to solving global problems and moving toward the sustainable development of human civilization.

As the main promising directions of the progressive development of international law for the transition from the spontaneous development of instability in international relations to the governability of contemporary processes and the sustainable development of human civilization, implying the development of its theoretical-legal, normative and institutional foundations, the following may be identified: human rights; freedom of conscience as the possibility of worldview choice; tolerance as equality and mutual respect; worldview neutrality of subjects of international law; reform and development of new principles of international law as its fundamental beginnings; the supremacy of international law on the basis of the principle of legal certainty; universalization of the principles and norms of international law; a non-polar system of international relations, that is, rejection of the "right of force" and transition to the "force of law"; and the increased role of the United Nations and other international organizations.

It should be especially emphasized that the realization of human rights is a necessary condition for reducing and overcoming the gap between power and society, for the evolutionary rather than revolutionary limitation of the power of some people over others, and ultimately for embodying human freedom as an objective requirement of contemporary globalizing social relations.

Human rights must be placed at the foundation of the progressive development of international law, which is called upon to form the legal basis for the transition from the spontaneous development of instability in international relations to the governability of contemporary processes and the sustainable development of human civilization.

Since the realization of human rights, as "the most essential opportunities for a person's development, inalienable properties determining the measure of his freedom", is the main criterion of the effectiveness of contemporary society, law and the state, this criterion is the principal one in the formation of a modern and prospective normative system.

One should agree with the view that "in a globalizing world, a necessary condition for the progressive and sustainable development of humanity is the recognition and implementation by states of extraterritorial obligations in relation to human rights. This obligation of states is conditioned by the indisputable imperative that fundamental human rights and freedoms are to be protected regardless of where individuals are located, whether within their own state or on the territory of another state: states must protect human rights and freedoms globally, especially from gross and systematic violations, in accordance with the principle of cooperation."

At the same time, the level of comprehension of contemporary global processes by the ruling elites of states does not yet allow for the creation of effective mechanisms for the realization of human rights at the domestic level, nor for effective cooperation in this area within the international sphere. This "clearly confirms the need to bring national legislation and law-enforcement practice of states into conformity with the new realities of contemporary human development and international-legal norms in the field of the promotion and protection of human rights and freedoms."

Thus, human rights may be named as the flagship branch of the emerging global law, called upon to eliminate the "distortions" of global governance in the sphere of social justice. It is precisely in human rights that social justice must be embodied, and human rights themselves are the embodiment of human freedom on the principles of equality.

No less important is the formation and development of an academic discussion environment on global studies in law and on the prospects for forming global law and education. Given the dependence of the formation of global principles and norms on the degree of their scholarly development, one may state the leading role of legal global studies in the context of the problem of forming a global governance system.

Another extremely important element should be the academic discipline "Global Jurisprudence", whose purpose is to train personnel for research and practical work under conditions of the emerging economic, informational, political and sociocultural integrity of the contemporary world.

Thus, global law should include the scholarly field of Legal Global Studies, the academic discipline Global Jurisprudence, a system of principles and norms regulating social relations on a planetary scale, as well as interactions between human beings and nature, and directed toward solving global problems and ensuring the sustainable development of civilization.

I believe that an adequate system of global governance and law must also be formed on the basis of the principles, norms and institutions of international law in interaction with domestic legal systems. To achieve this, however, they must undergo a whole series of evolutionary changes. Thus, in the context of the problem of forming global law as the basis of adequate global governance, one of the priorities must be the anticipatory development of global science and education, including legal education. On the basis of global studies, they must create the conceptual, normative and personnel foundation for the formation of global law as the basis of adequate global governance for sustainable development.

It is necessary to introduce new academic disciplines directed toward training personnel for research and practical work under conditions of the emerging economic, informational, political and sociocultural integrity of the contemporary world. In this context, it is necessary first of all to create a system for reproducing knowledge about global processes and problems, sustainable development, human rights and tolerance.

In forming educational programs, special attention should be given to human rights, the realization of which is a necessary condition for reducing and overcoming the gap between power and society, for the evolutionary rather than revolutionary limitation of the power of some people over others, and ultimately for embodying human freedom as an objective requirement of contemporary globalizing social relations. It should be especially emphasized that, given the importance of overcoming ethno-confessional barriers between people, which is a necessary condition for positive planetary integration, education must be worldview-neutral.

Thus, the principle of secularity, or, in my view, worldview neutrality, of the state and municipal education system is an important component of the principle of state secularity, as a condition for achieving tolerance and overcoming xenophobia, intolerance and discrimination. In the realities of the twenty-first century, denial and violation of the principle of worldview neutrality of states implies the use of religion for political purposes, the sacralization of power, and also contributes to turning democratic procedures into a farce and reproducing authoritarianism. In this context, worldview neutrality of states is the most important condition for the realization of human rights, as well as an antidote to the monopolization of power, social differentiation and ethno-confessional conflicts.

However, judging by all appearances, politicians both in Russia and throughout the world show little desire to change policy. No one seems eager to renounce such a profitable matter for power, but such a harmful matter for society, as the use of religious confessions for political needs. Moreover, the principle of state secularity as the most important guarantee of freedom of conscience is still not enshrined in the norms of international law. It is therefore unsurprising that even those states that declare their secularity are not secular in practice.

The relationship between the emerging global society and the traditional political structures of nation-states, based on the exploitation of divisive ethno-confessional principles, remains an extremely difficult problem. In particular, V. P. Malakhov and A. Kh. Abashidze have studied the problems of forming global civil society and its role in the international legal order. Among its elements they name "faith-based organizations, mainly membership-based religious organizations or organizations devoted to worship or the promotion of doctrine, or created for these purposes." In addition, according to these researchers, "global civil society, besides the elements identified in the classification, must also include interstate systems at the international level."

At the same time, it is obvious that international organizations composed of nation-states have absorbed the divisive ethno-confessional principles mentioned above, which have become a serious obstacle to positive integration for solving global problems. Under current conditions, at the level of the structure and distribution of power, new political leadership is needed, capable of integrating into the world community while organically combining national and global interests. "Politicians will have to change the priorities of their activity, at least equating planetary ecological interests with national interests."

A substantial factor of political disintegration is the exploitation by politicians of ethno-confessional divisive principles within relations between the state and different religious associations competing with one another, and in a broader sense within relations between politics and religion.

In turn, relations between politics and religion are interconnected with relations between science and religion, and may even be conditioned by them. In this context, it is relevant to study the influence of relations between science and religion on relations between politics and religion and, consequently, on the global prospects of human civilization, including the possibility of solving global problems.

Science and religion participate in forming the picture of the world in society and the worldview of each individual. The worldview need is a fundamental distinction between human beings and animals; together with freedom, it is one of the most important attributes and conditions of personal self-realization and, consequently, of the development of society. One may assume that the relationship between science and religion to a significant extent forms a kind of system-forming matrix of human civilization.

No less important and difficult is the task of "forming new thinking and a worldview corresponding to it among people who, for the most part, simply do not keep pace with the dynamics of the changes occurring around them. Many, especially adults, find it very difficult to get used to a new world, an open arena in which ideas, paradigms and values collide in free competition; it is still more difficult to leave the familiar little cave of established opinions, myths and ideologies."

In resolving these issues, a special role should be played by the formation of an effective legal mechanism for realizing the right of every individual to freedom of conscience, that is, the legal dimension of liberty of worldview choice, as the most important means of protecting the individual and society from the ideological domination of any doctrines and structures. It is precisely this legal mechanism that is a necessary condition for the formation of global thinking, because it is capable of limiting, if not preventing, the exploitation of ethno-confessional divisive principles for political purposes and the manipulation of individual, collective and mass consciousness.

Proceeding from the fact that the realization by a person of the need for worldview orientation is a way of becoming conscious of oneself in the infinite world, that is, the core that determines the specific essence of the human being as the basic element of social systems, freedom of conscience is the foundation of all human rights, a system-forming right, corresponding to the historical tendency toward an increase in the degrees of personal freedom.

Beyond possession of a long-term vision, the future depends on whether people can overcome division. "The further development of the species Homo sapiens already requires a planetary strategy." In this context, the creation of an effective legal mechanism for the realization by every person of the right to freedom of conscience will give every person, at the personal level, a chance to recognize themselves as part of a single planetary community and to overcome the attraction of ideologized groups, including those that call themselves religious. "The future belongs neither to individualism nor to collectivism, but to personalism, where the age-old conflict between the individual and society has a chance to be resolved on the basis of the affirmation of freedom." "Thus, it is not a rigid set of rules whose fulfilment guarantees 'salvation', but one's own conscience, knowledge and talent, if you will, one's own creativity combined with mercy and love for one's neighbor, that should determine the character of decisions made by a person. Without this, 'salvation', that is, futurity, is impossible."

Unlawful restrictions, and the reduction of every person's right to freedom of conscience to the choice of confessional preferences that are constantly engaged in a struggle for human souls, run directly counter to global tendencies, depriving society of freedom and humanity of a future. "A society equipped with unprecedented instruments of destruction and creation cannot return to the Middle Ages with impunity. To survive, it must develop and master a mentality adequate to its instrumental power, one that presupposes an extraordinarily high degree of tolerance, readiness for self-criticism and compromise. The acceptance by all regional cultures of common values and norms of coexistence, convergence and unification, is today not a pious wish but an imperative for preserving world civilization."

From this follows the urgency of developing both legal mechanisms for realizing each person's right to freedom of conscience and ideas about its significance and place in the system of contemporary law. At the same time, the principles and conceptual apparatus currently used in the sphere of freedom of conscience do not make it possible to create adequate international-legal standards and a domestic normative-legal basis for the full realization of this basic system-forming right. The definition and essential content of freedom of conscience as a legal category have in fact been distorted, creating from the outset insurmountable obstacles to the realization of the declared principles.

The creation of an effective system of legal guarantees of freedom of conscience on the basis of worldview neutrality of states and other subjects of international law is intended to limit the use of religion as ideology for political needs, to undermine the very foundation of totalitarianism, namely the possibility of not fully legal restrictions, control and suppression of the worldview sphere. In contemporary realities it is extremely necessary to recognize liberty of worldview choice and worldview neutrality of states as global universal human values.

Solving the global problems of humanity is possible only through integration, which implies a revision of the complex of national priorities and infrastructures and, possibly, the abandonment of state sovereignty or at least part of it. Moreover, solving global problems is impossible within a unipolar or even bipolar system of international relations. Even the formation of a multipolar system will not necessarily lead to the solution of the complex of global problems.

Yet even a multipolar system is impossible without overcoming intercivilizational tensions based on the combination of religion and politics and, consequently, on the suppression of liberty of worldview choice. In this connection, an effective legal mechanism for realizing the principle of worldview neutrality of states and other subjects of international law is a necessary factor in overcoming divisive principles that form the basis of the existence of traditional political structures and a condition for forming political leadership of nation-states capable of integrating into the world community.

The world has indeed changed and continues to change rapidly. The present and future of human civilization depend on the adequacy of responses to these changes. Accordingly, liberty of worldview choice and worldview neutrality of states are necessary conditions for the transformation of value orientations, the formation of global thinking and the transition to sustainable development of an integral non-polar world.

In this context, the formation of an adequate normative basis is not only a necessary condition for realizing liberty of worldview choice and worldview neutrality of states, but also one of the conditions for the survival of human civilization. Thus, one may agree with researchers who speak of the need for planetary-scale unification in order to solve global problems and move toward sustainable development. The most serious approaches are connected with the idea of forming global governance and global law as its foundation. At the same time, their foundation must be built through strengthening and modernizing international law and its universal institutions.

The diversity of models of global governance may be divided into two main groups: first, those directed toward limiting the power of some people over others on a planetary scale; and second, those presupposing the preservation and increase of the concentration of the power of some people over others on a planetary scale. In this context, approaches directed toward limiting power appear more consistent with the tasks of forming adequate global governance. Accordingly, the most promising system for governing global processes appears to be one that presupposes the strengthening of the influence of international organizations, including non-governmental organizations, and a further gradual transition to a non-hierarchical system.

Here, once again, it is necessary to note the need for anticipatory innovative development of science and education, including legal science and legal education, as the foundation of the system for governing global processes described above. It should be especially emphasized that, in searching for ways to resolve global problems, the most serious attention must be paid to global studies in the sphere of freedom of conscience and state secularity. "Without scientifically grounded solutions in this sphere, bloody conflicts will continue, and the transition to sustainable development will be impossible. The resolution of the global problems currently facing the world community, as well as the preservation and sustainable development of civilization, will depend on solving this problem. Further ignoring the principle of state secularity and the closely connected right to freedom of conscience will lead to wars and the destruction of human civilization." Thus, one should note the key significance of worldview neutrality of states and other subjects of international law for the search for scientifically grounded approaches to solving global problems and achieving the sustainable development of civilization.

The concept and content of the right to freedom of conscience were historically formed first as religious tolerance, then as the right to freedom of religion or confession, and finally as the right to free worldview choice. However, in fundamental international documents this right is fixed with reference to the legally undefined term "religion", which predetermined legal arbitrariness and mass violations of this system-forming right at the domestic level. As a result, freedom of thought, conscience and religion is declared in fundamental international documents, while in practice violations occur in all states, especially confessional ones.

We have substantiated the enormous significance of realizing freedom of conscience in the context of globalization and have proposed its innovative paradigm. Under conditions of intensifying global risks, an effective legal mechanism for realizing the right to freedom of conscience, with worldview neutrality as its guarantee, is a necessary condition for separating religious and political institutions and overcoming xenophobic divisive ethno-confessional principles exploited by domestic political structures, which to a significant extent obstruct the formation of an adequate system of global governance for sustainable development.

In addition, the realization of the possibility of free worldview choice is the main condition for weakening xenophobia, intolerance and violence on ethno-confessional grounds and for ensuring global security. The realization of the right to liberty of worldview choice is also a necessary condition for transforming value orientations, forming global thinking and moving toward the sustainable development of civilization. Thus, the right to liberty of worldview choice, as a system-forming right within the system of human rights, must occupy an important place in the global normative system, or law, in order to form global governance for overcoming global challenges and moving toward sustainable development.

Unlawful restrictions, and the reduction of every person's right to freedom of conscience to the choice of confessional preferences that are constantly engaged in a struggle for human souls, run directly counter to global tendencies, depriving society of freedom and humanity of a future. "A society equipped with unprecedented instruments of destruction and creation cannot return to the Middle Ages with impunity. To survive, it must develop and master a mentality adequate to its instrumental power, one that presupposes an extraordinarily high degree of tolerance, readiness for self-criticism and compromise. The acceptance by all regional cultures of common values and norms of coexistence, convergence and unification, is today not a pious wish but an imperative for preserving world civilization." From this follows the urgency of developing both legal mechanisms for realizing each person's right to freedom of conscience and ideas about its significance and place in the system of contemporary law and international relations.

At the same time, the principles and conceptual apparatus currently used in the sphere of freedom of conscience do not make it possible to create adequate international-legal standards and a domestic normative-legal basis for the full realization of this basic system-forming right. The definition and essential content of freedom of conscience as a legal category have in fact been distorted, creating from the outset insurmountable obstacles to the realization of the declared principles.

The creation of an effective system of legal guarantees of freedom of conscience at the world level is intended to limit the use of religion as ideology for political needs and to undermine the very foundation of totalitarianism: the possibility of extra-legal restrictions, control and suppression of the worldview sphere. In the realities of the twenty-first century, freedom of conscience must be understood as a global universal human value. Solving humanity's global problems is possible only through integration, which implies a revision of the complex of national priorities and infrastructures and, possibly, the abandonment of state sovereignty or at least part of it.

Taking into account the inevitable tendencies of globalization, which put forward demands for integration, the contemporary political system based on nation-states increasingly fails to meet the challenges of the new era. It should be noted, however, that it is precisely nation-states that serve as the starting point for forming a single world system called upon to solve, at a new level, the global problems facing humanity. This means that solving global problems is impossible within a unipolar or even bipolar system of international relations. Even the formation of a more balanced and viable multipolar system appears to be a necessary intermediate stage, but not a sufficient condition for solving the complex of global problems. Yet even a multipolar system is impossible without overcoming intercivilizational tensions based on the combination of religion and politics and, consequently, on the suppression of liberty of worldview choice.

In this connection, an effective legal mechanism for realizing the right to liberty of worldview choice and the principle of public-law worldview neutrality is a necessary factor in overcoming the divisive principles that form the basis of the existence of traditional political structures and a condition for forming political leadership of nation-states capable of integrating into the world community. The capacity of society to protect religion from use for political purposes and to avoid the incitement of interethnic and interconfessional conflicts as a means in the struggle for the redistribution of the world depends on the realization of the right to liberty of worldview choice for everyone, beyond divisive principles, through the formation of a corresponding legal mechanism. The adequacy of people's responses to the challenges of globalization also depends on the realization of this system-forming right and on overcoming the largely artificial political division; so too does the question of which conceptual model of globalization will prevail: the model of the "golden billion", that is, globalization of division and injustice, or the model of a just world with an equal right to a dignified life for every citizen of the planet. In turn, humanity's ability to solve global problems, and therefore the survival of humanity, depends on the victory of the just model.

Today the world has indeed changed and continues to change rapidly. The present and future of human civilization depend on the adequacy of responses to these changes. Accordingly, liberty of worldview choice, ensured by public-law worldview neutrality, is a necessary condition for transforming value orientations, forming global thinking and moving toward sustainable development of an integral non-polar world. In this context, the issue of increasing the adequacy and effectiveness of international and domestic principles and norms in this sphere is one of the conditions for the survival of human civilization.

REFERENCES TO CHAPTER 3

List of Sources Used

1. Robertson R., Lechner F. Modernization, Globalization and the Problem of Culture in World-Systems Theory // Theory, Culture & Society. 1985. No. 2. pp. 103-117.
2. Levitt T. The Globalization of Markets // Harvard Business Review. May-June 1983. No. 3. pp. 2-11. Electronic resource. URL: http://ksb.uek.krakow.pl/uploads/media/Levitt_The_Globalization_of_Markets_HBR_1983_01.pdf (accessed: 15.04.2019).
3. Beck U. What Is Globalization? Frankfurt am Main, 1998. pp. 28-44.
4. Giddens A. The Consequences of Modernity. Stanford: Stanford University Press, 1990. 186 p. pp. 63-71.
5. Sklair L. Sociology of the Global System. Hemel Hempstead, 1991. pp. 6-8.
6. Ohmae K. The Mind of the Strategist: The Art of Japanese Business. Moscow: Alpina Publisher, 2007. 224 p.
7. Ohmae K. The End of the Nation State: The Rise of Regional Economies. Free Press, 1995. 214 p.
8. Ohmae K. The Next Global Stage: Challenges and Opportunities in Our Borderless World. Wharton School Publishing, 2005. 312 p.
9. Chumakov A.N. The Metaphysics of Globalization: The Cultural and Civilizational Context. Monograph. 2nd ed., revised and expanded. Moscow: Prospekt, 2017. 496 p.
10. Chumakov A.N. Globalization: The Contours of an Integral World. Monograph. 3rd ed., revised and expanded. Moscow: Prospekt, 2017. 448 p. p. 24.
11. Global Studies: Encyclopedia / chief editors I.I. Mazur and A.N. Chumakov // Center for Scientific and Applied Programs 'Dialog'. Moscow: Raduga Publishing House, 2003. p. 181.
12. Ursul A.D. V.I. Vernadsky and the Noospheric Prospects of the Global Direction of Science // URL: http://gigabaza.ru/doc/41510.html (accessed: 17.07.2019).
13. Cheshkov M.A. Global Studies as Scientific Knowledge: Essays on Theory and the Categorical Apparatus. Moscow: Scientific and Educational Forum on International Relations, 2005. p. 12.
14. Cheshkov M.A. Global Studies as Scientific Knowledge: Essays on Theory and the Categorical Apparatus. Moscow: Scientific and Educational Forum on International Relations, 2005. p. 38.
15. Ivanov D.V. The Evolution of the Concept of Globalization // Telescope. 2002. No. 5. p. 1.
16. Trofimova A.Kh. Globalization: Processes of Formation and Development. Privatization and Globalization // Legal World. 2011. No. 10. p. 36.
17. Volkov V.V. Philosophy: Main Problems, Concepts, and Terms. Study Guide. Saint Petersburg: Lenand, 2015. 136 p.
18. Myasnikov V.A., Naydenova N.N., Tagunova I.A. Education in the Global Dimension. Monograph. Moscow: Institute of Theory and History of Pedagogy of the Russian Academy of Education, 2009. 512 p. p. 23.
19. Ilyin I.V. Global Studies in the Context of Political Processes: Abstract of Doctoral Dissertation in Political Science. Moscow, 2011. p. 7.
20. Yashkova T.A. Political Modernization under the Conditions of Global Transformational Challenges. Monograph. Moscow: MAKS Press, 2007. 255 p. p. 35.
21. Lukashuk I.I. International Law. General Part: Textbook for Students of Law Faculties and Universities. 3rd ed., revised and expanded. Moscow: Wolters Kluwer, 2005. p. 11.
22. Lukashuk I.I. International Law. General Part: Textbook for Students of Law Faculties and Universities. 3rd ed., revised and expanded. Moscow: Wolters Kluwer, 2005. p. 3.
23. Lazarev V.V., Saidov A.Kh. Patterns in the Development of Contemporary Law and Legal Global Studies // Bulletin of the International Institute of Economics and Law. 2011. No. 2 (2). pp. 85-91. p. 88.
24. Theory of State and Law. Textbook. 3rd expanded and supplemented edition. Edited by M.N. Marchenko. Moscow: Zertsalo Publishing House, 2000. 624 p.

25. Meshcheryakova O.M. Globalization of Law: A Philosophical-Legal Reflection // Russian Journal of Legal Studies. 2015. No. 2 (3). pp. 28-32.
26. Bekyashev K.A. Globalization and International Law // Eurasian Law Journal. 2014. No. 1.
27. Moiseev A.A. Current Aspects of International Law under Conditions of Globalization // Bulletin of RUDN University. Series: Legal Sciences. 2007. No. 2 (14). pp. 45-58.
28. Moiseev A.A. Supranationality in Contemporary International Law. Moscow: Scientific Book, 2007. 288 p. Moiseev A.A. The Phenomenon of Globalization and the International Community // Age of Globalization. 2015. No. 2 (16). pp. 59-72.
29. Zanibekov M.M. Current Problems of the Globalization of International Trade // Eurasian Law Journal. 2015. No. 4 (83). pp. 54-58.
30. Zelentsov A.B. Legal Globalization and Public Law // Contemporary Problems of Public Law in Russia and Foreign Countries. Collection of Scientific Works by Teachers, Postgraduate Students, and Degree Seekers of the Department of Administrative and Financial Law, Faculty of Law, RUDN University. Moscow, 2012. pp. 6-25.
31. Mamedov A.A. Contemporary Trends in the International-Legal Regulation of Banking under Conditions of Globalization // Eurasian Law Journal. 2013. No. 6 (61). pp. 37-40.
32. Mamedov A.A. International Norms-Principles and Economic Globalization // Bulletin of RUDN University. Series: Legal Sciences. 2014. No. 2. pp. 265-271.
33. Farkhutdinov I.Z. International Investment Law: Theory and Practice of Application / I.Z. Farkhutdinov. Moscow: Wolters Kluwer, 2005. p. 8.
34. Korshunov A.N. The Idea of Global Law: Philosophical and Methodological Aspects: Abstract of Candidate of Philosophical Sciences Dissertation. Rostov-on-Don, 2010. pp. 6-7.
35. Ibid.
36. Problems of Legal Understanding: Collective Monograph Based on the Results of the Third Alekseev Readings / executive editor V.D. Perevalov. Yekaterinburg: Publishing House of the Ural State Law University, 2018. 276 p. p. 15.
37. Cited work. p. 16.
38. Kostenko A.N. Social Naturalism as a Methodological Principle of Legal Global Studies // Electronic resource. URL: http://lib.chdu.edu.ua/pdf/naukpraci/politics/2007/69-56-42.pdf (accessed: 14.04.2019).
39. Ekimov A.I. On the Contemporary Model of Development of Russian Society // Economy. Law. Society. 2016. No. 4 (8). pp. 23-29. p. 27.
40. ILO Declaration on Social Justice for a Fair Globalization of June 13, 2008. Electronic resource. URL: http://www.un.org/ru/documents/decl_conv/declarations/pdf/fair_globalization.pdf (accessed: 17.04.2019).
41. Cheshkov M.A. Global Studies: Subject Matter, Problems, Prospects // Social Sciences and Contemporary World. 1998. No. 2. p. 129.
42. Cheshkov M.A. Global Studies as Scientific Knowledge: Essays on Theory and the Categorical Apparatus. Moscow: Scientific and Educational Forum on International Relations, 2005. p. 20.
43. King A., Schneider B. The First Global Revolution: A Report by the Council of the Club of Rome. Moscow: Progress-Pangeya, 1991. 344 p. p. 8.
44. Cited work. p. 152.
45. Philosophy: Encyclopedic Dictionary. Content. Edited by A.A. Ivin. Moscow: Gardariki, 2004. 1072 p.
46. Ibid.
47. Lukashuk I.I. Globalization, the State, and the 21st Century. Moscow, 2000. pp. 6-7.
48. Nizhnikov S.A. Philosophy: Course of Lectures. Study Guide for Universities. Moscow, 2006. 383 p. p. 21.
49. Chernikova N.V. Concepts of Global Education and Features of Their Implementation in the Practice of Russian Schools: Abstract of Candidate of Pedagogical Sciences Dissertation. Rostov-on-Don, 2005. 167 p. p. 5.

50. Malayan R.E. The United Nations in the Emerging System of Global Governance: Abstract of Candidate of Political Sciences Dissertation. Moscow, 2009. p. 3.
51. Farkhutdinov I.Z. The American Doctrine of the Preventive Military Strike: International-Legal Aspects. Moscow, 2017. 338 p.
52. Farkhutdinov I.Z. The American Doctrine of the Preventive Strike from Monroe to Trump: International-Legal Aspects. Moscow, 2018. 419 p.
53. Farkhutdinov I.Z. The Israeli Doctrine of Preventive Self-Defense and International Law // Eurasian Law Journal. 2016. No. 8. pp. 15-32.
54. Farkhutdinov I.Z. The Iranian Doctrine of Preventive Self-Defense and International Law // Eurasian Law Journal. 2017. No. 1. pp. 15-26.
55. Farkhutdinov I.Z. The Iranian Doctrine of Preventive Self-Defense and International Law (conclusion) // Eurasian Law Journal. 2017. No. 2. pp. 15-25.
56. Farkhutdinov I.Z. International Law and the U.S. Doctrine of Preventive Self-Defense // Eurasian Law Journal. 2016. No. 2. pp. 23-31.
57. Farkhutdinov I.Z. International Law on the Use of Military Force by a State against Non-State Actors // Eurasian Law Journal. 2016. No. 7. pp. 59-76.
58. Farkhutdinov I.Z. International Law on the Principle of Non-Use of Force or Threat of Force: History and the Present // Eurasian Law Journal. 2015. No. 11. pp. 34-38.
59. Farkhutdinov I.Z. International Law on the Self-Defense of States // Eurasian Law Journal. 2016. No. 1. pp. 91-100.
60. Podzigun I.M. Globalization and Global Problems: A Philosophical and Methodological Analysis: Doctoral Dissertation in Philosophy. Moscow, 2003. 384 p. p. 12.
61. Volkov V.V. Philosophy: Main Problems, Concepts, and Terms. Study Guide. Saint Petersburg: Lenand, 2015. 136 p.
62. Nazaretyan A.P. The Nonlinear Future: Mega-Historical, Synergetic, and Cultural-Psychological Preconditions for Global Forecasting. Moscow: MBA Publishing House, 2013. p. 33.
63. Cited work. p. 35.
64. Nizhnikov S.A. Philosophy: Course of Lectures. Study Guide for Universities. Moscow, 2006. 383 p.
65. Kochetov E.G. Main Characteristics of the Globalization Process and the Legal Dimension of the World // Journal of Russian Law. Moscow: Norma, 2003. No. 3. pp. 88-97. p. 88.
66. Philosophy: Encyclopedic Dictionary. Content. Edited by A.A. Ivin. Moscow: Gardariki, 2004. 1072 p.
67. Lukashuk I.I. Interaction between International and Domestic Law under Conditions of Globalization // Journal of Russian Law. 2002. No. 3. pp. 115-128.
68. Volkov V.V. Philosophy: Main Problems, Concepts, and Terms. Study Guide. Saint Petersburg: Lenand, 2015. p. 35.
69. Weber A.B. The Contemporary World and the Problem of Global Governance // Age of Globalization. 2009. No. 1. pp. 3-16.
70. Romasevich D.I. Global Problems of the Present and Russia's Environmental Policy: Candidate of Political Sciences Dissertation. Moscow, 2006. 148 p.
71. Our Common Future: Report of the World Commission on Environment and Development. Translated from English, edited by S.A. Evteev and R.A. Perelet. Electronic resource. URL: http://xn--80adbkckdfac8cd1ahpld0f.xn--p1ai/files/monographs/OurCommonFuture-introduction.pdf (accessed: 14.08.2019). pp. 30-43.
72. Khorin I.S. Globalization of Social Life: History and the Present // Knowledge. Understanding. Skill. 2005. No. 2. p. 118.
73. Podzigun I.M. Globalization and Global Problems: A Philosophical and Methodological Analysis: Doctoral Dissertation in Philosophy. Moscow, 2003. 384 p. p. 17.
74. Weber A.B. The Contemporary World and the Problem of Global Governance // Age of Globalization. 2009. No. 1. p. 16.
75. Facets of Globalization: Difficult Questions of Contemporary Development. Moscow: Alpina Publisher, 2003. p. 17.
76. Cited work. pp. 7-8.

77. Nemchuk A.A. Global Governance in the Contemporary World: A Political-Science Analysis: Doctoral Dissertation in Political Science. Moscow, 2004. 303 p. Cited work. p. 12.
78. Yashkova T.A. Political Modernization under the Conditions of Global Transformational Challenges. Monograph. Moscow: MAKS Press, 2007. 255 p. p. 26.
79. Lazarev V.V., Saidov A.Kh. Patterns in the Development of Contemporary Law and Legal Global Studies // Bulletin of the International Institute of Economics and Law. 2011. No. 2 (2). pp. 85-91. p. 88.
80. See: Integration Law in the Contemporary World: A Comparative-Legal Study / edited by S.Yu. Kashkin. Moscow, 2014. 112 p.
81. Bekyashev K.A. Globalization and International Law // Eurasian Law Journal. 2014. No. 1. pp. 25-29.
82. Our Global Neighborhood: Report of the Commission on Global Governance / translated from English by A.A. Obukhov; edited by A.P. Fomenko. Moscow: Ves Mir, 1996. 402 p.
83. Tsivatyi V.G. Legal Education and Law under Transformational Conditions of Polycentricity: Globalization and Institutional Discourses // Russian and International Law: The Common and the Particular. Proceedings of the All-Russian Scientific and Practical Conference in Memory of Professor F.M. Rudinsky. 2019. pp. 370-375.
84. Ilyin I.V. Global Studies in the Context of Political Processes: Doctoral Dissertation in Political Science. Moscow, 2011. 428 p.
85. Malayan R.E. The United Nations in the Emerging System of Global Governance: Abstract of Candidate of Political Sciences Dissertation. Moscow, 2009. 22 p. p. 3.
86. Cited work. p. 16.
87. Pavlenko V.B. Institutional Aspects of Global Governance of Political Processes: Doctoral Dissertation in Political Science. Moscow, 2008. p. 503.
88. Davydov V.M. New Centers of Power: Access to Mechanisms of Global Regulation // Global Governance in the 21st Century: Innovative Approaches / edited by Al.A. Gromyko. Moscow: Institute of Europe of the Russian Academy of Sciences; Nestor-Istoriya, 2013. pp. 54-55.
89. Barabanov O.N. Global Governance as a Topic for Scientific Analysis // Anti-Globalism and Global Governance: Reports, Discussions, Reference Materials. Moscow: MGIMO University, Ministry of Foreign Affairs of Russia, 2006. Electronic resource. URL: http://www.mgimo.ru/files/6195/global-anti[1].pdf (accessed: 23.09.2019).
90. Stetsko E.V. 'Global Governance' and the Role of Non-Governmental Organizations in Its Formation // Society. Environment. Development (Terra Humana). 2012. No. 4. p. 111.
91. Chumakov A.N. The Global World: The Problem of Governance // Age of Globalization. 2010. No. 1. pp. 3-15.
92. Ilyin I.V., Kaverin M.A. Issues of Transforming International Organizations into Institutions of Global Governance // Age of Globalization. 2014. No. 2. p. 37.
93. Romasevich D.I. Global Problems of the Present and Russia's Environmental Policy: Candidate of Political Sciences Dissertation. Moscow, 2006. 148 p.
94. Nemchuk A.A. Global Governance in the Contemporary World: A Political-Science Analysis: Doctoral Dissertation in Political Science. Moscow, 2004. p. 242.
95. Lukashuk I.I. International Law. General Part: Textbook for Students of Law Faculties and Universities. 3rd ed., revised and expanded. Moscow: Wolters Kluwer, 2005. p. 12.
96. For more detail, see: Lykov A.Yu. The World State as the Future of the International Community: Monograph / A.Yu. Lykov. Moscow: Prospekt, 2013. 248 p.
97. Lykov A.Yu. On the Problem of Integrating Legal Systems in the Formation of a World State // History of State and Law. 2013. No. 12. p. 16.
98. Lukashuk I.I. International Law. General Part: Textbook for Students of Law Faculties and Universities. 3rd ed., revised and expanded. Moscow: Wolters Kluwer, 2005. p. 23.

99. Prospects for Geopolitical Dynamics and the Interaction of Civilizations. Part 7 of the Global Forecast 'The Future of Civilizations' for the Period until 2050. Edited by Yu.V. Yakovets, A.I. Ageev, and T.T. Timofeev. Moscow: International Institute of Pitirim Sorokin-Nikolai Kondratieff, 2009. pp. 161-170.
100. Volkov V.V. Philosophy: Main Problems, Concepts, and Terms. Study Guide. Saint Petersburg: Lenand, 2015. 136 p.
101. Slaughter A.-M. A New World Order. Princeton and Oxford: Princeton University Press, 2004. 341 p. pp. 71-107.
102. Rosenau J.N., Czempiel E.-O. Governance Without Government: Order and Change in World Politics. Cambridge University Press, 1992. 311 p.
103. Ilyin I.V. Global Studies in the Context of Political Processes: Abstract of Doctoral Dissertation in Political Science. Moscow, 2011. 428 p.
104. Zuikov R.S. International Relations at the Beginning of the 21st Century: An Analysis of Criteria of Systemicity: Abstract of Candidate of Political Sciences Dissertation. Moscow, 2010.
105. Barabanov O.N. Global Governance as a Topic for Scientific Analysis // Anti-Globalism and Global Governance: Reports, Discussions, Reference Materials. Moscow: MGIMO University, Ministry of Foreign Affairs of Russia, 2006. Electronic resource. URL: http://www.mgimo.ru/files/6195/global-anti[1].pdf (accessed: 23.09.2019).
106. Smbatyan A.S. The Evolution of the International Legal Order: Universalization or Fragmentation? // Moscow University Bulletin. Series 11: Law. 2013. No. 6. pp. 48-58.
107. Cited work. p. 51.
108. Cited work. pp. 56-57.
109. Zhuzha D.Yu. Political Aspects of Global Governance of Natural Resources: Abstract of Candidate of Political Sciences Dissertation. Moscow, 2012. p. 15.
110. Zhuzha D.Yu. Political Aspects of Global Governance of Natural Resources: Abstract of Candidate of Political Sciences Dissertation. Moscow, 2012. p. 17.
111. Smorgunov L.V. Human Prospects in a Globalizing World / edited by V.V. Partsvaniya. Saint Petersburg: Saint Petersburg Philosophical Society, 2003. p. 159.
112. Lebedeva M.M. World Politics. 2nd ed., revised and expanded. Moscow: Aspect Press, 2007. 365 p. pp. 338-339.
113. Barabanov O.N. Global Governance as a Topic for Scientific Analysis. Electronic resource. URL: http://globalanti.risa.ru/reports.php?cat_id=31&doc_id=13 (accessed: 23.09.2019).
114. Korshunov A.N. The Idea of Global Law: Philosophical and Methodological Aspects: Abstract of Candidate of Philosophical Sciences Dissertation. Rostov-on-Don, 2010. p. 3.
115. Ursul A.D., Ursul T.A. Education for Sustainable Development: First Results, Problems, and Prospects // Sociodynamics. 2015. No. 1. pp. 11-74.
116. Lebedev S.A. Science in a Global World // Age of Globalization. 2012. No. 2 (10). pp. 145-151.
117. Farkhutdinov I.Z. International-Legal Regulation of Investment Relations: Theory and Practice: Doctoral Dissertation in Law. Moscow, 2006. 390 p.
118. Farkhutdinov I.Z. International or Global Law // International Lawyer. All-Russian Journal of International Law. 2004. No. 4. pp. 15-23.
119. Farkhutdinov I.Z. International Law under Conditions of Globalization // Law and Politics. 2003. No. 8. pp. 141-149.
120. Buryanov S.A., Krivenky A.I. Strategic Prospects for the Development of Law and Legal Education under Conditions of Contemporary Global Processes // Main Trends and Prospects in the Development of Contemporary Law: Proceedings of the Annual International Scientific Conference in Memory of Professor Felix Mikhailovich Rudinsky, April 19, 2018 / general editor Doctor of Legal Sciences, Professor D.A. Pashentsev. Moscow, 2018. pp. 247-251.
121. Global Processes and the Formation of Global Education (Interdisciplinary Study): Monograph in 2 books / authors: S.A. Buryanov, E.S. Demidenko, E.A. Dergacheva et al.; general editors S.A. Buryanov and A.I. Krivenky. Book 1. Moscow: Moscow City University, 2019. 200 p.
122. Global Processes and the Formation of Global Education (Interdisciplinary Study): A. Buryanov, A.V.

Zvonarev et al.; general editors S.A. Buryanov and A.I. Krivenky. Book 2.Moscow: Moscow City University, 2019. 276 p.
123. Lazarev V.V., Saidov A.Kh. Patterns in the Development of Contemporary Law and Legal Global Studies // Bulletin of the International Institute of Economics and Law. 2011. No. 2 (2). p. 91.
124. Abashidze A.Kh., Klishas A.A. Protection of Human Rights and Freedoms in Accordance with International Legal Standards under the New Realities of a Globalizing World // Bulletin of RUDN University. Series: Legal Sciences. 2013. No. 2. pp. 251-255. p. 251.
125. Cited work. p. 254.
126. Buryanov S.A. Freedom of Conscience as a Global Value: Toward Political Unity and the Solution of Global Problems // Age of Globalization. 2009. No. 1. pp. 136-151.
127. Kartashkin V.A. The UN Charter and the United Nations in a Globalizing World // International Lawyer. 2004. No. 1. pp. 4-8.
128. Pashentsev D.A. The Significance of Human Rights in Contemporary Conditions // Concepts for the Development of the Institution of Human Rights under Conditions of Globalization of Contemporary Law and Politics: Collection of Scientific Works Based on the Materials of an International Scientific and Practical Round Table. Executive editor N.N. Kuleshova. Ryazan, 2016. pp. 195-197.
129. Ivaneev S.V. Humanization of Education as a Source for Mastering Human Capital // Human Capital. 2015. No. 5 (77). pp. 123-127.
130. Civil Human Rights in Russia: Contemporary Problems of Theory and Practice [author collective including Pavel Viktorovich Anisimov et al.]; edited by F.M. Rudinsky. 3rd ed. Moscow, 2010. pp. 399-410.
131. Human Rights under Conditions of Globalization and Their Protection in Private International Law (Interdisciplinary Study): Collective Monograph in 2 books (edited by A.I. Krivenky). Moscow: Moscow City University, 2016. pp. 17-35.
132. Civil Human Rights in Russia: Contemporary Problems of Theory and Practice [author collective including Pavel Viktorovich Anisimov et al.]; edited by F.M. Rudinsky. 3rd ed. Moscow, 2010. 624 p. pp. 399-410.
133. Buryanov S.A. The Secular Nature of the State and Internationally Recognized Freedom of Conscience: Theoretical and Applied Study for 2015 - Early 2016. Moscow: Poligraf Servis, 2016. pp. 121-135.
134. Malakhov V.P., Abashidze A.Kh. International Civil Society: Essence and Current State // Bulletin of Moscow University of the Ministry of Internal Affairs of Russia. 2014. No. 11. pp. 164-167.
135. Moiseev N. The Environmental Background of Contemporary Politics // Social Sciences and Contemporary World. 1999. No. 4. p. 139.
136. Chumakov A.N. Cited work. p. 376.
137. Moiseev N. Morality and the Phenomenon of Evolution: The Ecological Imperative and the Ethics of the 21st Century // Social Sciences and Contemporary World. 1994. No. 6. p. 131.
138. Levitsky S.A. The Tragedy of Freedom. Posev Publishing House, 1984. p. 344.
139. Moiseev N. Morality and the Phenomenon of Evolution: The Ecological Imperative and the Ethics of the 21st Century // Social Sciences and Contemporary World. 1994. No. 6. p. 138.
140. Nazaretyan A.P. 'The Clash of Civilizations' and 'The End of History' // Social Sciences and Contemporary World. 1994. No. 6. p. 144.
141. Zuikov R.S. Problems in the Formation of a New System of International Relations from the Perspective of World-Systems and Civilizational Approaches // Space and Time in World Politics and International Relations. Proceedings of the 4th Convention of the Russian International Studies Association. In 10 vols. Vol. 6: New Trends in World Politics. Moscow: MGIMO University, 2007.
142. Buryanov S.A., Nikitaev D.M. The Problem of Freedom of Conscience as an Important Component of Global Studies // Global Processes and New Formats of Multilateral Cooperation: Collection of Scientific Works by Conference Participants / edited by I.V. Ilyin. Moscow: International Kondratieff Foundation, 2016. 282 p. pp. 163-167.

143. Buryanov S.A. Problems of Improving the Regulatory-Legal Framework for the Implementation of the Right to Freedom of Conscience in the Context of Globalization // State and Law. 2002. No. 10. pp. 26-31.
144. Buryanov S.A. A New Paradigm of Legal Regulation of Freedom of Conscience // Contemporary Law. 2003. No. 5. pp. 11-15.
145. Buryanov S.A. The Confessional Factor in the Context of the Problem of Ensuring Security: International-Legal and Domestic Aspects // Religions and the Security of the State and Society: Proceedings of a Public Conference. Moscow: Poligraf Servis, 2015. pp. 34-43.
146. Nazaretyan A.P. 'The Clash of Civilizations' and 'The End of History' // Social Sciences and Contemporary World. 1994. No. 6. p. 144.
147. Ivaneev S.V. Humanization of Education as a Source for Mastering Human Capital // Human Capital. 2015. No. 5 (77). pp. 123-127.
148. Buryanov S.A. The Right to Freedom of Conscience under Conditions of Global Processes and Challenges 4.0 // Eurasian Law Journal. 2023. No. 6 (181). pp. 18-21.
149. Buryanov S.A., Buryanov M.S. In Search of an Adequate World Order: The Future of the Rule of Law in the Era of Digital Globalization 4.0 // Age of Globalization. 2024. No. 2 (50). pp. 127-140. DOI: 10.30884/vglob/2024.02.10.
150. Freedom of Conscience: Problems of Theory and Practice. Monograph (edited by F.M. Rudinsky, S.A. Buryanov). Moscow: ZAO TF MIR, 2012. 1120 p.
151. Buryanov S.A. International Recognition of the Right to Freedom of Conscience and Problems of Its Implementation in the Russian Federation under Conditions of Contemporary Global Processes: Monograph. Moscow: Poligraf Servis, 2020. 624 p.
152. Civil Human Rights in Russia: Modern Problems of Theory and Practice / compiled by F.M. Rudinsky. Transaction Publishers, New Brunswick, New Jersey (U.S.A.), 2008. 458 p.
153. Buryanov S.A. State Worldview Neutrality in the Context of Deteriorating Imbalances in Globalization: The Case and Current State of Affairs in the Russian Federation from 2016 to the Beginning of 2017. Amsterdam, 2018. 228 p.
154. Buryanov S.A., Buryanov M.S., Nikitaev D.M. The Right to Freedom of Conscience under Conditions of Global Processes: Theory and Practice of Implementation in the Russian Federation (Master's Level). Monograph. Edited by S.A. Buryanov. Moscow: LitRes, 2022. 234 p.
155. Digital Human Rights under Conditions of Global Processes: Theory and Practice of Implementation: Monograph / M.S. Buryanov; scientific editor S.A. Buryanov. Moscow: RUSAINS, 2024. 148 p.

PART IV. INTERNATIONAL LAW BETWEEN PROTECTION AND POWERLESSNESS

CIVILIZATIONAL THESIS

"The inner forum is the last court of freedom."

International law protects human dignity most deeply when it protects the conditions in which free choice is formed. The law of the future must defend external rights and the inner space where convictions, doubts, conscience, and moral judgment arise.

CONCEPT BOX

CHAPTER 4

INTERNATIONAL LAW AND THE PROTECTION OF THE INNER WORLD

From fragmented guarantees to a wider legal framework

International human rights law already protects freedom of thought, conscience, religion, expression, dignity, equality, and non-discrimination.

Yet the existing legal language often remains fragmented. It protects separate elements of the person's inner life, while new forms of influence target the whole architecture of worldview formation.

Chapter 4 examines international legal principles, norms, and mechanisms related to freedom of conscience.

The chapter prepares the ground for a wider legal framework: liberty of worldview choice protected through public-law worldview neutrality.

CHAPTER IV

INTERNATIONAL LAW AND THE PROTECTION OF THE INNER WORLD

From fragmented guarantees to a wider legal framework

Chapter 4. International-Legal Consolidation and Protection of Human Rights in the Sphere of Freedom of Conscience

4.1. The Right to Freedom of Conscience in the System of International Protection of Human Rights

The protection of human rights occupies an important place in public international law. According to the authoritative view of V. A. Kartashkin, "at present human rights extend beyond national borders and are regulated not only by national law, but also by international law. For this reason, we now speak not only of the sovereignty of states, but also of the sovereignty of the individual, understanding sovereignty as the fundamental rights and freedoms of each person, protected by both national and international norms."

As A. Kh. Abashidze notes, the search for solutions to the problem of protecting the individual from abuses by unlimited state power and protecting the dignity of the human person in an unequal society gradually led to the formation of the institution of human rights protection. In this context, V. A. Kartashkin points to the complexity of the emergence of the principle of respect for and observance of human rights and fundamental freedoms in international law, since "until the beginning of the twentieth century human rights were regulated exclusively by domestic law, and states participating in international relations proceeded from the view that these issues belonged to their internal jurisdiction."

Internationally recognized human rights occupy a central place in the works of I. I. Lukashuk. It is no coincidence that this outstanding scholar, speaking of the need to form a legal system adequate to contemporary realities and directed toward solving global problems, names as its foundation "the principles of democracy, universally recognized human rights and the rule of law." In Lukashuk's entirely justified view, human rights are no longer an internal matter of states, and international control is not interference in domestic affairs. "The situation that has developed in this field is characterized by international instruments in such a way that questions relating to human rights, fundamental freedoms, democracy and the rule of law are international in character, since respect for these rights and freedoms constitutes one of the foundations of the international order." Moreover, "both international law and the constitutions of many states provide grounds for considering the right to a democratic system within a country to be one of the fundamental human rights. Otherwise human rights become an empty sound, since they can be secured only under conditions of democracy."

According to F. M. Rudinsky, "human rights are one variety of the subjective rights of the individual. In the modern understanding, human rights are the most essential possibilities of human development, inalienable attributes that determine the measure of human freedom." In many of his works, Rudinsky substantiates the idea of forming a new science - the science of human rights. V. A. Chetvernin notes that "by their nature, human rights are socially significant claims to a certain measure of freedom, compatible with the principle of formal equality, and capable of being universal demands of freedom." As O. I. Tiunov emphasizes, "in the broad sense, human rights are connected with the question of the relationship between the interests of society and the human being. Each person is part of society and must be regarded by society and the state as a person possessing value and the right to respect for their dignity. If one proceeds from the view that the purpose of society and the state is to affirm the principle of humanism, then the interests of the human being must be placed first."

M. M. Utyashev points to the distinction between freedom and right. "A right is a legitimate claim of a person to certain material and spiritual goods, the receipt of which he or she is empowered to claim; freedom is the possibility of acting at one's own discretion without any external interference or coercion, above all from power structures." At the same time, the scholar believes that "the long-proclaimed principle should become the practical norm of individual conduct: everything that is not prohibited by law is permitted."

E. A. Lukasheva believes that "the modern world cannot be imagined without human rights, which are based on the principles of freedom, equality and justice and have a universal character," and she speaks of the importance of universal and regional mechanisms for the protection of human rights. Speaking of the internationalization of these rights as a result of the adoption of international instruments, F. M. Rudinsky notes that they were "legally recognized by most states of the world, and a legal mechanism for their protection was formed." G. I. Tunkin also rightly notes that "the principle of respect for fundamental rights and freedoms has become one of the most important principles of international law. A new branch of international law is emerging, defining the obligations of states to ensure to all people, irrespective of race, language, religion or sex, fundamental rights and freedoms." In this context it should be especially emphasized that, in the competent view of V. A. Kartashkin, human rights do not belong exclusively to the domestic competence of states, whose sovereignty in this matter is substantially limited, while "the principles and norms relating to human rights are regarded by contemporary international law as jus cogens, operating erga omnes."

At the same time, other scholarly positions should also be noted. S. V. Chernichenko defends the view that the individual is endowed with rights and duties through the mechanism by which international norms are transformed into domestic law. Under conditions of globalization, the scholar emphasizes that the sphere of human rights belongs primarily to the internal competence of the state and opposes the use of human rights and freedom of conscience as instruments of political pressure. He considers that humanitarian intervention has no lawful basis in international law if it is carried out in circumvention of the UN Charter. In later works, the author stresses that international law should ensure a dialogue of cultures, not their forced unification.

In the view of T. A. Soshnikova, the legal mechanism for protecting rights and freedoms should be understood as "a set of legal norms, legal institutions and legal procedures by means of which the protection and restoration of legally protected human and civil rights and interests are carried out." At the same time, the role of the state mechanism within the structure of the above-mentioned mechanism for the protection of rights and freedoms is determined by the fact that all other forms and methods of protection are applied on the basis of, and in accordance with, legislation adopted by the state.

Accordingly, "the international protection of human rights consists in cooperation among states in two main directions: first, the creation of universal and regional international-legal norms in the field of human rights; second, the creation and operation of special mechanisms for monitoring their observance." Thus, the international protection of human rights, as a set of principles, norms and mechanisms directed toward the protection of the rights and fundamental freedoms of the individual, is recognized as one of the most important branches of contemporary public international law.

Wolfgang Benedek believes that "the desire to protect the human dignity of all people is the core of the concept of human rights. It places the human being at the center. It is based on a common universal system of values dedicated to the inviolability of life and provides the foundation for building a human rights system protected by internationally accepted norms and standards. During the twentieth century, human rights developed as a moral, political and legal basis and as guidance for the development of a world free from fear and want." All this indicates that "the aim and criterion of progressive social and state development is the human being, the reality of their rights and freedoms in combination with duties and responsibility for conduct," and also that "there is a close connection between states ensuring fundamental human rights and freedoms and maintaining international peace and security." Thus, "one of the most important results of the functioning of the universal system for the protection of human rights is the general recognition of the close interrelationship between development, security and the protection of human rights. At the same time, the realization of human rights is a necessary condition for ensuring development and security."

The materials of Oxfam International emphasize that respect for human rights will help lift people out of poverty and injustice, allow them to affirm their dignity and ensure sustainable development. Particular importance is attached to the fair distribution of the planet's resources and to counteracting the consequences of climate change in order to overcome hunger and ensure equal access to basic services, including health care and education.

The formation of this branch of international law took place over a long historical period, and it cannot be said that this process has been completed. Issues concerning respect for human rights were considered within the framework of the League of Nations and were included in bilateral and multilateral treaties of states. At a fundamentally new level, the protection of human rights was fixed in the documents of the United Nations (UN), of which most states are members, including the Russian Federation since 15 October 1945. According to V. A. Kartashkin, only with the creation of the United Nations did the stage of forming the modern system of international protection of human rights begin, with the UN Charter playing a special role in the development of cooperation among states. "The fact that the UN Charter, for the first time in international practice, fixed the principle of respect for human rights and obliged states to observe fundamental rights and freedoms received unanimous recognition." It is further emphasized that "the UN Charter is the first international-legal act that established an international organization one of whose purposes is based on universal recognition and respect for human rights. Provisions on human rights were also fixed in other articles of the UN Charter, namely Articles 2, 10, 13, 24, 34, 39, 55, 56, 60, 62, 68, 71, 75, 76, 87, 94 and 98, as well as in Article 34 of the Statute of the International Court of Justice, which is an integral part of the UN Charter." At the same time, unlike the League of Nations, "the UN Charter gives broad powers and provides the principal organs of the UN with substantial institutional possibilities in the promotion and protection of human rights for all. In addition to the general powers of the principal UN organs, auxiliary functional bodies were created in this field." The study by T. N. Neshataeva examines in detail the history of the creation of the UN system, as well as the mechanism by which the UN system and international organizations influence international law.

The UN Charter, which laid the foundation for modern international-legal protection and was adopted on 26 June 1945, speaks of the determination "to save succeeding generations from the scourge of war, which twice in our lifetime has brought untold sorrow to mankind, and to reaffirm faith in fundamental human rights, in the dignity and worth of the human person, in the equal rights of men and women and of nations large and small, and to establish conditions under which justice and respect for obligations arising from treaties and other sources of international law can be maintained, and to promote social progress and better standards of life in larger freedom." Special attention should be paid to the obligation of UN Member States "to achieve international cooperation in solving international problems of an economic, social, cultural or humanitarian character, and in promoting and encouraging respect for human rights and fundamental freedoms for all without distinction as to race, sex, language or religion."

Thus, "through the Charter, all UN Member States are legally bound in the struggle for the full realization of all human rights and freedoms," which in fact means the international recognition of the principle of respect for human rights. This principle was further developed in various international-legal instruments of both universal and regional character.

The Universal Declaration of Human Rights, adopted by UN General Assembly resolution 217 A (III) of 10 December 1948, "played a key role in the further development of the human rights mechanism and became the foundation on which the international system for protecting and implementing human rights was built." Article 1 of the Universal Declaration of Human Rights states that "all human beings are born free and equal in dignity and rights. They are endowed with reason and conscience and should act toward one another in a spirit of brotherhood." In accordance with Article 2, "everyone is entitled to all the rights and freedoms set forth in this Declaration, without distinction of any kind, such as race, colour, sex, language, religion, political or other opinion, national or social origin, property, birth or other status. Furthermore, no distinction shall be made on the basis of the political, jurisdictional or international status of the country or territory to which a person belongs, whether it is independent, trust, non-self-governing or under any other limitation of sovereignty."

Today, from the point of view of the United Nations, human rights are interdependent, indivisible, universal and inalienable, and states are obliged to respect, promote, protect and fulfil them without discrimination. "Under international law, states assume obligations to respect, protect and fulfil human rights. Respect for human rights means that the state does not interfere with the exercise of human rights and refrains from restricting rights. The obligation to protect human rights requires the state to prevent violations. Fulfilment of human rights obliges the state to guarantee the unhindered exercise of basic human rights."

However, the problems of the practical embodiment of human rights, and especially the limited possibility for the majority of the planet's population to enjoy social goods, make it possible to doubt the effectiveness of the above-mentioned UN approach. Moreover, the intensification of social differentiation in the modern world leads to the conclusion that the promotion, respect, protection and fulfilment of human rights, despite their significance and pleasing sound, do not in themselves guarantee the possibility for each person, without discrimination, to enjoy social goods for the purposes of development.

In particular, in 2019 Oxfam International noted that billionaire wealth had increased by 12 percent in the previous year, or by 2.5 billion dollars a day, while 3.8 billion people, constituting the poorest half of humanity, saw their wealth decline by 11 percent. The authors of the report point to the growing gap between rich and poor, which undermines the fight against poverty, damages the economy and fuels public anger throughout the world. This shows how governments aggravate inequality by underfunding public services, such as health care and education, while also taxing corporations and the wealthy inadequately and failing to curb tax evasion. It was also established that women and girls suffer most from growing economic inequality.

Under contemporary conditions, it appears necessary and timely to establish, as one of the fundamental principles of international law, the duty of its subjects to create effective conditions for the realization of human rights, including the very possibility sought here: the ability to enjoy the relevant goods.

In this context, F. M. Rudinsky identifies the following main stages of the realization of human rights: "Stage I - recognition of human rights in international covenants; Stage II - implementation of human rights in national legislation; Stage III - the emergence and existence of a citizen's right corresponding to a particular human right; Stage IV - the stage of readiness for realization; Stage V - direct realization, including the enjoyment of the personal good ensured by the corresponding right of the citizen."

At the same time, it should be noted that the recognition of human rights is preceded by the stage of scholarly and, in particular, theoretical-legal development, as well as inclusion in educational and awareness-raising programs at various levels of education. This level may conditionally be called the theoretical-legal level, and it is precisely this level that is system-forming in relation to the others, including those mentioned above. The theoretical-legal level largely predetermines the quality of the relevant principles and norms of international law, their implementation and the effectiveness of their application at domestic levels, as well as the possibility of protecting violated rights.

It seems necessary to listen to R. A. Podoprigora, who called for greater attention to the key conceptual apparatus in order to avoid extremely serious problems. "This is especially important for those states or legal systems where the main source of law is a normative legal act, since it initially sets the parameters that all participants in social relations must observe."

In other words, in relation to the subject of this study, the use of insufficiently correct or legally undefined key terms at the level of recognition and consolidation of the corresponding human rights in international covenants, as a rule, automatically and inevitably entails problems of implementation and law enforcement at domestic levels. In this case, it is hardly surprising that the possibility of enjoying the relevant goods, allegedly arising from internationally recognized human rights, in fact turns into fictitious good wishes.

With regard to the right to freedom of conscience, the theoretical problems mentioned above, including legally insufficiently defined formulations at the level of principles and norms of international law, the UN Charter, the International Bill of Human Rights and others, and consequently at the domestic level, predetermined substantial violations and the low effectiveness of human rights protection in the states of the modern world. It is clear that improvement should begin with overcoming theoretical-legal terminological problems and developing a legally correct conceptual and categorical apparatus that corresponds to the principle of the rule of law and modern legal technique.

After adequate theoretical-legal development, one may move to the next stage of international legal recognition in declarations and then to consolidation in legally binding forms of international law. This is followed by the stage of implementation of the principles and norms of international law, which presupposes the fulfilment by states of their international obligations through the inclusion of these principles and norms in their legal systems by means of a set of means, procedures, actions and methods. Here one should agree with I. I. Lukashuk, who considered that solving this question requires the existence of an appropriate mechanism in the form of a complex system of legal and organizational means.

This means that the implementation of legally correct principles and norms will contribute to an effective transition to the next stage, bringing closer the real possibility of enjoying goods in the interests of human development. If, however, the implemented principles and norms are not based on legally defined terms, the chances of realizing human rights are extremely small.

In the above context, the next extremely important stage in the realization of human rights is readiness for their embodiment at the domestic level, including the existence of a system of guarantees, including guarantees of protection from violations carried out not only at the domestic but also at the international level. At the same time, the system for protecting human rights can be effective only if the preceding stages have been successfully passed.

One should also agree with the view of scholars who consider the realization of human rights not only as an endless process, but also as a result in the form of obtaining, enjoying and disposing of a concrete good in the interests of the individual. At the same time, if we speak of forms of realization of human rights, realization may take place not only within legal relations, but also outside them. In particular, as F. M. Rudinsky notes, "it is also possible to distinguish such forms of realization of human rights as the use of rights or compliance with prohibitions, and the fulfilment of obligations provided by legal norms." In addition, "law enforcement, that is, the realization of legal norms by law enforcement bodies, should also be included among the forms of realization of human rights." V. O. Luchin likewise connects realization with implementation in life, noting that its mechanism "indicates orderliness and structure, subjects, objects and the nature of the links between them, and includes a variety of general social and legal factors, forms, methods, conditions and guarantees for the realization of norms in accordance with democratic procedures, the principles of legality and social justice." Speaking of forms of realization, the scholar identifies "observance, passive abstention from prohibited undesirable actions; execution, the obligatory performance of actions provided for by norms; use, permitted actions and the exercise of granted rights; and application, endowing some participants in relations with subjective rights, imposing corresponding duties on others and resolving questions of legal responsibility."

It is also necessary to mention the problem of understanding and implementing the prohibition of discrimination in international law. For example, O. V. Vashanova defines the principle of non-discrimination as "a universally recognized norm of international law consisting in the prohibition of distinctions in the legal status of people - their possession of unequal volumes of human rights and freedoms - provided that such a distinction: a) is based on natural or social characteristics of people, such as sex, race, religion and others, that do not affect the enjoyment of a specific right; b) is directed toward the violation of legal equality, that is, the equal possibility guaranteed by the state of exercising the same volume of human rights and freedoms."

At the UN level it is considered that "the principle of non-discrimination is interconnected with the principles of equality and freedom, applies to every person and extends to all human rights and freedoms, allowing no discrimination on any ground, whether sex, race, colour or any other ground." Analysis of the theory and practice of the international protection of human rights, also based on the principle of non-discrimination, shows that it presupposes equality of rights on the basis of respect for differences. Since the definitions of these differences do not always correspond to modern legal technique, in particular the principle of legal certainty, the effectiveness of the realization of the principle prohibiting discrimination is called into question. As a solution, I propose to understand the principle of non-discrimination as equality of rights irrespective of differences, provided that externally expressed conduct remains lawful.

Another problem is connected with the use of the term "standards" in relation to human rights. For example, in the view of E. G. Moiseev, the branch of international human rights law, the scholar considers this name more correct, "may conditionally be divided into two parts: the first includes questions connected with citizenship, its acquisition, change and loss, dual citizenship and statelessness; the rights of foreigners, stateless persons, refugees and forced migrants; the right of asylum. The second part of this branch includes international standards in the field of human rights, the activity of international organizations in ensuring and protecting human and civil rights and freedoms, and the policy and practice of the Russian Federation in this field."

As O. I. Tiunov notes, "international-legal standards are a variety of international norms that have a legal character, that is, they are part of the system of norms of international law. Consequently, we are speaking of standards that have international-legal significance. However, some international provisions, although they have the features of standards, have not yet been given legal force; they play the role of recommendations and are widely used unilaterally by particular state bodies." The scholar also gives a definition of universal international-legal standards, noting that these are "certain rules of conduct recognized by the states of the entire international community or by the predominant majority of them." It is specially stipulated that these standards include "one of the basic principles of international law - the principle of respect for human rights and fundamental freedoms, including freedom of thought, conscience, religion and belief."

F. M. Rudinsky likewise singles out the human rights fixed in the International Bill of Human Rights, considering that they "constitute international standards below which the level of ensuring these rights in individual states is impermissible." In the works of other authors, human rights standards directed toward the development of the principle of respect for human rights are usually understood as "generally recognized norms of state conduct that states must observe in legislation and practice in relation to all persons under their jurisdiction."

In addition, the authors mentioned above point to "regional standards developed by states of particular regions. There is no contradiction between universal and regional standards; rather, there is mutual influence. Regional standards complement universal standards, contribute to their strengthening and to the protection of human rights at the regional level. A separate group of human rights acts consists of universal international treaties aimed at suppressing crimes against humanity, war crimes, genocide and apartheid, and at combating gross and mass violations of human rights. A special group of universal treaties is formed by acts aimed at the additional protection of particular categories of persons, refugees, stateless persons, women, children and migrant workers."

"The application of international standards is not a right, but a constitutional duty of state bodies and officials," T. M. Pryakhina believes. G. I. Kurdyukov also considers that "standards are important for a universal understanding of the nature of human rights and freedoms and for the full fulfilment of obligations to respect and observe these rights. The conformity of human rights with international standards has been fixed in the concept of the human dimension."

At the same time, some scholars consider the use of the concept of "standards in the field of human rights", which are in substance recommendatory norms, not entirely correct. For example, according to A. Kh. Saidov, "it is more correct to speak not of international standards in the field of human rights fixed in international documents, but of universally recognized human rights. The term 'universally recognized human rights' more accurately reflects the role of international documents: they are developed at the inter-state level, are intended to regulate inter-state cooperation, and are addressed primarily to states that are able, within the limits of their jurisdiction, to ensure respect for human rights and that bear responsibility to their people and to the international community for the observance of human rights." The scholar believes that "the fixation of human rights in international documents makes them internationally universally recognized, that is, obligations of states. Thus, members of the international community agree not to fall below their level." This author also points to "three levels of sources of universally recognized human rights: 1. International sources of universally recognized human rights. 2. Regional sources of universally recognized human rights. 3. National sources of universally recognized human rights."

However, V. V. Ershov believes that "from the position of an integrative understanding of law, one may conclude that the law applied in Russia constitutes a single system of forms of international and Russian law, consisting of interdependent elements - forms of international and Russian law." In his works, the scholar substantiates the identification of the following forms of international law implemented in Russia: 1) fundamental or basic principles of international law; 2) international treaties; 3) customs of international law. He believes that "the source of international treaties is the agreement of states and other authorized subjects of international law. The source of the fundamental or general principles of international law is the activity of states and authorized specialized institutions or organizations, as well as custom as uniform and repeatedly repeated actions of subjects of international law aimed at protecting their interests and ensured by available legal and other forms of coercion." Indeed, the international level of human rights protection is closely interconnected with the domestic level, and "the question of the direction and nature of the influence of international law on the national law of Russia is one of the basic questions."

A number of scholars note that under conditions of globalization of social relations, the deepening interaction between international and national law, including in the field of human rights, is objective in character. At the same time, according to L. Kh. Mingazov, "the effectiveness of international-legal norms in many cases remains very low. In a number of areas of cooperation, they are sometimes violated. What is particularly important to note is that these violations occur in spheres connected with the maintenance of peace and security, the protection of human rights, environmental protection and so on. Norms of international law prohibiting the use of force in international relations, interference in the internal affairs of sovereign states, combating international terrorism and other crimes of an international character are insufficiently effective, or entirely ineffective." The insufficient effectiveness of international-legal principles and norms indicates that "the considerable potential embedded in international law is still far from fully realized in practice, and much remains to be done in its development and in increasing its effectiveness." The concept, criterion and conditions of effectiveness of international law norms are examined in the work of S. Yu. Marochkin. In particular, the co tion to law-making and to the quality of norms, as well as in relation to the operation and realization of norms.

Taking into account their significance under conditions of globalization of social relations, the extremely urgent problem of increasing the effectiveness of international principles and norms in the field of human rights, including the right to freedom of conscience, causes particular concern. In the context of the foregoing, it seems promising to study the role of institutions of international law not only in regulating international relations, but also in solving global problems that threaten the very existence of human civilization. The problem of the effectiveness of international-legal principles, norms and institutions in the sphere of freedom of conscience deserves special attention, since the ethno-confessional factor is one of the main obstacles on the path toward the integration of states for the solution of the problems mentioned above.

Summarizing the above, it should be noted that international law needs to fix legal obligations, rather than essentially recommendatory standards, of subjects of international law to form effective conditions for the realization of human rights, without discrimination, on the basis of equality of rights irrespective of differences, and directed toward making it possible to enjoy the relevant goods. Accordingly, the main stages of realization of human rights are: theoretical-legal development; consolidation in forms of international law; implementation in forms of domestic law; readiness for realization, including the existence of a system of guarantees, including protection from violations; and direct realization, meaning the possibility of enjoying the relevant goods without discrimination. This means that the contemporary international system for protecting human rights is in fact a necessary and important, but insufficient, component of the realization of human rights, including the right to freedom of conscience.

In addition to the main stages of realization of human rights, several auxiliary stages may be identified, arising from specific socio-political conditions: reflection within the system of education and enlightenment; perception and support by civil society; and support by the mass media.

Thus, the contemporary system for protecting the human right to freedom of conscience includes international and domestic subsystems interconnected through the mechanism of implementation and containing the corresponding principles, norms and mechanisms. In this context, it is necessary to note the need to move from a concept of protection to a concept of realization of human rights, which ultimately implies the possibility of enjoying the relevant goods in the interests of the development of every person irrespective of differences. In addition, it appears necessary to form a system of guarantees for the realization of the human right to freedom of conscience, above all the worldview neutrality of states and other subjects of law, as well as tolerance and resistance to intolerance.

4.2. International-Legal Principles and Norms for the Protection of the Right to Freedom of Conscience

4.2.1. Universal International-Legal Principles and Norms for the Protection of the Right to Freedom of Conscience

The universal international-legal principles and norms for the protection of human rights in the sphere of freedom of conscience include those fixed in universal international treaties and customs by the United Nations (UN) and its specialized agencies. Universal principles and norms are generally recognized, imperative and operate in relation to all states, since they regulate important and universal social relations, including the human right to freedom of conscience.

According to V. A. Kartashkin and E. A. Lukasheva, "the catalogue of human rights proclaimed in the Universal Declaration of Human Rights of 1948, based on European values, was proclaimed as universal. No fewer than 90 national constitutions adopted after 1948 contain a list of fundamental rights that either reproduce the provisions of the Declaration or were included under its influence. It truly expresses values without which the normal development of society based on the principles of freedom, the rule of law and human rights is impossible. The idea of the universality of human rights, formulated in the Universal Declaration of Human Rights of 1948 and other international-legal acts, was accepted as an axiom. For a number of years it did not raise doubts either as to its acceptability or as to the possibility of immediate realization in all countries of the modern world."

Thus, the United Nations, of which most states of the world are members, possesses planetary legitimacy. This means that the principles and norms of the UN are fundamental in the contemporary international-legal system for the protection of human rights, including in the sphere of freedom of conscience.

Principles and norms of the right to freedom of conscience are contained in a number of universal international legal documents, among which the following should be singled out: the UN Charter of 1945, the Universal Declaration of Human Rights of 1948, the International Covenant on Civil and Political Rights of 1966, the International Covenant on Economic, Social and Cultural Rights of 1966, the Convention on the Prevention and Punishment of the Crime of Genocide of 1948, the Convention against Discrimination in Education of 1960, the International Convention on the Elimination of All Forms of Racial Discrimination of 1965, the International Convention on the Suppression and Punishment of the Crime of Apartheid of 1973, the

Declaration on the Elimination of All Forms of Intolerance and of Discrimination Based on Religion or Belief of 1981, the Declaration of Principles on Tolerance of 1995 and others.

The UN Charter was signed on 26 June 1945 in San Francisco at the final meeting of the United Nations Conference on International Organization and entered into force on 24 October 1945. The Statute of the International Court of Justice forms an integral part of the Charter. Article 1 of the UN Charter defines the purposes of the Organization, among which human rights occupy an important place: 1) to maintain international peace and security and, to that end, to take effective collective measures for the prevention and removal of threats to the peace and for the suppression of acts of aggression or other breaches of the peace, and to bring about, by peaceful means and in conformity with the principles of justice and international law, adjustment or settlement of international disputes or situations which might lead to a breach of the peace; 2) to develop friendly relations among nations based on respect for the principle of equal rights and self-determination of peoples, and to take other appropriate measures to strengthen universal peace; 3) to achieve international cooperation in solving international problems of an economic, social, cultural or humanitarian character and in promoting and encouraging respect for human rights and fundamental freedoms for all without distinction as to race, sex, language or religion; and 4) to be a center for harmonizing the actions of nations in the attainment of these common ends.

The Universal Declaration of Human Rights

The Universal Declaration of Human Rights, adopted in December 1948 by UN General Assembly resolution 217 A (III), is not a legally binding international-legal document, but it is an authoritative source and model for the consolidation of human rights in the sphere of freedom of conscience. In this context, V. A. Kartashkin notes that "when assessing the legal force of its provisions, it should be taken into account that in international law, alongside treaty, custom also plays a significant role, formed as a result of the international practice of states and gradually recognized by them as a legal norm. The fundamental rights and freedoms proclaimed in the Universal Declaration are now regarded by states as legally binding customary or treaty norms. The constitutions of many countries directly refer to this document and include a number of its provisions. The principles and norms set out in the Declaration are constantly developing through the conclusion of new international agreements, which testifies to the realism of the document and its great significance for the development of interstate relations."

In accordance with Article 18 of the Universal Declaration of Human Rights, "everyone has the right to freedom of thought, conscience and religion; this right includes freedom to change his religion or belief, and freedom, either alone or in community with others and in public or private, to manifest his religion or belief in teaching, practice, worship and observance."

Article 19 states that "everyone has the right to freedom of opinion and expression; this right includes freedom to hold opinions without interference and to seek, receive and impart information and ideas through any media and regardless of frontiers."

It should be especially noted that the Universal Declaration refers to the provisions fixed in it as "a common standard of achievement for all peoples and all nations." In accordance with Article 20: "1. Everyone has the right to freedom of peaceful assembly and association. 2. No one may be compelled to belong to an association."

Article 26 speaks of everyone's right to education. In accordance with paragraph 2, "education shall be directed to the full development of the human personality and to the strengthening of respect for human rights and fundamental freedoms. It shall promote understanding, tolerance and friendship among all nations, racial and religious groups, and shall further the activities of the United Nations for the maintenance of peace."

Paragraphs 2 and 3 of Article 29 of the Universal Declaration of Human Rights define limitations in the exercise of rights and freedoms. "In the exercise of his rights and freedoms, everyone shall be subject only to such limitations as are determined by law solely for the purpose of securing due recognition and respect for the rights and freedoms of others and of meeting the just requirements of morality, public order and the general welfare in a democratic society. These rights and freedoms may in no case be exercised contrary to the purposes and principles of the United Nations."

In subsequent years, the provisions of the Universal Declaration of Human Rights in the sphere of freedom of conscience were developed in the human rights covenants, which, according to V. A. Kartashkin, "became the framework and foundation of the entire legal system in the field of human rights."

The International Covenant on Civil and Political Rights

The International Covenant on Civil and Political Rights was adopted and opened for signature, ratification and accession by UN General Assembly resolution 2200 A (XXI) on 16 December 1966 and entered into force on 23

March 1976. The Covenant is an international treaty binding on the states parties.

However, when it was adopted, sharp controversy arose over the key formulations fixing the right to freedom of conscience. "Taking into account the different attitudes toward religion among the states that participated in drafting the Covenant, Article 18 is of a compromise character. At the same time, at the insistence of the Soviet Union and a number of other countries, paragraph 3 of Article 18 of the Covenant stated that freedom to manifest religion or belief may be subject to limitations. Representatives of Western countries, however, insisted on including in the text the provision on the obligation of states to respect the freedom of parents and, where applicable, legal guardians to ensure the religious and moral education of their children in conformity with their own convictions (paragraph 4 of Article 18)."

Article 2 of the Covenant states that "each State Party to the present Covenant undertakes to respect and to ensure to all individuals within its territory and subject to its jurisdiction the rights recognized in the Covenant, without distinction of any kind, such as race, colour, sex, language, religion, political or other opinion, national or social origin, property, birth or other status."

Article 18 of the Covenant guarantees: "1. Everyone shall have the right to freedom of thought, conscience and religion. This right shall include freedom to have or to adopt a religion or belief of his choice, and freedom, either individually or in community with others and in public or private, to manifest his religion or belief in worship, observance, practice and teaching. 2. No one shall be subject to coercion which would impair his freedom to have or to adopt a religion or belief of his choice. 3. Freedom to manifest one's religion or beliefs may be subject only to such limitations as are prescribed by law and are necessary to protect public safety, order, health or morals or the fundamental rights and freedoms of others. 4. The States Parties to the present Covenant undertake to respect the liberty of parents and, when applicable, legal guardians to ensure the religious and moral education of their children in conformity with their own convictions."

Under Article 19, "everyone shall have the right to hold opinions without interference and to freedom of expression; this right shall include freedom to seek, receive and impart information and ideas of all kinds, regardless of frontiers, either orally, in writing or in print, in the form of art, or through any other media of his choice."

However, paragraph 3 of this article reminds us that "the exercise of the rights provided for in paragraph 2 of this article carries with it special duties and responsibilities." Here we are speaking of limitations, some of which lack clear legal criteria: "It may therefore be subject to certain restrictions, but these shall only be such as are provided by law and are necessary: a) for respect of the rights or reputations of others; b) for the protection of national security or of public order, or of public health or morals."

Article 24 of the Covenant guarantees protection of the rights of the child "without any discrimination as to race, colour, sex, language, religion, national or social origin, property or birth."

In accordance with Article 26, "all persons are equal before the law and are entitled without any discrimination to the equal protection of the law. In this respect, the law shall prohibit any discrimination and guarantee to all persons equal and effective protection against discrimination on any ground, such as race, colour, sex, language, religion, political or other opinion, national or social origin, property, birth or other status."

Article 27 provides additional guarantees for the protection of minorities. "In those states in which ethnic, religious or linguistic minorities exist, persons belonging to such minorities shall not be denied the right, in community with the other members of their group, to enjoy their own culture, to profess and practise their own religion, or to use their own language."

In accordance with the Optional Protocol, 102 states parties to the Covenant recognize the competence of the Human Rights Committee to consider confidential complaints received from citizens who consider themselves victims of violations of the human rights provided for in the Covenant.

To monitor implementation by states parties, and pursuant to Part IV (Articles 28-45) of the International Covenant on Civil and Political Rights, the UN Human Rights Committee was established. Among the powers of the Committee are the consideration of reports and complaints of violations, as well as the development of general comments of a recommendatory character.

In accordance with Article 40 of the Covenant, states parties, at the request of the Committee, "undertake to submit reports on the measures they have adopted which give effect to the rights recognized herein and on the progress made in the enjoyment of those rights." Reports are submitted to the Secretary-General of the United Nations, who transmits them to the Committee for consideration. "The reports shall indicate the factors and difficulties, if any, affecting the implementation of the present Covenant. The Secretary-General of the United Nations, after consultation with the Committee, may transmit to the specialized agencies concerned copies of those parts of the reports which may fall within their field of competence. The Committee shall study the reports submitted by the States Parties. It shall transmit its reports and such general comments as it may consider appropriate to the States Parties. The Committee may also transmit to the Economic and Social Council these comments along with copies of the reports received from States Parties."

The International Covenant on Economic, Social and Cultural Rights

The International Covenant on Economic, Social and Cultural Rights was adopted by General Assembly resolution 2200 A (XXI) of 16 December 1966 and entered into force on 3 January 1976. The states parties undertook to take steps, by all appropriate means, including in particular the adoption of legislative measures, to ensure progressively the full realization of the rights recognized in the Covenant. In particular, states undertook that "the rights enunciated in the present Covenant will be exercised without discrimination of any kind as to race, colour, sex, language, religion, political or other opinion, national or social origin, property, birth or other status."

Article 13 of the Covenant fixed the right of everyone to education, which "shall be directed to the full development of the human personality and the sense of its dignity, and shall strengthen respect for human rights and fundamental freedoms," and which "shall enable all persons to participate effectively in a free society, promote understanding, tolerance and friendship among all nations and all racial, ethnic or religious groups, and further the activities of the United Nations for the maintenance of peace."

In accordance with paragraph 3 of Article 13, the states parties to the Covenant undertake "to respect the liberty of parents and, when applicable, legal guardians to choose for their children schools other than those established by public authorities, which conform to such minimum educational standards as may be laid down or approved by the state, and to ensure the religious and moral education of their children in conformity with their own convictions."

It is necessary once again to turn to V. A. Kartashkin, who, noting the special role of the Covenants in the development of the architecture of human rights in the modern world, observes that "over fifty years seven further key treaties listed above were adopted, covering most human rights and freedoms. But for many years to come the Covenants will serve as the basis on which cooperation among states in the sphere of human rights continues to develop." The well-known scholar reminds us that "all states are obliged to undertake legislative, administrative and other measures that may prove necessary for the practical implementation of all rights fixed in these fundamental international treaties."

The Declaration on the Elimination of All Forms of Intolerance and of Discrimination Based on Religion or Belief

The Declaration on the Elimination of All Forms of Intolerance and of Discrimination Based on Religion or Belief is recommendatory in nature, but, in the opinion of D. Davis, it is "one of the most important international documents protecting religious rights and prohibiting intolerance or discrimination with regard to religion or belief." The document was adopted as General Assembly resolution 36/55 A (XXI) of 13 November 1981.

"The 1981 Declaration on the Elimination of All Forms of Intolerance and of Discrimination Based on Religion or Belief is a document that creates no legally binding obligation to implement its provisions and, accordingly, provides no supervisory mechanism. Instead, through the so-called extra-conventional mechanism, the Commission on Human Rights appointed a Special Rapporteur (an independent expert) to ensure the implementation of the provisions of the 1981 Declaration. The Special Rapporteur is entrusted with the duty of reporting annually to the Commission on the state of the right to freedom of religion and belief throughout the world."

The preamble notes the special role of freedom of thought, conscience, religion or belief, the disregard or violation of which "have brought, directly or indirectly, wars and great suffering to mankind."

Article 1 of the Declaration repeats the above-mentioned formulations concerning freedom of thought, conscience and religion. Pursuant to paragraph 1 of Article 2 of the Declaration, "no one shall be subject to discrimination by any state, institution, group of persons, or person on the grounds of religion or other belief."

Paragraph 2 of Article 2 defines the concept of "intolerance and discrimination based on religion or belief" as "any distinction, exclusion, restriction or preference based on religion or belief and having as its purpose or as its effect nullifying or impairing the recognition, enjoyment or exercise, on an equal basis, of human rights and fundamental freedoms."

Article 3 states: "Discrimination between human beings on the grounds of religion or belief constitutes an affront to human dignity and a disavowal of the principles of the Charter of the United Nations, and shall be condemned as a violation of the human rights and fundamental freedoms proclaimed in the Universal Declaration of Human Rights and enunciated in detail in the International Covenants on Human Rights, and as an obstacle to friendly and peaceful relations between nations."

Article 4 of the Declaration obliges all states "to take effective measures to prevent and eliminate discrimination on the grounds of religion or belief in the recognition, exercise and enjoyment of human rights and fundamental freedoms in all fields of civil, economic, political, social and cultural life. All states shall make all efforts to enact or rescind legislation where necessary to prohibit any such discrimination, and to take all appropriate measures to combat intolerance on the grounds of religion or other beliefs in this matter."

Article 5 of the Declaration protects children from any form of discrimination based on religion or belief and guarantees the right of parents or legal guardians to raise children in accordance with their convictions.

Article 6 specifies the content of the right to freedom of thought, conscience, religion or belief, which includes, in particular, "the following freedoms: (a) to worship or assemble in connection with a religion or belief, and to establish and maintain places for these purposes; (b) to establish and maintain appropriate charitable or humanitarian institutions; (c) to make, acquire and use to an adequate extent the necessary articles and materials related to the rites or customs of a religion or belief; (d) to write, issue and disseminate relevant publications in these areas; (e) to teach a religion or belief in places suitable for these purposes; (f) to solicit and receive voluntary financial and other contributions from individuals and institutions; (g) to train, appoint, elect or designate by succession appropriate leaders called for by the requirements and standards of any religion or belief; (h) to observe days of rest and celebrate holidays and ceremonies in accordance with the precepts of one's religion or belief; and (i) to establish and maintain communications with individuals and communities in matters of religion and belief at the national and international levels."

It should be noted that although the Declaration on the Elimination of All Forms of Intolerance and of Discrimination Based on Religion or Belief "does not have binding status, it is an official UN document and possesses a high level of moral suasion. The Declaration undoubtedly has a certain legal weight, since it calls for respect for standards protecting religious human rights. As B. Tahzib observed: "States regard the 1981 Declaration, or at least some of its provisions, as normative and as part of international law."

The Declaration on the Rights of Persons Belonging to National or Ethnic, Religious and Linguistic Minorities

The Declaration on the Rights of Persons Belonging to National or Ethnic, Religious and Linguistic Minorities, also recommendatory in nature, was adopted by General Assembly resolution 47/135 of 18 December 1992. Since minority issues are complex, and proceeding from the interests of political and social stability of states, this Declaration is directed toward the implementation of the principles contained in the following documents: the UN Charter, the Universal Declaration of Human Rights, the Convention on the Prevention and Punishment of the Crime of Genocide, the International Convention on the Elimination of All Forms of Racial Discrimination, the International Covenant on Economic, Social and Cultural Rights, the Declaration on the Elimination of All Forms of Intolerance and of Discrimination Based on Religion or Belief, the Convention on the Rights of the Child and others.

As A. Kh. Abashidze notes, "the aim of contemporary international law should be to define not the concept of a 'minority' in general, but only that part of it which falls under international protection. With this approach, the concept of 'minority' should be understood as a social phenomenon, and it ceases to be an ethnic or biological phenomenon." No less important is his observation that "the right of self-identification is the prerogative of minorities themselves and of persons belonging to them."

By adopting the Declaration on the Rights of Persons Belonging to National or Ethnic, Religious and Linguistic Minorities, the General Assembly reaffirmed that one of the fundamental purposes of the United Nations, as proclaimed in the Charter, is "to promote and encourage respect for human rights and fundamental freedoms for all, without distinction as to race, sex, language or religion," as well as faith "in fundamental human rights, in the dignity and worth of the human person, in the equal rights of men and women and of nations large and small."

According to Article 1, "states shall protect the existence and the national or ethnic, cultural, religious and linguistic identity of minorities within their respective territories and shall encourage conditions for the promotion of that identity," and shall also "adopt appropriate legislative and other measures to achieve those ends."

Articles 2-4 affirm the rights of persons belonging to national or ethnic, religious and linguistic minorities, hereinafter referred to as persons belonging to minorities, individually as well as in community with other members of their group and without any discrimination: "to enjoy their own culture, to profess and practise their own religion, and to use their own language, in private and in public, freely and without interference or any form of discrimination"; "to participate effectively in cultural, religious, social, economic and public life"; "to participate effectively in decisions on the national and, where appropriate, regional level concerning the minority to which they belong or the regions in which they live, in a manner not incompatible with national legislation"; "to establish and maintain their own associations"; "to establish and maintain, without any discrimination, free and peaceful contacts with other members of their group and with persons belonging to other minorities, as well as contacts across frontiers with citizens of other states to whom they are related by national or ethnic, religious or linguistic ties"; "to exercise fully and effectively all their human rights and fundamental freedoms without any discrimination and in full equality before the law"; and "to express their characteristics and to develop their culture, language, religion, traditions and customs, except where specific practices are in violation of national law and contrary to international standards."

Paragraphs 2 and 3 of Article 8 of the Declaration state that "the exercise of the rights set forth in the present Declaration shall not prejudice the enjoyment by all persons of universally recognized human rights and fundamental freedoms," and that "measures taken by states to ensure the effective enjoyment of the rights set forth in the present Declaration shall not prima facie be considered contrary to the principle of equality contained in the Universal Declaration of Human Rights."

Mechanisms for the protection of minority rights include the UN Commission on Human Rights and the ECOSOC Sub-Commission on Prevention of Discrimination and Protection of Minorities. In addition to increasing their effectiveness, A. Kh. Abashidze proposes the creation of a judicial body for the protection of minorities and the development of "a comprehensive programme of action that would cover measures to eliminate ethnic, linguistic and religious discrimination in combination with the promotion of the rights of representatives of ethnic, religious and linguistic minorities, on the basis of respect for the territorial integrity of states."

At the same time, certain research findings indicate that the special identification of the social group of "religious minorities" is not beyond dispute in contemporary conditions. Historically, the need to protect "religious minorities" was not in doubt. However, in connection with the formation of an international system for the protection of the rights of every person without discrimination, on the basis of equality and irrespective of differences, this approach requires improvement. It appears that the attempt to realize the worthy idea of protecting "religious minorities," under conditions in which fundamental principles and the conceptual apparatus remain insufficiently developed, leads to the ineffectiveness of the relevant norms. Moreover, from the point of view of modern legal technique, it appears debatable and creates preconditions for serious problems in the realization of freedom of conscience.

However, as A. Kh. Abashidze notes, "the difference between the concept of 'minority protection' and the formula of 'equality and non-discrimination' consists in the fact that the latter provides a formal guarantee of equal treatment for all individuals in the enjoyment of rights and the performance of obligations, whereas minority protection provides for special measures to ensure their real equality with the rest of the country's population."

To a certain extent one may agree that, while the realization of the right of everyone to freedom of conscience remains far away, religious and other minorities must be protected. Yet this must not be done at the expense of the realization of the right of every person to free worldview choice irrespective of worldview differences.

The Convention on the Rights of the Child

The Convention on the Rights of the Child was adopted by General Assembly resolution 44/25 of 20 November 1989 and is binding on the states parties. It entered into force on 2 September 1990. The Convention was preceded by the adoption in 1959 by the General Assembly of the non-binding Declaration of the Rights of the Child.

The states parties to the Convention undertook to respect and ensure all rights of the child without discrimination, and, according to Article 14: "1. States Parties shall respect the right of the child to freedom of thought, conscience and religion. 2. States Parties shall respect the rights and duties of the parents and, when applicable, legal guardians, to provide direction to the child in the exercise of their right in a manner consistent with the evolving capacities of the child. 3. Freedom to manifest one's religion or beliefs may be subject only to such limitations as are prescribed by law and are necessary to protect public safety, order, health or morals, or the fundamental rights and freedoms of others."

Article 20 provides for the right to special protection and assistance provided by the state for a child "temporarily or permanently deprived of their family environment, or in whose own best interests cannot be allowed to remain in that environment." In particular, "States Parties shall in accordance with their national laws ensure alternative care for such a child. Such care could include, inter alia, foster placement, kafalah of Islamic law, adoption or, if necessary, placement in suitable institutions for the care of children. When considering solutions, due regard shall be paid to the desirability of continuity in a child's upbringing and to the child's ethnic, religious, cultural and linguistic background."

In implementation of Part II of the Convention on the Rights of the Child, and for the purpose of examining the progress made by states parties in fulfilling the obligations undertaken under the Convention, the Committee on the Rights of the Child was established. "In order to foster the effective implementation of the Convention and to encourage international cooperation in the field covered by the Convention, the Committee may invite the specialized agencies, the United Nations Children's Fund and other competent bodies, as it may consider appropriate, to provide expert advice on the implementation of the Convention in areas falling within the scope of their respective mandates; the Committee may invite specialized agencies, the United Nations Children's Fund and other United Nations organs to submit reports on the implementation of the Convention in areas falling within the scope of their activities; the Committee shall transmit, as it may consider appropriate, to the specialized agencies, the United Nations Children's Fund and other competent bodies any reports from states parties that contain a request, or indicate a need, for technical advice or assistance, along with the Committee's observations and suggestions, if any, on these requests or indications; the Committee may recommend to the General Assembly that it request the Secretary-General to undertake studies on specific issues relating to the rights of the child; the Committee may make suggestions and general recommendations based on information received pursuant to Articles 44 and 45 of the Convention. Such suggestions and general recommendations shall be transmitted to any state party concerned and reported to the General Assembly together with comments, if any, from states parties."

Unfortunately, a number of provisions of the Convention on the Rights of the Child that concern questions of freedom of conscience contain the already mentioned fundamental scientific and theoretical contradictions that hinder their practical implementation.

Among the instruments adopted within the framework of the United Nations Educational, Scientific and Cultural Organization (UNESCO), special attention should be paid to the Convention against Discrimination in Education, adopted by the General Conference of UNESCO at its eleventh session on 14 December 1960, and the Declaration of Principles on Tolerance, approved by resolution 5.61 of the UNESCO General Conference of 16 November 1995.

The Convention against Discrimination in Education

The Convention against Discrimination in Education, which has a recommendatory character in the author's treatment of this context, was adopted by the General Conference of the United Nations Educational, Scientific and Cultural Organization at its eleventh session on 14 December 1960.

Paragraph 1 of the document deserves attention, since it defines the concept of "discrimination" as covering "any distinction, exclusion, limitation or preference which, being based on race, colour, sex, language, religion, political or other opinion, national or social origin, economic condition or birth, has the purpose or effect of nullifying or impairing equality of treatment in education and in particular: (a) depriving any person or group of persons of access to education of any type or at any level; (b) limiting any person or group of persons to education of an inferior standard; (c) establishing or maintaining separate educational systems or institutions for persons or groups of persons, except as provided in Section II of the present Recommendation; or (d) inflicting on any person or group of persons conditions incompatible with human dignity."

The Declaration of Principles on Tolerance

Since the realization of human rights in the sphere of freedom of conscience depends to a very significant degree on respect for the principle of equality, tolerance, understood as mutual respect irrespective of differences, plays an important role. The Declaration of Principles on Tolerance was approved by resolution 5.61 of the UNESCO General Conference of 16 November 1995.

Article 1 of the Declaration provides a modern and detailed definition of a key concept inseparably connected with freedom of conscience:

"1.1. Tolerance means respect, acceptance and appreciation of the rich diversity of our world's cultures, our forms of expression and ways of being human. It is fostered by knowledge, openness, communication and freedom of thought, conscience and belief. Tolerance is harmony in difference. It is not only a moral duty, it is also a political and legal requirement. Tolerance is the virtue that makes peace possible and contributes to the replacement of the culture of war by a culture of peace.

1.2. Tolerance is not concession, condescension or indulgence. Tolerance is, above all, an active attitude prompted by recognition of the universal human rights and fundamental freedoms of others. Under no circumstances can tolerance be used to justify infringements of these fundamental values. Tolerance is to be exercised by individuals, groups and states.

1.3. Tolerance is the responsibility that upholds human rights, pluralism, including cultural pluralism, democracy and the rule of law. Tolerance involves the rejection of dogmatism and absolutism and affirms the standards set out in international human rights instruments.

1.4. Consistent with respect for human rights, the practice of tolerance does not mean toleration of social injustice, or the abandonment or weakening of one's convictions. It means that one is free to adhere to one's own convictions and accepts that others may adhere to theirs. It means accepting the fact that human beings, naturally diverse in their appearance, situation, speech, behaviour and values, have the right to live in peace and to be as they are. It also means that one's views are not to be imposed on others."

Article 2 defines the main directions of state policy in achieving tolerance and counteracting intolerance:

"2.1. At the state level, tolerance requires just and impartial legislation, law enforcement and judicial and administrative process. Tolerance also requires that economic and social opportunities be made available to each person without any discrimination. Exclusion and marginalization can lead to frustration, hostility and fanaticism.

2.2. In order to achieve a more tolerant society, states should ratify existing international human rights conventions and, where necessary, draft new legislation to ensure equality of treatment and equality of opportunity for all groups and individuals in society.

2.3. For international harmony it is essential that individuals, communities and nations accept and respect the multicultural character of the human family. Peace is impossible without tolerance, and development and democracy are impossible without peace.

2.4. Intolerance may take the form of marginalization of vulnerable groups and their exclusion from social and political participation, as well as violence and discrimination against them. As the Declaration on Race and Racial Prejudice states, 'all individuals and groups have the right to be different' (Article 1.2)."

Article 3 reveals the social aspects, significance of tolerance and the strengthening of relevant cooperation in the modern globalizing world:

"3.1. Tolerance is more important than ever in the modern world. We live in an age marked by the globalization of the economy and by increasingly rapid mobility, communication, integration and interdependence, large-scale migrations and displacement of populations, urbanization and changing social patterns. Since every part of the world is characterized by diversity, escalating intolerance and conflict potentially threaten every region. This threat cannot be confined within national borders, for it is global in character.

3.2. Tolerance is necessary between individuals and at the level of the family and community. In schools and universities, through non-formal education, at home and in the workplace, a spirit of tolerance should be strengthened and attitudes of openness, mutual attention and solidarity should be formed. Means of communication can play a constructive role in facilitating free and open dialogue and discussion, disseminating the values of tolerance and explaining the danger of indifference toward groups and ideologies that preach intolerance and are gaining strength.

3.3. The UNESCO Declaration on Race and Racial Prejudice proclaims that special measures must be taken, wherever necessary, to ensure equality in dignity and rights for individuals and groups. In this connection, particular attention should be paid to vulnerable groups in adverse social or economic circumstances, so as to provide them with legal and social protection, especially in respect of housing, employment and health care, to respect the authenticity of their culture and values, and to facilitate, particularly through education, their social and occupational advancement and integration.

3.4. In order to address this global challenge, appropriate scientific research and networking are necessary to coordinate the work of the international community, including social-science analysis of the root causes of this phenomenon, effective countermeasures, and research and monitoring that support policy-making and standard-setting by member states."

Article 4 defines education as the most effective means of preventing intolerance. It provides that "education for tolerance begins with teaching people what their shared rights and freedoms are, in order to ensure the exercise of those rights, and with encouraging the will to protect the rights of others.

4.2. Education for tolerance should be regarded as an urgent imperative; for this reason, systematic and rational methods of teaching tolerance should be promoted, methods that address the cultural, social, economic, political and religious sources of intolerance that lie at the roots of violence and exclusion. Education policies and programmes should contribute to improving mutual understanding, strengthening solidarity and tolerance in relations among individuals, as well as among ethnic, social, cultural, religious and linguistic groups and nations.

4.3. Education in a spirit of respect should be directed toward countering influences that lead to fear and alienation toward others. It should help young people develop capacities for independent thinking, critical reflection and judgement based on moral values.

4.4. We declare our readiness to support and implement programmes of social-science research and education for tolerance, human rights and non-violence. This means that special attention must be paid to improving teacher training, curricula, the content of textbooks and lessons, and other educational materials, including new educational technologies, with the aim of educating caring and responsible citizens who are open to other cultures, capable of appreciating freedom, respecting human dignity and individuality, and preventing conflicts or resolving them by non-violent means."

Thus, states have undertaken to promote tolerance and non-violence, using for this purpose programmes and institutions in the fields of education, science, culture and communication.

In addition to the instruments examined above, certain aspects connected with freedom of conscience are addressed in the Convention on the Prevention and Punishment of the Crime of Genocide of 1948, the Convention relating to the Status of Refugees of 1951, the Convention relating to the Status of Stateless Persons of 1954, the Convention on the Elimination of All Forms of Discrimination against Women of 1979 and other universal instruments.

The Recommendation on Education for Peace and Human Rights, International Understanding, Cooperation, Fundamental Freedoms, Global Citizenship and Sustainable Development (2023)

The Recommendation on Education for Peace and Human Rights, International Understanding, Cooperation, Fundamental Freedoms, Global Citizenship and Sustainable Development (2023) deserves mention. It was adopted on 20 November 2023 at the forty-second session of the UNESCO General Conference in Paris. This document updated the 1974 Recommendation concerning Education for International Understanding, Co-operation and Peace and Education relating to Human Rights and Fundamental Freedoms. It adapts the previous framework to contemporary conditions, including digitalization, climate challenges, social inequality and increasing disinformation.

The main purpose of the 2023 Recommendation is to provide Member States with guidance in their efforts to embed in education, through a multilateral approach, the aims and objectives set forth in the Charter of the United Nations, the Constitution of UNESCO, the International Bill of Human Rights and other relevant international human rights instruments.

Several key aspects of the Recommendation should be noted. First, the document links different spheres: digital technologies, climate change, gender issues, fundamental freedoms, and media and information literacy. It recognizes that education in all its forms and at all levels, both in schools and beyond them, shapes perceptions of the world and attitudes toward other people, and can become a path toward building lasting peace.

The Recommendation describes the principles that should determine the transformation of education systems in the coming decades. These principles cover all aspects, from law and policy to curriculum development, teaching practice, learning environments and assessment.

The document defines the knowledge, skills, values, attitudes and behaviours that learners should acquire through education for life in peace. These include empathy, critical thinking, intercultural understanding and environmental stewardship.

The Recommendation includes areas in which stakeholders, such as policy-makers, teachers, non-formal educators and others, may apply guiding principles for change. For example, teachers may adapt lesson plans to integrate themes of peace, human rights and sustainable development, while local community leaders may advocate changes in policy and curricula.

Promoting the philosophy, principles and components of global citizenship education is recognized as important for preparing learners who value human dignity, cooperation and dialogue. This may include studying the impact of past and current events and conflicts, economic, social and political links between countries, and fostering empathy and respect for cultural diversity and diversity of opinions.

The promotion of gender equality and the recognition of its importance for the realization of the right to education for all are among the guiding principles of the Recommendation and reflect a global priority of UNESCO.

The document emphasizes the problems of disinformation and incitement to hatred, as well as the opportunities offered by new technologies in teaching and learning. Special attention is paid to critical thinking and to understanding the principles of digital safety, confidentiality and ethical online interaction.

The Recommendation stresses the need to create guaranteed safe and inclusive learning conditions, free from any violence, for all learners and educational communities.

Overall, the document serves as a global standard defining how education should be used to achieve lasting peace and promote human development. It also serves as a legal basis for Sustainable Development Goal (SDG) 4.7, which concerns education for sustainable development and global citizenship.

It should be especially emphasized that the 2023 Recommendation on Education for Peace and Human Rights, International Understanding, Cooperation, Fundamental Freedoms, Global Citizenship and Sustainable Development points to "the importance of preventing and suppressing wars, aggression and all forms of violence and violations of human rights, and instilling in each individual and community an understanding and acceptance of responsibility for maintaining, strengthening and defending peace." It also notes that "education should also contribute to activities aimed at combating racism, xenophobia and all acts and ideologies that incite hatred, all forms of intolerance, discrimination and violence." In particular, "education should be transformative, lay a solid foundation for literacy and numeracy, and promote the development of knowledge, skills, values, attitudes and behaviours," including: (a) analytical and critical thinking, namely the ability to question norms, practices and opinions, critically analyze and understand complex systems and multicultural environments, and understand power dynamics and interconnections among countries, populations and the natural environment, as well as between local, national, regional and global levels; (b) anticipatory skills, namely the ability to act as agents of change, assess and understand emerging and future opportunities and threats, and adapt to new possibilities in order to promote peaceful, just, equitable, inclusive, healthy and sustainable futures for all; (c) respect for diversity, namely the ability to understand, value and respect the equal dignity and rights of every person, irrespective of race, colour, descent, sex, age, language, religion, political opinion, national, ethnic or social origin, economic or social condition at birth, disability and any other grounds, as well as their needs, views and peaceful actions, and to value their diverse knowledge systems and learning activities; (d) self-awareness, namely the ability to recognize and critically reflect on one's personal values, perceptions and actions, to know, value and live in harmony with oneself, to understand and manage emotions, to feel and show empathy and respect for others and for one's role in the local, national, regional and global community; (e) a sense of connectedness and belonging to a common and diverse humanity and to planet Earth, namely understanding humanity as a global community sharing responsibility for a healthy planet and for respect for the needs and rights of one another, of other living beings and of nature itself; (f) empowerment, agency and resilience, namely motivation, confidence and the capacity to act and respond effectively, proactively, consciously and responsibly to challenges at local, national, regional and global levels, and to cope with risks, conflicting objectives and uncertainty; (g) decision-making skills, namely the ability to assess the impact of actions and make decisions using available information from diverse and reliable sources; (h) cooperation skills, namely the ability to express feelings and opinions effectively and constructively, participate in joint work, stakeholder planning and collaborative problem-solving marked by responsible and respectful conduct; (i) adaptive and creative skills, namely the ability to adapt, engage, create, innovate and succeed in a rapidly changing environment and in diverse and variable conditions so that new ideas can be translated into action; (j) citizenship skills, namely the ability to act ethically and responsibly and to participate fully in civic and social life in the digital age and in local, national and global contexts; and (k) skills for the peaceful resolution and transformation of conflicts, considerable extent, to the Framework Convention for the Protection of National Minorities.

Opinion No. 193 of the Parliamentary Assembly of the Council of Europe on Russia's Application for Membership of the Council of Europe

Opinion No. 193 of the Parliamentary Assembly of the Council of Europe concerned Russia's application for membership of the Council of Europe and was adopted in Strasbourg on 25 January 1996. As is known, the Russian Federation applied for membership of the Council of Europe on 7 May 1992. On 25 June 1992, the Committee of Ministers adopted Resolution 92(27), inviting the Parliamentary Assembly to provide an opinion pursuant to Statutory Resolution 51(30A).

The Opinion listed the obligations of the Russian Federation, including, among others: to return, within the shortest possible time, property belonging to religious organizations; to ensure that the application of the CIS Convention on Human Rights would in no way impede the procedures and guarantees of the European Convention on Human Rights; to adopt a law on alternative military service as soon as possible; and to ensure full freedom of religion.

On the basis of these obligations and agreements, the Parliamentary Assembly recommended that the Committee of Ministers invite the Russian Federation to become a member of the Council of Europe, grant the Russian Federation eighteen seats in the Parliamentary Assembly, and bring the resources and capacities of the Council of Europe into line with the consequences of this decision.

PACE Resolution No. 1278 of 2002, "Russia's Law on Religion"

Resolution No. 1278 of the Parliamentary Assembly of the Council of Europe, "Russia's Law on Religion," adopted in 2002, was devoted to criticism of several provisions of the 1997 Federal Law "On Freedom of Conscience and Religious Associations." The Resolution states that the law, although an acceptable basis for the activity of most religious associations, could nevertheless be improved. Although the Constitutional Court of the Russian Federation had already limited the application of the so-called "fifteen-year rule," which had initially seriously restricted the rights of religious groups unable to prove that they had existed on the territory of the Russian Federation for fifteen years before the entry into force of the new law, the complete abolition of this rule, in the opinion of a number of such groups, would significantly improve the legislative framework.

The Resolution further noted problems connected with the fact that local officials in some cases, although not systematically, discriminated against and freedom of conscience are not fully correct. The problem is complicated by the fact that unresolved scientific and theoretical problems concern fundamental principles and the conceptual apparatus."

E. V. Kiseleva points to the insufficient effectiveness of international legal norms and structures in the sphere of freedom of conscience, noting that "international law has reflected a number of elements of this right, with different correlations between the powers of the individual and the ability of states to restrict them." The author indicates "the need to end discriminatory practices that reflect particular aspects of human rights violations on the ground of religion, which must be brought to an end."

A. M. Solntsev observes "an intolerant attitude toward different religions in the modern world, where representatives of traditional confessions, including Christians and Muslims, are subjected to discrimination."

T. F. Akchurin points to the narrowing of the right to freedom of conscience, noting that in international legal principles and norms "freedom of conscience is reduced exclusively to freedom of religion." The researcher then proposes various ways of overcoming this problem: "to develop and propose for signature by states a new international legal treaty that would emphasize the right to freedom of conscience; to provide a fuller clarification and interpretation by the relevant international body of norms concerning religious human rights and freedoms; to devote special attention, when international courts consider international disputes concerning the protection of religious rights and freedoms, to issues of interpretation of these norms, and so on."

It is evident that the insufficient scholarly development of the problem of freedom of conscience, and especially the system-forming problems of the conceptual apparatus, excessive dependence on political interests and other factors, have a destructive impact on universal international legal principles and norms. Unresolved theoretical and legal problems in the field of freedom of conscience significantly reduce their effectiveness, if they do not render them ineffective altogether.

Universal international legal principles and norms for the protection of human rights in the sphere of freedom of conscience are not fully effective and require substantial reform on the basis of: a new paradigm of freedom of conscience as the possibility of worldview choice; the principle of the supremacy of international law, including legal certainty of key terms; the principle of worldview neutrality; and the principle of mutual respect irrespective of differences and the overcoming of xenophobia, intolerance and discrimination.

4.2.2. Regional, Interregional and Subregional International-Legal Principles and Norms for the Protection of the Right to Freedom of Conscience

Regional, interregional and subregional international-legal principles and norms for the protection of human rights in the field of freedom of conscience include those principles and norms that are enshrined in the relevant international treaties and customs. Unlike universal principles and norms, regional, interregional and subregional standards operate within specific historical, cultural and institutional spaces. They do not replace the universal level. Rather, they test whether universal guarantees can be made real within different legal civilizations and political traditions.

Thus, in addition to universal norms, there are regional, interregional and subregional international-legal principles and norms which, in the view of W. Benedek, may lead not only to precedent-setting cases interpreting and clarifying the provisions of human rights instruments, but also to changes in national law and practice.

At the same time, V. A. Kartashkin and E. A. Lukasheva point to the difficulties faced by the perception of human rights values among the peoples of a number of regions and cultures that belonged to other civilizations. They note that the process of promoting and protecting human rights must be universal in nature and carried out with respect for the historical, cultural and religious characteristics of peoples.

The relevant regional, interregional and subregional principles and norms in the field of freedom of conscience are contained in the acts of the following organizations and legal systems:

The Council of Europe: the Convention for the Protection of Human Rights and Fundamental Freedoms of 1950; Protocol No. 1 to the Convention for the Protection of Human Rights and Fundamental Freedoms of 1952; the European Social Charter of 1996; the Recommendation of the Committee of Ministers to member states on the European Prison Rules; the Framework Convention for the Protection of National Minorities of 1995; opinions, resolutions and recommendations of the Parliamentary Assembly of the Council of Europe, and other acts.

The European Union: the Charter of Fundamental Rights of the European Union of 2000 and other acts.

The Organization of American States: the American Declaration of the Rights and Duties of Man of 1948; the American Convention on Human Rights of 1969; the Inter-American Convention against All Forms of Discrimination and Intolerance of 2013, and other acts.

The African Union: the African Charter on Human and Peoples' Rights of 1981 and other acts.

The CSCE/OSCE: the Helsinki Final Act of the Conference on Security and Co-operation in Europe of 1975; the Concluding Document of the Madrid Meeting of 1980; the Vienna Concluding Document of 1989; the Document of the Copenhagen Meeting of the Conference on the Human Dimension of 1990; the Charter of Paris for a New Europe of 1990; the Document of the Moscow Meeting of the Conference on the Human Dimension of 1991; the Charter for European Security of 1999; decisions of the OSCE Ministerial Council, and other acts.

The Commonwealth of Independent States: the Declaration of the Heads of State of the CIS on International Obligations in the Field of Human Rights and Fundamental Freedoms of 1993; the CIS Convention on Human Rights and Fundamental Freedoms of 1995, and other acts.

The Association of Southeast Asian Nations: the ASEAN Human Rights Declaration of 2012 and other acts.

The Organization of Islamic Cooperation: the Declaration on Human Rights in Islam, known as the Cairo Declaration, of 1990; the Covenant on the Rights of the Child in Islam of 2004; the Plan of Action for the Advancement of Women of 2008, and other acts.

The League of Arab States: the Arab Charter on Human Rights of 2004 and other acts.

Non-governmental regional organizations: the Asia-Pacific Declaration of Human Rights of Individuals and Peoples of 1988; the Final Document of the International Islamic Non-Governmental Conference of 2005, "Leading Ulema of the Islamic World on Accusations of Unbelief and the Unity of Muslims"; the Universal Islamic Declaration of Human Rights and other documents.

We shall now consider the main regional, interregional and subregional international-legal principles and norms of human rights in the field of freedom of conscience.

The Convention for the Protection of Human Rights and Fundamental Freedoms

The European Convention for the Protection of Human Rights and Fundamental Freedoms, the foundational document of the Council of Europe, was signed on 4 November 1950 by representatives of the member states of the Council of Europe.

The initial text of the Convention was amended in accordance with Protocol No. 3, which entered into force on 21 September 1970, Protocol No. 5, which entered into force on 20 December 1971, and Protocol No. 8, which entered into force on 1 January 1990. It also includes the text of Protocol No. 2, which, pursuant to Article 5, paragraph 3, formed an integral part of the Convention from the date of its entry into force on 21 September 1970. All provisions that had been amended or added by these protocols were subsequently replaced by Protocol No. 11, which entered into force on 1 November 1998.

Russia ratified the Convention on 20 February 1998, and it entered into force for Russia on 5 May 1998. Thus, Russia recognized the jurisdiction of the European Court of Human Rights, which, despite many difficulties, has functioned as a real mechanism for the protection of the rights declared by the European Convention.

Article 9 of the European Convention provides that everyone has the right to freedom of thought, conscience and religion. This right includes freedom to change religion or belief, and freedom, either alone or in community with others, in public or in private, to manifest religion or belief in worship, teaching, practice and observance.

However, paragraph 2 of Article 9 establishes limitations on the exercise of this right. Freedom to manifest one's religion or beliefs may be subject only to such limitations as are prescribed by law and are necessary in a democratic society in the interests of public safety, for the protection of public order, health or morals, or for the protection of the rights and freedoms of others.

It should also be noted that Article 9 is closely connected with Article 10, on freedom of expression, Article 11, on freedom of assembly and association, and Article 14, on the prohibition of discrimination. Article 10 states that everyone has the right to freedom of expression. This right includes freedom to hold opinions and to receive and impart information and ideas without interference by public authority and regardless of frontiers.

Article 14 prohibits discrimination on grounds such as sex, race, colour, language, religion, political or other opinion, national or social origin, association with a national minority, property, birth or other status.

In addition, Article 2 of Protocol No. 1 to the European Convention, adopted in Paris on 20 March 1952, establishes the right to education, including respect by the state for the right of parents to ensure such education and teaching in conformity with their own religious and philosophical convictions.

The European Social Charter

The European Social Charter was adopted by the Council of Europe in Turin on 18 October 1961 and entered into force on 26 February 1965. Among its main purposes were the strengthening of European unity, the promotion of progress and the protection of ideals, including human rights. The governments that signed the Charter declared their determination to make every effort jointly to improve the standard of living and promote the social well-being of both urban and rural populations by means of appropriate institutions and actions.

Recommendation No. R (87) 3 of the Committee of Ministers to Member States on the European Prison Rules

Recommendation No. R (87) 3 of the Committee of Ministers to member states on the European Prison Rules, adopted by the Committee of Ministers of the Council of Europe on 12 February 1987, addresses the problem of realizing the right to freedom of conscience for prisoners. According to the basic principles of the Recommendation, deprivation of liberty must be carried out under material and moral conditions that ensure respect for human dignity and comply with the Rules.

The section of the Recommendation concerning prisoners' nutrition states that food should, as far as possible, correspond to the religion and cultural traditions of prisoners.

The Recommendation also contains a special section on religious and moral assistance to prisoners. It provides that every prisoner shall be allowed, as far as practicable, to satisfy the needs of their religious, spiritual or moral life by attending services or meetings provided in the institution and by possessing books or literature relating to such matters.

The Framework Convention for the Protection of National Minorities

The Framework Convention for the Protection of National Minorities was adopted by the Council of Europe in Strasbourg on 1 November 1995. Under Article 5, the parties undertake to promote the conditions necessary for persons belonging to national minorities to maintain and develop their culture and preserve the essential elements of their identity, including religion, language, traditions and cultural heritage.

Article 6 declares the need to encourage a spirit of tolerance and intercultural dialogue, and to take effective measures to promote mutual respect, understanding and cooperation among all persons living on their territory, irrespective of ethnic, cultural, linguistic or religious identity.

Under Article 7, the parties ensure respect for the rights of every person belonging to a national minority to freedom of peaceful assembly, freedom of association, freedom of expression, and freedom of thought, conscience and religion.

On the basis of Article 8, the parties undertake to recognize that every person belonging to a national minority has the right to manifest religion or belief and to establish religious institutions, organizations and associations.

According to Article 17, the parties undertake not to interfere with the right of persons belonging to national minorities to establish and maintain free and peaceful contacts across borders with persons lawfully residing in other states, in particular with those with whom they share an ethnic, cultural, linguistic or religious identity or a common cultural heritage.

The author has previously expressed the view that the special separation of the social group of "religious minorities" in the context of the realization of freedom of conscience is disputable. This observation also applies, to a considerable extent, to the Framework Convention for the Protection of National Minorities.

Opinion No. 193 of the Parliamentary Assembly of the Council of Europe on Russia's Application for Membership of the Council of Europe

Opinion No. 193 of the Parliamentary Assembly of the Council of Europe concerned Russia's application for membership of the Council of Europe and was adopted in Strasbourg on 25 January 1996. As is known, the Russian Federation applied for membership of the Council of Europe on 7 May 1992. On 25 June 1992, the Committee of Ministers adopted Resolution 92(27), inviting the Parliamentary Assembly to provide an opinion pursuant to Statutory Resolution 51(30A).

The Opinion listed the obligations of the Russian Federation, including, among others: to return, within the shortest possible time, property belonging to religious organizations; to ensure that the application of the CIS Convention on Human Rights would in no way impede the procedures and guarantees of the European Convention on Human Rights; to adopt a law on alternative military service as soon as possible; and to ensure full freedom of religion.

On the basis of these obligations and agreements, the Parliamentary Assembly recommended that the Committee of Ministers invite the Russian Federation to become a member of the Council of Europe, grant the Russian Federation eighteen seats in the Parliamentary Assembly, and bring the resources and capacities of the Council of Europe into line with the consequences of this decision.

PACE Resolution No. 1278 of 2002, "Russia's Law on Religion"

Resolution No. 1278 of the Parliamentary Assembly of the Council of Europe, "Russia's Law on Religion," adopted in 2002, was devoted to criticism of several provisions of the 1997 Federal Law "On Freedom of Conscience and

Religious Associations." The Resolution states that the law, although an acceptable basis for the activity of most religious associations, could nevertheless be improved. Although the Constitutional Court of the Russian Federation had already limited the application of the so-called "fifteen-year rule," which had initially seriously restricted the rights of religious groups unable to prove that they had existed on the territory of the Russian Federation for fifteen years before the entry into force of the new law, the complete abolition of this rule, in the opinion of a number of such groups, would significantly improve the legislative framework.

The Resolution further noted problems connected with the fact that local officials in some cases, although not systematically, discriminated against and persecuted representatives of certain religious associations whose communities were small in a given territory. Sometimes local officials recognized a dominant status for the Russian Orthodox Church, forcing other religious associations to obtain consent from local representatives of the Russian Orthodox Church for their projects, for example for the rental or construction of buildings for a church or mosque. Sometimes complaints by religious associations to competent authorities, including prosecutors' offices, concerning physical attacks or arson attacks against churches remained unanswered, forcing such associations to apply to courts. In addition, some local and regional departments of the Ministry of Justice refused registration or re-registration to certain religious associations, even though those associations had already been registered at the federal level. The Ministry of Justice of the Russian Federation appeared unable to control its local and regional departments in accordance with the principle of the rule of law, preferring that religious associations themselves seek judicial review of registration decisions instead of taking internal departmental measures to correct the situation.

Proceeding from this understanding of the situation in Russia, the Assembly recommended that the Russian authorities ensure a more uniform application of the law on religion throughout the Russian Federation; end unjustified discrimination against certain religious associations in the regions and localities; end the practice of privileged treatment of the Russian Orthodox Church by local officials; strengthen the role of the Ministry of Justice in resolving disputes between its local or regional officials and religious organizations before such disputes are transferred to court; refrain from confiscating property in cases where a religious association is liquidated by court decision before all legal remedies have been exhausted; establish an independent extra-judicial review body including representatives of religious associations; speed up parliamentary consideration of the draft law on alternative civil service; and guarantee regular broadcasting time for religious organizations in state media.

In our view, a not entirely adequate understanding of the Russian situation in the field of freedom of conscience, and of the reasons underlying it, led to the emergence of recommendations, some of which contradict the principle of state secularity.

PACE Recommendation No. 1556 of 2002, "Religion and Change in Central and Eastern Europe"

Recommendation No. 1556 of the Parliamentary Assembly of the Council of Europe, "Religion and Change in Central and Eastern Europe," adopted in 2002, was devoted to assessing the interaction between religion and society in the new circumstances created primarily by the collapse of the socialist system.

In particular, the Assembly recommended that the Committee of Ministers of the Council of Europe call upon the governments of member states, the European Union and the relevant authorities and organizations to bring domestic legislation into conformity with the European Convention on Human Rights, especially Article 9; guarantee legal personality to all churches, religious organizations, centres and associations whose activity does not violate human rights or international law; take effective measures to guarantee freedom for religious minorities, especially in Central and Eastern Europe; apply ordinary procedures under domestic law in proven cases of abuse of religious freedoms that harm society or the rights, freedoms and health of individuals; guarantee restitution of church property nationalized in the past, or fair compensation where restitution is impossible; offer mediation in conflicts and disputes between parties, provided that state bodies do not interfere in matters of dogma or internal religious disputes; request that European religions and Christian traditions be included in the preamble of a future European constitution as a foundation of human dignity, human rights and the ethical roots of European identity; develop information strategies, cultural exchanges and educational programmes; cooperate with church authorities in fields such as the restoration of historical buildings and religious education; include information about major European religious cultures and customs in school curricula; support NGOs working for mutual understanding among religious groups; ensure equal access to the media, education and culture for representatives of all religious traditions; encourage centres for interfaith relations and cultural heritage exchanges; promote student, academic and cultural exchanges on the ethical, moral and cultural values of European religions; develop cultural routes in Europe and with neighbouring countries; supply public libraries with publications on various religious traditions; and support research aimed at identifying common roots of European cultures and improving understanding of how they interact and complement one another.

Alongside the repetition of general declarations, the PACE document draws attention by using divisions that lack clear legal criteria, such as "religious majority" and "religious minority," as well as non-legal references to "traditional churches of the region," "major European religious cultures," "European religions" and "Christian traditions."

The recommendation to include "European religions" and "Christian traditions" in the preamble of the European constitution "as the basis of the dignity of the human person and human rights and the ethical roots of European identity" appears to contradict the principles of freedom of conscience and to be insufficiently thought through.

In addition to the above, the following also highly disputable recommendatory documents of the Parliamentary Assembly of the Council of Europe should be mentioned: Recommendation 1720 (2005), "Education and Religion"; Resolution 1510 (2006), "Freedom of Expression and Respect for Religious Beliefs"; Resolution 1535 (2007), "Threats to the Lives and Freedom of Expression of Journalists"; Recommendation 1805 (2007), "Blasphemy, Religious Insults and Hate Speech against Persons on Grounds of their Religion"; Recommendation 1804 (2007), "State, Religion, Secularity and Human Rights"; Resolution 1763 (2010), "The Right to Conscientious Objection in Lawful Medical Care"; Resolution 1846 (2011), "Combating All Forms of Discrimination Based on Religion"; Recommendation 1987 (2011), "Combating All Forms of Discrimination Based on Religion"; Resolution 1928 (2013), "Safeguarding Human Rights in Relation to Religion and Belief and Protecting Religious Communities from Violence"; Resolution 1992 (2014), "Protecting Minors against the Harmful Influence of Sects"; Resolution 2076 (2015), "Freedom of Religion and Living Together in a Democratic Society"; Recommendation 2080 (2015), "Freedom of Religion and Living Together in a Democratic Society," and others.

In our view, despite the positive wishes and mostly good intentions behind these documents, compliance by states with recommendations based on not entirely correct principles and conceptual language, including concepts that imply non-legal selection among religious associations and sometimes even contain elements of xenophobia, cannot lead to the affirmation of the principles of freedom of conscience for every person.

The Charter of Fundamental Rights of the European Union

The Charter of Fundamental Rights of the European Union, established in the context of the European Union created on 7 February 1992 in Maastricht, was adopted in Nice on 7 December 2000. Its preamble refers to the decision of the peoples of Europe to build a peaceful future together on the basis of common values: human dignity, freedom, equality, solidarity, democracy and the rule of law, while respecting the diversity of the cultures and traditions of the peoples of Europe, as well as the national identities of the member states and the organization of their public authorities at national, regional and local levels.

The Charter reaffirms, with due regard for the powers and tasks of the European Community and the European Union and for the principle of subsidiarity, the rights that arise in particular from the constitutional traditions and common international obligations of the member states, the Treaty on European Union and the Treaties establishing the European Communities, the European Convention for the Protection of Human Rights and Fundamental Freedoms, the Social Charters adopted by the Community and by the Council of Europe, and the case-law of the Court of Justice of the European Communities and the European Court of Human Rights.

Article 10 of the Charter declares freedom of thought, conscience and religion. Everyone has the right to freedom of thought, conscience and religion. This right includes freedom to change religion or belief, and freedom, either alone or in community with others and in public or in private, to manifest religion or belief in worship, teaching, practice and observance. The right to conscientious objection is recognized in accordance with the national laws governing its exercise.

Article 14 states that the freedom to found educational establishments with due respect for democratic principles, and the right of parents to ensure the education and teaching of their children in conformity with their religious, philosophical and pedagogical convictions, shall be respected in accordance with the national laws governing the exercise of such freedom and right.

Under Article 21 of the Charter, any discrimination based on any ground such as sex, race, colour, ethnic or social origin, genetic features, language, religion or belief, political or any other opinion, membership of a national minority, property, birth, disability, age or sexual orientation is prohibited. Within the scope of application of the treaties, and without prejudice to their special provisions, any discrimination on grounds of nationality is also prohibited. Article 22 further emphasizes that the European Union respects cultural, religious and linguistic diversity.

Thus, in its foundational documents, the European Union confirmed its commitment to human rights, including in the field of freedom of conscience, as well as to the prohibition of discrimination and the encouragement of diversity.

Overall, the position of the European Union absorbed the spirit, the letter and the fundamental problems of universal international-legal principles and norms, as well as those of the European community itself.

The American Convention on Human Rights

The American Convention on Human Rights was adopted on 22 November 1969 and entered into force in 1978 within the framework of the Organization of American States, which had been created in 1948. To supervise the implementation of the Convention, the Inter-American Court of Human Rights was established to examine complaints concerning violations of the rights enshrined in the Convention.

The American Convention on Human Rights enshrined the right to freedom of conscience and religion, as well as the right of parents, and where applicable guardians, to provide religious and moral education for their children or wards in accordance with their own convictions.

Article 12 of the Convention provides that everyone has the right to freedom of conscience and religion. This right includes freedom to maintain or change one's religion or beliefs, and freedom to profess or disseminate one's religion or beliefs, either individually or together with others, in public or in private. No one shall be subject to restrictions that might impair freedom to maintain or change religion or beliefs. Freedom to manifest religion and beliefs may be subject only to limitations prescribed by law and necessary to protect public safety, order, health or morals, or the rights and freedoms of others. Parents and guardians have the right to provide religious and moral education for their children or wards in accordance with their own convictions.

Article 13 further provides that everyone has the right to freedom of thought and expression. This includes freedom to seek, receive and impart information and ideas of all kinds, regardless of frontiers, orally, in writing, in print, in artistic form, or by any other medium of one's choice. The exercise of this right shall not be subject to prior censorship, but may be subject to subsequent liability expressly established by law and necessary to ensure respect for the rights or reputation of others, or to protect national security, public order, or public health or morals. The right of expression may not be restricted by indirect methods or means, including abuse of public or private controls over newsprint, broadcasting frequencies, equipment used in the dissemination of information, or any other means tending to impede communication and the circulation of ideas and opinions. Prior censorship may be allowed only for public entertainments and solely to regulate access for the moral protection of children and adolescents. Any propaganda for war and any advocacy of national, racial or religious hatred that constitutes incitement to unlawful violence or other similar illegal action shall be considered an offence punishable by law.

Thus, the American Convention confirmed the aspiration to consolidate, within its hemisphere and within democratic institutions, a system of personal freedom and social justice based on respect for human rights. The nature of this document was undoubtedly influenced by the spiritual and historical heritage of the American Declaration of Independence of 1776 and the Bill of Rights of 1791. It should also be noted that the American Declaration of the Rights and Duties of Man of 1948 preceded the American human rights system, the Inter-American Democratic Charter was adopted in 2001, and the Inter-American Convention against All Forms of Discrimination and Intolerance of 2013 reaffirmed the prohibition of religious intolerance and discrimination.

These regional principles and norms in the field of freedom of conscience generally correspond to universal principles and norms, including their shortcomings.

The African Charter on Human and Peoples' Rights

The member states of the Organization of African Unity, since 2002 the African Union, adopted the African Charter on Human and Peoples' Rights in June 1981. In 2004, the 1998 Protocol establishing the African Court on Human and Peoples' Rights entered into force. The African Commission on Human and Peoples' Rights supervises the implementation of the Charter, including by examining violations.

Under Article 1 of the African Charter, the member states of the Organization of African Unity that are parties to the Charter recognize the rights, duties and freedoms enshrined in it and undertake to adopt legislative and other measures to give effect to them. Under Article 2, every individual is entitled to the rights and freedoms recognized and guaranteed in the Charter without distinction of any kind, including race, ethnic group, colour, sex, language, religion, political or other opinion, national or social origin, fortune, birth or other status.

Freedom of conscience and the free practice of religion are guaranteed by Article 8. No one may, subject to law and order, be subjected to measures restricting the exercise of these freedoms. Under Article 9, everyone has the right to express and disseminate opinions within the law. Article 17, together with the right to education, grants everyone the right freely to take part in the cultural life of their community. The same article states that the promotion and protection of morals and traditional values recognized by the community shall be the duty of the state.

The documents of the African Union declare freedom of conscience and prohibit discrimination on the basis of religion. However, compared with universal standards, the range of rights in this field is somewhat narrower. One may assume that in this way the African Charter on Human and Peoples' Rights marks the dignity of the historical traditions and values of African civilization, which should be reflected in the formation and content of human rights concepts.

The Final Act of the Conference on Security and Co-operation in Europe, Helsinki, 1 August 1975

The Organization for Security and Co-operation in Europe, originally the CSCE, traces its origin to the meeting in Helsinki on 3 July 1973, which continued in Geneva from 18 September 1973 to 21 July 1975 and concluded in Helsinki on 1 August 1975 with the adoption of the Final Act.

The Final Act adopted in Helsinki on 1 August 1975 contains Section VII, "Respect for human rights and fundamental freedoms, including freedom of thought, conscience, religion or belief." It states that the participating states will respect human rights and fundamental freedoms, including freedom of thought, conscience, religion or belief, for all without distinction as to race, sex, language or religion. They will promote and encourage the effective exercise of civil, political, economic, social, cultural and other rights and freedoms, all of which derive from the inherent dignity of the human person and are essential for that person's free and full development. Within this framework, the participating states will recognize and respect the freedom of the individual to profess and practise, alone or in community with others, religion or belief, acting in accordance with the dictates of their own conscience.

The Final Act of the Conference on Security and Co-operation in Europe, Madrid, 1983

The Final Act of the Conference on Security and Co-operation in Europe, adopted in Madrid in 1983, confirmed the obligations of the participating states to recognize, respect and take the measures necessary to ensure the freedom of the individual to profess and practise, alone or in community with others, religion or belief in accordance with the dictates of conscience. The participating states also undertook to give favourable consideration to requests by religious communities of believers, practising or prepared to practise their faith within the constitutional framework of their states, for the granting of the status provided in their countries for religious denominations, institutions and organizations.

The Final Act of the Conference on Security and Co-operation in Europe, Vienna, 1989

In the Final Act of the Conference on Security and Co-operation in Europe adopted in Vienna in 1989, the participating states once again confirmed that they would respect human rights and fundamental freedoms, including freedom of thought, conscience, religion or belief, for all, without distinction as to race, sex, language or religion. They reaffirmed the universal significance of human rights and fundamental freedoms, respect for which is an essential factor for peace, justice and security and is necessary for the development of friendly relations and cooperation among them and among all states.

The states also expressed their determination to guarantee the effective exercise of human rights and fundamental freedoms, all of which derive from the dignity inherent in the human person and are essential for free and full development. They recognized that all civil, political, economic, social, cultural and other rights and freedoms are of primary importance and must be fully realized by all appropriate means.

To ensure the freedom of the individual to profess and practise religion or belief, the participating states undertook, among other things, to take effective measures to prevent and eliminate discrimination against individuals or communities on the grounds of religion or belief in all spheres of civil, political, economic, social and cultural life; to ensure genuine equality between believers and non-believers; to promote a climate of mutual tolerance and respect among believers of different communities and between believers and non-believers; to grant, upon request, recognition of the status provided in their countries to communities of believers acting within the constitutional framework of their states; and to respect the right of religious communities to establish and maintain freely accessible places of worship or assembly, organize themselves according to their own hierarchical and institutional structure, choose and replace their personnel according to their requirements and standards, and solicit and receive voluntary financial and other contributions.

The participating states also undertook to consult with religious denominations, institutions and organizations in order to achieve a better understanding of the needs of religious freedoms; to respect the right of everyone to give and receive religious education in the language of their choice, individually or together with others; to respect, in this context, the freedom of parents to ensure the religious and moral education of their children in accordance with their own convictions; to allow the training of religious personnel in appropriate institutions; to respect the right of believers and religious communities to acquire, possess and use sacred books, religious publications and other items and materials related to the practice of religion or belief; to allow religious denominations, institutions and organizations to produce, import and distribute religious publications and materials; and to give favourable consideration to the interest of religious communities in participating in public dialogue, including through the mass media.

The participating states further recognized that the exercise of the above-mentioned rights relating to freedom of religion or belief may be subject only to such limitations as are prescribed by law and consistent with their obligations under international law. They undertook to ensure, in their laws and administrative regulations and in their application, the full and effective exercise of freedom of thought, conscience, religion or belief.

Document of the Copenhagen Meeting of the Conference on the Human Dimension of the CSCE, Copenhagen, 29 June 1990

The Document of the Copenhagen Meeting of the Conference on the Human Dimension of the CSCE, adopted in Copenhagen on 29 June 1990, confirmed the following provisions.

Everyone has the right to freedom of expression, including the right to communication. This right includes freedom to hold opinions and to receive and impart information and ideas without interference by public authorities and regardless of frontiers. The exercise of this right may be subject only to such restrictions as are prescribed by law and are consistent with international standards. In particular, no limitations shall be imposed on access to, and use of, means of reproducing documents of any kind, while respecting intellectual property rights, including copyright.

Everyone has the right to freedom of thought, conscience and religion. This right includes freedom to change religion or belief and freedom to manifest religion or belief, either alone or in community with others and in public or in private, through worship, teaching, practice and observance. The exercise of these rights may be subject only to such limitations as are prescribed by law and consistent with international standards.

The participating states also took note of the recognition by the United Nations Commission on Human Rights of the right of everyone to refuse military service on grounds of conscience. They noted the measures recently adopted by a number of participating states to allow exemption from compulsory military service on the basis of conscientious objection, as well as the activity of certain non-governmental organizations in this field.

They agreed to consider introducing, where this had not yet been done, various forms of alternative service compatible with the reasons for conscientious objection. Such service, as a rule, would not be linked to combat units, or would be civilian in character, socially useful and non-punitive. The participating states also agreed to make information on this issue available to the public and to continue considering relevant issues within the framework of the Conference on the Human Dimension.

The Copenhagen Document also contains guarantees for the protection of national minorities, with an important reservation: belonging to a national minority is a matter of a person's individual choice, and no disadvantage may arise from the exercise of that choice. Persons belonging to national minorities have the right freely to express, preserve and develop their ethnic, cultural, linguistic or religious identity and to maintain and develop their culture in all its aspects, without being subjected to any attempt at assimilation against their will.

Considerable attention is paid to countering xenophobia, intolerance and discrimination. The participating states clearly and unconditionally condemned totalitarianism, racial and ethnic hatred, antisemitism, xenophobia, discrimination against anyone, and persecution on religious and ideological grounds. In this context, they also recognized the special problems of Roma communities.

The participating states declared their firm intention to intensify efforts to combat these phenomena in all their forms. They undertook to adopt effective measures, including legislation where necessary, to protect against any acts that constitute incitement to violence against persons or groups on grounds of national, racial, ethnic or religious discrimination, hostility or hatred, including antisemitism; to protect persons or groups threatened by discrimination, hostility or violence because of their racial, ethnic, cultural, linguistic or religious identity; to promote understanding and tolerance at national, regional and local levels, especially in education, culture and information; to strengthen respect for different civilizations and cultures; to recognize the right of every person to effective remedies; and to consider accession to international instruments dealing with discrimination and recognition of international mechanisms allowing states and individuals to submit communications concerning discrimination to international bodies.

The Charter of Paris for a New Europe, Paris, 21 November 1990

The Charter of Paris for a New Europe, as the concluding document of the Conference on Security and Co-operation in Europe, was adopted in Paris on 21 November 1990. In the document, the heads of state and government of the CSCE participating states optimistically proclaimed a new era of democracy, peace and unity.

The era of confrontation and division of Europe had ended. The participating states declared that their relations would henceforth be founded on mutual respect and cooperation. Europe was liberating itself from the legacy of the past. The courage of men and women, the strength of peoples and the power of the ideas of the Helsinki Final Act had opened a new era of democracy, peace and unity in Europe. The document described the time as one of realizing the hopes and expectations that had lived in the hearts of peoples for decades: firm commitment to democracy based on human rights and fundamental freedoms; prosperity through economic freedom and social justice; and equal security for all countries. The ten principles of the Final Act were proclaimed as a guiding star toward this desired future, just as they had illuminated the path toward improved relations over the previous fifteen years.

The participating states further confirmed that, without discrimination, everyone has the right to freedom of thought, conscience, religion and belief, freedom of expression, freedom of association and peaceful assembly.

They also confirmed that the ethnic, cultural, linguistic and religious identity of national minorities would be protected, and that persons belonging to national minorities have the right freely to express, preserve and develop that identity without discrimination and in full equality before the law. The states also undertook to ensure that everyone has access to effective national or international remedies against violations of rights.

Document of the Moscow Meeting of the Conference on the Human Dimension of the CSCE, Moscow, 3 October 1991

The Document of the Moscow Meeting of the Conference on the Human Dimension of the CSCE was adopted in Moscow on 3 October 1991. In particular, the participating states recognized the need to ensure respect for the rights of migrant workers and their families lawfully residing in participating states, and especially their right freely to express their national, cultural, religious and linguistic characteristics. The exercise of such rights may be subject only to such limitations as are prescribed by law and consistent with international standards.

The participating states also condemned all acts of discrimination on grounds of race, colour and ethnic origin, as well as intolerance and xenophobia against migrant workers. In accordance with national legislation and international obligations, they undertook to take effective measures to promote tolerance, understanding, equality of opportunity and respect for the fundamental human rights of migrant workers. They also undertook, where they had not yet done so, to prohibit acts constituting incitement to violence based on national, racial, ethnic or religious discrimination, hostility or hatred.

In addition, the participating states recognized that effective human rights education contributes to combating intolerance, religious, racial and ethnic prejudice and hatred, including against Roma, as well as xenophobia and antisemitism.

The Charter for European Security, Istanbul, November 1999

The Charter for European Security was adopted in Istanbul in November 1999. The participating states reaffirmed that respect for human rights and fundamental freedoms, democracy and the rule of law occupies a central place in the OSCE's comprehensive concept of security.

The states undertook to confront threats to security such as violations of human rights and fundamental freedoms, including freedom of thought, conscience, religion or belief, and manifestations of intolerance, aggressive nationalism, racism, chauvinism, xenophobia and antisemitism. They stressed that the protection and promotion of the rights of persons belonging to national minorities are essential factors for democracy, peace, justice and stability within and between participating states. Full respect for human rights, including the rights of persons belonging to national minorities, is an end in itself and can strengthen, rather than undermine, territorial integrity and sovereignty. Various concepts of autonomy, and other approaches consistent with OSCE principles, may serve as means of protecting and developing the ethnic, cultural, linguistic and religious identity of national minorities within an existing state. The participating states condemned violence against any minority and undertook to promote tolerance and build pluralistic societies in which everyone, regardless of ethnic origin, enjoys full equality of opportunity.

Ministerial Council Decision No. 4/03: Tolerance and Non-Discrimination

Following the meetings of the Ministerial Council in March 2003, Ministerial Council Decision No. 4/03, "Tolerance and Non-Discrimination," was adopted.

Recognizing that respect for human rights and fundamental freedoms, democracy and the rule of law form the basis of the OSCE's comprehensive concept of security, the Decision continued the work initiated at the OSCE Conference on Racism, Xenophobia and Discrimination held in Vienna on 4-5 September 2003. It instructed the Permanent Council to continue discussing ways to expand OSCE and participating-state efforts to promote tolerance and non-discrimination in all fields. It encouraged participating states to collect and maintain reliable information and statistics on hate crimes, including violent manifestations of racism, xenophobia, discrimination and antisemitism, and to inform the OSCE Office for Democratic Institutions and Human Rights about legislation concerning crimes motivated by intolerance and discrimination.

The Decision instructed ODIHR, in cooperation with the UN Committee on the Elimination of Racial Discrimination, the European Commission against Racism and Intolerance, the European Monitoring Centre on Racism and Xenophobia, and relevant NGOs, to serve as a collection point for information and statistics provided by participating states and to report regularly on these matters. It recognized the need to combat hate crimes that may be encouraged by racist, xenophobic and antisemitic propaganda on the Internet, while fully respecting freedom of information and freedom of expression.

The Decision declared the importance of freedom of thought, conscience, religion and belief and condemned any discrimination and violence, including against any religious groups or individual believers. It committed participating states to ensure and promote the freedom of the individual to profess and practise religion or belief, alone or in community with others, where necessary through transparent and non-discriminatory laws, regulations, practices and policies. It recommended that participating states seek assistance from ODIHR and its expert panel on freedom of religion or belief. It also emphasized the importance of continuing and strengthening interfaith and intercultural dialogue in the interests of greater tolerance, mutual respect and understanding.

The OSCE/CSCE documents confirmed the commitment of participating states to freedom of thought, conscience and religion as enshrined in universal international-legal principles and norms. Participating states also undertook obligations to counter violations in the field of freedom of conscience, xenophobia, intolerance and discrimination. It is positive that fundamental human rights are considered a necessary condition of peace, and that their violations are treated as threats to security in Europe. However, the key conceptual and categorical apparatus of these norms very substantially hinders their effective implementation.

4.3. International Mechanisms for the Protection of the Right to Freedom of Conscience

4.3.1. Universal Mechanisms for the Protection of the Right to Freedom of Conscience

The universal United Nations system for the promotion and protection of human rights, including the right to freedom of conscience, consists of several subsystems: the principal Charter-based organs and subsidiary bodies; specialized agencies established under the UN Charter; and treaty, or convention-based, bodies created under specific human rights treaties.

Accordingly, under Article 7 of the UN Charter, the principal organs empowered in the field of the promotion and protection of human rights include: the General Assembly, the Security Council, the Economic and Social

Council, the Trusteeship Council, the International Court of Justice, and the Secretariat headed by the Secretary-General.

1. The General Assembly

The General Assembly is one of the six principal organs of the United Nations. It is the only body in which all Member States have equal representation: one country, one vote. All 193 Member States are represented in this unique forum, where, in accordance with the UN Charter, joint action is discussed across the widest range of international issues, including development, peace and security, international law and human rights.

In this context, the Third Committee of the General Assembly, dealing with social, humanitarian and cultural issues, considers a broad range of questions, including social development and human rights. These include the advancement of women, the protection of children, the situation of Indigenous peoples, the treatment of refugees, the promotion of fundamental freedoms through the elimination of racism and racial discrimination, and the right of peoples to self-determination.

In general, the activity of the General Assembly in protecting the right to freedom of conscience has so far been limited mainly to discussion, research and recommendations. These tools have not yet been sufficient to increase the effectiveness of the relevant principles and norms or to bring the enjoyment of the corresponding goods closer to the populations of UN Member States. It therefore seems appropriate for the General Assembly to develop and adopt a conceptual document on the development of international cooperation in the realization of the right to liberty of worldview choice through a system of legal guarantees.

Such guarantees should include the principle of the rule of international law, including legal certainty in key terms; the principle of worldview neutrality; the principle of mutual respect irrespective of differences; and the overcoming of xenophobia, intolerance and discrimination. These approaches could be used by the General Assembly in its law-making activity, as well as in the exercise of its supervisory powers in interaction with the system of treaty bodies. In ordinary human language, this means that the world cannot keep treating freedom of conscience as a ceremonial phrase for declarations while leaving its practical implementation to political weather and administrative improvisation.

2. The United Nations Security Council

The UN Security Council plays a key role in maintaining peace and security on the planet. It consists of fifteen members. Member States agree, in accordance with the Charter, to accept and carry out the decisions of the Security Council. The Council has the leading role in determining the existence of a threat to peace, a breach of peace, or an act of aggression. It calls upon parties to a dispute to settle it by peaceful means and recommends methods or terms of settlement.

In certain cases, the Security Council may resort to sanctions or even authorize the use of force in order to maintain or restore international peace and security. Its powers include investigation, mediation, dispatching missions, appointing special envoys, and requesting the Secretary-General to use his good offices. The Council may order a ceasefire, send military observers or peacekeeping forces, and, where these measures prove insufficient, adopt enforcement measures such as economic sanctions, arms embargoes, financial sanctions and restrictions, travel bans, severance of diplomatic relations, blockade, or even collective military action.

Given the exceptional importance of the realization of the right to freedom of conscience for international peace and security, especially for overcoming ethno-confessional conflicts, it would be appropriate to consider establishing an international tribunal for the prosecution of persons responsible for large-scale violations of the right to freedom of conscience that threaten international peace and security.

This proposal is not a decorative legal gesture. Where worldview persecution feeds mass violence, ethnic cleansing, terrorist recruitment, or state-sponsored campaigns of hatred, the issue ceases to be merely "domestic." It becomes a question of international order. A world that criminalizes some forms of mass violence but treats systematic destruction of inner freedom as a secondary matter leaves the architecture of peace with a rather embarrassing hole in the wall.

3. The Economic and Social Council

The Economic and Social Council, or ECOSOC, promotes collective action for building sustainable peace. Within the UN development system, it performs an important coordinating and analytical role: it reviews operational trends, aligns global norms and supports progress towards sustainable development.

In the field examined here, ECOSOC's existing contribution is mainly indirect. It helps connect human rights, development, social inclusion, education, tolerance and non-discrimination. This is important because the right to freedom of conscience cannot be realized only through prohibitions. It requires social conditions: education, equality, institutional trust, access to cultural life, and protection against exclusion. A person whose worldview is formally "free" but socially punished, administratively blocked, or economically marginalized remains free mostly on paper, which is humanity's favourite place to store rights it has not yet learned to implement.

In this regard, it seems necessary to strengthen ECOSOC's role in coordinating development-oriented measures for the realization of liberty of worldview choice, including within the Sustainable Development Goals, education policies, anti-discrimination strategies and national human rights action plans.

4. The Trusteeship Council

In 1994, the Trusteeship Council fulfilled its historical mission related to the realization of the right of peoples to self-determination in relation to territories that later became sovereign states or acquired the status of commonwealth or association with the United States. Although the Council no longer performs its original functions in the same practical sense, its historical role is important for the present study.

The experience of decolonization shows that the right to self-determination concerns not only territory and political status. It is also connected with the ability of peoples and individuals to preserve their identity, culture, language, religion, convictions and worldview. The unfinished task of the present age is to translate this lesson into the post-digital context, where domination may be exercised without formal colonization, through information infrastructures, algorithmic dependence, cognitive pressure and ideological capture.

5. The International Court of Justice

The International Court of Justice functions on the basis of its Statute, which forms part of the UN Charter, and its Rules. It is the principal judicial organ of the United Nations. Its activity serves one of the main purposes of the UN: to bring about, by peaceful means and in conformity with the principles of justice and international law, the adjustment or settlement of international disputes or situations that might lead to a breach of the peace.

The Court is located in The Hague and is open to states, not individuals. This fact significantly limits its direct role in the protection of the individual right to freedom of conscience. However, the Court may still play an important systemic role. It can clarify the content of international obligations concerning freedom of thought, conscience and religion, non-discrimination, minority rights, the prohibition of persecution, and state responsibility for internationally wrongful acts.

The development of advisory opinions or contentious cases involving large-scale violations of worldview freedom could help strengthen the legal certainty of this field. At present, however, the potential of the Court remains insufficiently used in relation to freedom of conscience, worldview neutrality and the protection of the inner sphere of the human person.

6. The Secretariat and the Secretary-General

In the UN Charter, the Secretary-General is described as the chief administrative officer of the Organization. The Charter authorizes the Secretary-General to bring to the attention of the Security Council any matter which, in his opinion, may threaten the maintenance of international peace and security.

In particular, the Secretary-General appoints special representatives dealing with grave human rights violations, including the Special Representative on Children and Armed Conflict, the Special Representative on Sexual Violence in Conflict, and the Special Representative on Violence against Children. These mandates show that the Secretariat can become an institutional voice for vulnerable groups and for human rights issues requiring urgent global attention.

In the field of freedom of conscience and liberty of worldview choice, a comparable institutional development appears necessary. The increasing scale of worldview manipulation, digital propaganda, persecution of religious and non-religious minorities, ideological discrimination and algorithmic control over belief formation suggests the need for stronger attention within the UN system. In the future, this could include a special representative or high-level envoy on liberty of worldview choice, cognitive autonomy and worldview neutrality in analog and digital environments.

Such a mandate could prepare thematic reports, coordinate with treaty bodies and special procedures, collect information on systemic violations, and contribute to the development of international legal standards. It could also draw attention to situations where the destruction of worldview freedom becomes an early warning sign of mass violence, authoritarian consolidation or geopolitical escalation.

The Human Rights Council

The Human Rights Council, established in 2006, replaced the UN Commission on Human Rights, which had operated for sixty years. It became the key independent intergovernmental body of the United Nations responsible for human rights.

The Council uses several mechanisms. The Universal Periodic Review assesses the fulfilment of international obligations by Member States. Special procedures involve independent experts acting under mandates established by the Council and supported by the Office of the United Nations High

Commissioner for Human Rights. The Council may also create commissions of inquiry, fact-finding missions and other investigative mechanisms in response to serious violations.

In addition, there is a complaint procedure that allows individuals and organizations to bring patterns of gross and reliably attested violations of human rights to the attention of the Council. In theory, this system creates an important channel for addressing violations of freedom of conscience. In practice, its effectiveness depends heavily on political will, the quality of evidence, the persistence of civil society, and the readiness of states to cooperate. So yes, once again, the most delicate human rights problems are placed into a mechanism that depends on states behaving like adults. A bold institutional experiment, clearly.

In 1986, the Commission on Human Rights, now the Human Rights Council, by resolution 1986/20 decided to establish the mandate of the Special Rapporteur on Religious Intolerance. In 2000, the title was changed to Special Rapporteur on Freedom of Religion or Belief, a formulation subsequently confirmed by General Assembly resolution 55/97.

From the beginning, this mandate has played a significant role in identifying obstacles to the realization of freedom of religion or belief. It examines situations of intolerance, discrimination and violence, communicates with governments, undertakes country visits, prepares thematic reports and contributes to the interpretation of international standards.

In accordance with Human Rights Council resolution 6/37, the Special Rapporteur on freedom of religion or belief is mandated to promote measures at the national, regional and international levels to ensure the promotion and protection of the right to freedom of religion or belief; identify existing and emerging obstacles to the enjoyment of this right and make recommendations on ways and means to overcome them; continue efforts to examine incidents and governmental actions inconsistent with the provisions of the Declaration on the Elimination of All Forms of Intolerance and of Discrimination Based on Religion or Belief; and recommend remedial measures as appropriate.

Under the powers of the mandate, the Special Rapporteur sends communications to states concerning cases that may represent violations of, or obstacles to, the enjoyment of freedom of religion or belief; conducts fact-finding country visits; and submits annual reports to the Human Rights Council and the General Assembly on activities, trends and methods of work.

However, the very name and conceptual scope of the mandate reveal the central problem of the entire field. Freedom of religion or belief is broader than protection of religious organizations, but narrower than the full idea of liberty of worldview choice. The mandate can address atheists, agnostics, non-believers, converts and persons facing coercion in matters of belief, yet the conceptual language still remains tied to the older framework of "religion or belief." A new stage of legal development requires the explicit recognition of worldview choice as the broader category.

The Working Group on Arbitrary Detention

The Working Group on Arbitrary Detention is another important special procedure of the Human Rights Council. Its thematic mandate directly provides for the consideration of individual complaints. This means that its activity is based on the right of people throughout the world to submit complaints concerning unlawful detention.

The experts also receive information from human rights organizations and from families of persons who have become victims of arbitrary detention. The Working Group is called upon to investigate cases of deprivation of liberty imposed arbitrarily or inconsistently with international standards, including cases where detention is connected with the exercise of freedom of thought, conscience, religion or belief.

The Human Rights Council encourages the Working Group, in carrying out its mandate, to work in cooperation and dialogue with all stakeholders in cases submitted to it, especially with states providing information that deserves due attention; to coordinate with other Council mechanisms, other competent UN bodies and treaty bodies; and to use its mandate in a way that strengthens the protection of individuals from arbitrary state power.

Thus, the thematic mandate of the Working Group directly provides for the consideration of individual applications. It is therefore one of the practical universal mechanisms capable of responding where persons are deprived of liberty for their convictions, religious affiliation, conscientious objection, non-belief, minority status or refusal to accept an imposed ideology.

At the same time, the Working Group is still a reactive mechanism. It intervenes after harm has occurred. It cannot replace the creation of legal, institutional and educational guarantees that would prevent such violations in the first place. This is precisely why the present book insists on moving from fragmented protection to a systemic architecture for the realization of worldview freedom.

The Office of the United Nations High Commissioner for Human Rights

The Office of the United Nations High Commissioner for Human Rights, headed by the High Commissioner, is a subsidiary organ of the UN Secretariat.

The official UN materials emphasize the leading role of the High Commissioner in promoting and protecting human rights.

The Office interacts with UN bodies and mechanisms, provides them with advisory and other human rights support, and integrates human rights concerns into all areas of the Organization's work, including socio-economic development, peace and security, peacekeeping and humanitarian action. It also supports treaty bodies and special procedures, provides technical assistance to states, monitors human rights situations and contributes to the development of standards.

In the sphere of freedom of conscience, the Office has an important coordinating and analytical role. It can support the work of treaty bodies, special rapporteurs and working groups; collect information on violations; assist states in reforming legislation; and encourage the application of international standards. Its activity is especially significant where violations of freedom of conscience are connected with discrimination, persecution, hate speech, arbitrary detention, education, minority rights and the digital environment.

Nevertheless, the available instruments remain insufficient. The Office may condemn violations and provide expert assistance, but without a clearer legal standard of liberty of worldview choice and worldview neutrality, many violations remain conceptually blurred. They are often scattered between freedom of religion, freedom of expression, minority rights, education, privacy, political rights and anti-discrimination norms. The result is a fragmented map of protection where the central object, the human capacity to form a worldview freely, remains unnamed.

The International Law Commission

The International Law Commission was established to promote the progressive development of international law and its codification. Its function is not to protect individual applicants directly, but to develop legal concepts, draft articles, studies and proposals capable of shaping the future of international law.

This role is of direct relevance to the topic of the present study. The development of a future international legal framework on the right to liberty of worldview choice and the principle of worldview neutrality would require careful conceptual work: defining terms, identifying state obligations, clarifying responsibility, establishing limits on permissible restrictions, and linking analog and digital forms of interference.

The International Law Commission could, in the future, contribute to the codification or progressive development of principles concerning the protection of worldview freedom, cognitive autonomy, mental integrity and the neutrality of public power in the field of meaning. Otherwise, international law will continue patching holes after each technological and ideological disaster, like a tailor repairing a parachute after the fall.

Specialized Agencies of the United Nations

The specialized agencies of the United Nations also have significance for the protection of freedom of conscience, although often indirectly. Their mandates concern education, culture, science, labour, health, communication and social development. These areas shape the real conditions under which freedom of conscience may either flourish or remain a beautiful phrase framed on the wall of an institution that no one listens to.

UNESCO

UNESCO has particular importance because the right to freedom of conscience is closely connected with education, culture, science, information and tolerance. The Director-General of UNESCO has powers in promoting the realization of human rights within the Organization's mandate. According to established practice, communications concerning alleged violations of human rights within UNESCO's fields of competence may be considered through the relevant procedures.

Within this framework, UNESCO's competence covers rights connected with education, science, culture and information. These include the right to education, the right to share in scientific advancement and its benefits, the right freely to participate in cultural life, and the right to information, including freedom of opinion and expression.

Thus, the right to freedom of conscience falls within UNESCO's competence mainly in the context of educational and cultural human rights. This includes the formation of tolerance, respect for cultural diversity, the prevention of discrimination in education, and the development of curricula that support human dignity and critical thinking. UNESCO's role is especially significant because worldview freedom is not produced by legal prohibitions alone. It also requires an educational environment in which a person can compare, doubt, interpret, and choose.

At the same time, UNESCO's mandate does not by itself solve the legal uncertainty surrounding freedom of conscience. Educational and cultural approaches must be linked with a clearer legal concept of liberty of worldview choice. Without this link, tolerance education risks becoming a noble ritual repeated in stems and digital platforms continue shaping minds through invisible hierarchies of acceptable belief.

The Subsystem of Treaty Bodies

The subsystem of treaty bodies occupies a central place in the universal protection of human rights. These bodies monitor the implementation of specific human rights treaties. They examine state reports, adopt concluding observations, issue general comments, and, where the relevant procedures are accepted, consider individual communications.

For the right to freedom of conscience, the UN Human Rights Committee is particularly important because it monitors the implementation of the International Covenant on Civil and Political Rights. Article 18 of the Covenant protects freedom of thought, conscience and religion. Article 19 protects freedom of opinion and expression. Article 27 concerns the rights of persons belonging to minorities to enjoy their own culture, profess and practise their own religion, and use their own language.

General Comment No. 10 of the Human Rights Committee: Freedom of Opinion, Article 19

General Comment No. 10 of the Human Rights Committee, adopted in 1983, concerned freedom of opinion under Article 19. It required protection of the right to hold opinions without interference. With respect to this right, the Covenant permits no exception or restriction.

This position is extremely important for the present study. Freedom of opinion is not merely the right to express a finished view. It begins earlier, in the inner domain where a person forms a judgment. If this inner process is subjected to coercion, manipulation or technological steering, freedom of expression becomes only the final, already damaged stage of a deeper freedom.

General Comment No. 34 of the Human Rights Committee: Freedoms of Opinion and Expression, Article 19

At its 102nd session in Geneva, held from 11 to 29 July 2011, the Human Rights Committee adopted General Comment No. 34 on freedoms of opinion and expression, replacing General Comment No. 10.

General Comment No. 34 confirms the absolute nature of the freedom to hold opinions. It emphasizes that all forms of opinion are protected, including political, scientific, historic, moral or religious opinions. It also stresses that harassment, intimidation or stigmatization of a person, including arrest, detention, trial or imprisonment, for opinions held, constitutes a violation of Article 19.

This interpretation has profound significance for liberty of worldview choice. If freedom to hold opinions is absolute, then the state may not claim authority over the internal convictions of a person. The state may regulate external conduct within lawful limits, but it may not own the inner architecture of belief. This is exactly where the principle of worldview neutrality becomes indispensable.

General Comment No. 22 of the Human Rights Committee: Freedom of Thought, Conscience and Religion, Article 18

General Comment No. 22 remains one of the most important interpretive documents in this field. It concerns Article 18 of the Covenant and clarifies the content of freedom of thought, conscience and religion.

The Committee stresses that Article 18 protects theistic, non-theistic and atheistic beliefs, as well as the right not to profess any religion or belief. The terms "belief" and "religion" are to be broadly construed. Article 18 is not limited in its application to traditional religions or to religions and beliefs with institutional characteristics or practices analogous to those of traditional religions.

The Committee notes that the freedom to "have or to adopt" a religion or belief necessarily entails the freedom to choose a religion or belief, including the right to replace one's current religion or belief with another or to adopt atheistic views, as well as the right to retain one's religion or belief. Article 18(2) prohibits coercion that would impair the right to have or adopt a religion or belief, including the use of threat of physical force or penal sanctions to compel believers or non-believers to adhere to religious beliefs and congregations, to recant their religion or belief, or to convert.

The Committee also considers that Article 18(4) permits instruction in public schools in subjects such as the general history of religions and ethics, provided that such instruction is given in a neutral and objective way. The freedom of parents or legal guardians to ensure the religious and moral education of their children in conformity with their own convictions must be respected.

It is especially important that the Committee states that public education including instruction in a particular religion or belief is inconsistent with Article 18(4) unless non-discriminatory exemptions or alternatives are available that accommodate the wishes of parents and guardians.

The Committee also gives an adequate interpretation of situations in which a state has an official or traditional religion, or where adherents of a particular religion constitute the majority of the population. Such circumstances must not impair the enjoyment of any rights under the Covenant, including Articles 18 and 27, nor result in discrimination against adherents of other religions or non-believers.

This position is one of the strongest points in the existing universal system. It approaches, though still does not fully formulate, the principle of worldview neutrality. The state may have history, culture and traditions, but it must not convert them into a legal hierarchy of human worth. Tradition may explain a society. It cannot become a license for discrimination.

General Comment No. 23 of the Human Rights Committee: Rights of Minorities, Article 27

General Comment No. 23 concerns the rights of persons belonging to minorities under Article 27 of the Covenant. The Committee notes that Article 27 establishes and recognizes a right conferred on persons belonging to minorities which is distinct from, and additional to, all the other rights which they, in common with everyone else, are already entitled to enjoy under the Covenant.

In the context of freedom of conscience, Article 27 is relevant because religious minorities may require protection not only against direct persecution, but also against assimilation, exclusion, unequal access to institutions and denial of cultural or religious identity.

At the same time, general comments on Article 27 that touch upon issues of freedom of conscience are still based partly on older approaches. They tend to focus on minorities as groups with cultural, religious or linguistic characteristics, while the modern problem is broader. A person may require protection not only as a member of a traditional minority, but also as an individual exercising worldview choice outside established collective categories. This includes non-believers, converts, persons with mixed identities, secular worldview communities, and individuals resisting ideological pressure.

Thus, the existing framework remains important but incomplete. It protects some visible communities better than it protects the invisible process of choosing, changing or refusing a worldview.

General Comment No. 24 of the Human Rights Committee: Reservations or Declarations

General Comment No. 24, adopted in 1994, concerns reservations or declarations made by states in relation to the Covenant and the First Optional Protocol. It sets out principles of international law relating to reservations, by which their acceptability and meaning must be determined. It also addresses the role of states parties in relation to reservations made by others, as well as the role of the Committee itself.

The Committee recognizes that the possibility of entering reservations may encourage a state that has difficulties guaranteeing all the rights under the Covenant nevertheless to accept the treaty's general obligations. Reservations can perform a useful function by allowing states to adapt specific elements of their domestic law to the obligations under the Covenant.

However, the Committee also stresses that the absence of a prohibition on reservations does not mean that any reservation is permissible. The question of reservations to the Covenant and the First Optional Protocol is governed by international law, including Article 19(3) of the Vienna Convention on the Law of Treaties. Where a reservation is not prohibited by the treaty, a state may formulate it only if it is not incompatible with the object and purpose of the treaty.

The Committee's position is important for the right to freedom of conscience. Reservations that undermine the core content of freedom of thought, conscience and religion, or that permit discrimination on religious or worldview grounds, should be treated as incompatible with the object and purpose of the Covenant. Otherwise, states could sign universal human rights instruments with one hand and quietly remove their substance with the other. A charming legal magic trick, but not one that should be accepted as international law.

In this context, reports to the Committee should contain information on what actions have been taken to review or withdraw reservations. This is especially important where reservations affect Article 18, Article 19 or Article 27 of the Covenant.

The Practical Limits of Universal Treaty Protection

The Human Rights Committee has made important efforts to protect individuals and religious communities, including in situations of persecution, discrimination and unlawful interference. These efforts deserve approval. Yet their effectiveness remains limited where states ignore views, delay implementation, or treat international procedures as inconvenient paperwork rather than binding moral and legal commitments.

For example, attempts by the Committee to protect individual religious associations have not always produced effective results. Cases involving religious communities, including situations connected with occupied or disputed territories, demonstrate that interim measures and views of the Committee may be politically disregarded or implemented only partially.

The Office of the High Commissioner for Human Rights has also spoken against persecution of Jehovah's Witnesses in Russia, whose organizations were banned as extremist by a decision of the Supreme Court of the Russian Federation of 20 April 2017. The UN Working Group on Arbitrary Detention adopted opinions concerning the unlawfulness of criminal prosecution in such cases. Yet, in response to one such UN position, domestic proceedings and penalties continued. This illustrates the central weakness of the international system: recognition of a violation does not automatically restore freedom.

The universal mechanisms are therefore indispensable, but insufficient. They offer interpretation, monitoring, recommendations, communications, interim measures and public pressure. What they do not yet provide is a fully operational architecture for the realization of the right to liberty of worldview choice.

The problem is not merely the lack of mechanisms. The deeper problem is the lack of a unified conceptual standard. The current system distributes related issues between freedom of religion or belief, freedom of opinion and expression, minority rights, education, non-discrimination, arbitrary detention and cultural rights. Each mechanism sees part of the picture. The human being, meanwhile, needs protection of the whole inner domain in which conscience, thought, belief, doubt, identity and worldview are formed.

That is why the next stage of international law should move from the protection of separate fragments toward the recognition of a comprehensive right to liberty of worldview choice, ensured by the principle of public-law worldview neutrality of states and other subjects of public power, including in the digital environment.

4.3.2. Regional, Interregional and Subregional Mechanisms for the Protection of the Right to Freedom of Conscience

Regional Mechanisms for the Protection of the Right to Freedom of Conscience

The relevant regional mechanisms for the protection of human rights in the sphere of freedom of conscience include those mechanisms that have been formed within the respective regions of the world. These are, first of all, the European mechanisms, including the instruments of the Council of Europe and the European Union, the Inter-American mechanisms, and the African mechanisms. The European mechanisms are of particular interest in this study, since the Russian Federation was a participant in them for a significant period.

The European system for the protection of human rights may be understood as a body of legal norms contained in conventions and agreements adopted within the framework of the Council of Europe, the Organization for Security and Co-operation in Europe and other European human rights organizations, incorporated into the national legislation of European states, as well as the activity of the Council of Europe and the European Court of Human Rights aimed at ensuring compliance with European convention-based norms for the protection of human rights.

European Mechanisms for the Protection of the Right to Freedom of Conscience

Mechanisms of the Council of Europe

The Council of Europe was established on 5 May 1949. For Russia, the Statute of the Council of Europe entered into force on 28 February 1996. According to Article 1 of the Statute, the aim of the Council of Europe is to achieve greater unity among its members for the purpose of safeguarding and realizing the ideals and principles which are their common heritage, and facilitating their economic and social progress. This aim is pursued through the organs of the Council of Europe by discussing questions of common concern, concluding agreements and taking joint action in economic, social, cultural, scientific, legal and administrative matters, as well as by maintaining and further realizing human rights and fundamental freedoms.

Under Article 3 of the Statute, every member of the Council of Europe must accept the principle of the rule of law and the principle that all persons within its jurisdiction shall enjoy human rights and fundamental freedoms. Each member must also collaborate sincerely and effectively in the realization of the aim of the Council of Europe as defined in Chapter I of the Statute.

Thus, the Council of Europe has historically set itself the task of building a Europe without dividing borders, on the basis of democracy, the rule of law and human rights. Its governing body is the Committee of Ministers. Under its authority operate expert bodies that prepare decisions on matters within the Committee's competence: the conclusion of conventions and agreements, adoption of recommendations to member states, admission of new member states, development of cooperation programmes, supervision of the execution of judgments of the European Court of Human Rights and other matters.

The Parliamentary Assembly is the consultative body of the Council of Europe and consists of representatives of the parliaments of member states. On the basis of reports prepared by its members, the Parliamentary Assembly adopts resolutions and recommendations, including on issues touching upon freedom of conscience. The consultative bodies of the Council of Europe also include the European Commission for Democracy through Law, known as the

Venice Commission, and the Congress of Local and Regional Authorities of Europe.

In S. A. Gorshkova's monograph, the Council of Europe system for the protection of human rights is examined in detail, including the question of the conformity of Russian legislation with the Convention for the Protection of Human Rights and Fundamental Freedoms, as well as the application of European human rights standards in the Russian Federation.

As G. E. Lukyantsev notes, the regional system for the protection of human rights created under the European Convention of 1950 is an integral institutional organism. It includes a control mechanism composed of the two bodies of the Convention, namely the European Commission and the European Court, and two institutions of the Council of Europe, namely the Committee of Ministers and the Secretary General of the organization, endowed with functions of control over the implementation of the Convention's provisions.

In 1976, the Committee of Ministers established the Steering Committee for Human Rights, composed of experts representing the member states of the Council of Europe. The Steering Committee carries out the Council of Europe's intergovernmental work in the field of human rights, taking into account, in particular, the legal standards of the Council of Europe and the relevant case law of the European Court of Human Rights. It advises the Committee of Ministers and provides legal expertise on all questions falling within its competence.

The Committee of Experts on the System of the European Convention on Human Rights, operating under the authority of the Steering Committee for Human Rights, conducts intergovernmental work aimed at strengthening human rights protection by increasing the effectiveness of the control mechanism of the European Convention on Human Rights and supporting the implementation of the Convention at national level. Like the Steering Committee itself, this body is a plenary committee composed of experts representing the member states of the Council of Europe, as well as other participants and observers defined by its terms of reference. It exchanges views with various stakeholders within the Convention system, including the Registry of the Court, the Secretariat of the Parliamentary Assembly and the Department for the Execution of Judgments of the European Court of Human Rights.

At the summit of heads of state and government of the Council of Europe in 1993, the European Commission against Racism and Intolerance was established. It is a human rights monitoring body specializing in issues related to combating racism, discrimination on grounds such as "race," ethnic or national origin, colour, citizenship, religion, language, sexual orientation and gender identity, as well as xenophobia, antisemitism and intolerance.

An important role in promoting the ideal of human rights is played by the Commissioner for Human Rights of the Council of Europe. The Commissioner provides advice and raises awareness of human rights issues by publishing thematic documents and organizing events and seminars. Human rights recommendations are most often addressed to all member states in the form of recommendations, opinions, reports and issue papers published by the Commissioner. The Commissioner also organizes workshops, seminars and lectures to examine specific human rights problems, and the Commissioner's Office participates in wider human rights debates, including major conferences.

In January 2018, Dunja Mijatovic was elected Commissioner for Human Rights by the Parliamentary Assembly of the Council of Europe. Before her, the office was held by Nils Muiznieks from 2012 to 2018, Thomas Hammarberg from 2006 to 2012, and Alvaro Gil-Robles from 1999 to 2006. Since January 2024, the Commissioner for Human Rights of the Council of Europe has been Michael O'Flaherty, an Irish lawyer and human rights specialist.

The European Committee for Social Cohesion, Human Dignity and Equality was established to coordinate the intergovernmental work of the Council of Europe in the fields of social cohesion, dignity, equality and non-discrimination. It is also intended to advise the Committee of Ministers on all matters in these areas. Its work is directed toward ensuring the full realization of and access to all human rights, including social and economic rights, for all members of society; reducing people's vulnerability; and strengthening the foundations of resilient democratic societies.

This is to be achieved through the promotion of solidarity, the protection of human dignity, equality and equal opportunities for all, and the prevention and suppression of violence and discrimination on any grounds. The Committee, relying on the relevant Council of Europe strategies and cross-cutting perspectives, is intended to improve the implementation of relevant standards and policies in member states and, where necessary, cooperate with other structures and bodies.

Among the mechanisms mentioned above, the key role in protecting human rights in the field of freedom of conscience belongs to the European Court of Human Rights, created on the basis of the European Convention on Human Rights of 1950. The rest of the system matters greatly, but the Court is the place where abstract guarantees meet a concrete violated person. This is precisely why its practice is so important: it shows whether the solemn vocabulary of human rights can survive contact with administrative reality.

The European Court of Human Rights

The European Court of Human Rights is an international judicial body which, under specific conditions, may examine applications from individuals who claim that their rights guaranteed by the European Convention on Human Rights have been violated. The protected rights are set out directly in the text of the Convention itself, as well as in Protocols Nos. 1, 4, 6 and 7, which have been ratified only by some of the states.

The European Court of Human Rights was established in Strasbourg by the member states of the Council of Europe in 1959 in order to deal with alleged violations of the Convention for the Protection of Human Rights and Fundamental Freedoms. The procedure before the Court consists of two principal stages: first, the determination of the admissibility of the application; and second, the examination of the merits of the case. These stages are preceded by correspondence between the applicant and the Registry of the European Court, up to the point at which the applicant is informed that the application has been communicated to the government for observations.

The question of admissibility, that is, whether the application meets the necessary formal criteria, is decided by a Committee of judges or by a Chamber of the Court. At the merits stage, the Chamber decides whether one of the rights listed in the European Convention has been violated.

A judgment of the European Court of Human Rights is binding on states that have recognized its jurisdiction and may include monetary compensation for the harm caused. Over the long history of the Court's practice, the author notes that there had been no recorded case of a member state of the Council of Europe refusing to execute a judgment of the Court. At the same time, applicants must be prepared for the fact that more than five years may pass between lodging an application and obtaining a judgment. The growing flow of applications has led to an overload of the system. A mechanism designed to protect the person thus risks becoming one more test of endurance for the very person it is meant to protect. Apparently even justice sometimes enjoys standing in line.

Article 9 of the European Convention for the Protection of Human Rights and Fundamental Freedoms guarantees to "everyone" the "right to freedom of thought, conscience and religion." This right includes freedom to change one's religion or belief and freedom, either alone or in community with others and in public or private, to manifest one's religion or belief in worship, teaching, practice and observance.

Article 9 of the European Convention is closely connected with Article 10 on freedom of expression, Article 11 on freedom of assembly and association, and Article 14 on the prohibition of discrimination. It should also be noted that Article 2 of Protocol No. 1 to the European Convention, adopted in Paris on 20 March 1952, secures the right to education, including respect by the state for the right of parents to ensure such education and teaching in conformity with their own religious and philosophical convictions.

The Convention for the Protection of Human Rights and Fundamental Freedoms of 1950 entered into force for the Russian Federation on 5 May 1998. Thus, Russia recognized the jurisdiction of the European Court of Human Rights, and citizens of the country who had exhausted all national remedies received the right to apply to the Court with complaints. In particular, the insufficient effectiveness of the judicial branch in Russia led to a number of cases concerning the protection of freedom of religion being declared admissible and examined on the merits by the European Court of Human Rights.

The study reveals the insufficient effectiveness of the European Court mechanism, which itself requires improvement. In addition to reducing the time required for the examination of cases, it is necessary to resolve the question of reopening and reviewing cases at the domestic level in connection with judgments of the European Court.

To a certain extent, one may agree with Yu. V. Samovich, who argues that the functioning of the international mechanism for the protection of human rights is one of the real manifestations of the effectiveness of the international legal order and, above all, of the basic postulate of international law: pacta sunt servanda. The practical expression of this mechanism, namely the bringing of domestic legislation and practice into conformity with judgments of the European Court of Human Rights and taking account of recommendations made by universal supervisory bodies, demonstrates the existence of an operative enforcement mechanism within the international community.

However, the analysis of the case law of the European Court of Human Rights in cases connected with the realization of the right to freedom of conscience confirms the author's hypothesis that the solution to the problem of realizing rights in this field lies not only in law enforcement. First of all, it requires deep scholarly rethinking of the problem and improvement of the normative and legal framework.

Judgments of the European Court of Human Rights concerning the Russian Federation

Since Russia recognized the jurisdiction of the European Court of Human Rights, the Court has delivered a significant number of judgments finding violations of the 1950 European Convention by Russia. In 2018, 12.15 thousand applications from Russia were submitted to the European Court of Human Rights, compared with 7.96 thousand in 2017. At that time, 11.75 thousand applications against Russia were awaiting examination, amounting to 20.9 percent of the total number, half of which had been communicated, namely 5.235 thousand applications.

A detailed study of the scope and content of freedom of thought, conscience and religion, guaranteed in particular by Article 9 of the European Convention as interpreted in the practice of the European Court and the former European Commission of Human Rights, is contained in the guide by J. Murdoch. In particular, he notes that from 1959 to 2010, violations of Article 9 were found in 35 judgments: nine against Greece, five against Russia, four against Bulgaria, three each against Latvia, Moldova, Turkey and Ukraine, and one each against Austria, Georgia, Poland, San Marino and Switzerland. By the beginning of 2019, according to the Court's data, 75 violations of Article 9 of the European Convention had been recorded; ten of them were in cases against Russia, with higher figures only against Greece, with thirteen cases, and Turkey, with twelve cases. It should also be noted that the first case concerning a violation of Article 9, Kokkinakis v. Greece, was examined in 1993.

It seems appropriate to agree with researchers who emphasize the significance of realizing the right to freedom of conscience and the need to study the relevant practice of the European Court of Human Rights, especially in relation to the Russian Federation, since judgments delivered against one's own state acquire precedent-forming significance. Judgments delivered against foreign states should likewise not be ignored by domestic courts; they should be analysed and taken into account.

This is all the more important because the insufficient effectiveness of the judicial branch in Russia led to a number of cases concerning the protection of freedom of religion being declared admissible by the European Court of Human Rights. We shall now examine several Russian cases that were considered by the Court on the merits.

For example, on 5 October 2006, the European Court of Human Rights granted the application of the Moscow branch of the Salvation Army, lodged in 2001 and declared admissible on 24 June 2004, concerning the refusal of the national authorities to re-register it as a legal entity. The Court held that in the case of the Moscow branch of the Salvation Army there had been a violation of Article 11 of the European Convention, concerning freedom of assembly and association, interpreted in the light of Article 9, concerning freedom of thought, conscience and religion. In addition, the Court held that the Russian Federation must pay the Moscow branch of the Salvation Army 10,000 euros in compensation for non-pecuniary damage.

The judgment in the case of Kuznetsov and Others v. Russia, concerning the disruption of a religious meeting of Jehovah's Witnesses and the unfair judicial proceedings in their case, was delivered by the European Court of Human Rights on 11 January 2007. The application, No. 184/02, had been lodged on 17 December 2001 and declared admissible on 9 September 2004. The applicants complained that in 2000 representatives of state authorities and law enforcement bodies had unlawfully disrupted a religious service, and that the lease agreement for the premises used for worship had been terminated. They also complained of the denial of a fair trial.

On 16 April 2000, the chair of the Human Rights Commission under the Governor of the Chelyabinsk Region, E. V. Gorina, accompanied by police officers, attempted in a rough manner to disrupt the religious service. Then, under pressure from representatives of the authorities, on 17 April the director of School No. 85 announced that the lease agreement between the school and the community would be terminated "in connection with certain violations committed by the school administration during the conclusion of the agreement." The prosecutor's office refused to initiate a criminal case concerning the actions of the authorities, while during the court proceedings the believers were repeatedly referred to as a "totalitarian sect."

The European Court of Human Rights unanimously held that there had been a violation of Articles 6 and 9 of the European Convention, namely the right to a fair trial and freedom of thought, conscience and religion. The Court ordered that Mr. Kuznetsov, and through him all the applicants, should receive 30,000 euros as compensation for non-pecuniary damage and 60,544 euros for costs and expenses.

Another judgment of the European Court was delivered on 5 April 2007 in the case brought by the Church of Scientology of Moscow against Russia. The Church was later liquidated by court decision in 2015. The Court unanimously found a violation of Articles 9 and 11 of the European Convention. The religious organization complained to the European Court of Human Rights about the refusal of the authorities to re-register it as a legal entity after the entry into force of the federal law on freedom of conscience. The document notes that the believers had applied more than ten times to the Moscow Justice

Department for re-registration without success, although the organization had first been officially registered on 25 January 1994. The European Court of Human Rights ordered Russia to pay 10,000 euros in compensation for non-pecuniary damage and 15,000 euros for costs. The organization was later liquidated by decision of the Moscow City Court in 2015, and that status was confirmed by the Supreme Court of the Russian Federation. At present, the legal entity has been excluded from the register of active organizations.

On 26 July 2007, the European Court of Human Rights in Strasbourg delivered a judgment in favour of the Evangelical Christian Church "Grace of Christ," operating in the town of Chekhov in the Moscow Region. The Court found that there had been a violation of Article 11 of the Convention, concerning freedom of assembly and association, considered in the light of Article 9, concerning freedom of thought, conscience and religion. The Court awarded the applicant 6,000 euros in compensation for non-pecuniary damage. The case arose from application No. 10519/03 against the Russian Federation, lodged before the Court under Article 34 of the Convention by Russian citizen Petr Ivanovich Barankevich on 23 February 2003. By decision of 20 October 2005, the Court declared the application admissible.

On 12 February 2009, the European Court of Human Rights delivered a judgment in case No. 2512/04, Nolan and K. v. Russia. The judgment states that the Russian side's ban on Patrick Nolan's entry into Russia in 2002, when he was a missionary of the Unification Church, violated several articles of the European Convention on Human Rights, including Article 9 on freedom of religion. Russian officials had claimed that Mr. Nolan's activity as a foreign missionary "threatened the security of the country." The European Court held that Article 9 of the Convention does not allow restrictions on freedom of religion on the ground of "national security." In the Court's view, the absence of national security from the list of permissible grounds for restricting rights under Article 9 reflects the primary importance of religious pluralism as "one of the foundations of a democratic society" under the European Convention. Consequently, national security interests could not justify the measures taken by Russian officials. The Court also held that Russia had violated Article 38 of the Convention, concerning the obligation to furnish all necessary facilities for the examination of the case; Article 8, concerning respect for private and family life; and Article 1 of Protocol No. 7, concerning procedural safeguards relating to the expulsion of aliens.

On 1 October 2009, the European Court of Human Rights examined applications by religious groups, Nos. 76836/01 and 32782/03, concerning the refusal to register them as religious organizations. The Court held that Russia had violated Articles 9 and 11 of the European Convention for the Protection of Human Rights and Fundamental Freedoms. The European Court concluded that the religious groups in question had the right to be registered as religious organizations under Russian law. In the winter of 2010, the Russian Federation attempted to challenge this judgment, but in March the judgment in the case of Kimlya and Others v. Russia, concerning organizations in Nizhnekamsk and Surgut, became final.

On 10 June 2010, the European Court of Human Rights delivered its judgment in case No. 302/02, Religious Community of Jehovah's Witnesses in Moscow v. Russia. The organization was later banned in Russia as extremist by the decision of the Supreme Court of the Russian Federation of 20 April 2017. The Court declared admissible the applicants' claims concerning the liquidation of the applicant community and the ban on its activities, the applicant community's claim concerning the refusal to re-register it, the claims of religious discrimination, and the applicant community's claim concerning the excessive length of the judicial proceedings in the liquidation case.

The European Court of Human Rights also decided that the authorities of the Russian Federation should pay the applicants 20,000 euros as compensation for non-pecuniary damage and 50,000 euros for costs and expenses connected with the application to the Court. The Court concluded that the interference with the applicants' rights to freedom of religion and association had been unjustified. The national courts had not provided relevant and sufficient reasons showing that the applicant community had coerced members into breaking up families, violated the rights and freedoms of its members or third persons, induced followers to suicide or refusal of medical care, infringed the rights of parents and children, or encouraged members to refuse to perform legally established duties.

The sanction imposed by the Russian courts was extremely severe because of the inflexibility of national legislation and was not proportionate to any legitimate aim pursued. Accordingly, there had been a violation of Article 9 of the European Convention taken together with Article 11. On 9 September 2010, the Russian Federation appealed this judgment and requested that the case be referred to the Grand Chamber of the European Court. On 13 December 2010, it became known that a panel of the Grand Chamber had refused Russia's request for review. This made final a case that had lasted approximately fifteen years.

In the 1990s, criminal proceedings against the community were initiated four times and discontinued four times on rehabilitating grounds. Later, the prosecutor's office brought a civil action in the Golovinsky District Court seeking liquidation of the Moscow community. In 2001 the court dismissed the claim, but the cassation instance returned the case for a new examination; three years later, the Golovinsky District Court ruled against the religious community in Moscow. The believers challenged the actions of the Russian authorities before the European Court, which unanimously found that the rights of the ten-thousand-member religious community in Moscow had been grossly violated and ordered individual and general measures to "restore, as far as possible, the situation existing before the violation."

The reference to "general measures" indicated that more was expected of the Russian side than merely the payment of compensation and the restoration of the community's registration, as Vasily Kalin explained. Yet these hopes were not fulfilled, since in 2017 all religious organizations and the Administrative Centre in Russia were banned, and their members were subjected to new persecution, including allegations of torture. In this context, it is difficult to avoid the conclusion that the judgments of the European Court concerning Russia have not fully contributed to the effective protection of the right to freedom of conscience.

By 2019, the total number of applications by the above-mentioned believers against Russia exceeded forty. In particular, in 2018 the following applications were lodged against the Russian Federation before the European Court of Human Rights: Markin and Trofimov v. Russia, No. 20519/18; Suvorov and Others v. Russia, No. 29779/18; Polyakovy v. Russia, No. 32532/18; Klimov v. Russia, No. 40768/18; Puyda and Others v. Russia, No. 42412/18; Britvin and Levchuk v. Russia, lodged on 1 November 2018; and Bazhenov and Others v. Russia, lodged on 30 November 2018. These applications challenged the warning on the inadmissibility of "extremist" activity, the decision of the Supreme Court banning the activities of the Administrative Centre in Russia and local communities as "extremist," the liquidation of all these organizations and the transfer of their property to the state, as well as violations of believers' rights in connection with their detention.

It also became known in 2018 that the European Court of Human Rights had communicated applications Nos. 10188/17 and 3215/18. The total compensation claims in these applications exceeded six billion roubles, or 79,215,679 euros. The applications were lodged in connection with the above-mentioned decision of the Russian Supreme Court to liquidate and ban the activities of all registered organizations and to transfer all their property, above all all houses of worship, to the state. In their compensation claim, the believers emphasized that their principal demand against the Russian Federation was the return of confiscated real estate throughout Russia. The Court declared applications Nos. 10188/17 and 3215/18 admissible and indicated that it would examine them as a matter of priority.

In March 2018, the Russian Government submitted its observations to the Strasbourg Court on the merits of application No. 10188/17, emphasizing that the Supreme Court's decision and the appellate ruling that upheld it "do not assess the doctrine and do not contain a restriction or prohibition on individually professing the above-mentioned teaching." Yet less than a month after this statement, mass searches and arrests of believers began in Russia, which causes particular concern as a direct consequence of the Supreme Court's decision. In December 2018, it became known that Russia had submitted observations to the European Court on the applications, rejecting all substantive claims of the applicants.

A particularly revealing example in this context is the instruction of the European Court of Human Rights to the Government of the Russian Federation of 26 February 2019 to send Sergey Loginov, one of the believers who complained of torture in the Investigative Committee of Russia in Surgut, for a medical examination. Doctors were to determine whether he needed treatment and whether his condition was compatible with continued detention. The application Loginov and Others v. Russia was lodged with the Court within a short time and was assigned No. 10618/19.

Thus, the situation appears paradoxical and deeply alarming: as believers lodge applications with the European Court and win cases against Russia, their organizations are outlawed, while hundreds of believers are prosecuted under Article 282.2 of the Criminal Code of the Russian Federation, concerning the organization of the activities of a religious organization in respect of which a liquidation decision has been adopted in connection with extremist activity, and also under Article 282.3 of the Criminal Code, concerning the financing of extremist activity.

By 1 November 2019, the total number of the above-mentioned believers with the status of "suspect" was 72, with the status of "accused" 145, with the status of "defendant" 57, and with the status of "convicted person" 26. Among them, 46 were in custody, four were in custody in absentia, meaning wanted persons, 23 were under house arrest, 22 were subject to a preventive measure in the form of a prohibition on certain actions, and 189 were subject to another preventive or coercive measure, such as an undertaking not to leave or an obligation to appear. In total, 122 cases under Article 282.2 of the Criminal Code had been initiated against members of the religious organization in 47 regions of Russia. The largest number of cases, 24, had been initiated in the Khanty-Mansi Autonomous Okrug-Yugra. Fourteen cases had been opened in

Khabarovsk Krai, 21 in Primorsky Krai, 12 in Kirov Region, and 14 in the Jewish Autonomous Region. In other regions, fewer than ten cases had been initiated. It should be recalled that all organizations of Jehovah's Witnesses were banned in Russia in mid-2017. After that, searches with significant use of force were carried out against followers of this confession in many regions of the country, and criminal cases began to be initiated against them.

The study allows us to assume that fast and effective protection of violated human rights in the field of freedom of conscience should not be expected from the European Court of Human Rights alone. First, between the violation of the right and examination of the case on the merits, six or seven years often passed. Second, even where the European Court delivered a judgment in favour of the applicant, this did not in itself guarantee execution by the respondent state. Here the limit of the judicial model becomes visible with uncomfortable clarity: a court may recognize the violation, but recognition itself does not automatically rebuild the legal environment that produced the violation in the first place.

The Execution of Judgments of the European Court of Human Rights and the Limits of the Judicial Model

For example, after the judgment of the European Court of Human Rights had been delivered, the representative of the believers, A. Lychkin, applied to the Nikulinsky District Court of Moscow seeking a review, on the basis of newly discovered circumstances, of the judicial decision of 2001. The court, however, refused. It relied, among other things, on the argument that a judgment of the European Court of Human Rights was not included in the list of grounds established by Article 392 of the Code of Civil Procedure of the Russian Federation for reviewing final judicial decisions on the basis of newly discovered circumstances. The Moscow City Court upheld that decision. The believers then lodged a complaint with the Constitutional Court of the Russian Federation, challenging, in particular, the constitutionality of Part 2 of Article 392 of the Code of Civil Procedure. The Constitutional Court refused to examine the complaint.

The ruling stated that the provisions of Part 2 of Article 392 of the Code of Civil Procedure of the Russian Federation, in themselves, could not be regarded in that case as violating the constitutional rights of the religious association, and therefore the complaint, since it did not meet the admissibility criteria, could not be accepted for consideration by the Constitutional Court. The text also emphasized that assessing whether the actions of authorized federal bodies were sufficient to remedy a specific situation, as well as assessing the legality of decisions taken by registration authorities, could not be carried out within constitutional proceedings, because such questions were connected with the establishment of factual circumstances, which did not fall within the competence of the Constitutional Court of the Russian Federation. In practical terms, this meant that the European Court, having awarded 10,000 euros in compensation against the Russian Federation in favour of the church, did not itself have the power to compel the respondent state to re-register the applicant.

At the same time, the recommendations of the Committee of Ministers of the Council of Europe concerning the reopening of cases at the domestic level following judgments of the European Court stated that the respondent state is obliged to take measures ensuring, as far as possible, restoration of the situation in which the applicant had found themselves before the state violated the Convention. In this context, the weakness of the mechanism becomes visible: recognition of a violation is only the beginning; the legal system must still be capable of restoring the right in reality. Thus, the ECHR mechanism clearly requires improvement. Changes are also necessary at the Russian national level.

On 26 February 2010, a judgment was delivered in the case concerning the constitutionality of Articles 392 and 393 of the Code of Civil Procedure of the Russian Federation, considered in connection with other provisions of Section IV of that Code, in particular the provisions on the review of final judicial decisions. The Constitutional Court of the Russian Federation obliged courts of general jurisdiction to accept civil cases for review on the basis of judgments of the European Court of Human Rights. It should be added, however, that this judgment of the Constitutional Court did not mean that such cases would be reviewed automatically.

On 29 October 2010, V. D. Zorkin, Chairman of the Constitutional Court of the Russian Federation, published an article in which he stated that the review of judicial decisions following ECHR judgments was reliably ensured in Russia. According to him, the last gap in the implementation of ECHR judgments had been eliminated by the Constitutional Court's ruling of 26 February 2010 concerning the constitutionality of Article 392 of the Code of Civil Procedure. He stressed that every ECHR judgment establishing a violation of human rights entails two consequences: first, a financial sanction against the state; and second, a decision by domestic authorities on restoring the violated right. The Constitutional Court's approach, according to Zorkin, was that in every case an application for review of domestic judicial decisions must be accepted for consideration, while the court in each specific case must determine whether everything necessary has been done to restore the right, including whether monetary compensation awarded by the European Court is sufficient.

At the same time, Zorkin noted that the introduction of many approaches and positions of the European Court into the fabric of the Russian legal space was far from simple, since most law-enforcement bodies were simply not ready to apply those approaches and positions directly as binding. He then referred to the European Court's recent judgment concerning the unlawfulness of the ban on gay pride marches in Moscow, which, in his view, "break the cultural, moral and religious code," as well as to the proceedings before the ECHR on the complaint of the "united opposition," which questioned the results of the 2003 State Duma elections. From this he drew the following thesis: the principles of state sovereignty and supremacy of the Constitution in the Russian legal system belong to the foundations of Russia's constitutional order. The Convention, as an international treaty of Russia, forms part of its legal system, but it does not stand above the Constitution. Article 15 of the Constitution establishes the priority of an international treaty over statutory law, but not over the Constitution. The monopoly on interpreting constitutional provisions and revealing the constitutional meaning of a law belongs to the Constitutional Court. Therefore, the interpretation of the Constitution given by the highest judicial body of the state cannot be overcome by interpreting the Convention, because the legal force of the Convention does not surpass the legal force of the Constitution.

The topic was developed further at a meeting held on 11 December 2010 at the Gorki residence between the President of Russia, D. A. Medvedev, and judges of the Constitutional Court headed by V. D. Zorkin. In particular, Zorkin stated that problems do indeed arise, but not problems of whether or not to execute judgments of the Constitutional Court, as the issue is sometimes presented. Rather, the problems concern the boundary between national jurisdiction and the jurisdiction, for example, of the Strasbourg Court. It is one thing when a specific dispute about a right is resolved between citizen Ivanov and the Russian Federation; it is another when a judgment contains so-called recommendations of a general character, including conclusions that a law is defective and must be repealed or amended. Then the question arises: who puts the final dot over the "i"? Is it the Russian Federation itself, determining concrete measures of a general nature, or are we dealing with supranational justice that replaces constitutional justice? The problem, he said, is not as simple as it may first appear.

In response, D. A. Medvedev effectively confirmed Zorkin's doubts concerning the limits of jurisdiction and the binding nature of ECHR judgments. He stated that, in contacts with foreign partners and European institutions, Russia should proceed above all from the scope of powers that it had transferred to the European Court when concluding the relevant agreements and signing the relevant laws. In his view, Russia had never transferred that part of its sovereignty which would allow any international court or foreign court to issue judgments changing national legislation.

As a result, on 15 February 2011, Judge Elena Novikova of the Golovinsky District Court of Moscow refused to grant the religious community's application for review of that court's judgment of 26 March 2004, which the European Court had found unlawful. "This is an obvious disregard for the judgment of the European Court and an unwillingness to execute it properly," explained lawyer Artur Leontyev. "The rights of the believers have not been restored. We will appeal this decision."

An interim outcome of the above-mentioned discussion came in July 2015 with the adoption of the highly controversial Judgment of the Constitutional Court of the Russian Federation of 14 July 2015 No. 21-P. The case concerned the constitutionality of provisions of the Federal Law "On Ratification of the Convention for the Protection of Human Rights and Fundamental Freedoms and the Protocols thereto," provisions of the Federal Law "On International Treaties of the Russian Federation," provisions of the Code of Civil Procedure, the Arbitrazh Procedure Code, the Code of Administrative Judicial Procedure and the Code of Criminal Procedure, in connection with a request from a group of deputies of the State Duma.

In particular, the Constitutional Court held that the relevant interconnected provisions did not contradict the Constitution of the Russian Federation, because, according to their constitutional and legal meaning in the system of existing legal regulation, they ensured the application of the Convention for the Protection of Human Rights and Fundamental Freedoms as a multilateral international treaty of the Russian Federation and as a component of its legal system. They also ensured the execution of judgments adopted on the basis of the Convention by the European Court of Human Rights, acting as a subsidiary inter-state judicial body in specific cases where all constitutionally established domestic judicial remedies had been exhausted.

The Constitutional Court further explained that, when a court of general jurisdiction or an arbitrazh court reviews a case following an ECHR judgment establishing that rights and freedoms had been violated in Russia through the application of a law or certain provisions thereof, and if that domestic court concludes that the question of applying the relevant law can be resolved only after confirming its conformity with the Constitution, it must apply to the

Constitutional Court to review that law's constitutionality. In any event, a domestic court reviewing a final judicial act at the request of a person whose application to the European Court led to a judgment finding a violation of the Convention must suspend the proceedings and refer the question to the Constitutional Court if the violation resulted from provisions of Russian legislation applied in that person's case.

The judgment also stated that state bodies responsible for ensuring Russia's performance of its international treaty obligations may apply to the Constitutional Court if they conclude that it is impossible to execute an ECHR judgment against Russia because the part of the judgment requiring individual or general measures is based on an interpretation of the Convention that diverges from the Constitution of the Russian Federation. If the Constitutional Court concludes that an ECHR judgment, in so far as it is based on an interpretation of the Convention contradicting the Constitution, cannot be executed, then that judgment is not subject to execution in that part.

At the same time, the Court added that this does not exclude the competence of the federal legislator, proceeding from the requirements of the Constitution and taking into account the legal positions expressed by the Constitutional Court, to create a special legal mechanism under which the Constitutional Court may decide whether, in light of the principles of the supremacy and highest legal force of the Constitution, a judgment delivered on an application against Russia by the European Court of Human Rights can or cannot be executed, including in relation to general measures.

The court decisions of 2016 then effectively determined that judgments of the European Court of Human Rights delivered on applications against the Russian Federation could be executed only on the basis of the supremacy of the Constitution of the Russian Federation. The Constitutional Court was thus placed in a position where it could assess the possibility of executing such judgments.

Federal Constitutional Law No. 7-FKZ of 14 December 2015 introduced amendments to the Federal Constitutional Law "On the Constitutional Court of the Russian Federation." These amendments gave the Constitutional Court the power, at the request of a federal executive body vested with competence in the field of protecting the interests of the Russian Federation when complaints against Russia are examined by an inter-state body for the protection of human rights and freedoms, to decide whether a judgment of such an inter-state body may be executed. The law also supplemented the relevant procedures by adding a special chapter on cases concerning the possibility of executing decisions of inter-state bodies for the protection of human rights and freedoms.

In fact, this Federal Constitutional Law gave the Constitutional Court of Russia the authority to determine whether judgments of the European Court of Human Rights and other inter-state bodies should be executed.

The Russian Federation subsequently ceased to be a member of the Council of Europe and left the jurisdiction of the European Court of Human Rights; the Court stopped examining complaints concerning violations occurring after 16 September 2022. Under Federal Law No. 183-FZ, the Russian Federation does not execute ECHR judgments that entered into force after 15 March 2022, and the Supreme Court removed references to the European Convention, replacing them with references to United Nations and CIS documents.

As a result, it should be noted that the mechanism for protecting the right to freedom of conscience through the European Court of Human Rights was the most effective of the existing human rights mechanisms, both at the regional and the universal levels. Yet, in conditions of systemic problems of legal recognition, reflected in the extremely low effectiveness of principles and norms of international law as well as domestic legal systems, the European Court was unable to exert a decisive influence on the realization of the right to freedom of conscience in Europe, including Russia.

European Union

The European Union was established by the Treaty on European Union of 7 February 1992, signed in Maastricht by the leaders of Belgium, Denmark, Germany, Greece, Spain, France, Ireland, Italy, Luxembourg, the Netherlands, Portugal, the United Kingdom and Northern Ireland. The Treaty marked a new stage of European integration. The European Union was founded on the basis of the European Community.

In particular, the preamble refers to the determination to promote, through joint efforts, the development of democracy based on the fundamental rights recognized in the constitutions and laws of the Member States, in the Convention for the Protection of Human Rights and Fundamental Freedoms and in the European Social Charter, above all the rights of freedom, equality and social justice.

As E. S. Alisevich notes, truly important human rights provisions appeared in the law of the European Communities and the European Union with the adoption of the new version of the Treaty on European Union on 2 October 1997, the Amsterdam Treaty. Human rights were included among the founding principles of the European Union, and respect for them became a condition for accession to the Union. The Amsterdam Treaty introduced provisions on monitoring respect by EU Member States for the Union's fundamental principles, including respect for human rights, as well as measures of responsibility for violations of those principles. Article 13 prohibited discrimination on grounds of sex, race, ethnic and religious affiliation, and, for the first time, on grounds of age and sexual orientation. Article 7 of the Treaty on European Union was later supplemented by the Nice Treaty of 11 December 2000. In addition to establishing the existence of a serious breach by a state of the principles of the European Union and adopting measures against that state, it provided for a preventive procedure where there is a clear risk of a serious violation of human rights by a Member State, including the preparation of recommendations and continuous monitoring by the Council of the European Union.

With the adoption of the Charter of Fundamental Rights of the European Union in 2000, the Charter came to occupy one of the key positions in the European Union's human rights system. Its purpose is to strengthen the protection of fundamental rights within the European Union in light of changes in society, social progress and scientific and technological development.

In 2007, the European Union Agency for Fundamental Rights was established by Council Regulation (EC) No. 168/2007. It is a decentralized expert institution within the EU framework. The Agency carries out its tasks, as defined in its founding regulation, through specific projects and activities within thematic areas of a five-year multiannual framework programme, generally corresponding to different chapters of the Charter of Fundamental Rights. The Agency collects relevant and timely data and information, exchanges evidence-based views and advice with policymakers and decision-makers, raises awareness of rights and promotes fundamental rights through effective communication, and works with a wide range of stakeholders from the local to the international level through targeted assistance and deep expertise. This is what makes it a European centre of fundamental rights expertise.

The thematic areas of the Agency's work are defined by five-year multiannual frameworks. At its session of 7-8 December, the Council of the European Union adopted a decision establishing the multiannual framework for the European Union Agency for Fundamental Rights for 2018-2022. Since 1 January 2018, the Agency's priority themes have included: victims of crime and access to justice; equality and discrimination on any ground, such as sex, race, colour, ethnic or social origin, genetic features, language, religion or belief, political or any other opinion, membership of a national minority, property, birth, disability, age, sexual orientation or nationality; the information society, in particular respect for private life and protection of personal data; judicial cooperation except in criminal matters; migration, borders, asylum and integration of refugees and migrants; racism, xenophobia and related intolerance; rights of the child; and Roma integration and social inclusion. These frameworks are implemented in addition to the work of other EU bodies, the Council of Europe and other international organizations dealing with human rights.

In general, through data collection and analysis in the European Union, and through cooperation with EU institutions, EU Member States and other organizations at the international, European and national levels, the European Union Agency for Fundamental Rights is designed to help make fundamental rights a reality for everyone living in the European Union.

Inter-American Mechanisms for the Protection of the Right to Freedom of Conscience

The Inter-American human rights protection system began to take shape after the Second World War. It emerged as a heterogeneous formation, bringing together countries with different values, cultures, languages and traditions: industrially developed states of North America with long-standing traditions of federalism, self-government, a developed civil society and independent human rights institutions on the one hand, and countries of the Caribbean basin and the South American continent on the other, where throughout the twentieth century periods of instability and tension frequently arose, dictators came to power, and military regimes were established with systematic violations of human rights. The Organization of American States plays a key role in the Inter-American system of human rights protection.

Organization of American States

The Organization of American States was established on 30 April 1948 in Bogota, Colombia. It is considered the successor of the Pan-American Union, created in 1889. Under Article 2 of the Charter of the Organization of American States of 1948, in order to implement the principles on which it is founded and to fulfil its regional obligations under the Charter of the United Nations, the Organization sets the following essential purposes: to strengthen the peace and security of the continent; to promote and consolidate representative democracy, with due respect for the principle of non-intervention; to prevent possible causes of difficulties and to ensure the peaceful settlement of disputes that may arise among Member States; to provide for common action in the event of aggression; to seek the solution of political, juridical and economic problems that may arise among them; to promote, by cooperative action, their economic, social and cultural development; and to achieve an effective limitation of conventional weapons that will make it possible to devote the largest amount of resources to the economic and social development of Member States.

It should be especially noted that the OAS Charter of 1948 states that the American States proclaim the fundamental rights of the individual without distinction as to race, nationality, creed or sex.

To implement the purposes of the OAS, the following principal organs were established: the General Assembly; the Meeting of Consultation of Ministers of Foreign Affairs; the Councils of the Organization; the Inter-American Juridical Committee; the Inter-American Commission on Human Rights; the General Secretariat; and specialized conferences and organizations. Chapter XVI of the Charter, concerning the Inter-American Commission on Human Rights, provides in Article 111 that the Commission's principal function is to promote the observance and protection of human rights and to serve as a consultative organ of the Organization in this field. Its structure, competence and procedure, as well as those of other bodies dealing with these matters, are determined by the American Convention on Human Rights.

The American Declaration of the Rights and Duties of Man was also adopted in 1948, and in 1960 the Inter-American Commission on Human Rights was established to implement its provisions. Since 1970, on the basis of amendments to the Charter, the Commission has been an organ of the OAS. The American Convention on Human Rights was adopted on 22 November 1969 and entered into force in 1978. To supervise the implementation of the Convention, the Inter-American Court of Human Rights was established in 1979. The American Convention on Human Rights secures the right to freedom of conscience and religion, as well as the right of parents and, where applicable, guardians, to provide for the religious and moral education of their children or wards in accordance with their own convictions. The Inter-American Convention against All Forms of Discrimination and Intolerance of 2013 prohibits religious discrimination.

African Mechanisms for the Protection of the Right to Freedom of Conscience

African mechanisms are based on a number of foundational documents and on a system of bodies intended to implement them. First of all, this is the African Union.

African Union

The African Union was established as an international intergovernmental organization on 9 July 2002. It is the successor to the Organization of African Unity, founded in 1963.

The Constitutive Act of the African Union of 2000 defines its objectives as follows: to achieve greater unity and solidarity between the African countries and the peoples of Africa; to defend the sovereignty, territorial integrity and independence of its Member States; to accelerate the political and socio-economic integration of the continent; to promote and defend common African positions on issues of interest to the continent and its peoples; to encourage international cooperation, taking due account of the Charter of the United Nations and the Universal Declaration of Human Rights; to promote peace, security and stability on the continent; to promote and protect human and peoples' rights in accordance with the African Charter on Human and Peoples' Rights and other relevant human rights instruments; to establish the necessary conditions enabling the continent to play its rightful role in the global economy and in international negotiations; to promote sustainable development at the economic, social and cultural levels and the integration of African economies; to promote cooperation in all fields of human activity in order to raise the living standards of African peoples; to coordinate and harmonize policies between existing and future regional economic communities for the gradual attainment of the objectives of the African Union; to advance the development of the continent by promoting research in all fields, especially science and technology; and to work with relevant international partners in the eradication of preventable diseases and the promotion of good health on the continent.

The governing body of the African Union is the Assembly, composed of Heads of State and Government of Member States. The Secretariat and the Commission of the African Union ensure the functioning of the organization. In February 2009, a decision was adopted to transform the African Union Commission into the African Union Authority.

In June 1981, the Member States of the Organization of African Unity adopted the African Charter on Human and Peoples' Rights. The Charter emphasizes the value of historical traditions and the values of African civilization, which should be reflected in the formation and content of human rights concepts. In 2004, the 1998 Protocol establishing the African Court on Human and Peoples' Rights entered into force. The African Commission on Human and Peoples' Rights plays a leading role and monitors the implementation of the Charter, including by examining human rights violations.

An important feature of the Commission's work is the existence of special procedures that make it possible, within mandates approved by resolutions, to respond relatively promptly to information concerning human rights violations. The mandates of special procedures require mandate-holders to monitor, advise and inform the public about the human rights situation in particular countries or territories, known as country mandates, or about mass human rights violations in Africa as a whole, known as thematic mandates. Researchers point to several problems in the Commission's work: too much time passes between the submission of complaints and their consideration; the Commission makes insufficient use of its powers to refer cases to the African Court; it does not adopt and examine reports of special rapporteurs in a timely manner; it lacks a mechanism for monitoring compliance with its recommendations; many states fail to fulfil their reporting obligations under the Charter, thereby depriving the Commission of the opportunity to review human rights situations in African states; states often lack the political will to implement the Commission's recommendations; and cooperation between the African Union and the African Commission is also poorly implemented.

The African Committee of Experts on the Rights and Welfare of the Child should also be included among African human rights protection mechanisms. It operates under the African Charter on the Rights and Welfare of the Child of 1990, which entered into force in 1999. The Committee supervises the fulfilment of obligations by states parties through the submission of reports. In addition, when considering state reports, the Committee may request additional information from civil society organizations. The African Charter on the Rights and Welfare of the Child allows any person or group of persons to submit a complaint to the Committee concerning violations of children's rights.

The key role in the African human rights protection system belongs to the African Charter on Human and Peoples' Rights of 1981, with its two protocols, and to the African Commission on Human and Peoples' Rights. The African Court on Human and Peoples' Rights and the African Committee of Experts on the Rights and Welfare of the Child also play a certain role.

Thus, European regional mechanisms include the instruments of the Council of Europe, including the European Convention for the Protection of Human Rights and Fundamental Freedoms, the former European Commission of Human Rights, the European Court of Human Rights, the Committee of Ministers, the Parliamentary Assembly, the Secretary General of the Organization, the Secretariat, the Steering Committee for Human Rights, the European Commission against Racism and Intolerance, the Council of Europe Commissioner for Human Rights, the European Committee for Social Cohesion, Human Dignity and Equality, the European Commission for Democracy through Law, known as the Venice Commission, the Congress of Local and Regional Authorities of Europe and other bodies, as well as the instruments of the European Union, including the Charter of Fundamental Rights of the European Union and the European Union Agency for Fundamental Rights. Regional mechanisms also include the Inter-American mechanisms, namely the American Convention on Human Rights, the Organization of American States, the Inter-American Commission on Human Rights and the Inter-American Court of Human Rights, and the African mechanisms, namely the African Charter on Human and Peoples' Rights, the African Union, the African Commission on Human and Peoples' Rights, the African Court on Human and Peoples' Rights and the African Committee of Experts on the Rights and Welfare of the Child. Of particular interest are the European mechanisms in which the Russian Federation participated.

REFERENCES TO CHAPTER 4

List of Sources Used

1. Kartashkin V.A. Human Rights and International Security // Proceedings of the Institute of State and Law of the Russian Academy of Sciences. 2013. No. 1. pp. 33-59. p. 43.
2. International Protection of Human Rights: Textbook / executive editor A.Kh. Abashidze. Moscow: RUDN University, 2017. 466 p. p. 10.
3. Cited work. pp. 89-90.
4. Lukashuk I.I. International Law. General Part: Textbook for Students of Law Faculties and Universities. 3rd ed., revised and expanded. Moscow: Wolters Kluwer, 2005. p. 152.
5. Cited work. p. 159.
6. Buryanov S.A. The Future of International Law under Conditions of the Globalization of Social Relations through the Prism of the Creative Legacy of Igor Ivanovich Lukashuk // Eurasian Law Journal. 2016. No. 7 (98). pp. 77-81.
7. Civil Human Rights in Russia: Contemporary Problems of Theory and Practice (edited by F.M. Rudinsky). Moscow: ZAO TF MIR, 2006. p. 13.
8. Rudinsky F.M. The Science of Human Rights and Problems of Constitutional Law. Moscow: ZAO TF MIR, 2006.
9. Chetvernin V.A. Introduction to the Course of the General Theory of Law and the State. Moscow: Institute of State and Law of the Russian Academy of Sciences, 2003. 204 p. pp. 19-20.
10. Constitutional Rights and Freedoms of the Human Person and Citizen in the Russian Federation. Edited by O.I. Tiunov. Moscow: Norma, 2005. p. XIV.
11. Utyashev M.M. Human Rights and Freedoms in the System of Universal Human Values // Bulletin of Bashkir University. Vol. 14. No. 3-1. 2009. pp. 1244-1250.
12. Human Rights. Executive editor corresponding member of the Russian Academy of Sciences E.A. Lukasheva. Moscow: Norma, 2004. p. 1.
13. Ibid. p. 4.
14. Civil Human Rights in Russia: Contemporary Problems of Theory and Practice (edited by F.M. Rudinsky). Moscow: ZAO TF MIR, 2006. p. 13.
15. Tunkin G.I. Theory of International Law. General editor L.N. Shestakov. Moscow: Zertsalo, 2006. p. 68.
16. Kartashkin V.A. Human Rights and Principles of International Law in the 21st Century: Monograph. Moscow: Norma: INFRA-M, 2018. 148 p. p. 115.
17. Chernichenko S.V. The Individual and International Law. Moscow: International Relations, 1974. 166 p.
18. Chernichenko S.V. Theory of International Law: in 2 volumes. Moscow: NIMP, 1999. Vol. 1: Contemporary Theoretical Problems. 336 p.; Vol. 2: Old and New Theoretical Problems. 526 p.
19. Chernichenko S.V. The Contours of International Law. General Issues. Moscow: Scientific Book, 2014. 592 p.
20. Rudinsky F.M., Gavrilova Yu.V., Krikunova A.A., Soshnikova T.A. Economic and Social Rights of the Human Person and Citizen: Contemporary Problems of Theory and Practice. General editor F.M. Rudinsky. Moscow: Human Rights, 2009. p. 283.
21. International Law: Textbook / executive editors Yu.M. Kolosov and E.S. Krivchikova. 2nd ed., revised and expanded. Moscow: International Relations, 2005. p. 532.
22. Understanding Human Rights: Manual on Human Rights Education / edited by W. Benedek and M. Nikolova. Graz: European Training and Research Centre for Human Rights and Democracy; Russian version: Warsaw: Homework, 2005. p. 14.
23. Vitruk N.V. General Theory of the Legal Status of the Individual. Moscow: Norma, 2008. p. 443.
24. Tunkin G.I. Theory of International Law. General editor L.N. Shestakov. Moscow: Zertsalo, 2006. p. 68.
25. Koneva A.E. International-Legal Aspects of Strengthening the System of Human Rights Treaty Bodies: Candidate of Legal Sciences Dissertation. Moscow, 2015. 235 p. p. 3.

26. Our Commitment to Human Rights // Oxfam International. Electronic resource. URL: https://www.oxfam.org/en/our-commitment-human-rights (accessed: 1.10.2019).
27. International Protection of Human Rights: Textbook / executive editor A.Kh. Abashidze. Moscow: RUDN University, 2017. 466 p. Cited work. p. 93.
28. Cited work. p. 117.
29. Cited work. p. 120.
30. See: Neshataeva T.N. International Organizations and Law. New Trends in International-Legal Regulation. Moscow: Delo, 1998. 272 p.
31. Ibid. p. 66.
32. Azarov A., Reuter W., Hufner K. Protection of Human Rights: International and Russian Mechanisms. Moscow: Moscow School of Human Rights, 2000. p. 11.
33. Ibid. p. 12.
34. Electronic resource. URL: https://www.ohchr.org/RU/Issues/Pages/WhatareHumanRights.aspx (accessed: 29.09.2019).
35. Billionaire Fortunes Grew by $2.5 Billion a Day Last Year as Poorest Saw Their Wealth Fall // Oxfam International. Electronic resource. URL: https://www.oxfam.org/en/pressroom/pressreleases/2019-01-18/billionaire-fortunes-grew-25-billion-day-last-year-poorest-saw (accessed: 1.10.2019).
36. Rudinsky F.M. The Science of Human Rights and Problems of Constitutional Law. Moscow: ZAO TF MIR, 2006. p. 651.
37. Podoprigora R.A. The State and Religious Organizations: Administrative-Legal Issues. Almaty: Arkaim, 2002. p. 45.
38. Civil Human Rights in Russia: Contemporary Problems of Theory and Practice (edited by F.M. Rudinsky). Moscow: ZAO TF MIR, 2006. p. 441.
39. Rudinsky F.M. The Science of Human Rights and Problems of Constitutional Law. Moscow: ZAO TF MIR, 2006. p. 107.
40. Luchin V.O. The Constitution of the Russian Federation: Problems of Implementation. Moscow: UNITI-DANA, 2002. p. 63.
41. Luchin V.O. Cited work. pp. 87-108.
42. Vashanova O.V. The Principle of Non-Discrimination of the Individual in International Law: Candidate of Legal Sciences Dissertation. Moscow, 2004. 232 p. p. 7.
43. Electronic resource. URL: https://www.ohchr.org/RU/Issues/Pages/WhatareHumanRights.aspx (accessed: 29.09.2019).
44. International Public Law: Textbook / L.P. Anufrieva, D.K. Bekyashev, K.A. Bekyashev, V.V. Ustinov et al.; executive editor K.A. Bekyashev. 4th ed., revised and expanded. Moscow: TK Welby; Prospekt Publishing House, 2007. p. 227.
45. Constitutional Rights and Freedoms of the Human Person and Citizen in the Russian Federation. Edited by O.I. Tiunov. Moscow: Norma, 2005. p. 535.
46. Cited work. p. 536.
47. Rudinsky F.M. The Science of Human Rights and Problems of Constitutional Law. Moscow: ZAO TF MIR, 2006. p. 209.
48. International Law: Textbook / executive editors Yu.M. Kolosov and E.S. Krivchikova. 2nd ed., revised and expanded. Moscow: International Relations, 2005. p. 535.
49. Ibid. p. 536.
50. Ibid. p. 539.
51. Ibid. p. 539.
52. Pryakhina T.M. Ensuring Interaction between International and National Law by the Constitutional Court of the Russian Federation. Saratov: Nauka Information Center, 2010. p. 27.
53. International Law and Soviet Legislation. Collective Monograph (edited by G.I. Kurdyukov). Kazan: Kazan University Publishing House, 1991. p. 23.
54. Saidov A.Kh. Universally Recognized Human Rights: Study Guide / A.Kh. Saidov; edited by I.I. Lukashuk; Academic Law University. Moscow: MZ Press, 2004. p. 64.
55. Cited work. p. 65.

56. Ershov V.V. Current Problems of the Expert Activity of the Public Chamber of the Russian Federation in Light of Contemporary Legal Understanding, Lawmaking, and Law Enforcement // Russian Justice. 2009. No. 5. p. 105.
57. Ershov V.V. International and Domestic Law from the Standpoints of Legalism and an Integrative Understanding of Law // Russian Justice. 2011. No. 8 (64). p. 6.
58. Ibid. p. 3.
59. Mingazov L.Kh. The Effectiveness of Norms of International Law (Theoretical Problems): Abstract of Doctoral Dissertation in Law. Kazan, 2000. p. 7.
60. Ibid. pp. 7-8.
61. Marochkin S.Yu. The Problem of the Effectiveness of Norms of International Law. Scientific editor G.V. Ignatenko. Irkutsk: Irkutsk University Publishing House, 1988. pp. 7-21.
62. Kartashkin V.A., Lukasheva E.A. Human and Citizen Rights and Freedoms: Universalism, Regionalism, Realities // State and Law. 2010. No. 7. p. 38.
63. V.A. Kartashkin: On the 50th Anniversary of the Human Rights Covenants. Interview with Vladimir Alekseevich Kartashkin, Doctor of Legal Sciences, Professor, Honored Lawyer of the Russian Federation // Eurasian Law Journal. 2016. No. 2 (93). pp. 8-15.
64. Rowe M. The Right to Freedom of Religion or Belief. Study Guide. Minnesota: Human Rights Center, University of Minnesota, 2003. Electronic resource. URL: http://www1.umn.edu/humanrts/russian/edumat/studyguides/Rreligion.html (accessed: 10.11.2019).
65. De Zayas A., Moller J.Th. The United Nations Human Rights Committee Case Law 1977-2008: A Handbook. N.P. Engel Publishers, Kehl am Rhein, 2009.
66. Davis D. The Evolution of Religious Freedom as a Universal Human Right: A Study of the Role of the 1981 UN Declaration on the Elimination of All Forms of Intolerance and of Discrimination Based on Religion or Belief // Protection of the Individual from Discrimination: Reader. In 3 vols. Vol. 3. Moscow: New Justice, 2006. 360 p. p. 198.
67. Rowe M. The Right to Freedom of Religion or Belief. Study Guide. Minnesota: Human Rights Center, University of Minnesota, 2003 // Electronic resource. URL: http://www1.umn.edu/humanrts/russian/edumat/studyguides/Rreligion.html (accessed: 21.11.2019).
68. Davis D. The Evolution of Religious Freedom as a Universal Human Right: A Study of the Role of the 1981 UN Declaration on the Elimination of All Forms of Intolerance and of Discrimination Based on Religion or Belief // Protection of the Individual from Discrimination: Reader. In 3 vols. Vol. 3. Moscow: New Justice, 2006. 360 p. p. 200.
69. Abashidze A.Kh. Problems of International-Legal Protection of Minorities: Doctoral Dissertation in Law. Moscow, 1997. 58 p. p. 7. p. 11.
70. Cited work. p. 20.
71. Abashidze A.Kh. Problems of International-Legal Protection of Minorities: Doctoral Dissertation in Law. Moscow, 1997. 58 p. p. 23.
72. Abashidze A.Kh. Problems of International-Legal Protection of Minorities: Doctoral Dissertation in Law. Moscow, 1997. 58 p. p. 7.
73. Kartashkin V.A. Human Rights and Principles of International Law in the 21st Century: Monograph. Moscow: Norma: INFRA-M, 2018. 148 p. p. 107.
74. Drinan Robert F. Can God and Caesar Coexist? Balancing Religious Freedom and International Law. New Haven: Yale University Press, 2004. 266 p.
75. Hordecki B. On Problems with the Opacity of the Concept of 'Freedom of Conscience and Religion' in United Nations Documents // Russian Political Science. 2016. No. 1. pp. 137-146.
76. Baboshina E.V., Kosov G.V., Stankevich G.V. A Theoretical-Legal Analysis of Universal and Regional International Documents on Freedom of Conscience, Belief, and Religion // Caspian Region: Politics, Economy, Culture. 2018. No. 4 (57). pp. 76-85. p. 83.
77. Kiseleva E.V. The Concept of Discrimination as Applied to the Right to Freedom of Thought, Conscience, and Religion: International-Legal Norms and the Practice of Universal and Regional International Human Rights Structures // Eurasian Law Journal. 2019. No. 1 (128). pp. 29-34. pp. 32-33.

78. Solntsev A.M. Certain Issues of Religious Discrimination in the Contemporary World // Eurasian Law Journal. 2019. No. 1. pp. 25-28.
79. Akchurin T.F. The Right to Freedom of Conscience: The International-Legal Aspect // Russian Law Online. 2018. No. 3. pp. 19-28.
80. Understanding Human Rights: Manual on Human Rights Education / edited by W. Benedek and M. Nikolova. Graz: European Training and Research Centre for Human Rights and Democracy; Russian version. Warsaw: Homework, 2005. p. 30.
81. Kartashkin V.A., Lukasheva E.A. Human and Citizen Rights and Freedoms: Universalism, Regionalism, Realities // State and Law. 2010. No. 7. p. 38.
82. International Public Law: Collection of Documents / compiled and with a foreword by K.A. Bekyashev and D.K. Bekyashev. Moscow: Prospekt, 2009. 1200 p.
83. Recommendation of the Parliamentary Assembly of the Council of Europe No. 1556, 'Religion and Change in Central and Eastern Europe' of 2002 // Council of Europe website: www.coe.int; URL: http://www.coe.int/t/r/parliamentary_assembly/%5BRussian_documents%5D/%5B2002%5D/%5BAvril2002%5D/Rek_1556.asp
84. Kartashkin V.A., Lukasheva E.A. Human and Citizen Rights and Freedoms: Universalism, Regionalism, Realities // State and Law. 2010. No. 7. p. 39.
85. Decision No. 4/03: Tolerance and Non-Discrimination // OSCE Human Dimension Commitments. Vol. II. Chronological Compilation of Documents, 2nd edition. Warsaw, Poland: Poligrafus Andrzej Adamiak, 2006. pp. 316-319.
86. Kartashkin V.A., Lukasheva E.A. Human and Citizen Rights and Freedoms: Universalism, Regionalism, Realities // State and Law. 2010. No. 7. pp. 39-40.
87. League of Arab States (Reference Information) // Official Website of the Ministry of Foreign Affairs of the Russian Federation. Electronic resource. URL: http://www.mid.ru/international_organizations/-/asset_publisher/km9HkaXMTium/content/id/290490 (accessed: 7.11.2019).
88. Buryanov S.A. The Fourth International Scientific Congress 'Global Studies 2015': Prospects for the Development of International Law under Conditions of the Globalization of Social Relations // Eurasian Law Journal. 2015. No. 10. p. 351.
89. Cited work. p. 350.
90. Kartashkin V.A., Lukasheva E.A. Human and Citizen Rights and Freedoms: Universalism, Regionalism, Realities // State and Law. 2010. No. 7. pp. 38-39.
91. Samovich Yu.V. The International Mechanism for the Protection of Human Rights and Freedoms: Universal and European Aspects: Candidate of Legal Sciences Dissertation. Yekaterinburg, 1999. 217 p.; Gorshkova S.A. Council of Europe Human Rights Standards and Russian Legislation. Monograph. Moscow: NIMP, 2001. 348 p.
92. Electronic resource. URL: https://www.un.org/ru/ga/ (accessed: 8.10.2019).
93. Electronic resource. URL: https://www.un.org/securitycouncil/ru (accessed: 8.10.2019).
94. Electronic resource. URL: https://www.un.org/ru/sections/what-we-do/protect-human-rights/index.html (accessed: 8.10.2019).
95. Electronic resource. URL: https://www.un.org/ru/sections/what-we-do/protect-human-rights/index.html (accessed: 8.10.2019).
96. Electronic resource. URL: https://www.un.org/ecosoc/ru/ (accessed: 8.10.2019).
97. Electronic resource. URL: https://www.un.org/ecosoc/ru/ (accessed: 8.10.2019).
98. Electronic resource. URL: https://www.un.org/ru/icj/ (accessed: 8.10.2019).
99. Electronic resource. URL: https://www.un.org/ru/sections/about-un/secretariat/index.html (accessed: 8.10.2019).
100. Electronic resource. URL: https://www.un.org/sg/ru (accessed: 8.10.2019).
101. Electronic resource. URL: https://www.un.org/ru/sections/what-we-do/protect-human-rights/index.html (accessed: 8.10.2019).
102. Electronic resource. URL: https://www.un.org/ru/chronicle/article/21872 (accessed: 8.10.2019).
103. Electronic resource. URL: https://www.un.org/ru/sections/what-we-do/protect-human-rights/index.html (accessed: 8.10.2019).

104. Electronic resource. URL: https://spinternet.ohchr.org/_layouts/15/SpecialProceduresInternet/ViewAllCountryMandates.aspx?Type=TM (accessed: 8.10.2019).
105. Electronic resource. URL: https://www.ohchr.org/RU/Issues/FreedomReligion/Pages/FreedomReligionIndex.aspx (accessed: 8.10.2019).
106. Ghanea N. Freedom of Religion or Belief: An International Law Commentary / N. Ghanea, H. Bielefeldt, M. Wiener. Oxford: Oxford University Press, 2016. 640 p.; Ghanea N. Human Rights, the UN and the Baha'is in Iran / N. Ghanea. Liverpool: Liverpool University Press, 2021. 656 p.
107. Electronic resource. URL: https://www.ohchr.org/RU/Issues/FreedomReligion/Pages/FreedomReligionIndex.aspx (accessed: 8.10.2019).
108. Electronic resource. URL: https://www.ohchr.org/RU/Issues/FreedomReligion/Pages/FreedomReligionIndex.aspx (accessed: 8.10.2019).
109. Electronic resource. URL: https://www.ohchr.org/RU/Issues/FreedomReligion/Pages/FreedomReligionIndex.aspx (accessed: 8.10.2019).
110. Electronic resource. URL: https://www.ohchr.org/RU/Issues/FreedomReligion/Pages/FreedomReligionIndex.aspx (accessed: 8.10.2019).
111. Electronic resource. URL: https://www.ohchr.org/EN/Issues/FreedomReligion/Pages/Annual.aspx (accessed: 8.10.2019).
112. Ghanea N. Freedom of Religion or Belief and the Right to Life: Report of the Special Rapporteur: A/HRC/52/38 / Nazila Ghanea; UN Human Rights Council. Geneva: UN, 2023. 25 p.
113. Working Group on Arbitrary Detention. Fact Sheet No. 26. Electronic resource. URL: https://www.ohchr.org/documents/publications/factsheet26ru.pdf (accessed: 04.10.2019).
114. Working Group on Arbitrary Detention. Electronic resource. URL: https://www.ohchr.org/EN/Issues/Detention/Pages/WGADIndex.aspx (accessed: 04.10.2019).
115. Electronic resource. URL: https://www.un.org/ru/sections/what-we-do/protect-human-rights/index.html (accessed: 8.10.2019).
116. Electronic resource. URL: https://www.ohchr.org/RU/AboutUs/Pages/Mandate.aspx (accessed: 8.10.2019).
117. Electronic resource. URL: https://www.un.org/ru/sections/what-we-do/uphold-international-law/index.html (accessed: 8.10.2019).
118. International Protection of Human Rights: Textbook / executive editor A.Kh. Abashidze. Moscow: RUDN University, 2017. 466 p. pp. 158-159.
119. Electronic resource. URL: https://ru.unesco.org/about-us/introducing-unesco (accessed: 8.10.2019).
120. Electronic resource. URL: https://ru.unesco.org/about-us/procedure104 (accessed: 8.10.2019).
121. Electronic resource. URL: https://ru.unesco.org/about-us/procedure104 (accessed: 8.10.2019).
122. Electronic resource. URL: https://ru.unesco.org/about-us/procedure104 (accessed: 8.10.2019).
123. Electronic resource. URL: https://ru.unesco.org/themes/zashchita-svobody-vyrazheniya-mneniy (accessed: 8.10.2019).
124. Davis D. The Evolution of Religious Freedom as a Universal Human Right: A Study of the Role of the 1981 UN Declaration on the Elimination of All Forms of Intolerance and of Discrimination Based on Religion or Belief // Protection of the Individual from Discrimination: Reader. In 3 vols. Vol. 3. Moscow: New Justice, 2006. 360 p. pp. 198-199.
125. Rowe M. The Right to Freedom of Religion or Belief. Study Guide. Minnesota: Human Rights Center, University of Minnesota, 2003 // Electronic resource. URL: http://www1.umn.edu/humanrts/russian/edumat/studyguides/Rreligion.html (accessed: 17.10.2019).
126. International Protection of Human Rights: Textbook / executive editor A.Kh. Abashidze. Moscow: RUDN University, 2017. 466 p. pp. 247-248.
127. Buryanov S.A. International-Legal Documents in the Sphere of Freedom of Conscience and the Practice of Their Application in the Russian Federation: Theoretical and Applied Study for 2011. Moscow: Russian Helsinki Group, 2012. 244 p.
128. Gorshkova S.A. Council of Europe Human Rights Standards and Russian Legisla

129. Lukyantsev G.E. The Regional System of Human Rights Protection within the Council of Europe: Using the Example of the 1950 European Convention for the Protection of Human Rights and Fundamental Freedoms: Candidate of Legal Sciences Dissertation. Moscow, 1998. 190 p. p. 161.
130. Council of Europe. Electronic resource. URL: https://www.coe.int/web/human-rights-intergovernmental-cooperation/ (accessed: 21.10.2019).
131. Council of Europe. Electronic resource. URL: https://www.coe.int/en/web/human-rights-intergovernmental-cooperation/echr-system (accessed: 21.10.2019).
132. Council of Europe. Electronic resource. URL: https://www.coe.int/en/web/european-commission-against-racism-and-intolerance/home (accessed: 21.10.2019).
133. Council of Europe. Electronic resource. URL: https://www.coe.int/en/web/commissioner/thematic-work (accessed: 21.10.2019).
134. Council of Europe. Electronic resource. URL: https://www.coe.int/en/web/cddecs/ (accessed: 21.10.2019).
135. Samovich Yu.V. The International Mechanism for the Protection of Human Rights and Freedoms: Universal and European Aspects: Candidate of Legal Sciences Dissertation. Yekaterinburg, 1999. 217 p. p. 197.
136. Murdoch J. Protecting the Right to Freedom of Thought, Conscience and Religion under the European Convention on Human Rights. Council of Europe Human Rights Handbooks Series. Voronezh: Elist LLC, 2014. 163 p.
137. Cited work. p. 8.
138. Annual Report of the European Court of Human Rights 2018. Electronic resource. URL: https://www.echr.coe.int/Documents/Annual_report_2018_ENG.pdf (accessed: 21.11.2019).
139. Nikolaev A.M., Folosyan M.E. Certain Problems in the Practice of the European Court of Human Rights in the Area of Freedom of Thought, Conscience, and Religion // Issues of Russian and International Law. 2016. No. 9. pp. 184-195.
140. Nikolaev A.M., Davtyan M.K. Execution of Judgments of the European Court of Human Rights and the Inter-American Court of Human Rights: A Comparative Analysis // Journal of Foreign Legislation and Comparative Law. 2018. No. 4 (71). pp. 40-46.
141. Nikolaev A.M., Folosyan M.E. Freedom of Conscience and Religion in the System of Individual Rights and Freedoms of the Human Person and Citizen in the Russian Federation // Contemporary Problems of Social Work. 2015. No. 1. pp. 107-111.
142. Nikolaev A.M. The European Convention for the Protection of Human Rights and Fundamental Freedoms and Constitutional Human Rights and Freedoms // Science and Education. 2011. No. 11 (17). pp. 85-88.
143. Bezrukova L. The Constitutional Court of the Russian Federation Allowed Civil Cases to Be Reviewed if a Judgment of the European Court of Human Rights Has Been Issued in Them // Rossiyskaya Gazeta. February 26, 2010.
144. Zorkin V.D. The Limits of Concession // Rossiyskaya Gazeta, Federal Issue No. 5325 (246). October 29, 2010.
145. Meeting with Judges of the Constitutional Court // Official Website of the President of the Russian Federation. Electronic resource. URL: http://www.kremlin.ru/events/president/transcripts/comminity_meetings/9792 (accessed: 21.11.2019).
146. International Protection of Human Rights: Textbook / executive editor A.Kh. Abashidze. Moscow: RUDN University, 2017. 466 p. pp. 286-287.
147. European Union Agency for Fundamental Rights. Electronic resource. URL: https://fra.europa.eu/en/about-fra/what-we-do/areas-of-work (accessed: 19.10.2019).
148. European Union Agency for Fundamental Rights. Electronic resource. URL: https://fra.europa.eu/en/about-fra (accessed: 19.10.2019).
149. European Union Agency for Fundamental Rights. Electronic resource. URL: https://fra.europa.eu/en/about-fra/what-we-do/areas-of-work/multi-annual-framework-2018 (accessed: 19.10.2019).

150. International Protection of Human Rights: Textbook / executive editor A.Kh. Abashidze. Moscow: RUDN University, 2017. 466 p. p. 297.
151. Kartashkin V.A., Lukasheva E.A. Human and Citizen Rights and Freedoms: Universalism, Regionalism, Realities // State and Law. 2010. No. 7. p. 39.
152. Abashidze A.Kh., Solntsev A.M. The Anniversary of the African Charter on Human and Peoples' Rights // Eurasian Law Journal. 2012. No. 2 (45). pp. 22-25. p. 23.
153. Abashidze A.Kh., Solntsev A.M. The Anniversary of the African Charter on Human and Peoples' Rights // Eurasian Law Journal. 2012. No. 2 (45). pp. 22-25. p. 25.
154. Kete K.J. The Activities of the African Committee of Experts on the Rights and Welfare of the Child // Issues of Russian and International Law. 2016. No. 9. pp. 175-183. p. 180.
155. Buryanov S.A. International-Legal Documents in the Sphere of Freedom of Conscience and the Practice of Their Application in the Russian Federation: Theoretical and Applied Study for 2011. Moscow, 2012. 244 p.
156. Organization for Security and Co-operation in Europe. Electronic resource. URL: https://www.osce.org/whatistheosce (accessed: 19.10.2019).
157. Office for Democratic Institutions and Human Rights. Electronic resource. URL: https://www.osce.org/ru/odihr/elections/13702?download=true (accessed: 19.10.2019).
158. Office for Democratic Institutions and Human Rights. Advisory Council on Freedom of Religion or Belief. Electronic resource. URL: https://www.osce.org/ru/odihr/25455?download=true (accessed: 19.10.2019).
159. See: Guidelines for Review of Legislation Pertaining to Religion or Belief. OSCE/ODIHR. Warsaw, 2005. Electronic resource. URL: https://www.osce.org/ru/odihr/13994?download=true (accessed: 19.10.2019).
160. Guidelines on the Legal Personality of Religious or Belief Communities, 2015, OSCE/ODIHR. Warsaw, 2015. Electronic resource. URL: https://www.osce.org/ru/odihr/145431?download=true (accessed: 19.10.2019).
161. Members of the Advisory Council on Freedom of Religion or Belief of the OSCE Office for Democratic Institutions and Human Rights (ODIHR): Jolanta Ambrosewicz-Jacobs (Poland), Cole Durham (United States), Anastasia Crickley (Ireland), Silvio Ferrari (Italy), and others.
162. Buryanov S.A. International-Legal Documents in the Sphere of Freedom of Conscience and the Practice of Their Application in the Russian Federation: Theoretical and Applied Study for 2011. Moscow, 2012. 244 p.
163. Office for Democratic Institutions and Human Rights. Annual Report 2018. Electronic resource. URL: https://www.osce.org/ru/node/431876?download=true (accessed: 19.10.2019).
164. All materials of the meeting can be found at: www.osce.org/odihr/hdim_2018.
165. Commonwealth of Independent States. Electronic resource. URL: http://www.ved.gov.ru/sng/ (accessed: 21.10.2019).
166. International Protection of Human Rights: Textbook / executive editor A.Kh. Abashidze. Moscow: RUDN University, 2017. 466 p. p. 358.
167. Association of Southeast Asian Nations. Electronic resource. URL: https://asean.org/asean/about-asean/overview/ (accessed: 21.10.2019).
168. International Protection of Human Rights: Textbook / executive editor A.Kh. Abashidze. Moscow: RUDN University, 2017. 466 p. p. 360.
169. Association of Southeast Asian Nations. Electronic resource. URL: http://tass.ru/info/3290905 (accessed: 21.10.2019).
170. A Human Rights Commission Has Been Established under the Organisation of Islamic Cooperation // Credo.Press Portal. Electronic resource. URL: https://credo.press/121829/ (accessed: 21.11.2019).
171. Organisation of Islamic Cooperation. Electronic resource. URL: http://www.oic-iphrc.org/en/legal/ (accessed: 28.02.2017).
172. International Protection of Human Rights: Textbook / executive editor A.Kh. Abashidze. Moscow: RUDN University, 2017. 466 p. p. 364.
173. Goryan E.V. The Islamic Concept of Human Rights and the Progressive Development of International Law // International Law. 2015. No. 3. pp. 91-101.
174. Abashidze A.Kh.. Solntsev A.M.. Vorobvev D.V. The Competence of the Organi[illegible] ights Protection //

Bulletin of Volgograd State University. Series 5: Jurisprudence. 2013. No. 3. pp.213-219. p. 117.

175. Al-Ali Naser Abdel Rahim. Universalism and the Islamic Concept of Human Rights and Freedoms: Candidate of Legal Sciences Dissertation. Moscow, 1999. 188 p. p. 170.

176. Ibid.

177. Al-Ali Naser Abdel Rahim. Universalism and the Islamic Concept of Human Rights and Freedoms: Candidate of Legal Sciences Dissertation. Moscow, 1999. 188 p. p. 178.

178. Al-Ali Naser Abdel Rahim. Universalism and the Islamic Concept of Human Rights and Freedoms: Candidate of Legal Sciences Dissertation. Moscow, 1999. 188 p. p. 179.

179. Ibrahim Ahmed Haj Youssef Abdalla. International and National Mechanisms for the Protection of Human Rights in the Middle East: Candidate of Legal Sciences Dissertation. Moscow, 1999. 192 p. pp. 8-9.

180. League of Arab States (Reference Information) // Official Website of the Ministry of Foreign Affairs of the Russian Federation. Electronic resource. URL: http://www.mid.ru/international_organizations/-/asset_publisher/km9HkaXMTium/content/id/290490 (accessed: 26.10.2019).

181. Ibrahim Ahmed Haj Youssef Abdalla. International and National Mechanisms for the Protection of Human Rights in the Middle East: Candidate of Legal Sciences Dissertation. Moscow, 1999. 192 p. p. 14.

182. Kartashkin V.A., Lukasheva E.A. Human and Citizen Rights and Freedoms: Universalism, Regionalism, Realities // State and Law. 2010. No. 7. pp. 38-39.

183. Cited work. p. 350.

184. Amini A.R. The Political System of Afghanistan and Prospects for State Secularity // PolitBook. 2018. No. 2. pp. 93-103.

185. Buryanov S.A. The Right to Freedom of Conscience under Conditions of Global Processes and Challenges 4.0 // Eurasian Law Journal. 2023. No. 6 (181). pp. 18-21.

186. Freedom of Conscience: Problems of Theory and Practice. Monograph (edited by F.M. Rudinsky, S.A. Buryanov). Moscow: ZAO TF MIR, 2012. 1120 p.

187. Buryanov S.A. International Recognition of the Right to Freedom of Conscience and Problems of Its Implementation in the Russian Federation under Conditions of Contemporary Global Processes: Monograph. Moscow: Poligraf Servis, 2020. 624 p.

188. Civil Human Rights in Russia: Modern Problems of Theory and Practice / compiled by F.M. Rudinsky. Transaction Publishers, New Brunswick, New Jersey (U.S.A.), 2008. 458 p.

189. Buryanov S.A. State Worldview Neutrality in the Context of Deteriorating Imbalances in Globalization: The Case and Current State of Affairs in the Russian Federation from 2016 to the Beginning of 2017. Amsterdam, 2018. 228 p.

PART V. RUSSIA AS A CASE, A SYMPTOM, AND A WARNING

CIVILIZATIONAL THESIS

"The abuse of meaning is one of the oldest forms of power."

The Russian case is examined as a warning about the fragility of legal guarantees, state neutrality, and freedom of conscience. It shows how conflicts of meaning become institutional conflicts, and how the language of security can turn the worldview sphere into an object of control.

CONCEPT BOX

CHAPTER 5

RUSSIA AS A LEGAL CASE

International obligations and domestic legal reality

The Russian Federation provides a significant case for examining the gap between international human rights obligations and domestic legal reality.

Chapter 5 analyzes the constitutional, legislative, and institutional framework related to freedom of conscience, religion, and state neutrality.

The Russian case is treated not as an isolated national issue, but as a legal symptom of a wider civilizational problem.

When a state struggles to maintain neutrality in the worldview sphere, the rights of the person become vulnerable to administrative, ideological, and institutional pressure.

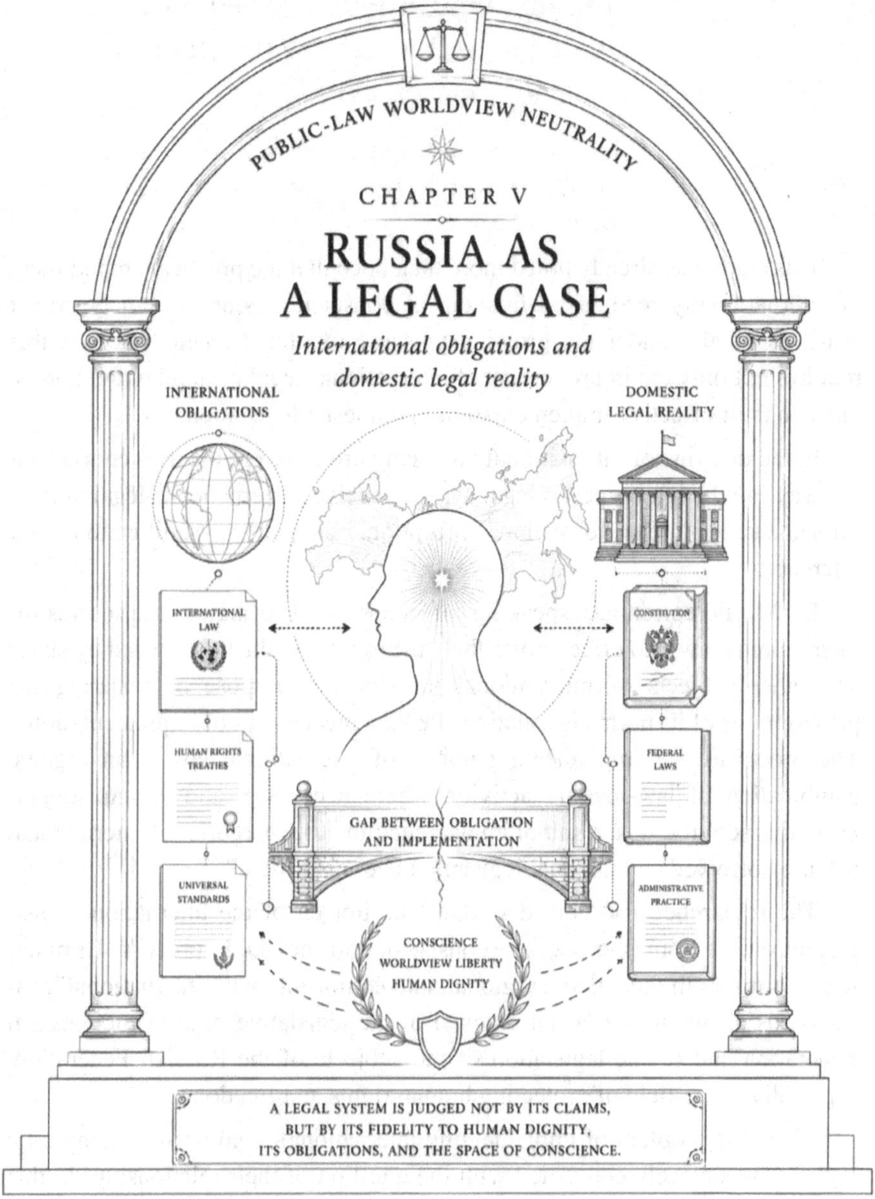

Between promise and practice stands the person.
Law must serve conscience, not suppress it.

Chapter 5. International Obligations and Domestic Law of the Russian Federation in the Field of the Right to Freedom of Conscience

5.1. Certain Theoretical and Legal Issues of Implementing International-Legal Principles and Norms on the Right to Freedom of Conscience in Domestic Legislation

This study has already noted more than once that the practical embodiment of internationally recognized freedom of conscience is an extremely urgent issue, especially under the growing pressure of global challenges. Yet this requires not only the improvement of international-legal principles and norms, but also their effective implementation in domestic legislation.

In the doctrine of international law, there are different theories concerning the way in which a domestic legal system receives international-legal norms: implementation, transformation, reception, adaptation, unification and reference.

E. M. Pereplesnina, speaking of theories of bringing the norms of international law into effect, notes that "the totality of the means used by states and other subjects of international law for the purpose of realizing the prescriptions of its norms is usually called the mechanism of implementation... The procedure for implementing norms of international law is an organic combination of law-making activity, where necessary, and organizational-executive activity. As a result of implementation, the categories of international law are borrowed and national legislation is transformed."

The researcher also considers that "the impact of the international legal system on the domestic one is a consistent and inevitable process. Bringing legal norms, both federal and regional, into conformity with international legal standards entails a significant renewal of the legislative base of the Russian Federation and of the legislation of the subjects of the Russian Federation, especially in the field of protecting human rights and freedoms."

Thus, the problem of implementing international-legal norms in domestic legislation is closely connected with the question of their relationship. In this context, the monist and dualist theories of the relationship between international and domestic law are well known. The monist theory presupposes the existence of a single legal system combining national, or domestic, law and international law. Within this theory, in turn, there arose the theory of the priority of domestic law, associated with Hegel, Lasson and others, and the theory of the priority of international law, associated with Kelsen and others.

As a consequence, the principle of secularity, or worldview neutrality, of the state, as one of the most important guarantees for the realization of freedom of conscience, has not received proper development in theory or law. In particular, it has not been consolidated as such in the norms of international law. Even states that declare secularity practise interference in the activities of religious organizations and unlawful control over the worldview sphere in general.

There also exist confessional states, whose legal systems are based on religious doctrines. For example, "religious states include the Vatican and a number of Muslim states, such as Iran and Saudi Arabia, where Islam permeates all spheres of state, social and personal life; where there is no separation between the state and the religious community, religion and politics, law and religion; and where state-legal and religious-moral institutions are closely intertwined under the dominance of the clergy. The main features of such states are: the dominant role of Muslim law in the system of sources of law and jurisdiction; the obligation of heads of state to implement the principles of Islam; the existence of special religious-political bodies supervising the compliance of political decisions and legislation with the requirements of Islam; appointment to state offices only of Muslims; and prohibition or significant restrictions on the activity of non-Islamic religious organizations and non-Muslim religious practice."

This means that in confessional states human rights violations may be justified by religious doctrines, which is reflected, for example, in the Cairo Declaration on Human Rights in Islam of 1990 and the Universal Islamic Declaration of Human Rights of 1981.

It is obvious that the creation of a global legal system is extremely problematic without the realization of the principles of freedom of conscience in the broadest possible legal understanding. It is precisely the realization of the mentioned principles of freedom of conscience and state secularity that should gently and evolutionarily separate religion and politics, thereby reducing ethno-confessional contradictions that hinder the positive integration of states.

It should be noted that resolving theoretical and legal problems will assist the effective implementation of internationally recognized freedom of conscience in the domestic legislation of states. This, in turn, will contribute to the evolutionary separation of religion and politics, the reduction of ethno-confessional contradictions, the positive integration of states and sustainable development.

5.2. The Problem of Implementing International-Legal Principles and Norms on the Right to Freedom of Conscience in the Domestic Legislation of the Russian Federation

Having chosen the path of building a law-governed state at the end of the 1980s, the Russian Federation undertook to bring its national legislation into conformity with international-legal principles and norms.

The Declaration of the Rights and Freedoms of Man and Citizen was adopted by the Supreme Soviet of the RSFSR on 22 November 1991 in order to affirm human rights and freedoms, and human honour and dignity, as the highest values of society and the state. It specifically emphasized the need to bring the legislation of the RSFSR into conformity with standards of human rights and freedoms generally recognized by the international community.

Under Article 14 of the Declaration, everyone is guaranteed "freedom of conscience, religion, and religious or atheist activity." The same provision further states that "everyone has the right freely to profess any religion or not to profess any, to choose, hold and disseminate religious or atheist beliefs, and to act in accordance with them, provided that the law is observed."

Article 3 guarantees equality of rights and freedoms "regardless of race, nationality, language, social origin, property and official status, place of residence, attitude to religion, beliefs, membership of public associations, and other circumstances."

Article 15 provides that "every citizen of the RSFSR whose convictions conflict with military service has the right to replace it with the performance of alternative civil duties in accordance with the procedure established by law."

Particular attention should be paid to the fact that the Declaration expressly enshrined freedom of atheist activity and mentioned the right freely to choose, hold and disseminate atheist beliefs and to act in accordance with them. This wording is found neither in international legal standards nor in later Russian legal documents.

The Constitution of the Russian Federation was adopted by popular vote on 12 December 1993. Its preamble speaks of affirming human rights and freedoms, civil peace and accord, and of the awareness of the multinational people of the Russian Federation as part of the world community.

Part 1 of Article 17 of the Russian Constitution states that "in the Russian Federation, the rights and freedoms of man and citizen shall be recognized and guaranteed in accordance with the generally recognized principles and norms of international law and in accordance with this Constitution." Article 18 provides that "the rights and freedoms of man and citizen shall have direct effect. They character, and therefore states are not entitled to deviate from the model of conduct fixed in them even by mutual agreement. The treaty formulation of this category of norms does not deprive them of their essential nature, and they are binding on states that are not parties to the multilateral treaty. The difficulty of applying generally recognized principles and norms of international law is caused by the absence in international law of a single act listing those principles and norms."

In its Plenum Resolution No. 5 of 10 October 2003, "On the Application by Courts of General Jurisdiction of Generally Recognized Principles and Norms of International Law and International Treaties of the Russian Federation," the Supreme Court of Russia interpreted the legal categories under consideration as follows: "Generally recognized principles of international law should be understood as fundamental imperative norms of international law accepted and recognized by the international community of states as a whole, deviation from which is inadmissible. Generally recognized principles of international law include, in particular, the principle of universal respect for human rights and the principle of good-faith fulfilment of international obligations. A generally recognized norm of international law should be understood as a rule of conduct accepted and recognized by the international community of states as a whole as legally binding. The content of the indicated principles and norms of international law may be disclosed, in particular, in the documents of the United Nations and its specialized agencies."

In addition, the Resolution confirms that "international treaties of the Russian Federation, along with generally recognized principles and norms of international law, are part of its legal system, under Article 15, part 4, of the Constitution of the Russian Federation and Article 5, part 1, of the Federal Law 'On International Treaties of the Russian Federation'. International treaties concluded by the USSR that remain in force, in relation to which the Russian Federation continues to exercise the international rights and obligations of the USSR as the continuator state of the Union of Soviet Socialist Republics, are also part of the legal system of the Russian Federation... Where difficulties arise in interpreting generally recognized principles and norms of international law or international treaties of the Russian Federation, courts are advised to use the acts and decisions of international organizations, including United Nations bodies and its specialized agencies, and also to contact the Legal Department of the Ministry of Foreign Affairs of the Russian Federation and the Ministry of Justice of the Russian Federation, for example, in order to clarify issues connected with the duration of an international treaty, the composition of states participating in a treaty and the international practice of its application."

Attention should also be paid to the definition of the constitutional prescription contained in Article 15, paragraph 4, formulated by E. M. Pereplesnina for the purposes of accessible application in Russia's legal system: "Generally recognized principles and norms of international law are a legal standard included in the Russian legal system, embodied in concrete legal norms of international legal acts and ratified international treaties of the Russian Federation, and subject to application for the creation of favourable domestic conditions for ensuring the life activity of the human being and citizen in the Russian Federation and for protecting their rights and freedoms. International treaties of the Russian Federation are written agreements of the Russian Federation with other subjects of international law, at the basis of which, regardless of the subject matter of regulation, lie generally recognized principles and norms of international law."

The combination of the principle of supremacy and the highest legal force of the constitution with the binding character of international-legal norms is also one of the important questions. "The mechanism for the formation of the foreign-policy will of a democratic state is marked by greater transparency, which makes it possible to determine the degree of its good faith in fulfilling the international obligations it has assumed, and also to identify in due time facts of their violation. Accordingly, by making demands upon the basic law, international-legal norms fix the duty of the government and state authorities to observe the constitution and act in a manner compatible with the law, according to the legally established constitutional order. A substantial divergence between the constitution and generally recognized principles and norms of international law may place the state in political isolation. Power becomes ineffective and costly, forced to compensate for the costs of refusing international cooperation from internal reserves. For Russia, such a problem does not exist. Its Constitution meets all international standards, and on a number of issues provides higher standards than those contained in international documents."

Questions of the international legal personality of federal authorities and authorities of the subjects of the Russian Federation are addressed in Articles 71 and 72 of the Constitution of Russia. Under Article 71 of the Constitution, the jurisdiction of the Russian Federation includes "k) the foreign policy and international relations of the Russian Federation, international treaties of the Russian Federation, issues of war and peace; l) foreign economic relations of the Russian Federation." Accordingly, under Article 72, "the joint jurisdiction of the Russian Federation and the subjects of the Russian Federation includes: ... o) coordination of the international and foreign economic relations of the subjects international treaties of the Russian Federation." Thus, the Constitution of the Russian Federation establishes the priority of the federal centre in foreign policy and international relations.

S. Yu. Marochkin speaks of the importance of, and points to numerous problems in, the interaction between international and domestic law: "The current constitutional principle concerning generally recognized principles and norms of international law and international treaties of the Russian Federation as part of its legal system, as well as the priority of international treaties over laws in the event of conflict, is not an abstract and declarative provision. In the almost ten years after the adoption of the Constitution, it has had, and continues to have, a cardinal influence on legislation, doctrine, law-realization practice, above all judicial practice, and on the country's legal system as a whole; therein lies its enduring significance. During this period, legislation has changed significantly and is now built on different approaches to the role and significance of the norms of international law in the country's legal system. Practically every code and many federal laws are accompanied by scholarly-practical commentaries that include recommendations on the application of norms of international law in the context of the code or law. Practice is steadily increasing. Previously, the use of international treaties could be found mainly in the materials of court cases, and even then far from always when it was necessary. Now a rare ruling or determination of the Constitutional Court of the Russian Federation is made without reference to norms of international law. Increasingly, all types of courts, other bodies of state authority, local self-government bodies, enterprises and firms, practising lawyers and advocates, statesmen and citizens refer to them in their activities. At the same time, the main task, namely the creation of the necessary legal, organizational and even psychological conditions for the country's legal system to be 'tuned' to the coexistence and coordinated operation of norms of international and Russian law, can hardly be considered solved. Sectoral legislation does not always adequately regulate, in accordance with the constitutional principle, the question of its relationship with international law. Theory has not formulated answers to all fundamental questions. Scholarly-practical commentaries to codes and federal laws rarely provide a correct and complete interpretation of the relevant provisions. Accordingly, judicial and other law-applying practice concerning the application of international law develops inconsistently and unevenly. As for legislation, codes and laws do not help disclose the content of the constitutional principle; they mainly repeat it, whereas at this level it is clearly necessary to specify it." He adds: "There is an urgent need, repeatedly noted in theory and practice, to adopt a law on the procedure for the operation and application in the legal system of the Russian Federation of international norms and acts, including treaties, resolutions and decisions of international organizations and decisions of international courts."

It is largely necessary to agree with researchers on the urgent need to resolve existing problems in the interaction between international and domestic norms. It has to be admitted that, despite objective necessity, there has not yet been any substantial movement toward the formation of a unified legal and governance system.

At the same time, it also appears evident that under contemporary globalization of social relations, which is characterized not merely by convergence but by the interpenetration and interdependence of interactions among individuals, communities and states, the strengthening of interaction among legal systems has an objective character. All the more so because, in view of global problems, the survival of civilization depends on the formation of a capable global law.

This means that, as V. V. Ershov rightly believes, "state sovereignty may be limited at the national level by the state itself, by other states or by supranational bodies."

In my view, state sovereignty must ultimately be limited in favour of the individual, and this is already gradually finding normative expression in international human rights documents.

Accordingly, the question of the international legal capacity of the individual should be considered in the context of the globalization of social relations through the prism of human rights and freedoms. "The denial of the international legal personality of individuals is, first, based on legalism, on the restriction of 'all' law to national normative legal acts, to 'legislation'. Second, an individual who is in certain legal relations both with state authorities and with natural and legal persons bears certain duties and responsibilities established not only by national law, but also by international law. Third, the argument of opponents of the international legal personality of the individual that the latter does not create norms and principles of international law appears, at the very least, debatable. In this case, within the framework of concrete legal relations and a particular legal status, their subjects, including both states and individuals, act not as law-making bodies or persons, but as parties to certain legal relations possessing rights and duties provided for by previously established national or international law. Fourth, the arguments of opponents of the international legal capacity of the individual clearly contradict many articles of the Constitution of the Russian Federation, for example Article 15, part 4; Article 17, part 1; Article 18; and Article 46, part 3."

In the context of the present study, it is no less important to define the concept of the realization of a subjective right, or freedom, since the realization of the principles of freedom of conscience depends on their level of development, their appropriate consolidation in the legal system, the presence of legal guarantees and, finally, objective social conditions.

As I. V. Rostovshchikov rightly notes, the realization of a subjective right, or freedom, may be viewed "both as a process of achieving a result through its stages and as the final outcome: the actual acquisition by a citizen of a concrete benefit, and the use and disposal of it."

F. M. Rudinsky identifies the following forms of realization of human rights: "By forms of realization of human rights, we mean their exercise within legal relations or outside legal relations. It is also possible to identify such forms of realization of human rights as the use of rights or the observance of prohibitions, and the fulfilment of duties provided for by legal norms. Law application should also be included among the forms of realization of human rights, that is, the realization of legal norms by law-enforcement bodies."

The author agrees with I. I. Lukashuk and other scholars who point to the objective tendency toward the interpenetration of international law and the internal law of states, leading to the formation of a global legal system or supersystem. "Within it, national legal systems actively interact with one another, with regional systems and with general international law. It is based on the principle of democracy, which makes it possible to coordinate different legal systems. Such a system has nothing in common either with a world state and corresponding law, or with similar concepts."

Here the remark of V. V. Ershov seems important. He believes that "the concepts traditionally used by Russian and foreign specialists, such as 'interaction', 'harmonization' and 'convergence', do not reflect a scientifically grounded contemporary concept of the relationship between international and domestic law."

Indeed, under the globalization of social relations it is more correct to speak of a tendency toward the merging and interpenetration of the corresponding legal systems. As I. A. Konyukhova notes, "the process of convergence of norms of international and constitutional law manifests itself most vividly in the contemporary period precisely in the sphere of regulating and ensuring rights and freedoms. In this regard, both a general and a Russian tendency is revealed: ever closer interpenetration of international-legal and constitutional-legal provisions, ensured by the authority of these norms and by an adequate mechanism for their protection."

For the effective implementation of internationally recognized freedom of conscience in domestic legislation, fundamental importance belongs to the legal understanding of the phenomenon of freedom of conscience and to the formation of a correct conceptual apparatus connected with that understanding, based on modern legal technique. It is difficult here not to agree with the Kazakhstani scholar R. A. Podoprigora, who said that "inattention to the conceptual apparatus leads to serious problems in the activities of both state bodies and religious organizations. This is especially important for those states or legal systems where the main source of law is a normative legal act, since it initially sets the parameters that all participants in social relations must observe."

At the same time, it should be noted that "important, but only initial, steps have been taken in the formation of a global legal system. The process will be long and complex. It should not be understood simplistically as meaning the standardization of national legal systems... We are speaking of creating such a global legal system as would ensure the normal functioning both of interconnected national societies and of a world order capable of creating conditions for solving global problems. It will rest on the principles of democracy, generally recognized human rights and the rule of law."

In this vein, it should be emphasized that a serious obstacle to the creation of a global legal system lies in divisive ethno-confessional principles based on the union of religion and politics and underlying national states. In Russia in particular, the sacralization of power and the dominance of a state religion formed part of the political and legal tradition for a long historical period. As K. E. Sigalov and A. A. Prokuratov note, "official religiosity denied freedom as such, replacing it with a cast of state thinking. Unofficial religiosity was connected with the moral self-improvement of the person and orientation toward absolute values. The world of unofficial religiosity is impractical, though elevated. Freedom is a value here, but only spiritual and intellectual freedom is at stake, and even that had to be suffered through; social, political and, even more so, legal freedom was not at issue."

As a consequence, the principle of secularity, or worldview neutrality, of the state, as one of the most important guarantees for the realization of freedom of conscience, has not received proper development in theory or law. In particular, it has not been consolidated as such in the norms of international law. Even states that declare secularity practise interference in the activities of religious organizations and unlawful control over the worldview sphere in general.

There also exist confessional states, whose legal systems are based on religious doctrines. For example, "religious states include the Vatican and a number of Muslim states, such as Iran and Saudi Arabia, where Islam permeates all spheres of state, social and personal life; where there is no separation between the state and the religious community, religion and politics, law and religion; and where state-legal and religious-moral institutions are closely intertwined under the dominance of the clergy. The main features of such states are: the dominant role of Muslim law in the system of sources of law and jurisdiction; the obligation of heads of state to implement the principles of Islam; the existence of special religious-political bodies supervising the compliance of political decisions and legislation with the requirements of Islam; appointment to state offices only of Muslims; and prohibition or significant restrictions on the activity of non-Islamic religious organizations and non-Muslim religious practice."

This means that in confessional states human rights violations may be justified by religious doctrines, which is reflected, for example, in the Cairo Declaration on Human Rights in Islam of 1990 and the Universal Islamic Declaration of Human Rights of 1981.

It is obvious that the creation of a global legal system is extremely problematic without the realization of the principles of freedom of conscience in the broadest possible legal understanding. It is precisely the realization of the mentioned principles of freedom of conscience and state secularity that should gently and evolutionarily separate religion and politics, thereby reducing ethno-confessional contradictions that hinder the positive integration of states.

It should be noted that resolving theoretical and legal problems will assist the effective implementation of internationally recognized freedom of conscience in the domestic legislation of states. This, in turn, will contribute to the evolutionary separation of religion and politics, the reduction of ethno-confessional contradictions, the positive integration of states and sustainable development.

5.2. The Problem of Implementing International-Legal Principles and Norms on the Right to Freedom of Conscience in the Domestic Legislation of the Russian Federation

Having chosen the path of building a law-governed state at the end of the 1980s, the Russian Federation undertook to bring its national legislation into conformity with international-legal principles and norms.

The Declaration of the Rights and Freedoms of Man and Citizen was adopted by the Supreme Soviet of the RSFSR on 22 November 1991 in order to affirm human rights and freedoms, and human honour and dignity, as the highest values of society and the state. It specifically emphasized the need to bring the legislation of the RSFSR into conformity with standards of human rights and freedoms generally recognized by the international community.

Under Article 14 of the Declaration, everyone is guaranteed "freedom of conscience, religion, and religious or atheist activity." The same provision further states that "everyone has the right freely to profess any religion or not to profess any, to choose, hold and disseminate religious or atheist beliefs, and to act in accordance with them, provided that the law is observed."

Article 3 guarantees equality of rights and freedoms "regardless of race, nationality, language, social origin, property and official status, place of residence, attitude to religion, beliefs, membership of public associations, and other circumstances."

Article 15 provides that "every citizen of the RSFSR whose convictions conflict with military service has the right to replace it with the performance of alternative civil duties in accordance with the procedure established by law."

Particular attention should be paid to the fact that the Declaration expressly enshrined freedom of atheist activity and mentioned the right freely to choose, hold and disseminate atheist beliefs and to act in accordance with them. This wording is found neither in international legal standards nor in later Russian legal documents.

The Constitution of the Russian Federation was adopted by popular vote on 12 December 1993. Its preamble speaks of affirming human rights and freedoms, civil peace and accord, and of the awareness of the multinational people of the Russian Federation as part of the world community.

Part 1 of Article 17 of the Russian Constitution states that "in the Russian Federation, the rights and freedoms of man and citizen shall be recognized and guaranteed in accordance with the generally recognized principles and norms of international law and in accordance with this Constitution." Article 18 provides that "the rights and freedoms of man and citizen shall have direct effect. They shall determine the meaning, content and application of laws, the activity of legislative and executive authorities and local self-government, and shall be ensured by justice."

Article 2 of the Basic Law states that "the human being, his rights and freedoms are the highest value. The recognition, observance and protection of the rights and freedoms of man and citizen shall be the duty of the state."

Russia's declared commitment to generally recognized principles and norms of international law is also reflected in part 4 of Article 15 of the 1993 Constitution: "The generally recognized principles and norms of international law and international treaties of the Russian Federation shall be a component part of its legal system. If an international treaty of the Russian Federation establishes rules other than those provided by law, the rules of the international treaty shall apply."

Under paragraph 3 of Article 5 of the Federal Law "On International Treaties of the Russian Federation," "provisions of officially published international treaties of the Russian Federation that do not require the adoption of domestic acts for their application shall operate in the Russian Federation directly. Appropriate legal acts shall be adopted for the implementation of other provisions of international treaties of the Russian Federation."

In addition, one should take into account the statement contained in the note of the Ministry of Foreign Affairs of the Russian Federation to heads of diplomatic missions dated 13 January 1992, according to which "the Russian Federation continues to exercise the rights and fulfil the obligations arising from international treaties concluded by the USSR."

Parts 1 and 2 of Article 46 of the Russian Constitution guarantee everyone judicial protection of his rights and freedoms, while decisions and actions, or inaction, of state authorities, local self-government bodies, public associations and officials may be appealed in court. Under part 3 of the same article, "everyone shall have the right, in accordance with international treaties of the Russian Federation, to apply to interstate bodies for the protection of human rights and freedoms if all available domestic remedies have been exhausted."

In this context, V. V. Ershov concludes that "the Constitution of the Russian Federation substantially reflects, in many respects, mutually exclusive concepts of dualism and moderate monism."

As O. I. Tiunov notes, under contemporary social relations "constitutional norms on human rights and the corresponding norms of international law are converging. This convergence takes place within the framework of coordinating norms containing Russia's international obligations in the field of human rights and norms concerning guarantees of human rights in domestic normative legal acts."

It is particularly emphasized that "both in international and domestic regulation, norms on human rights are interconnected, inseparable from one another, and must be applied in the context of one another's content."

The Constitution of the Russian Federation also confirmed, as a legal foundation, such civilized norms as ideological diversity (part 2 of Article 13), the secularity of the state, equality before the law and the separation of religious associations from the state (Article 14), equality of human and civil rights and freedoms regardless of attitude to religion or convictions, and the prohibition of any forms of restriction of citizens' rights on grounds of religious affiliation (part 2 of Article 19).

Under Article 28, "everyone shall be guaranteed freedom of conscience and freedom of religion, including the right individually or jointly with others to profess any religion or not to profess any, freely to choose, hold and disseminate religious and other beliefs and to act in accordance with them." Article 56 indicates that the provisions of Article 28 may not be restricted even under a state of emergency.

Part 2 of Article 29 emphasizes that "propaganda or agitation inciting religious hatred and enmity shall not be permitted. Propaganda of religious superiority shall be prohibited." In addition, part 3 of Article 59 provides that "a citizen of the Russian Federation, if his convictions or religious belief conflict with military service, as well as in other cases established by federal law, shall have the right to replace it with alternative civil service."

Article 30, which guarantees the right of association, also deserves attention, as does part 2 of Article 55, which establishes that "laws abolishing or diminishing the rights and freedoms of man and citizen shall not be issued in the Russian Federation."

At the same time, the application of this provision directly depends on the degree to which the principles of the relevant rights and freedoms have been developed. It must be admitted that, with regard to the right to freedom of conscience in the Russian Federation, this provision does not work properly. Laws that, in one way or another, infringe the rights and freedoms of the individual are adopted both at the federal level and at the level of the constituent entities of the Russian Federation.

Unfortunately, many constitutional formulations are rather vague, and in some respects even more abstract than the corresponding norms of international law. They do not fully reflect the essence of the right to freedom of conscience and, as a consequence, place its realization in dependence on actual religious policy and on the real pattern of relations between the state and religious associations.

Moreover, under part 3 of Article 55, "the rights and freedoms of man and citizen may be restricted by federal law only to the extent necessary for the protection of the foundations of the constitutional order, morality, health, the rights and lawful interests of other persons, and for ensuring the defence of the country and the security of the state."

These restrictions are insufficiently precise and excessively broad. As has already been noted above, it would have been enough to indicate, as the boundary of freedom of conscience, the rights and freedoms of others and lawful conduct.

In contemporary Russia, due to violations of the principle of state secularity in the form of confessional preferences on the part of the authorities, the constitutional formulations mentioned above are, from the outset, condemned to arbitrary interpretation by state structures. They create a legal space for a not always justified "special" and restrictive law-making.

Overall, the Constitution of the Russian Federation corresponds to international-legal documents in the field of freedom of conscience. A point of fundamental importance is the declaration of the secularity of the state, which is one of the most important guarantees of freedom of conscience and which is not enshrined as such in international-legal norms.

However, it should be noted that the constitutional formulations enshrining the right to freedom of conscience and freedom of religion are not entirely correct. A legally precise definition of state secularity is absent both from the text of the Basic Law and from other Russian legislation.

Thus, the principles of freedom of conscience were legally consolidated in the Constitution and other legislation of Russia, as well as in various international-legal documents, including those which, under Article 15 of the Constitution of the Russian Federation, "are a component part of its legal system."

The Federal Law "On Freedom of Conscience and Religious Associations" was adopted by the State Duma of the Russian Federation on 19 September 1997 and entered into force on 1 October. The discriminatory essence of that law was not especially hidden at the time, as is shown by the words of V. I. Zorkaltsev, then Chairman of the Committee for Affairs of Public and Religious Organizations and one of the authors of the law, spoken immediately before the vote.

On 19 September 1997, at a meeting of the State Duma of the Russian Federation, V. I. Zorkaltsev stated before the vote on the adoption of the law: "And yet I shall recall the essence of this law. It consists in the fact that the law creates a barrier to religious expansion into Russia, prevents the development of totalitarian sects, restricts the activity of foreign missionaries, and, at the same time, creates conditions for the activity of our traditional religions and confessions... We are confident that the practice of applying this law will help solve the problems that now face society, the state and the Church... I would also like to refer to the noteworthy fact that all confessions whose representatives objected to certain provisions of the law have their governing centres abroad. I say this for those who today consider our law unacceptable and intend to vote against it. And I want to ask the question: whose side are you on, dear colleagues?"

Immediately after the adoption of the law, many lawyers stated that it was discriminatory in nature and inconsistent with international-legal norms. The main objections concerned the following provisions: the establishment of a fifteen-year probationary period for state registration of a religious association; restriction of the territorial activity of religious associations; prohibition of foreign citizens and stateless persons residing in the territory of the Russian Federation from creating religious associations and restriction of this right for Russian citizens; dependence of foreign religious organizations, or their representative offices, on the will of Russian religious associations; and other similar provisions. The Federal Law "On Freedom of Conscience and Religious Associations" consolidated confessional preferences of the state in its preamble, divided religious associations into religious organizations and religious groups in a legally incorrect way (Articles 6-8), substantially toughened the procedure for establishing religious organizations (Articles 9-12), and restricted the activity of foreign religious organizations (Article 13). The grounds for a judicial decision to liquidate a religious association were also significantly broadened (Article 14). A particularly repressive innovation was the introduction of the institution of prohibiting the activity of a religious organization (paragraph 7 of Article 14).

Since its adoption, this federal law has repeatedly been subjected to severe criticism. In particular, in the conclusion of 25 March 1999 of the Commissioner for Human Rights in the Russian Federation, O. O. Mironov, on the verification of the conformity of the Federal Law "On Freedom of Conscience and Religious Associations" with the international-legal obligations of the Russian Federation, it was recognized that a number of provisions of the law contradicted principles enshrined in multilateral international treaties to which Russia is a party.

That conclusion of the Commissioner for Human Rights, in substance, subjected the Federal Law "On Freedom of Conscience and Religious Associations" to rather severe criticism. However, the Commissioner's resolution that the law could be brought into conformity with international law and Russia's international treaties, taking into account the comments made, remained unrealized.

The Constitutional Court of the Russian Federation has repeatedly considered the Federal Law "On Freedom of Conscience and Religious

Associations." In my view, however, in its determinations and judgment the Constitutional Court did not give the law a proper legal assessment and avoided recognizing as unconstitutional the norms challenged in the complaints. In practice, such decisions do not prevent religious discrimination, although they may formally satisfy individual applicants. The "constitutional-legal meaning" identified by the Constitutional Court replaces the literal interpretation of the Federal Law without changing its unconstitutional essence.

In this context, I believe that the Federal Law "On Freedom of Conscience and Religious Associations" is an obstacle to the realization of freedom of conscience in Russia. The attack on freedom of conscience and on the secularity of the state that began with its adoption continues and expands today.

Moreover, the very existence of a special law devoted to the regulation of the activity of religious associations appears doubtful. The experience of applying the Federal Law "On Freedom of Conscience and Religious Associations" has shown that transforming legislation on freedom of conscience into sectoral religious-confessional legislation, with special rights for believers and a special status for their associations, resulted in a serious defeat in the struggle for religious freedom and an attack on human rights. By creating a legal basis for the realization of rights only for "believers," that is, for a part of society, the state in fact deprives everyone of rights. The suppression of liberty of worldview choice and the division of religions into grades begins precisely with the division into "believers" and "non-believers," through the substitution of everyone's freedom of conscience with freedom of confessions for believers.

At the federal level, in addition to the Federal Law "On Freedom of Conscience and Religious Associations," there are more than one hundred normative legal acts affecting various aspects connected with the activity of religious associations.

According to the classification proposed by A. V. Pchelintsev and V. V. Ryakhovsky, there are several conditional groups of laws and other legal acts containing norms that regulate the relevant social relations. In particular, there are a number of normative acts providing for certain restrictions on the exercise of the right to freedom of conscience for particular categories of citizens, due to the specificity of the places where they are held. These include the Penal Enforcement Code of the Russian Federation, the Federal Law "On the Detention of Suspects and Accused Persons," and the Federal Law "On the Status of Servicemen." The financial and economic activity of religious associations is regulated by the Civil Code of the Russian Federation, the Tax

Code of the Russian Federation, the Federal Law "On Payment for Land," and other acts.

Very controversial and not entirely effective provisions are contained in the Criminal Code of the Russian Federation and in the Code of Administrative Offences of the Russian Federation.

The Federal Law of 25 July 2002 No. 114-FZ "On Countering Extremist Activity" and the amendments to other legislative acts made in connection with its adoption deserve separate mention. First of all, attention is drawn to the absence in this law of a definition of the key concept "extremism." Moreover, for some reason the legislator identifies the undefined concept of "extremism" with the concept of "extremist activity." Against the background of the absence of the key definition of "extremism," the definitions of "extremist organization" and "extremist literature" look no less strange, doubtful and uncertain.

The use of the undefined key concept "extremism," and therefore the legally incorrect derivative concepts "extremist organization" and "extremist literature," contradicts such principles as the recognition, observance and protection of the rights and freedoms of man and citizen, as well as the principle of legality.

Nevertheless, in September 2008 President of the Russian Federation D. A. Medvedev decided to establish a department of the Ministry of Internal Affairs for countering "extremism." The department for countering extremism, originally the Department for Countering Extremism of the Ministry of Internal Affairs and later the Main Directorate for Countering Extremism, became one of the central instruments of this policy. In practice, the vagueness of the concept created a convenient space for arbitrary enforcement against religious and worldview minorities, civic activists and critics of state-confessional policy. When the law is unclear, the law enforcer becomes almost a co-author of the norm. As usual, this is a dangerous promotion.

The problem is intensified by the fact that anti-extremist legislation is used in the sphere of freedom of conscience not only against actual violence, incitement or threats, but also against texts, associations, religious literature and forms of missionary or educational activity. In such cases, the legal boundary between preventing public danger and suppressing an undesirable worldview becomes blurred.

The same logic is visible in the use of provisions on insulting the feelings of believers. These norms are also based on legally uncertain concepts and therefore contradict the principle of legal certainty. They enable selective restriction of expression, artistic activity, public criticism and civic protest, especially where such criticism concerns privileged religious organizations or violations of the constitutional secularity of the state.

The cumulative effect of these legislative mechanisms is the emergence of a special restrictive regime around the worldview sphere. Formally, it is justified by the need to fight extremism, protect public order, protect morality or preserve religious peace. In substance, however, it often creates unequal legal conditions for different religious and worldview communities, gives advantages to associations close to the authorities and narrows the space of constitutional freedom.

Thus, the implementation of international-legal principles and norms on freedom of conscience in Russian domestic legislation remains contradictory. On the one hand, the Constitution and several legislative provisions reproduce the language of international human rights law. On the other hand, special legislation, anti-extremist norms, missionary restrictions, property regulation and administrative control have formed a system that often operates in the opposite direction.

This contradiction is not merely technical. It concerns the very structure of the legal order. A state may solemnly recognize international standards and even include them in its constitutional text. But if ordinary legislation and law-enforcement practice turn these standards into exceptions, permissions and filters, the right becomes decorative. The citizen sees the right in the Constitution, but meets the restriction in daily life. This is a familiar human invention: to declare freedom at the entrance and install a turnstile immediately after it.

It may therefore be concluded that the problem of implementing international-legal principles and norms on the right to freedom of conscience in the domestic legislation of the Russian Federation is not reducible to the absence of norms. The problem lies in the internal inconsistency of the legal system, the vagueness of key concepts, the persistence of confessional preferences, the underdevelopment of the principle of state secularity and the substitution of freedom of conscience with the administrative management of religious life.

5.3. International Obligations and the Transformation of Russian Legislation in the Field of the Right to Freedom of Conscience

The transformation of Russian legislation in the field of freedom of conscience has been taking place under the formal influence of international obligations and constitutional guarantees. At the same time, the real direction of this transformation has often diverged from the logic of expanding human rights. Instead of developing a universal legal regime for the freedom of worldview choice of every person, legislation has increasingly moved toward special regulation of religious associations, control over missionary activity, restrictions connected with "extremism," property preferences, and selective support for religious organizations close to state power.

The Federal Law of 26 September 1997 No. 125-FZ "On Freedom of Conscience and Religious Associations" became the basic act in this field. From the standpoint of constitutional principles and international obligations, it was problematic from the moment of its adoption. It fixed a confessional hierarchy in the legal language of the state, gave symbolic priority to "traditional" religions, created complicated and unequal procedures for religious associations, and opened the door for administrative interference in religious and worldview life.

Subsequent amendments did not remove this structural defect. On the contrary, many of them deepened it. The law gradually became not a guarantee of freedom of conscience, but a mechanism for managing religious activity, religious property, missionary work, pilgrimage, education, foreign participation, reporting duties and interaction with public authorities.

A significant line of legislative transformation concerned property. The Federal Law "On the Transfer to Religious Organizations of Religious Property in Federal or Municipal Ownership" of 2010 continued to be implemented. In practice, this entailed the transfer of large volumes of public property, including cultural heritage objects, to religious organizations useful to the authorities. According to independent expert estimates, in Moscow alone more than twenty thousand objects belonging to the state could be transferred to the Russian Orthodox Church of the Moscow Patriarchate; across the country, the number of such objects may be counted in the hundreds of thousands.

In substance, this law allows state property, including cultural heritage monuments, to be transferred on arbitrary grounds to religious associations favoured by the authorities, including associations to which the property had never belonged. This raises not only questions of property law, but also the deeper constitutional problem of the neutrality of the state in the worldview sphere.

Another important line of changes concerned labour and financial regulation. The Federal Law of 27 December 2018 No. 542-FZ "On Amendments to the Labour Code of the Russian Federation concerning the Establishment of Specific Features of the Extension of Regional Agreements on the Minimum Wage to Religious Organizations" made it possible for religious organizations, on the basis of Article 345.1 of the Labour Code, not to comply with the general requirements concerning the minimum wage.

Legislation on freedom of conscience in a number of constituent entities of the Russian Federation also does not correspond to the Russian Constitution or to international-legal documents. In recent years, some constituent entities of the Russian Federation have adopted new unconstitutional laws on "missionary activity" that do not correspond to the principle of legal certainty or to the constitutional secularity of the state.

In July 2019, Federal Law No. 170-FZ of 3 July 2019 "On Amendments to the Federal Law 'On the Fundamentals of Tourist Activity in the Russian Federation' and the Federal Law 'On Freedom of Conscience and Religious Associations'" was adopted. In particular, Article 2 supplemented Federal Law No. 125-FZ of 26 September 1997 with Article 17.1 on pilgrimage activity.

Under that provision, a pilgrim is defined as a natural person who undertakes a journey, or trip, to visit places of religious veneration and objects of religious purpose located in the territory of the Russian Federation or beyond its borders in order to participate in religious rites and ceremonies. Pilgrims have the right to undertake and organize their own pilgrimage trips directly, as well as through religious organizations or organizations created by religious organizations that carry out pilgrimage activity. Participation of pilgrims in worship services and other religious rites and ceremonies at places of religious veneration is carried out in accordance with the internal rules of religious organizations.

A journey undertaken for recreation and entertainment, for income-generating activity at the place of temporary stay, or for purposes other than those specified in the law cannot be recognized as a pilgrimage trip.

Pilgrimage activity is defined as the activity of religious organizations in organizing pilgrimage trips and in establishing, maintaining and developing international links and contacts for the purpose of organizing such trips. The activity of religious organizations in organizing pilgrimage trips includes arranging accommodation for pilgrims, meals for pilgrims with due regard to requirements provided by the internal rules of religious organizations, transport services, escorting of pilgrims, including services of guides, interpreter-guides and instructor-guides, and other activities connected with organizing pilgrimage trips.

Religious organizations received the exclusive right to carry out pilgrimage activity directly or by creating organizations whose main purpose is pilgrimage activity. Such activity may be carried out by religious organizations on a paid or unpaid basis, with or without the involvement of legal entities and individual entrepreneurs entitled to carry out tour-operator or travel-agent activity.

Thus, Federal Law No. 170-FZ of 3 July 2019 was, in substance, aimed at establishing the monopoly of the "main" religious organizations in the field of pilgrimage. Travel agencies can now organize "pilgrimage" only through the leadership of the relevant religious organizations.

Another line of transformation concerned the digital information sphere. On 12 April 2018, bill No. 223849-7 "On Amendments to Article 10-1 of the Federal Law 'On Information, Information Technologies and the Protection of Information'" was adopted in the first reading and prepared for the second reading. Among other things, it proposed to require an "organizer of dissemination of information on the Internet" with more than two million users to create, on the territory of the Russian Federation, a representative office of the social network operator and to restrict access to, or delete, upon a user's request, information disseminated in the social network that was clearly aimed at propaganda of war, incitement of national, racial or religious hatred and enmity, false information or information defaming another person's honour, dignity or reputation, or other information for the dissemination of which criminal or administrative liability is provided, within twenty-four hours from receipt of the request. In the second reading, the bill was adopted on 22 December 2020 and later became part of Federal Law No. 530-FZ of 30 December 2020.

Federal Law No. 530-FZ of 30 December 2020 obliged social networks with more than 500,000 daily users independently to identify and block content aimed at inciting religious hatred and enmity. Platforms are given one day to make decisions on user complaints or monitoring results, and in disputed cases access to the material must be restricted during the period of Roskomnadzor's examination. Thus, the law transferred the function of operational censorship of inter-confessional conflicts from the state to the administrations of Internet resources.

The key amendments and additions to Federal Law No. 125-FZ "On Freedom of Conscience and Religious Associations" over the last decade were made on 30 March and 6 July 2016; 5 February 2018; 1 May, 3 July and 2 December 2019; 5 and 30 April and 11 June 2021; 28 June and 5 and 29 December 2022; 24 July and 2 November 2023; 6 April, 8 August and 28 December 2024; 31 July 2025; and 20 February 2026.

During this period, Law No. 125-FZ changed unevenly, but in substance the amendments followed several lines: property issues of religious organizations; missionary activity; pilgrimage; educational activity; "anti-extremist" and "anti-terrorist" restrictions; registration procedures; and, in 2025-2026, protection of religious symbolism and clarification of rules on the establishment and income-generating activity of religious organizations.

On 30 March 2016, the law introduced a special norm on disposal of real estate belonging to a religious organization. Transactions with such real estate were made dependent on the written consent of the body of the religious organization authorized by its charter. In substance, this strengthened internal control over alienation of worship-related and other significant real estate.

On 6 July 2016, within the so-called "Yarovaya package," the law was supplemented by a block of provisions regulating missionary activity. The law introduced a special definition of missionary activity, established rules for its implementation, and provided administrative liability for violations of those rules. This was one of the most significant restrictive turns in the legal regulation of religious activity. The notion of missionary activity was formulated broadly, which created risks for ordinary religious communication, distribution of literature, public discussion of beliefs and activity of small religious communities.

On 5 February 2018, amendments connected with voluntary and volunteer activity were adopted. They affected the interaction between religious organizations and volunteers and clarified certain aspects of voluntary activity within religious communities. Yet the very discussion around these amendments showed how deeply religious organizations close to the authorities were involved in the legislative process.

On 1 May 2019, amendments affected questions of educational activity and the relationship between spiritual educational organizations and general educational regulation. On 3 July 2019, the already mentioned amendments on pilgrimage activity were adopted. On 2 December 2019, changes were introduced into provisions related to the activity of religious organizations and their reporting or registration obligations.

On 5 April 2021, a new package of amendments strengthened the requirements imposed on founders, participants and leaders of religious organizations, including restrictions concerning persons whose activities were regarded as connected with extremism or terrorism. On 30 April 2021, additional changes were made in relation to the legal status and registration of religious organizations. On 11 June 2021, further amendments clarified particular rules on religious organizations and their interaction with public authorities.

On 28 June 2022, the law was amended in connection with the extension of certain requirements to public authorities of a federal territory. The amendment also clarified that the activity of authorities and officials must not be accompanied by public religious rites and may not be used to shape citizens' attitude to religion.

On 29 December 2022, Article 19 was amended, primarily paragraph 3. The amendment was connected with the further linking of the status of spiritual educational organizations to general educational legislation and federal requirements for educational programmes.

On 24 July 2023, in the first paragraph of paragraph 9 of Article 8, the period for notifying the registration authority of changes in information about a religious organization was increased from three to seven working days. This amendment is procedural, but in practice it is noticeable.

On 2 November 2023, Article 16 was supplemented by paragraph 6. The amendment concerned the mechanism for restricting access to information disseminated in violation of the law, in connection with the religious activity of centralized religious organizations and their authorized persons.

On 6 April 2024, in a number of provisions, including Articles 7, 8, 11, 14 and 27, earlier formulations concerning the body making a registration decision were replaced by more precise wording: "federal body of state registration" and its territorial body.

On 8 August 2024, the second paragraph of paragraph 3 of Article 9 was amended. A rule was introduced under which a foreign citizen or stateless person whose information is included in the register of controlled persons may not be a founder of a religious organization.

On 28 December 2024, paragraph 1.1 of Article 7, paragraph 3 of Article 9 and paragraph 4 of Article 25 were amended. In the relevant provisions, after words concerning countering the financing of "terrorism," the words "and extremist activity" were added. In other words, the anti-money-laundering and control block was expanded to cover "extremist" activity as well.

On 31 July 2025, paragraph 6.1 was added to Article 3. It prohibited the use of images and other reproductions of places of worship, other religious objects and official heraldic signs without the religious symbols historically present on them, except where historical accuracy so requires or where the reproduction of the symbols would lead to their desecration. This amendment concerns the protection of religious symbolism in visual reproduction.

On 20 February 2026, amendments affected Articles 8, 10, 23 and 24.1. Paragraph 4 of Article 8 was supplemented by a provision that a centralized religious organization may be created by another centralized religious organization; paragraph 2 of Article 10 clarified the content of a charter, including the subject, purposes and directions for the use of property and funds;

Article 23 was set out in a new version and systematically regulated income-generating activity of religious organizations; Article 24.1 received a technical amendment.

Thus, at the law-making level, since the adoption in 1997 of the unconstitutional Federal Law "On Freedom of Conscience and Religious Associations," which also fails to correspond to modern legal technique, its repressive potential has continued to grow. As already noted, norms inconsistent with the principle of legal certainty are regularly introduced into this act. On the one hand, they mainly serve the interests of religious associations close to state power; on the other hand, they are aimed at unlawfully restricting the activity of any other religious associations regarded as "non-traditional" by the authorities, supporters of state secularity, and, in practice, almost anything else. Regulation of "missionary activity," countering "extremism" and similar grounds serve as pretexts for mass persecution.

It is highly indicative that in 2018 lawyers of certain religious organizations participated in the drafting of six federal laws adopted in the interests of that organization. This was reported on 30 January 2019 at the round table "The Role of Traditional Religions in the Development of Civil Society" in the State Duma of the Russian Federation, held within the framework of the Christmas Parliamentary Meetings, by the head of the Legal Service of the Russian Orthodox Church of the Moscow Patriarchate, Abbess Ksenia (Chernega), according to TASS. She recalled that President Vladimir Putin had signed Federal Law No. 15 of 5 February 2018 on amendments to certain legislative acts concerning volunteer activity. According to her, that law regulated voluntary and religious activity in religious communities. She stated that the original version of the act had contained a number of provisions violating the principle of non-interference of the state in the activity of religious associations, and that those provisions had been removed. In particular, at the level of subordinate acts it had been proposed to regulate the conditions and procedure for interaction between state bodies and religious communities in all areas of volunteer activity, including physical and missionary activity. Mandatory inclusion of information about all volunteers, including monastics, in a unified information system was also considered unacceptable.

Among other laws in the drafting of which the Russian Orthodox Church of the Moscow Patriarchate participated was the law concerning the transformation of the federal state unitary enterprise Russian Post into a joint-stock company. At first glance, this law did not relate to the religious sphere at all. However, according to the same statement, an amendment was introduced prohibiting the transfer into private ownership of religious-purpose property on the balance sheet of Russian Post after its corporatization. The Church also participated in the development of a legislative act amending the Civil Code of the Russian Federation concerning unauthorized constructions. Amendments were proposed in order to remove unauthorized constructions of religious significance from the effect of the relevant prohibition. Other legislative acts included amendments to the Tax Code and the law on accounting, fully exempting all religious organizations from submitting accounting reports regardless of whether they had tax obligations, as well as amendments to the Labour Code establishing specific features of the regional minimum wage agreement for religious organizations. This made it easier for religious organizations to refuse to join regional minimum wage agreements and allowed them to pay employees below the minimum wage.

If religious organizations close to the authorities receive support, the rights of others are unlawfully restricted. In particular, in addition to "anti-extremist" legislation, norms on protecting the "feelings of believers," which also fail to correspond to the principle of legal certainty, are used as a legal instrument for persecuting supporters of state secularity. In recent years, the number of cases applying these norms has significantly increased, including in connection with growing public opposition to violations of the constitutional secularity of the state. Special attention should be paid here to the movement against the unlawful construction of places of worship in parks and public gardens, the most resonant example being the case of Yekaterinburg.

For many years, a working group in the Federation Council of the Russian Federation continued its not entirely constitutional activity in preparing proposals to improve Russian legislation in the field of protecting Russian citizens from fraudulent actions of "sects." Its core consisted of state officials and representatives of the Russian Orthodox Church of the Moscow Patriarchate.

The lower chamber of the Federal Assembly also did not remain apart from unconstitutional activity. Within the walls of the legislative body of a secular state, the Inter-Factional Deputy Group for the Protection of Christian Values continued to meet, while the first deputy chairman of the Synodal Department for Relations between the Church, Society and the Media of the Russian Orthodox Church of the Moscow Patriarchate worked as an adviser to the Speaker of the State Duma.

Thus, the problems of developing legislation on freedom of conscience largely flow from the problems of observing the principle of state secularity by the chambers of the legislative body of the Federal Assembly of Russia, namely the State Duma and the Federation Council.

According to M. O. Shakhov, speaking at the Second All-Russian Forum "Law. Religion. State" on 21-22 November 2019 in the Civic Chamber of the Russian Federation, current legislation on freedom of conscience and religious associations contains a number of gaps and uncertainties. This may be qualified as conscious legislative silence, sometimes called by legal scholars a qualified or intentional gap in law. The researcher reasonably believes that for decades the legislator has deliberately avoided taking responsibility for filling gaps in religious legislation and specifying excessively abstract terms and expressions, shifting this function and responsibility to the law enforcer and thereby giving him excessive freedom to interpret and apply legal norms. In other words, uncertainty of legal regulation creates the impossibility of an unambiguous interpretation of the legal norm, which ultimately leads to an unjustified expansion of the discriminatory powers of the law enforcer.

It may be concluded that legislative initiatives of recent years, contrary to the constitutional principle of state secularity, are aimed at strengthening a state "confessional vertical" in the narrow interests of the confessional bureaucracy of "traditional confessions" and ruling "elites." In this context, I believe that legislative norms inconsistent with the principle of legal certainty, that is, based on legally undefined concepts, predetermine arbitrariness, inequality and human rights violations in the sphere of internationally recognized freedom of conscience. Ultimately, it must be recognized that the evolution of Russian legislation in the field of freedom of conscience does not fully correspond to Russia's international obligations.

REFERENCES TO CHAPTER 5

List of Sources Used

1. Pereplesnina E.M. Universally Recognized Principles and Norms of International Law and International Treaties of the Russian Federation in the Russian Legal System. Petrozavodsk: Verso, 2009. p. 49.
2. Ibid. p. 64.
3. Khizhnyak V.S. The Constitutional-Legal Mechanism of Interaction between the Domestic Law of the Russian Federation and International Law: Doctoral Dissertation in Law. Saratov, 2008. 476 p. p. 18.
4. Khizhnyak V.S. The Constitutional-Legal Mechanism of Interaction between the Domestic Law of the Russian Federation and International Law: Doctoral Dissertation in Law. Saratov, 2008. 476 p. p. 15.
5. Ershov V.V. International and Domestic Law from the Standpoints of Legalism and an Integrative Understanding of Law // Russian Justice. No. 8 (64). 2011. p. 8.
6. Lukashuk I.I. The Interaction of International and Domestic Law under Conditions of Globalization // Journal of Russian Law. 2002. No. 3. pp. 115-128.
7. Kartashkin V.A. Human Rights in International and Domestic Law. Executive editor E.A. Lukasheva. Moscow: Publishing House of the Institute of State and Law of the Russian Academy of Sciences, 1995. p. 131.
8. Constitutional Rights and Freedoms of the Human Person and Citizen in the Russian Federation. Edited by O.I. Tiunov. Moscow: Norma, 2005. p. 542.
9. Pryakhina T.M. Constitutional Problems of the Interaction between International and National Law: Study Guide. Edited by V.O. Luchin. Moscow: Russian State Social University Publishing House, 2006. pp. 11-12.
10. Pereplesnina E.M. Universally Recognized Principles and Norms of International Law and International Treaties of the Russian Federation in the Russian Legal System. Petrozavodsk: Verso, 2009. p. 29.
11. Pryakhina T.M. Constitutional Problems of the Interaction between International and National Law: Study Guide. Edited by V.O. Luchin. Moscow: Russian State Social University Publishing House, 2006. p. 10.
12. The Constitution of the Russian Federation. Moscow: Legal Literature, 1995.
13. Marochkin S.Yu. Legislative, Doctrinal, and Practical Assimilation of the Constitutional Principle on Norms of International Law in the Russian Legal System // Application of International Human Rights Treaties in the Legal System of the Russian Federation: Proceedings of the All-Russian Scientific and Practical Conference. Yekaterinburg, 2003. p. 60.
14. Ibid. p. 63.
15. Ershov V.V. The Relationship between International and Domestic Law under Conditions of Globalization // Russian Justice. No. 6 (62). 2011. p. 6.
16. Civil Human Rights in Russia: Contemporary Problems of Theory and Practice (edited by F.M. Rudinsky). Moscow: ZAO TF MIR, 2006. p. 441.
17. Rudinsky F.M. The Science of Human Rights and Problems of Constitutional Law. Moscow: ZAO TF MIR, 2006. p. 107.
18. Lukashuk I.I. The Interaction of International and Domestic Law under Conditions of Globalization // Journal of Russian Law. No. 3, 2002. p. 115.
19. Ershov V.V. The Relationship between International and Domestic Law under Conditions of Globalization // Russian Justice. No. 6 (62). 2011. p. 6.
20. Konyukhova I.A. Sources of International and Constitutional Law in the Regulation of Human and Citizen Rights and Freedoms: International, Foreign, and Domestic Experience // International-Legal Standards in Constitutional Law: Collection of Scientific Works. Part I / Russian Academy of Justice; INION RAS; executive editor I.A. Konyukhova. Moscow, 2006. 376 p. pp. 108-138.
21. Podoprigora R.A. The State and Religious Organizations: Administrative-Legal Issues. Almaty: Arkaim, 2002. p. 45.
22. Lukashuk I.I. The Interaction of International and Domestic Law under Conditions of Globalization // Journal of Russian Law. No. 3, 2002. p. 120.
23. Sigalov K.E. The Sacralization of Power as an Immanent Feature of the Russian Political-Legal Tradition // Legal Traditions. Zhidkov Readings: Proceedings of the International Scientific Conference. Moscow, 2014. pp. 235-239.

24. Sigalov K.E., Prokuratov A.A. The Legal Environment and the Specific Features of Russian Legal Consciousness in the Context of Historical Epochs // Law and Education. 2005. No. 4. pp. 246-259.
25. Mozgovoy S.A. Confessional (Clerical) State // Religious Studies / Encyclopedic Dictionary. Moscow: Academic Project, 2006. p. 538.
26. Ershov V.V. The Relationship between International and Domestic Law under Conditions of Globalization // Russian Justice. No. 6 (62). 2011. p. 8.
27. Constitutional Rights and Freedoms of the Human Person and Citizen in the Russian Federation. Edited by O.I. Tiunov. Moscow: Norma, 2005. p. 590.
28. Expert Opinion on the Federal Law 'On Freedom of Conscience and Religious Associations' // Religion and Law. 1997. No. 1. pp. 21-23.
29. Buryanov S.A. Freedom of Conscience in Russian Scholarship, Lawmaking, and Law Enforcement // Law and Politics. 2001. No. 7. p. 31.
30. Protection of Human Rights: Collection of Documents 1998-2000. Moscow, 2001. pp. 419-423.
31. Religious Associations. Freedom of Conscience and Religion. Normative Acts. Judicial Practice. Moscow: Jurisprudence, 2001. pp. 5-6.
32. Issues Concerning the Transfer of Religious-Purpose Property Were Discussed at a Meeting of the State Duma Committee // Federal Agency for State Property Management. Electronic resource. URL: https://www.rosim.ru/press/news/319931 (accessed: 14.12.2019).
33. Kulagin M.A. Property Rights of Religious Organizations: Abstract of Candidate of Legal Sciences Dissertation. Moscow, 2013. p. 19.
34. Buryanov S.A. Implementation of the Constitutional Freedom of Conscience and Freedom of Religion in the Russian Federation: Monograph. Moscow: ZAO TF MIR, 2009. 288 p.
35. Freedom of Conscience: Problems of Theory and Practice. Monograph (edited by F.M. Rudinsky, S.A. Buryanov). Moscow: ZAO TF MIR, 2012. 1120 p.
36. Opinion of the Commissioner for Human Rights in the Russian Federation 'On Reviewing the Compliance of the Federal Law "On Freedom of Conscience and Religious Associations" with the International-Legal Obligations of the Russian Federation.' Electronic resource. URL: http://www.consultant.ru/document/cons_doc_LAW_22805/ (accessed: 21.11.2019).
37. Federal Law No. 374-FZ of July 6, 2016 'On Amendments to the Federal Law " On Countering Terrorism" and Certain Legislative Acts of the Russian Federation Regarding the Establishment of Additional Measures to Counter Terrorism and Ensure Public Security' // Rossiyskaya Gazeta, Federal Issue No. 7017 (149). July 8, 2016. Electronic resource. URL: https://rg.ru/2016/07/08/antiterror-dok.html (accessed: 17.12.2019).
38. Zubarev E. Pity and Irony against the Golden Censer. It Is No Longer Possible to Restore Russia's Former Secular Status without Revolutionary Reconstruction: No One Voluntarily Gives Up a Sinecure // Rosbalt. Electronic resource. URL: http://www.rosbalt.ru/blogs/2013/05/01/1124348.html (accessed: 17.12.2019).
39. Buryanov S.A. Analysis of New Provisions of Russian Legislation in the Sphere of Freedom of Conscience from the Standpoint of International Law // Human Rights in International and National Law: Collection of Scientific Articles Dedicated to the 10th Anniversary of the Department of International Law and Human Rights of the Law Institute of Moscow City Pedagogical University. Moscow: Human Rights, 2015. pp. 102-114.
40. Freedom of Conscience: Problems of Theory and Practice. Monograph (edited by F.M. Rudinsky, S.A. Buryanov). Moscow: ZAO TF MIR, 2012. 1120 p.
41. Buryanov S.A. International Recognition of the Right to Freedom of Conscience and Problems of Its Implementation in the Russian Federation under Conditions of Contemporary Global Processes: Monograph. Moscow: Poligraf Servis, 2020. 624 p.
42. Civil Human Rights in Russia: Modern Problems of Theory and Practice / compiled by F.M. Rudinsky. Transaction Publishers, New Brunswick, New Jersey (U.S.A.), 2008. 458 p.

CONCEPT BOX

CHAPTER 6

WHEN WORLDVIEW NEUTRALITY FAILS

Legal uncertainty, selective enforcement, and symbolic inequality

The violation of freedom of conscience rarely appears as a single dramatic event. More often, it develops through legal uncertainty, selective enforcement, privileged institutions, informal pressure, discrimination, and symbolic inequality.

Chapter 6 examines the problems that arise when the principle of state worldview neutrality remains incomplete or inconsistent.

The chapter shows that freedom of conscience requires more than formal recognition. It needs clear legal guarantees, neutral institutions, effective remedies, and protection from worldview discrimination.

Without these safeguards, public authority can silently reshape the conditions of belief, doubt, loyalty, and civic belonging.

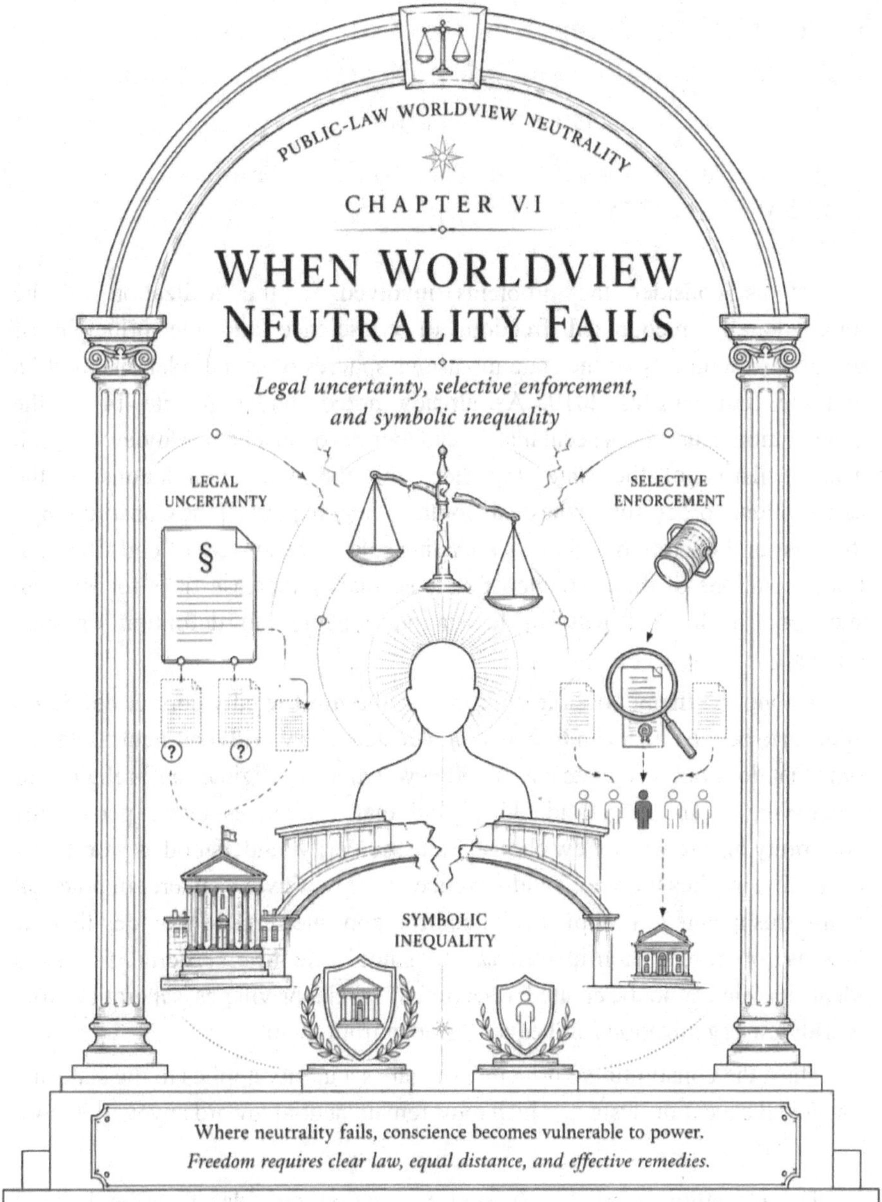
PUBLIC-LAW WORLDVIEW NEUTRALITY
CHAPTER VI
WHEN WORLDVIEW NEUTRALITY FAILS
Legal uncertainty, selective enforcement, and symbolic inequality
LEGAL UNCERTAINTY
SELECTIVE ENFORCEMENT
§
SYMBOLIC INEQUALITY
Where neutrality fails, conscience becomes vulnerable to power.
Freedom requires clear law, equal distance, and effective remedies.

Chapter 6. Problems in the Realization of the Internationally Recognized Right to Freedom of Conscience and the Principle of Worldview Neutrality of the State in the Russian Federation

6.1. Problems in the Realization of the Right to Freedom of Conscience and the Principle of Worldview Neutrality of the State in Various Spheres of Social Relations

Let us consider the problems involved in the realization of the internationally recognized freedom of conscience and the principle of worldview neutrality of the state in various spheres of social relations in 2018 and the first half of 2019. As already noted, under Article 14 of the Constitution, Russia is a secular state and therefore must be worldview-neutral. The secularity of the state, together with the other foundations of the constitutional order, forms the basis for the other chapters of the Constitution of the Russian Federation and plays a leading role in the system of constitutional law. Violations of the principle of state secularity increase risks for Russian statehood itself. Yet Russian legislation contains no definition of state secularity.

Among the many possible approaches, the most legally correct definition appears to be the understanding of a secular state as a worldview-neutral state: a state that does not recognize any worldview, whether religious, anti-religious or otherwise, as an official ideology, and that guarantees every person the possibility of free worldview choice. State secularity, understood as worldview neutrality, implies the state's indifference in the worldview sphere. In practical terms, this means refusal of special control, non-interference provided that the law is observed, non-identification, since scientific criteria for such identification cannot be created, rejection of special privileges, separation from worldview organizations and equal distance from them.

Thus, the constitutional principle of state secularity applies to the state and municipal education system, which must remain neutral toward any worldview.

Problems in Observing the Neutrality of the Legislative Branch

The Constitution of the Russian Federation provides that the Federal Assembly, the Parliament of the Russian Federation, is the representative and legislative body of the Russian Federation (Article 94). Under Article 95, the Federal Assembly consists of two chambers: the Federation Council and the State Duma. The Federation Council includes two representatives from each constituent entity of the Russian Federation, one from the representative body and one from the executive body of state power. The State Duma consists of 450 deputies.

Bills are introduced in the State Duma. The right of legislative initiative belongs to the President of the Russian Federation, the Federation Council, members of the Federation Council, deputies of the State Duma, the Government of the Russian Federation, and legislative representative bodies of the constituent entities of the Russian Federation. The right of legislative initiative also belongs to the Constitutional Court of the Russian Federation, the Supreme Court of the Russian Federation and the Higher Arbitration Court of the Russian Federation on matters within their competence (Article 104).

Federal laws are adopted by the State Duma and, within five days, submitted for consideration to the Federation Council (Article 105). An adopted federal law is sent within five days to the President of the Russian Federation, who signs and promulgates it within fourteen days (Article 107).

In accordance with Article 4 of Federal Law No. 184-FZ of 6 October 1999, "On the General Principles of the Organization of Legislative (Representative) and Executive Bodies of State Power of the Constituent Entities of the Russian Federation," the legislative representative body of state power of a constituent entity of the Russian Federation is the permanent, highest and sole body of legislative authority of that constituent entity. The number of deputies of the legislative representative body is established by the constitution or charter of the constituent entity and is determined depending on the number of voters registered within its territory.

Article 5 of the same federal law defines the main powers of the legislative representative body of a constituent entity of the Russian Federation. These powers include legislative regulation on matters within the jurisdiction of the constituent entity and on matters of joint jurisdiction of the Russian Federation and its constituent entities, within the limits of the authority of the relevant constituent entity.

The right of legislative initiative in the legislative representative body of a constituent entity of the Russian Federation belongs to deputies, the highest official of the constituent entity, that is, the head of the highest executive body of state power of the constituent entity, and representative bodies of local self-government. The constitution or charter of a constituent entity may also grant the right of legislative initiative to other bodies, to members of the Federation Council of the Federal Assembly of the Russian Federation representing the legislative and executive bodies of state power of the given constituent entity, to public associations, and also to citizens residing in the territory of that constituent entity (Article 6).

The laws of a constituent entity of the Russian Federation and resolutions of the legislative representative body of state power of the constituent entity are adopted by a majority vote and are then sent by that body for promulgation to the highest official of the constituent entity, the head of the highest executive body of state power, within the period established by the constitution or charter and by the law of the relevant constituent entity (Articles 7 and 8).

It follows from the Constitution of Russia and current legislation that the exercise of legislative power in the Russian Federation must have a secular and worldview-neutral character.

However, monitoring makes it possible to conclude that the transformation of legislation on freedom of conscience in the Russian Federation, in violation of the Constitution of Russia and current legislation, has been influenced by certain arrangements between the "party of power," together with its sparring partners, and the Russian Orthodox Church of the Moscow Patriarchate. Other religious organizations controlled by the authorities, usually described as "traditional," are often used to legitimize initiatives that are not fully constitutional.

The relationship between the legislative body of the state and one religious organization, contrary to legislation, is very close. Joint events speak for themselves. In recent years, every winter the State Duma of the Russian Federation has hosted Christmas Parliamentary Meetings with the participation of deputies and representatives of the Russian Orthodox Church of the Moscow Patriarchate.

For example, in winter 2014, among other legally questionable initiatives, deputies proposed compiling a special list of theological universities in foreign countries where ideas of "religious extremism" were allegedly preached and where students from Russia should not be allowed to study. This was stated by the then First Deputy Chair of the State Duma Committee on Public Associations and Religious Organizations.

"We insist that it is necessary to compile a list of foreign universities that provide religious education and from which young men later return here, to Russia, and engage in unlawful activity. They receive the basics of non-traditional forms of interpreting Islam in certain educational institutions in Saudi Arabia, Pakistan, Qatar, Turkey and Egypt; these universities are known."

In his view, representatives of the Ministry of Education and Science and Russian security services should participate in drawing up a federal list of non-recommended universities. As Markelov noted, if such a document were adopted, those students who nevertheless decided to study at a university from the "black list" could face restrictions on entering the territory of Russia. He described this as a preventive measure that would make it possible, in some way, to organize the education received by Russian citizens, mainly representatives of the Caucasus republics, who go abroad to study. The statement was made at a section of the Christmas Readings devoted to the topic "Spiritual and Moral Education at School as Prevention of Religious Extremism."

In turn, the Deputy Chair of the Duma Committee clarified to RIA Novosti that when students from Russia are taught "non-traditional" forms of Islam, the matter concerns, to a greater extent, the spread of Salafism. According to him, this represents a real threat because a loose interpretation of the Qur'an allegedly leads to the replenishment of the ranks of so-called "forest brothers" by young men.

In 2015, the discussion about the role of Orthodoxy in a secular state intensified inside the walls of the State Duma of the Russian Federation. In the spring, a prominent deputy called for a "more serious" attitude toward attacks on the Church and personally on the Patriarch. This was stated by the then head of the Duma Committee on Culture, film director Stanislav Govorukhin, as TASS reported on 25 March. He said that attacks had begun against His Holiness the Patriarch of Moscow and All Rus': first that he wore the wrong watch, then that he arrived in the wrong car. According to him, all such attacks in literature and art should be treated very seriously.

As an example, the director referred to the punk group's action in the Cathedral of Christ the Saviour. In his view, something similar had occurred during the February Revolution of 1917. Unlike the beginning of the twentieth century, he argued, present activity was aimed at "undermining the pillars on which the state rests." He added: "We are not Charlie; we will not allow anyone to mock or deride either Jesus Christ, or the Prophet Muhammad, or any shrine of any people." At the same time, the head of the Duma Committee acknowledged that representatives of the Church had often been negative characters in Russian literature.

In this connection, a proposal was made to amend Article 13 of the Constitution, which prohibits the establishment of a state ideology, by adding provisions on the spiritual sovereignty of the state. The deputy noted that the 1993 Constitution specifies the state sovereignty of the Russian Federation, but does not mention spiritual sovereignty. According to him, everyone remembers that the unsuccessful implementation of the "orange scenario" began with a massive attack on the Church. He argued that the spiritual unity of Russians had been underestimated and that today "our positions and values" had to be clearly stated. “We have nothing to be ashamed of. On the contrary, this step will unite our people even more strongly and allow us to give a worthy response to modern challenges. Russia is a unique country, and we must be proud of it,” he said.

Then, at the end of spring, in May 2015, Elena Mizulina, then head of the State Duma Committee on Family, Women and Children, proposed teaching the school course “Fundamentals of Religious Cultures and Secular Ethics” from the primary grades through senior classes in all secondary schools in Russia, Interfax reported. She stated at a round table in Kazan that the course “Fundamentals of Orthodox Culture” was already part of the school curriculum, but was limited to two educational modules in the fourth and fifth grades. She said that they would seek to ensure that schoolchildren began learning the fundamentals of religious culture already in primary school.

It should also be recalled that in April 2012 State Duma deputies Sergey Gavrilov from the Communist Party and Sergey Popov from United Russia initiated the creation, within Parliament, of an association of State Duma deputies and Federation Council members who recognized the significance of Christianity in the history, culture and tradition of the Russian people and other indigenous peoples of Russia, and its “invaluable contribution to world culture.”

Alongside the initiators, the group included, among others, Deputy Chair of the Committee on International Affairs Valentina Tereshkova, member of the Committee on CIS Affairs and Relations with Compatriots Andrey Skoch, head of the Committee on Security and Anti-Corruption Irina Yarovaya, head of the Committee on Constitutional Legislation and State Building Vladimir Pligin, First Deputy Chair of the Committee on International Affairs Andrey Klimov, and others.

In spring 2014 it became known that a working group had been created in the State Duma of the Russian Federation to study the issue of a legal formulation of the concept of a “sect,” Interfax-Religion reported on 4 March. State Duma deputy Mikhail Markelov stated in an interview published by the Orthodox newspaper Krestovsky Most that a working group had been created in the State Duma to resolve this problem, and that a joint appeal would be prepared by deputies, representatives of the Ministry of Justice, the Prosecutor General’s Office and other agencies, so that the Supreme Court would define “sect.” In his view, the matter concerned “destructive religious organizations” that actively operate in different regions, exert psychological influence on people, conduct unlawful “healing” sessions, often damage health and alienate the property of credulous followers in their own favor.

In autumn 2015, another meeting took place in the State Duma of the inter-factional deputy group for the protection of Christian values, headed by Communist Party member Sergey Gavrilov. After discussing the suffering of Christians in Syria and praying, the deputies focused on the familiar "struggle against external and internal enemies." Having condemned "absolute freedom of speech," they proposed to "burn out with a red-hot iron" non-traditional religious confessions and called for strengthening the "fight against sodomy online." Toward the end, the discussion also turned to the need to amend the Constitution, although the parliamentarians left that task to the next convocation of the State Duma.

On 8 December 2015, a round table entitled "Sects and Destructive Cults as a Challenge to Russia's National Security" was held in the State Duma of the Federal Assembly of the Russian Federation. The round table was organized by the State Duma Committee on Public Associations and Religious Organizations and the Synodal Missionary Department of the Russian Orthodox Church. It was attended by the First Deputy Chair of the State Duma Committee on Public Associations, M. Yu. Markelov; Deputy Chair of the Synodal Missionary Department, Hegumen Serapion (Mitko); President of the St Irenaeus of Lyon Centre for Religious Studies and professor at the Orthodox St Tikhon Humanitarian University, A. L. Dvorkin; head of the Centre for the Geography of Religion of the Synodal Department for Church and Society Relations, R. A. Silantiev; representatives of synodal departments; and invited experts.

Opening the meeting, M. Yu. Markelov noted that the discussion on the declared topic, involving experts, lawyers, representatives of the Ministry of Justice, the Ministry of Internal Affairs, other law-enforcement agencies and public organizations, had long been taking place in the State Duma. It was mainly built around the question of whether a separate law should be created against destructive religious cults and elements and whether the concept of "sect" should be legally fixed, or whether that would be excessive and it would be enough to improve a number of provisions of existing regulatory legal acts.

On 17 January 2016, seventy-six Baptist Christians sent a letter to the President of the Russian Federation, V. V. Putin, with copies to the Minister of Justice of the Russian Federation, A. V. Konovalov, and the Chair of the State Duma, S. E. Naryshkin. In that letter, they expressed concern about changes taking place in the state in the sphere of freedom of conscience and religion. They referred to the round table held on 8 December 2015 in the State Duma on "Sects and Destructive Cults as a Challenge to Russia's National Security," organized by the State Duma Committee on Public Associations and Religious

Organizations and the Synodal Missionary Department of the Russian Orthodox Church.

The authors of the letter emphasized that the participants of the meeting were convinced that "sects" represented a danger for Russia and had to be fought. The presiding deputy, M. Yu. Markelov, acknowledged that at present there was no legal definition of "sect": it was not legally enshrined and there was no definition from the Supreme Court. Yet none of the speakers proposed any variant of such a definition. In other words, it remained unclear by what principle religious associations were to be classified as "sects" against which a struggle should be waged. The question of clarifying terminology so that innocent people would not suffer in the process of "fighting sects" was not raised at the round table. The Baptist believers therefore drew attention to the danger that the practical application of the proposals voiced at that round table could create.

However, there are no grounds for hoping that freedom of conscience will be protected by Russian legislators. On 28 January 2016, the State Duma Committee on Public Associations and Religious Organizations held another meeting within the framework of the Fourth Parliamentary Meetings of the Twenty-Fourth International Christmas Educational Readings. The event was attended by State Duma deputies, representatives of so-called "traditional" religious organizations, ministries and agencies, and public figures. According to Nilov, the conclusions at that time were as follows: "There is a common understanding of how state-church relations should develop. The general line has been determined; adjustments are introduced depending on what problems and new challenges arise."

This means that instead of implementing the principles of state neutrality and freedom of conscience, the legislative body of Russia intends to continue serving unconstitutional relations between the authorities and the dominant Church.

Problems in Observing the Neutrality of the Executive Branch

In accordance with the Constitution of the Russian Federation, executive power in the Russian Federation is exercised by the Government of the Russian Federation, which consists of the Chair of the Government, Deputy Chairs of the Government and federal ministers (Article 110).

The Chair of the Government of the Russian Federation is appointed by the President of the Russian Federation with the consent of the State Duma (Article 111). Under Article 112 of the Constitution, the Chair of the Government, no later than one week after appointment, submits proposals to the President of the Russian Federation on the structure of federal executive bodies and proposes candidates for the positions of Deputy Chairs of the Government and federal ministers. The activity of a number of federal ministries, federal services and federal agencies is directly managed by the President of the Russian Federation. The President also directs the federal services and federal agencies subordinate to those federal ministries. The Government of the Russian Federation manages the federal ministries, federal services and federal agencies subordinate to those ministries. In addition, the Government directs a number of federal services and federal agencies.

According to Article 77 of the Constitution, the system of bodies of state power of republics, territories, regions, cities of federal significance, the autonomous region and autonomous districts is established independently by the constituent entities of the Russian Federation, in accordance with the foundations of the constitutional order of the Russian Federation and the general principles for organizing representative and executive bodies of state power established by federal law. Federal executive bodies, for the exercise of their powers, may create their territorial bodies and appoint the relevant officials (Article 78).

Thus, a system of executive bodies headed by the highest executive body of state power of the constituent entity is established in each constituent entity of the Russian Federation. Under the Constitution, within the jurisdiction of the Russian Federation and the powers of the Russian Federation on matters of joint jurisdiction of the Russian Federation and its constituent entities, federal executive bodies and executive bodies of the constituent entities form a single system of executive power in the Russian Federation. The structure of executive bodies of state power of a constituent entity is determined by the highest official of the constituent entity, the head of its highest executive body, in accordance with the constitution or charter of that constituent entity.

The institution of representatives of the President of the Russian Federation in federal districts deserves separate mention. In accordance with the Regulation on the Plenipotentiary Representative of the President of the Russian Federation in a Federal District, approved by Decree of the President of the Russian Federation No. 849 of 13 May 2000, the plenipotentiary representative is an official representing the President within the relevant federal district. The representative ensures the realization of the constitutional powers of the head of state within that district. The plenipotentiary representative is a federal civil servant and is part of the Administration of the President of the Russian Federation. He or she is appointed and dismissed by the President upon the proposal of the Head of the Presidential Administration, is directly subordinate to the President and is accountable to him.

In this context, the subjects of executive power in the Russian Federation are: the President of the Russian Federation; the Government of the Russian Federation; federal authorities, including federal ministries, federal services and federal agencies and their territorial subdivisions; the highest officials of the constituent entities of the Russian Federation, that is, the heads of the highest executive bodies of state power of the constituent entities; and state authorities of the constituent entities of the Russian Federation.

The constitutional principle of state secularity (Article 14) applies to the entire system of state authorities and is one of the most important guarantees for the realization of freedom of conscience (Article 28). Violations of the constitutional principle of state secularity predetermine problems in the realization of freedom of conscience.

Alongside the Constitution of the Russian Federation, the Federal Law "On the State Civil Service of the Russian Federation" prohibits the use of official powers in the interests of religious associations and other organizations, and also prohibits civil servants from publicly expressing their attitude toward such associations and organizations in their official capacity, unless this falls within their official duties (subparagraph 13, paragraph 1, Article 17). It also prohibits the creation, within state bodies, of structures of religious associations or assistance in creating such structures (subparagraph 14, paragraph 1, Article 17).

In general, one may agree with V. N. Linkin, who considers the principle of secularity of the state service to be a system-forming principle of a secular state. Accordingly, the necessary conditions for the secularity of the state service include: support for the constitutional order and compliance with legislation in the sphere of freedom of conscience and religion; inadmissibility of any generally binding religion within the system of public service; a ban on establishing any direct or indirect restrictions or advantages depending on beliefs or attitude toward religion when entering or serving in public service; a ban on using official position in the interests of religious associations or demonstrating one's attitude toward them; a ban on religious associations performing separate functions of state authorities, other state bodies, state institutions and local self-government bodies; a ban on religious associations participating in elections to state authorities and local self-government bodies; inadmissibility of discrimination and of violations by civil servants, in their activity, of the principle of equality of citizens regardless of beliefs or attitude toward religion; inadmissibility of accompanying the activity of state authorities and local self-government bodies with public religious rites and ceremonies; secular professional ethics of civil servants; rejection of the practice of recognizing particular religious holidays as days off; and absence of religious elements in state and municipal symbols of authority, including coat of arms, anthem and flag.

In recent years, numerous facts have been identified that give grounds to suspect violations of the relevant legal norms.

Problems in Observing the Neutrality of the President of the Russian Federation and the Chair of the Government of the Russian Federation

For many years, the highest officials have stood with candles under the lenses of federal television cameras at Christmas and Easter services in the Cathedral of Christ the Saviour and other churches of the Russian Orthodox Church of the Moscow Patriarchate.

High-ranking officials not only publicly participate in religious events; they also openly state that the secular state of Russia is indebted to the Russian Orthodox Church of the Moscow Patriarchate and promise to repay that debt. In response to such demonstrations of confessional preference, certain public groups sometimes respond with counter-initiatives, even rather unusual ones. For example, a media statement by the "National Committee +60" of 29 December 2015 literally said the following: in connection with a flood of instructions from active citizens after the publication of a book of Vladimir Putin's quotations, the "National Committee +60" appealed to First Deputy Head of the Presidential Administration V. V. Volodin with an urgent request to satisfy the passionate desire of millions of supporters of the course of the repeatedly elected President of the Russian Federation to have a pocket version of a collection of aphorisms by the national leader.

The statement continued that for sixteen years, by the will of the people and of the Heavenly Powers, the "wheel" of government of the Russian state had been in the reliable hands of the "great fellow countryman." It suggested that the time had come to publish the complete works of VVP in sixteen volumes as a treasury of wisdom and practical experience. The authors believed that the leaders of the Presidential Administration should issue a strict circular requiring the presence, in the offices of senior officials from the rank of general upward, of new symbols of statehood. In their view, to fill an official office with the spirit of patriotism, it was necessary to have a book of quotations, a complete collected works and a full-length ceremonial portrait of VVP, and for Orthodox believers also "The Gospel According to Vladimir." It was suggested that Vyacheslav Volodin and Bishop Tikhon (Shevkunov) would be able, within a short time, to prepare an agreed text of the Gospel and a prayer to VVP, which were intended to become the spiritual support of the people of Great Russia. The statement also expressed confidence that the Bishops' Council of the Russian Orthodox Church would officially approve the results of the work of the national leader's loyal comrades and that Russians would receive a priceless gift of faith and hope in difficult times.

In July 2015, information appeared that a government order had amended the plan of events devoted to the memory of Prince Vladimir. The total amount of expenditures was increased by 240 million rubles, from 528.9 million to 769.3 million. The new version added previously unaccounted expenditures for the installation of a monument to Prince Vladimir, 94 million rubles from extra-budgetary sources, and increased funding for a number of events.

From the standpoint of worldview neutrality, this is important not only as a matter of budgetary allocation, but also as an indicator of the direction in which public symbolism is being organized. When state resources, official memory policy and a particular religious-historical narrative begin to merge, the line between cultural heritage and confessional preference becomes dangerously thin. In legal terms, that line cannot be replaced by ceremonial enthusiasm.

In recent years, expert analysis has also pointed to the special role of presidential grants. After the reform of the grant-distribution system in 2013, the state began to allocate far more money to grants. Large grants for "Orthodox projects" were won by organizations either directly subordinate to the Moscow Patriarchate, such as the religious organization of the Department of Religious Education and Catechesis of the Russian Orthodox Church, or close to the Russian Orthodox Church, including the Centre for Religious Studies, Orthodox Youth, the Union of Orthodox Citizens and others.

As a rule, "Orthodox grants" indirectly or directly supported official state rhetoric. A grant of 2 million rubles was won by the Orthodox youth information agency "Faith and Deed" to form patriotic Orthodox bloggers on the basis of its resource. Eight million rubles were allocated for the educational readings "Prince Vladimir: The Civilizational Choice of Rus'." In total, in 2013–2015 at least sixty-three grants on Orthodox themes were issued for a total amount exceeding 256 million rubles, as emphasized in the report. The authors considered that this could be assessed as hidden state support for the Russian Orthodox Church and structures close to it.

Presidential grants to NGOs, as expert analysis has shown, became an instrument for supporting pro-government organizations and projects backed by the highest officials of the country. The Russian Orthodox Church of the Moscow Patriarchate turned out to be a significant beneficiary of the grant program. The sharp decline in funding from Western foundations, which began in the early 2000s, led to the underfunding of the third sector. Many NGOs, the authors of the study recalled, simply fell out of "international networks" where they had regularly received financial and informational support and were forced to reorient themselves toward interaction with the authorities.

The introduction of state support for such NGOs solved, on the one hand, the problem of crisis in the third sector. On the other hand, it simultaneously brought NGOs closer to the state and increased governmental control over the public sphere. Since 2013, the size of the fund for NGO needs has grown sharply, and competitions themselves began to be held several times per year, experts noted. The abrupt change in state policy was a direct consequence of the tense atmosphere in foreign and domestic policy. The state was forced to spend more money on NGOs because of the tightening of control over foreign funding of public organizations, primarily because of the law on foreign agents. But funds were mainly allocated to NGOs loyal to the authorities.

On 26 January 2016, Alexander Beglov, the Presidential Plenipotentiary Representative in the Central Federal District, speaking at a section meeting within the framework of the Fourth Christmas Parliamentary Meetings held by the Russian Orthodox Church of the Moscow Patriarchate in the Federation Council, reported on preparations for the celebration of the millennium of Russian monasticism on Mount Athos in Greece. In particular, Beglov stated that the main repair and restoration works at the facilities of the Russian St Panteleimon Monastery on Athos and other events were nearing completion. He said that the stage of preparation for the celebration of the millennium of Russian monasticism on Athos had practically been completed.

According to the plenipotentiary representative, this stage was connected primarily with repair and restoration works at the facilities of the Russian St Panteleimon Monastery, its sketes and cells, the construction of a church, the installation of a monument to St Silouan the Athonite in his native village in Lipetsk Region, and the development of comprehensive programs for jubilee celebrations at the international, all-Russian and regional levels.

At this point the problem is no longer limited to isolated ceremonial gestures. It concerns the broader model of state behavior in which public authorities, public budgets, symbolic policy, grants, educational initiatives and administrative resources increasingly converge around one dominant religious institution. In a secular and worldview-neutral state, such convergence is not a harmless tradition. It is a structural legal problem. It narrows the space of equal freedom of conscience and turns neutrality from a constitutional principle into a decorative phrase.

Problems in Observing the Neutrality of Federal Ministries, Federal Services and Federal Agencies

The unconstitutional position of ministries and agencies appears to be a natural consequence of the broader political line set "from above." In such a system, lower executive bodies rarely invent their own model of worldview policy. They reproduce the signal they receive from the centre, and that signal increasingly blurs the constitutional boundary between the state and selected religious institutions.

Serious questions arise, for example, in connection with the implementation of the Federal Target Programme "Strengthening the Unity of the Russian Nation and the Ethnocultural Development of the Peoples of Russia (2014-2020)," approved by Resolution No. 718 of the Government of the Russian Federation of 20 August 2013.

As early as November 2014, it became known that the Russian Orthodox Church of the Moscow Patriarchate would receive about 2 billion rubles from the federal budget for the creation of spiritual and educational centres. A total of 958 million rubles from the federal budget was to be allocated in 2015 for the creation of such centres. These funds were allocated under the federal target programme "Strengthening the Unity of the Russian Nation" and were intended for "non-profit organizations in the field of spiritual and educational activity." The original version of the programme, approved by the Government in 2013, provided for measures aimed at maintaining interethnic peace. Yet in March 2014 a new line appeared in the programme, providing funding for spiritual and educational non-profit organizations. In May, the first 1.1 billion rubles were distributed among twenty-three such organizations. According to the list of winners published by the Ministry of Regional Development, which distributed the funds, in all but one case approximately 25-30 million rubles were received by organizations founded by local Orthodox dioceses in the Tver, Saratov and Irkutsk regions, as well as in Dagestan, Mordovia, North Ossetia and other regions. The only exception was the Nizhny Novgorod autonomous non-profit organization Khramostroitel, whose founders were formally "citizens of Russia," although its director, Alexander Glazkin, was directly connected with the Russian Orthodox Church as a bishop of the Lyskov Diocese. All recipients of subsidies had been registered shortly before the competition, from December 2013 to February 2014, and the names of twenty-one of the twenty-three organizations contained the words "construction" or "restoration."

In January 2015, the head of the Russian Ministry of Culture reported that in the preceding year the ministry had allocated more than 6 billion rubles for the restoration of church monuments. Speaking at the Twenty-Third

International Christmas Readings in the Kremlin Palace, the minister stated that more than 6 billion rubles from the budget had been allocated for the restoration of 255 cultural heritage monuments in the use of the Russian Orthodox Church. According to him, active work was also being carried out to transfer objects formerly belonging to the Church, and allegedly “not always used by secular owners in the best way,” to the Moscow Patriarchate. As a particular achievement, he referred to the relocation of the Literary Museum from the ancient Vysokopetrovsky Monastery in central Moscow. He described the freeing of the premises as mutually beneficial: the museum received a new building, while the monastery would be seriously restored and revived.

At the end of 2015, it became known that the Ministry of Culture of the Russian Federation would participate in organizing festive events connected with the 1000th anniversary of the Russian presence on Mount Athos, to be celebrated in September 2016. The Government approved a plan of major celebrations for the anniversary of the Russian presence on the Holy Mountain of Athos. According to the published information, these events were to be held “for the purpose of restoring the cultural and spiritual heritage of the Russian St Panteleimon Monastery on Athos,” with the participation of both the Russian Orthodox Church of the Moscow Patriarchate and the Ministry of Culture. The plan included international academic conferences in Moscow, St Petersburg, Athens and the St Panteleimon Monastery, devoted to the Russian contribution to Athonite heritage. Educational, cultural and awareness-raising events were also planned in theological and secular schools, universities and institutes of the Russian Academy of Sciences. The organizers further intended to erect a monument to St Silouan the Athonite in the village of Shovskoye in the Lipetsk Region, where he was born, and to complete restoration of the church dedicated to him. The jubilee programme also included bringing Orthodox relics to the Russian Federation, organizing pilgrimages of abbots of Russian Orthodox monasteries to the Holy Mountain, and other events.

On 18 June 2015, an Agreement on Cooperation between the Russian Orthodox Church and the Ministry of Health of the Russian Federation was signed in the Patriarchal residence in Chisty Lane in Moscow. The document was signed by Patriarch Kirill of the Russian Orthodox Church of the Moscow Patriarchate and Minister of Health V. I. Skvortsova. According to the Patriarch, interaction between the Church and the Ministry of Health was actively developing because “more and more areas are being identified where the Church and health care could work together in order to treat people and support a healthy lifestyle.” The Minister of Health, in turn, stated that the interaction between the Ministry and the Church in recent years had helped implement “very ian citizens, including projects in palliative care, protection of motherhood and childhood, assistance to the homeless and support for people in difficult life situations.

At the same time, alongside the creation of conditions for the activity of the Russian Orthodox Church of the Moscow Patriarchate in medical organizations, including conditions for religious services, other rites and ceremonies, catechetical and educational activity, dissemination of religious literature and audio and video materials, and charitable work, the agreement also contained provisions that may be read as restricting competitors. In particular, the Ministry of Health and the Russian Orthodox Church were to work together against unlawful folk medicine and "healing," including the activity of "pseudo-religious structures" and "pseudo-scientific practices." From the point of view of worldview neutrality, the danger lies not only in cooperation itself, but in the fact that one religious institution receives a privileged interpretive role in relation to other forms of belief, healing practice and worldview life.

Problems in Observing the Neutrality of Executive Authorities in the Constituent Entities of the Russian Federation

Within the system commonly described as the "power vertical," the not entirely constitutional position of the highest officials becomes a model for appointed heads of constituent entities of the Russian Federation and for lower-level officials. Privileges granted to the so-called "main" confessions are inevitably accompanied by unlawful restrictions on other confessions, often labelled as "sects." Accordingly, in regions with a significant Muslim population, the preferences of the authorities extend to "traditional" representatives of Islam, while those regarded by the authorities as "non-traditional," often called "Wahhabis," become objects of pressure and persecution.

In Moscow, with state support, twenty-two churches and seventy-nine wooden chapels had been built under the programme for the construction of Orthodox churches, according to Vladimir Resin, State Duma deputy and curator of the project. He further stated that thirty more churches were under construction and that 148 land plots had been formalized for the Russian Church, four of them in the newly incorporated territories of Moscow.

In the summer of 2015, the Governor of the Moscow Region was approved as chair of the board of trustees for the restoration of the Joseph-Volokolamsk Monastery in the Volokolamsk District of the Moscow Region. Opening the meeting, Patriarch Kirill proposed electing the Governor to this position, noting that there would probably be no objections. The head of the Moscow Region stated that he would try to unite "all the healthiest forces" around the restoration and support of the monastery, adding that the work of the board of trustees had special significance in the year marking the 500th anniversary of the death of St Joseph of Volokolamsk.

In the same period, it became known that the authorities of St Petersburg would transfer the Annunciation Church, which at that time housed the Museum of Urban Sculpture, to the Alexander Nevsky Lavra of the Russian Orthodox Church of the Moscow Patriarchate. The museum exhibition was to be moved to the dilapidated Utkina Dacha on the Okhta River; under the law, this could take up to six years. In addition, the Russian Orthodox Church claimed the Church of St Nicholas, then housing the Museum of the Arctic and Antarctic, as well as several churches and a number of apartment buildings on Nevsky Prospekt. In the longer term, the museum complex known as St Isaac's Cathedral, including the Church of the Saviour on Spilled Blood, the Sampsonievsky Cathedral and the Smolny Cathedral, could also come under the control of the religious organization.

The Church of the Annunciation of the Most Holy Mother of God at 31-33 Fifth Sovetskaya Street was also to be transferred to the Russian Orthodox Church. The draft order prepared by the property relations committee provided that the transfer should be completed by July 2016. The archive of design and cartographic documentation was to be moved to Tambovskaya Street. The building, with an area of 2,500 square metres, housed the Central State Archive of Scientific and Technical Documentation of St Petersburg, which was to be relocated by June of the following year. In the same year, 28.7 million rubles were allocated for major repair and equipment of the new premises.

As early as April 2015, it became known that the Smolny Cathedral of St Petersburg, previously part of the State Museum-Monument "St Isaac's Cathedral," would be transferred to the Russian Orthodox Church of the Moscow Patriarchate. The director of the museum-monument told journalists that the museum was transferring the cathedral to the diocese because "such is the command of the law," although the deadlines did not suit the museum. The grand pianos, harpsichords and digital organ were to be removed by the end of January 2016. According to him, the instruments and the choir intended for concerts were to be placed temporarily in Petrikirche on Nevsky Prospekt, while a number of legal questions remained unresolved. Premises for the museum component of the Smolny Cathedral were being prepared on Dumskaya Street, in a building occupied by Sberbank. The preparation of the concert hall could take about three years, and the museum intended to request funding from the city budget for this purpose.

On 25 January 2015, a ceremony was held in the Smolny Cathedral for the transfer of the key to the western gates of the church to the Russian Orthodox Church. This eighteenth-century historical relic was handed over by Vice-Governor Alexander Govorunov to Archbishop Amvrosy of Peterhof, rector of the St Petersburg Theological Academy, and to Archpriest Pyotr Mukhin, the rector of the cathedral. The event took place after the liturgy on the day of St Tatiana, the heavenly patron of students. The archbishop also performed a minor consecration of the cathedral. At the ceremony, the vice-governor declared that after its consecration in 1835 the cathedral had become the church of all educational institutions, and that for the first time in modern Russian history people were congratulating one another on Student's Day "not in a museum, but in a church." In his words, the church had returned to its true purpose.

On 1 September 2015, Archimandrite Porfiry, abbot of the monastery and director of the local museum-reserve, stated during the restoration of the Solovetsky Monastery of the Transfiguration that it was planned to restore one hundred of its objects at a cost of 1 billion rubles per year. He emphasized that it was not a question of restoring all existing monuments, as this would be too expensive and difficult and, most importantly, not always justified. In his view, it was sufficient to restore the main hundred objects, and such a task could be solved only within the framework of a state programme, which was already being implemented. The level of annual allocations, he said, had reached 1 billion rubles, and he expressed hope that financial and economic conditions would make it possible to maintain such a scale in the coming years.

In Smolensk, by order of the head of the administration of the Promyshlenny District, measures were taken to organize a mass audience in connection with the service of Patriarch Kirill in the cathedral. Heads of enterprises, organizations and institutions of the district were invited to ensure attendance at the Patriarchal service on 30 August 2015. The notice requested that responsible persons be appointed for participation in the event and that the time and number of participants be clarified in advance with the district administration. This example is important not because of its ceremonial form, but because it illustrates the administrative organization of participation in a religious event. When officials begin to mobilize attendance at a religious service through the channels of public administration, neutrality becomes not a legal principle, but a decorative inscription on paper.

In 2015, a mosque named after Abdulhamid-afandi was opened in Makhachkala and presented by regional authorities as a source for spreading "traditional" Islam. The acting head of Dagestan, Ramazan Abdulatipov, stated at the opening ceremony that the republic, where many outstanding theologians had lived, had always been considered enlightened in religious terms, and that the opening of the mosque should become another source that would spread only good. He added that there were already more than one and a half thousand mosques in the republic and that their number would grow, but that the mosques should be filled with Muslims, not fanatics.

The President of Tatarstan, in turn, stated that Tatarstan's children should have an understanding of all "traditional" religions. In his annual address to the State Council of Tatarstan, he declared that the inhabitants of Tatarstan, and above all children, should have an idea of all traditional religions, as this was the basis for public harmony and trust. He also argued that the system of religious education should take maximum account of the ethnoconfessional composition of the population. At the same time, he warned that those who, under the cover of religion, sought to bring radical views and extremist actions into the republic should expect a tough response.

These examples allow one to conclude that, contrary to the Constitution of Russia, regional authorities are following a trajectory of violating neutrality and reproducing a dangerous federal religious policy built on confessional distinctions.

Problems in Observing the Neutrality of the Judiciary

Under Article 18 of the Constitution of the Russian Federation, human and civil rights and freedoms are directly applicable. They determine the meaning, content and application of laws, the activity of legislative and executive authorities and local self-government, and are ensured by justice. Article 19 declares that all are equal before the law and the court.

Under Article 118 of the Constitution, justice in the Russian Federation is administered only by courts. Judicial power is exercised through constitutional, civil, administrative and criminal proceedings. Articles 125-127 define the place of the higher courts in the judicial system and the position of the courts subordinate to them. Article 125 establishes the special status of the Constitutional Court of the Russian Federation as the judicial body called upon to interpret the Constitution and to review the constitutionality of laws and other legal acts. Although no courts are directly subordinate to the Constitutional Court, its decisions are significant for all judicial bodies and influence judicial practice as a whole. Article 126 provides that the Supreme Court of the Russian Federation is the highest judicial body in civil, criminal, administrative and other cases within the jurisdiction of courts of general jurisdiction; it exercises judicial supervision over their activity in the procedural forms established by federal law and provides clarifications on matters of judicial practice.

In practice, however, the Russian judicial system has also been seen issuing decisions that are difficult to describe as fully just and that appear to be substantially conditioned by the systemic clerical ideologization of state institutions.

There are grounds to believe that clerical ideologization may have influenced the decision of the Moscow Arbitration Court under which the rehabilitation centre for seriously ill children, Detstvo, was obliged to transfer one of its two buildings to a women's monastery subordinate to Patriarch Kirill of the Russian Orthodox Church of the Moscow Patriarchate. The medical centre, where seriously ill children from across the country receive treatment, lost more than half a hectare of land and, with it, one of its two buildings, including two departments, a clinic and a school.

Higher judicial instances ultimately left the complaint against the judgment of the court of first instance "without movement," thereby preserving the requirement to transfer the land plot and medical building to the Exaltation of the Cross Jerusalem Stauropegial Women's Monastery of the Russian Orthodox Church in the Moscow Region.

The influence of clerical ideologization may also be observed in numerous decisions based on the application of the Federal Law "On Countering Extremist Activity" and other "anti-extremist" legislation in relation to religious associations. Some examples are discussed below. No less revealing is the decision of the Supreme Court of the Russian Federation, which in spring 2015 refused to review the legally questionable sentence in the case of the punk group.

It may be assumed that problems in the neutrality of the judiciary have had a significant impact on the intensification of persecution both against opponents of violations of the principle of state neutrality and against religious associations independent of the authorities. These questions will be examined in greater detail below.

The State and Municipal Education System

In addition to the constitutional establishment of the principle of secularity, the Federal Law "On Freedom of Conscience and Religious Associations" of 26 September 1997 No. 125-FZ confirms that the state does not interfere in a citizen's determination of their attitude toward religion and religious affiliation, or in the upbringing of children by parents or persons replacing them in accordance with their beliefs and with due regard for the child's right to freedom of conscience and freedom of religion. The same law provides that the state ensures the secular character of education in state and municipal educational institutions. It also provides that the state regulates the granting of tax and other benefits to religious organizations and provides financial, material and other assistance to religious organizations in ensuring the teaching of general educational disciplines in educational organizations created by religious organizations in accordance with Russian legislation on education.

Article 5 of the same law states that everyone has the right to receive religious education of their choice, individually or together with others; that the upbringing and education of children are carried out by parents or persons replacing them with due regard for the child's right to freedom of conscience and religion; that religious organizations may create educational organizations in accordance with their statutes and Russian education legislation; and that, upon written request of parents or persons replacing them and with the consent of children studying in state or municipal educational organizations, such organizations may, on the basis of a decision of their collegiate governing body and by agreement with the founders, give a religious organization the opportunity to teach children religion outside the educational programme. Religious associations may also carry out religious teaching and religious upbringing of their followers in the manner established by Russian legislation and in forms determined by their internal rules. At the same time, the law directly states that religious teaching and religious upbringing are not educational activity.

The Federal Law "On Education in the Russian Federation" No. 273-FZ of 29 December 2012 also contains provisions establishing the principle of secularity of the state and municipal education system. Article 5 guarantees the right to education regardless of sex, race, nationality, language, origin, property, social or official status, place of residence, attitude toward religion, beliefs, membership in public associations and other circumstances.

In this context, the legal content of the secularity of education in state and municipal educational institutions in Russia presupposes several elements: ensuring constitutional freedom of conscience and religion; separation of worldview associations, whether religious, anti-religious or otherwise, from the state and municipal education system and their equality before the law; inadmissibility of establishing any worldview, religious, anti-religious or other, as mandatory or official in state and municipal educational institutions; prohibition on worldview and other public associations interfering in the administration of state or municipal educational institutions; prohibition on combining the educational process with compulsion to study a particular worldview, including religion, atheism or religious culture, or with compulsion to renounce a worldview; and inadmissibility of compulsory religious rites in state and municipal educational institutions, except during free time.

Yet in recent years, under pressure from a multi-level clerical lobby, the Ministry of Education and Science of the Russian Federation was forced to move toward an unlawful expansion of confessional-oriented teaching and the adoption of corresponding acts whose legality is highly questionable.

At the beginning of 2018, information appeared about Methodological Recommendations on Issues Related to the Resocialization of Adolescents Subjected to Destructive Psychological Influence by Supporters of Religious-Extremist and Terrorist Ideology, attached to a letter of the Ministry of Education and Science of 23 November 2017. Among other things, this document contained provisions that are highly debatable from a constitutional-law point of view and were based on legally undefined terms such as "extremism," "destructive psychological influence," "pseudo-religious groups" and similar expressions, as well as on unspecified "data from sociologists."

In winter 2018, state officials, teachers and representatives of the Russian Orthodox Church participated in the Twenty-Sixth Christmas Readings. As usual, participants, at public expense, demanded the expansion of the study of Orthodox culture in state schools. The Minister of Education and Science announced on 24 January, at the opening of the readings entitled "Moral Values and the Future of Humanity," that work on expanding and deepening the school course "Fundamentals of Religious Cultures and Secular Ethics" was continuing and that a working group was active.

In 2018, the Ministry of Education and Science was divided into the Ministry of Science and Higher Education and the Ministry of Education, but the above-mentioned tendencies of doubtful constitutionality continued.

It should be recalled that from 1 September 2012 the course "Fundamentals of Religious Cultures and Secular Ethics" was introduced in all general education schools in Russia. It is studied throughout the fourth grade at one hour per week and consists of six modules: "Fundamentals of Orthodox Culture," "Fundamentals of Islamic Culture," "Fundamentals of Jewish Culture," "Fundamentals of Buddhist Culture," "Fundamentals of World Religious Cultures," and "Fundamentals of Secular Ethics." Although schoolchildren and their parents formally have the right to choose one of the modules, in practice this rule is far from always implemented.

In autumn 2018, it became known that the Ministry of Education had allocated 40 million rubles to "increase the effectiveness" of the comprehensive course "Fundamentals of Religious Cultures and Secular Ethics." The ministry intended to analyse how this course was being implemented in the Russian Federation and to develop a concept for improving its effectiveness. At the same time, the task was set to form a "single positive information field" around the implementation of the educational module. Earlier, Patriarch Kirill had called for an increase in the volume of teaching of "Fundamentals of Religious Cultures" in schools, arguing that the subject was taught "for the sake of ticking a box," since one hour per week for one year could not provide serious knowledge.

In May 2019, the Ministry of Education approved a new Federal List of Textbooks recommended for use in general education organizations. In the sections devoted to the subject areas "Fundamentals of Religious Cultures and Secular Ethics" and "Fundamentals of the Spiritual and Moral Culture of the Peoples of Russia," changes were made, including the addition of new textbooks on Orthodox culture that created the possibility of teaching Orthodox culture at the level of basic general education, from the fifth to the ninth grades. The Synodal Department prepared recommendations on educational and methodological support for teaching subjects, courses and modules of Orthodox culture in connection with the new Federal List. These recommendations were intended for employees of diocesan departments of religious education and catechesis, church organizations and specialists interacting with state and municipal educational organizations on issues of educational and methodological support and the development of Orthodox culture teaching in schools.

At the beginning of 2019, it became known that deputies of the State Duma supported expanding the study of Orthodox culture. Participants in a round table in the State Duma entitled "Spiritual and Moral Education in the Russian School: Current State, Problems and Prospects for Development," held as part of the Seventh Christmas Parliamentary Meetings, concluded that the existing block of school spiritual and moral disciplines should consist predominantly of "Fundamentals of Orthodox Culture" and should be taught throughout school, from the second to the tenth grade. Regions where this was not happening were to be "taken note of," and priests of the Russian Orthodox Church were to participate in the education of schoolchildren in general education schools. One deputy stated that only 13 per cent of schoolchildren in rural dioceses and 20 per cent in urban dioceses had chosen "Fundamentals of Orthodox Culture," and that this was "very little." He proposed recommending that the Ministry of Education administratively oblige school leadership to expand the study of Orthodox culture.

Under the guise of cultural studies and similar labels, the clerical ideologization of the state education system continues. On 6 November 2018, at the Smolensk Regional Institute for the Development of Education, a solemn ceremony was held to award diplomas to graduates of professional retraining courses under the additional professional programme "Cultural Studies. Orthodox Culture." Twenty-three teachers of educational organizations in the Smolensk Region received diplomas granting them the right to professional activity in the field of cultural studies and spiritual and moral education. The ceremony emphasized the special significance of the programme for teachers and the relevance of Orthodox culture for the system of spiritual and moral education of children and youth in the region. The contribution of teachers from educational organizations of the Smolensk Metropolia, including the Smolensk Orthodox Theological Seminary and the Smolensk Interdiocesan Orthodox Theological School, was especially noted.

On 24 May 2019, at Moscow Pedagogical State University, advanced training courses were completed for heads and employees of diocesan departments of religious education, directors and teachers of Orthodox gymnasiums, and teachers of "Fundamentals of Orthodox Culture." The programme concerned the subject area "Fundamentals of the Spiritual and Moral Culture of the Peoples of Russia" in the modern school, including content, methodology and organization of teaching. It included classes on psychological and pedagogical aspects of modern childhood and on the teaching of Orthodox culture in continuity with the course on religious cultures and secular ethics. Participants were introduced to the Moscow experience of implementing the subject area, to the regulatory framework for teaching religious cultures, and to best practices of general education organizations and Orthodox schools. The courses were conducted by the Synodal Department of Religious Education and Catechesis and the Institute of Social and Humanitarian Education of Moscow Pedagogical State University under a cooperation agreement.

In May 2019, the Department of Religious Education and Catechesis of the Yekaterinburg Diocese completed a cycle of lectures for methodological associations of teachers of "Fundamentals of Religious Cultures and Secular Ethics" and "Fundamentals of the Spiritual and Moral Culture of the Peoples of Russia." Unlike previous years, when the programme covered all modules of the comprehensive course, the emphasis in that academic year was placed on teaching the module "Fundamentals of Orthodox Culture." Lecturers considered such topics as "The Church," "The Sacrament of Communion," "Orthodoxy on God's Judgment," and "Biblical and Gospel Parables." Teachers al ulture module and interactive forms of education and upbringing. The final meeting took place at the Novo-Tikhvin Women's Monastery, where teachers were given a tour of the monastery and told its history. During the 2018/19 academic year, nine seminars were held, attended by ninety teachers from Yekaterinburg and towns of the region.

At the end of 2018, a scandal occurred at School No. 43 in the Zavodskoy District of Saratov. On 14 December, the school was expected to receive the icon of the Mother of God "Deliverer from Troubles." According to available information, the school principal, Tamara Abramova, received a call from Alexander Blatman, deputy head of the district administration, who reprimanded her because the school had allegedly failed to notify the municipality about the planned event. After this, the principal was hospitalized in intensive care in the cardiology department of one of the city hospitals. The school had several Cossack classes, and its website reported children's participation in drill reviews, lessons of courage and "sabre-cutting competitions." In October, "following the traditions of spiritual and moral education," pupils of the Cossack classes attended a liturgy on the Feast of the Protection of the Mother of God. In that year, the school was to become the first point of stay in the Saratov Diocese for the icon "Deliverer from Troubles," normally kept in the church of the village of Tashla in the Samara Region.

In Kazan, local police requested information about Muslim schoolgirls wearing hijabs. A lawyer from Tatarstan wrote about this on Facebook, referring to an appeal from the Kazan Department of the Ministry of Internal Affairs. Information about the addressee of the document was hidden, but the lawyer stated that the police request had been sent to teachers in Kazan. Police asked teachers to provide "characterizing materials" on schoolgirls wearing hijabs and on their families. They also requested access to documents concerning conversations between school psychologists and the girls in order to determine which branch of Islam they followed. These requests were justified by "official necessity." The lawyer argued that by this request the Kazan police were attempting to collect personal data of children unlawfully, without parental consent, and were violating the constitutional right of citizens to freedom of conscience and religion.

The violation of the constitutional principle of secularity in the state education system is systemic in some state universities. For example, on 25 September 2018, an agreement on cooperation was signed in the meeting hall of the Academic Council of VGIK between the Moscow Theological Academy and the All-Russian State Institute of Cinematography. Representatives of the Academy included its rector, Archbishop Amvrosy of Vereya, and other clergy and religious media representatives; the Institute was represented by its rector and senior academic staff, including heads of animation and multimedia programmes. Such agreements are presented as cultural cooperation, but legally they raise the same question: where does cultural dialogue end, and where does privileged confessional influence within a state educational institution begin?

In September 2018, students of St Petersburg universities were, "by tradition," driven to participate in a procession of the Russian Orthodox Church of the Moscow Patriarchate held on 12 September in the centre of St Petersburg for the day of St Alexander Nevsky. Students of the St Petersburg State University of Technology and Design told journalists that they had been removed from classes and brought to the procession. Journalists also found students from the Herzen State Pedagogical University and the St Petersburg State University of Cinema and Television who had not come voluntarily. Media reports also indicated that residents of a student campus in St Petersburg were once again being compelled to attend the procession in a "voluntary-compulsory" manner.

It should also be recalled that in October 2015 the Presidium of the Higher Attestation Commission under the Ministry of Education and Science approved the passport of the "academic" specialty "theology," and in May 2017 the first dissertation in this area was defended before a joint dissertation council amid protests from the scientific community. In summer 2018, the master's programme of the Moscow Theological Academy of the Russian Orthodox Church received state accreditation. The certificate of accreditation for the master's programme in Theology was ceremonially handed to the rector of the Academy by the head of Rosobrnadzor, Sergey Kravtsov, after the order on reissuing the accreditation certificate for the Moscow Theological Academy had been signed.

For many years, state universities in Russian regions, including the Altai Territory, Omsk, Tver and Ivanovo regions, maintained departments and faculties of theology based on local dioceses of the Russian Orthodox Church. This contradicts the Constitution of the Russian Federation and the Federal Law "On Education in the Russian Federation," which provide for the neutral character of the educational process in state educational structures. By 2018, more than fifty universities in Russia had either departments of theology or theological educational programmes of various levels, according to Metropolitan Hilarion, rector of the Church-wide Postgraduate and Doctoral School. At the conference "Theology and Progress," held in January 2018 under his chairmanship, a proposal was made to establish a scientific and educational theological association. The initiative was presented by the rector of the National Research Nuclear University MEPhI, who stated that values inseparably connected with religion provide a powerful instrument for educating youth and that the worldview basis laid by theology serves as an "anchor principle" in training technical specialists.

From 2018 to 2026, the teaching of confessional-oriented disciplines in Russia moved from an optional experiment to a tightly regulated part of the educational standard. Several stages and features of this process can be distinguished.

From 2018 to 2022, the course "Fundamentals of Religious Cultures and Secular Ethics" in the fourth grade was mandatory, but the principle that parents could choose one of six modules formally remained. The most popular modules continued to be "Secular Ethics" and "World Religious Cultures," although the share of "Fundamentals of Orthodox Culture" steadily grew in the central regions. The subject "Fundamentals of the Spiritual and Moral Culture of the Peoples of Russia" in grades five to nine existed formally, since schools implemented it in different ways: some introduced a separate lesson, while others included the topics in history or social studies courses.

From 1 September 2023, changes came into force that made "Fundamentals of the Spiritual and Moral Culture of the Peoples of Russia" a full mandatory subject in middle school, grades five to nine. Unlike the earlier course on religious cultures and secular ethics, this subject does not divide pupils into confessional groups. Pupils study common values, cultural monuments and ethical norms of all the peoples of Russia in one classroom.

By 2026, the system of confessional-oriented education had changed further. From 1 September 2026, a new subject, "Spiritual and Moral Culture of Russia," is introduced to replace the previous course. It will be studied in grades five to seven. Its teaching is linked to Presidential Decree No. 809, "On the Approval of the Foundations of State Policy for the Preservation and Strengthening of Traditional Russian Spiritual and Moral Values," where the emphasis is placed on traditional values, including patriotism and civic identity. By 2025, unified federal work programmes and digital materials had been prepared, which reduced the subjectivity of teaching.

Thus, while in 2018 the formal emphasis was placed on freedom of choice between religious and secular modules, by 2026 the focus had shifted toward forming civic unity through the prism of traditional values. The study of religions is now presented as a way of understanding the history and culture of multinational Russia.

The unconstitutional interaction between the Russian Orthodox Church of the Moscow Patriarchate and security-related educational institutions deserves separate mention and will be discussed in more detail below.

As a result, violations of the principles of worldview neutrality and freedom of conscience in the state and municipal education system are directed toward a worldview split. They are unlawful, grossly violate the Constitution of Russia, and ultimately create threats to Russia's national security. They have also already led to a decline in the quality of education and to violations of the rights of parents and children to upbringing and education in accordance with their own convictions.

The Armed Forces and Other Security Formations of the State

The study has revealed violations of state neutrality and freedom of conscience in the Armed Forces and other security structures of the state. It should be noted that, in addition to the Armed Forces, the Russian Federation has no fewer than twenty state agencies that, on entirely lawful grounds, possess armed formations.

In the Armed Forces, other troops and military formations, freedom of conscience must be exercised in accordance with the Constitution of the Russian Federation, the Federal Law "On Freedom of Conscience and Religious Associations," and the Federal Law "On the Status of Servicemen."

In practice, however, for many years, in violation of the Russian Constitution and current legislation, and with the encouragement of the authorities, the confessions serving them and interested actors, places of worship have been opened and operated in the Armed Forces and other security formations of the Russian Federation under conditions of military discipline. Religious rites and ceremonies are conducted there, and awards of the Russian Orthodox Church of the Moscow Patriarchate are presented to employees of the state's security structures.

On 25 January 2018, the Military Academy of the General Staff of the Armed Forces of the Russian Federation hosted the scientific and practical conference "The Spiritual Foundations of Military Service to the Fatherland: Moral Values and Patriotic Education," as well as a meeting of the Military Section of the Twenty-Sixth International Christmas Educational Readings. Representatives of branches and arms of the Armed Forces of the Russian Federation, military academies and schools located in Moscow and the Moscow Region, the Main Directorate for Work with Personnel of the Armed Forces of the Russian Federation and other central bodies of military administration, as well as full-time assistants to commanders for work with believing servicemen, took part in the Military Section.

Following the work of the section, a resolution was adopted. It stated that the Military Section of the Twenty-Sixth International Christmas Educational Readings considered moral values and patriotic education to be priority areas in forming the spiritual foundations of service to the Fatherland. It also emphasized that in wars and military conflicts the religious factor is increasingly exploited and used to radicalize the actions of opposing sides, both in cases of external aggression and in efforts to destabilize the internal situation of the state. The resolution expressed concern about the spread of neo-paganism in certain military formations and in the cultural and information space. It stated that intensified work in the sphere of the spiritual enlightenment of servicemen, aimed at preventing and countering the spread of various destructive and radical movements, including Islamic and neo-pagan movements, remains a relevant task of pastoral service in the troops.

The resolution further described the ministry of military clergy at the current stage of development of the Armed Forces as an integral component of the educational process of servicemen, based on the traditional values of Orthodoxy. It presented the personality of the military priest as a key element of pastoral service in the troops. In that connection, it called for continued work to improve the qualifications of clergy providing pastoral care to servicemen, both on a full-time and non-staff basis; for the regular and prompt transmission of pastoral experience in military service; for continued interaction between modern church media and communications and the media of the Ministry of Defence of the Russian Federation in forming moral values among servicemen; for the development of spiritual and educational work using modern technical means; and for further promotion of the websites Pobeda.ru and Kapellan.ru, as well as the Internet radio station Pobeda of the Synodal Department, oriented toward the spiritual and moral education of servicemen.

The same resolution stated that the arrangement of military churches, chapels, prayer rooms and libraries of Orthodox literature should be regarded as one of the most significant components of pastoral activity in the troops. At the same time, the organization of worship services in military churches was named a priority task for the command staff of military units and formations, assistants to commanders for work with believing servicemen, and heads of diocesan military departments. Participants in the Military Section petitioned the hierarchy of the Russian Orthodox Church and the leadership of the Armed Forces of the Russian Federation to support all the initiatives listed above. They also expressed confidence that the persistent continuation and development of work on the moral and patriotic education of servicemen, and on forming a Christian worldview among soldiers, would have a significant influence on the formation of cohesive military collectives ready to perform complex training and combat-training tasks.

At the beginning of 2019, an appeal to the President of the Russian Federation was published on Change.org, demanding an end to the unconstitutional practice of introducing obscurantism into the Armed Forces of the Russian Federation. The appeal stated, in particular, that the construction of the Main Cathedral of the Armed Forces of the Russian Federation, called the "Main" cathedral, was a short-sighted state initiative because it created preconditions for the strengthening and aggravation of hidden inter-confessional and interethnic struggle, capable of manifesting itself in dangerous forms of national and religious hostility and thereby violating Article 13 of the Constitution of the Russian Federation. The authors of the appeal considered it appropriate to reconsider the emerging practice and its legislative regulation in light of contemporary realities and the need to counter interethnic and inter-confessional conflicts, in order to create real conditions for the recognition and guarantee of human and civil rights and freedoms in accordance with generally recognized principles and norms of international law and the Constitution of the Russian Federation.

It should be recalled that the decision to introduce the institution of military priests was adopted in 2009, contrary to the Russian Constitution. At that time, President D. A. Medvedev supported the proposal of leaders of religious associations loyal to the authorities to re-establish the institution of military clergy. The Regulation on the Organization of Work with Believing Servicemen of the Armed Forces of the Russian Federation was approved by the Minister of Defence of the Russian Federation on 24 January 2010. This Regulation establishes the procedure for interaction between military administration bodies and religious associations in relation to the realization by believing servicemen of their rights to freedom of conscience and freedom of religion.

In 2018, the first training and methodological gathering was held for clergy providing pastoral care to employees of the National Guard in Moscow and the Moscow Region, on the basis of the Synodal Department for Cooperation with the Armed Forces and Law Enforcement Agencies. The purpose of the gathering was to study new approaches to organizing work with believing employees of the Federal Service of National Guard Troops of the Russian Federation, as well as the forms and methods of military clergy in the spiritual education of National Guard servicemen.

Confessionally oriented events and religious ceremonies of the Russian Orthodox Church of the Moscow Patriarchate, whose constitutionality is highly doubtful, regularly take place in leading military universities: the Military Academy of the General Staff and the Military University of the Ministry of Defence of the Russian Federation. In addition, priests of the Russian Orthodox Church of the Moscow Patriarchate regularly undertake professional development at the Military University of the Ministry of Defence of the Russian Federation.

On 21 June 2018, a chapel of the Great Martyr George the Victorious was consecrated at the Military University of the Ministry of Defence of the Russian Federation in Moscow. Since 2014, a house church in honour of the Holy Right-Believing Prince Daniel of Moscow has operated at the same university.

It should also be recalled that since 2017 there has been a Cooperation Agreement between the Russian Orthodox Church of the Moscow Patriarchate and the Ministry of the Russian Federation for Civil Defence, Emergencies and Elimination of Consequences of Natural Disasters. Since 1 September 2017, the course "Fundamentals of Orthodox Culture" has been introduced in all educational institutions of the Russian Emergencies Ministry system.

From 2018 to 2026, interaction between the Russian Orthodox Church of the Moscow Patriarchate and the security structures of the Russian Federation moved from the format of spiritual care to systemic integration into educational and upbringing processes. The status of full-time military priests, or assistants to commanders for work with believers, was finally consolidated in departmental universities of the Ministry of Defence, the National Guard and the Ministry of Internal Affairs. The Main Cathedral of the Armed Forces, opened in 2020, became an ideological and educational centre. On its premises and in the adjacent "Road of Memory" complex, mandatory excursions and solemn oaths are held for cadets and military pupils.

Courses overlapping with the subject matter of "Fundamentals of the Spiritual and Moral Culture of the Peoples of Russia" were introduced into the curricula of security-sector universities, but with an emphasis on the "protection of traditional values" and "spiritual security." Particular attention is paid to the history of holy warriors such as Alexander Nevsky, Dmitry Donskoy and Fyodor Ushakov. Framework agreements between the Patriarch and the leadership of ministries, including the Ministry of Defence, the Ministry of Internal Affairs and the Ministry of Emergency Situations, were also renewed. They imply the participation of priests in certification commissions, professional development courses for educational officers and the prevention of destructive behaviour.

After 2022, the emphasis in Russian security-sector universities shifted toward the concept of sacred duty, understood as the highest moral obligation of the individual to serve the Fatherland, based on patriotism, honour and loyalty to the oath. This task is implemented through the study of history, rituals, problem-based learning and the personal example of mentors. The religious component has become an integral part of military-patriotic education aimed at forming readiness for self-sacrifice. Religion provides a value foundation by presenting the defence of the Fatherland as a sacred duty and strengthening readiness for self-sacrifice through ideas of the spiritual meaning of service and the immortality of the soul. In contemporary practice, this is implemented through cooperation between the army and religious organizations, the activity of military priests and symbolic practices, such as assigning the names of saints to units, which are intended to strengthen the morale of servicemen.

In my view, these trends in the "interaction" between "traditional religious associations" and the security ministries contribute to the creation and reproduction of an atmosphere of xenophobia, intolerance and fragmentation on worldview grounds. In the end, they lead to persecution both within the collectives of security structures and at the state level. These same trends also obstruct the establishment of tolerance, the democratization of military and law-enforcement collectives, the exercise of civilian control and, consequently, the construction of a law-governed state and the formation of civil society in Russia.

Places of Detention and Imprisonment

The right of persons held in custody and persons serving criminal sentences in the form of imprisonment to freedom of belief, conscience and religion is enshrined in the Constitution of the Russian Federation, the Federal Law "On Freedom of Conscience and Religious Associations," the Penal Enforcement Code of the Russian Federation, the Federal Law "On the Detention of Suspects and Accused Persons," the Internal Regulations of Correctional Institutions, the Rules for Serving Criminal Sentences by Convicted Servicemen and other regulatory legal acts.

In accordance with Article 14 of the Penal Enforcement Code of the Russian Federation, convicted persons are guaranteed freedom of conscience and freedom of religion. They have the right to profess any religion or to profess none, and freely to choose, hold and disseminate religious beliefs and act in accordance with them.

The exercise of the right to freedom of conscience and freedom of religion is voluntary. At the same time, it must not violate the internal regulations of the institution executing the sentence or infringe the rights of other persons.

At the request of persons sentenced to arrest or imprisonment, clergy belonging to duly registered religious associations are invited, according to the choice of the convicted persons. In institutions executing sentences, convicted persons are permitted to perform religious rites and to use religious objects and religious literature. For these purposes, the administration of the relevant institution allocates appropriate premises.

Seriously ill convicted persons, as well as persons sentenced to death before the execution of the sentence, must, at their request, be provided with the opportunity to perform all necessary religious rites with the invitation of clergy.

Under Federal Law No. 103-FZ of 20 April 2015 "On Amendments to Article 14 of the Penal Enforcement Code of the Russian Federation and Certain Legislative Acts of the Russian Federation," personal meetings between convicted persons and clergy were permitted, including private meetings. However, contrary to the Constitution of the Russian Federation, the provisions of this law are directed toward developing cooperation within the penal enforcement system only with "traditional confessions," which creates prerequisites for discrimination and violations of the rights of all those who do not fit within that concept.

As of the end of 2018, institutions of the penal enforcement system had 705 facilities used for religious rites and ceremonies, including 624 buildings and structures allocated for persons professing Orthodoxy, 62 for persons professing Islam, 11 for persons professing Buddhism, 3 for persons professing Catholicism and others. There were also more than 670 premises provided to suspects, accused persons and convicted persons for religious rites and ceremonies and for the use of religious objects and literature, including 345 premises for persons professing Orthodoxy, 274 for persons professing Islam, 15 for persons professing Judaism and others. According to updated 2024 data from the Federal Penitentiary Service of Russia, 1,508 facilities, including buildings, structures and premises, were functioning in the penal correctional system for the conduct of religious rites and ceremonies.

However, the closest and most constitutionally questionable relations within the Federal Penitentiary Service of Russia have developed specifically with the Russian Orthodox Church of the Moscow Patriarchate. In particular, the new Cooperation Agreement between the Federal Penitentiary Service of Russia and the Russian Orthodox Church of the Moscow Patriarchate, signed on 22 February 2011, and preceded by an earlier agreement dating back to 1999, states that it is concluded in order to satisfy the spiritual needs of employees of the penal enforcement system and members of their families, cadets of educational institutions of the Federal Penitentiary Service of Russia, and to ensure the constitutional rights to freedom of conscience and freedom of religion of persons sentenced to imprisonment and persons held in custody.

Among other things, this agreement defines obligations of the Federal Penitentiary Service of Russia that hardly require commentary: to assist in the construction of Orthodox churches and chapels and to allocate premises for prayer rooms, based on existing possibilities in institutions of the penal enforcement system; to support the distribution of printed materials, including newspapers, journals and books of spiritual, moral and patriotic content, religious literature and religious objects in institutions of the penal enforcement system, in accordance with established rules and restrictions; and, based on existing possibilities and with due regard for the specific regime of institutions, to assist in organizing the development of religious education for persons sentenced to imprisonment and persons held in custody, and to facilitate the realization of their rights to freedom of conscience and religion with the participation of clergy of the Russian Orthodox Church.

By 2026, cooperation between the Federal Penitentiary Service and the Russian Orthodox Church of the Moscow Patriarchate can be described as systemic integration. Full-time positions for clergy have appeared in penal institutions, educational projects have been launched, and interdepartmental cooperation has been strengthened. At the same time, the balance between spiritual care and respect for the rights of convicted persons to freedom of conscience depends on the development of interreligious councils under the Federal Penitentiary Service. Some opportunities for systemic work are also received by representatives of Islamic organizations loyal to the authorities and by certain other confessions.

Against the background of rapprochement with confessions close to the authorities, however, control over "non-systemic" believers has intensified through mechanisms of preventive registration. Order No. 17 of the Ministry of Justice of Russia of 26 January 2026 approved a new Procedure for the Preventive Registration of Suspects, Accused Persons and Convicted Persons in Correctional Institutions and Pre-Trial Detention Facilities. In particular, Muslims who do not belong to spiritual administrations loyal to the authorities, and whose religious practice is treated by the administration as suspicious, fall under special control. In practice, this results in the seizure of literature, prohibitions on rites and regular disciplinary penalties, which human rights defenders describe as discrimination on the basis of belief. Moreover, being placed on such a register limits a convicted person's chances of parole and mitigation of regime, creating an instrument of pressure on "non-systemic" believers. Thus the system clearly separates state-approved "traditional" believers from risk groups subject to harsh isolation and supervision.

Media and Mass Communications

Serious violations have also been identified in the sphere of media and mass communications. In recent years, this sphere in Russia has undergone a profound transformation characterized by intensified ideological control and the use of hate speech. Federal television channels and state media holdings have regularly produced content in which any religious movements not traditional for the authorities are branded as "destructive sects." Associations such as "Islamic extremism" or "religious terrorism" have been introduced into the information field and consolidated there, shifting responsibility from specific offenders onto entire worldview groups.

Religious minorities, especially those with foreign centres, are presented in the media as "agents of influence," which then justifies harsh administrative measures and searches conducted under the language of "spiritual security." Reliable information about the activity of religious associations is replaced by propaganda clichés, while representatives of discriminated groups are almost deprived of the opportunity publicly to refute accusations or defend their reputation.

Such information policy has provoked an increase in xenophobia and everyday violence, since aggressive rhetoric in the media legitimizes intolerance in the eyes of ordinary citizens. The screen does not merely describe hostility; it trains it, repeats it, normalizes it and then pretends to be surprised when society learns the lesson. Apparently, even prejudice now needs a broadcast schedule.

In my view, the growth of violence based on intolerance in society follows primarily from a state policy saturated with xenophobia, including its information component.

6.2. The Problem of Violations of the Right to Freedom of Conscience in Relation to Religious Associations and Their Members

In recent years, numerous violations of the internationally recognized right to freedom of conscience have been identified in relation to religious associations and their members. These violations are connected, above all, with breaches of the constitutional principle of the secular state. An analysis of the situation of specific religious currents shows, first of all, systematic unlawful persecution of Muslims, alternative Orthodox churches, and a growing unlawful pressure by the authorities on believers belonging to Protestant communities, pagans and a number of new religious movements.

Among the typical problems faced by religious associations and individual believers, the following should be noted: unlawful liquidation or attempted liquidation and problems with registration; unlawful restrictions on the distribution of texts, printed publications and online materials; difficulties in acquiring premises and land for the construction of places of worship; unlawful obstacles to holding public events; unlawful restrictions on the activity of religious associations whose leaders are foreign citizens; unlawful interference by state authorities and local self-government bodies in the activity of religious associations; and the dissemination of false and defamatory information about the activity of religious associations. Some religious organizations have also faced the blocking of their websites.

Among the numerous negative consequences of violations of the constitutional secularity of the state, particular attention should be paid to violence and vandalism motivated by xenophobia and intolerance on the grounds of religion or belief.

At the law-enforcement level, the unprecedented intensification of persecution by state security structures, as well as administrative pressure against religious associations and their members, is especially striking.

In the summer of 2018, the Presidential Council of the Russian Federation for the Development of Civil Society and Human Rights expressed concern about the persecution of believers. This was stated in a letter from the then Head of the Council, Mikhail Fedotov. Fedotov reported that the situation surrounding law-enforcement practice in criminal cases concerning extremism had been discussed on 19 June at a meeting of the Council's Presidium. As a result of that meeting, it was decided to hold a joint meeting with the Presidential Council for Cooperation with Religious Organizations. In addition, on 9 June, the Human Rights Council appealed to Prosecutor General Yuri Chaika with a request to verify the legality and justification of criminal prosecution of believers. Earlier, representatives of human rights organizations had issued a statement in defence of believers, signed by about one hundred people. The statement was sent to Mikhail Fedotov and to the Commissioner for Human Rights in the Russian Federation, Tatiana Moskalkova.

In the spring of 2018, human rights defenders in Saint Petersburg issued a statement condemning the persecution of citizens on religious grounds and calling for the repeal of the latest amendments to the article on "insulting the feelings of believers" (Article 148 of the Criminal Code of the Russian Federation). The statement, published on the official website of the Legislative Assembly of Saint Petersburg, noted that after the adoption of amendments to criminal legislation, known as the "Yarovaya package," prosecutions of citizens on the basis of their religious self-identification had significantly intensified. In various religious communities, unjustified searches were being carried out, religious literature was being seized, buildings were being confiscated and then destroyed, and in a number of cities decisions on permits for construction or transfer of religious-purpose real estate had been delayed for years.

At the end of 2018, during a meeting of the Human Rights Council, President Vladimir Putin expressed surprise at the persecution of believers, calling it "complete nonsense," and promised to study the issue together with the Chairman of the Supreme Court, V. Lebedev.

The Russian judicial system in cases connected with the realization of the right to freedom of conscience, like other state protection mechanisms, to put it mildly, has not had a positive influence on the situation in this sphere.

As in previous years, in March 2018 the Constitutional Court of the Russian Federation issued another decision that can hardly be called fully constitutional. It concerned Sergey Stepanov's complaint against Part 4 of Article 5.26 of the Code of Administrative Offences of the Russian Federation, which deals with missionary activity carried out in violation of legislation on freedom of conscience, freedom of religion and religious associations. Stepanov had been fined 10,000 roubles; the Tambov District Court of the Tambov Region upheld the ruling, after which he appealed to the Constitutional Court. In his complaint, he challenged the constitutionality of the provisions of the Law of the Russian Federation "On Freedom of Conscience and Religious Associations" that define missionary activity and determine the persons entitled to carry out missionary activity on behalf of a religious organization, as well as Part 4 of Article 5.26 of the Code of Administrative Offences. Under that provision, missionary activity carried out in violation of legal requirements entails an administrative fine for citizens from five thousand to fifty thousand roubles, and for legal entities from one hundred thousand to one million roubles. Both provisions were introduced into legislation in 2016 as part of the so-called "Yarovaya package." The Constitutional Court refused to accept Stepanov's complaint, finding that the provisions he listed could not be regarded as violating his constitutional rights. Moreover, despite the absence of clear legal definitions, the Constitutional Court called for a distinction between "missionary activity" and "confessional activity," although this distinction was formulated in a legally problematic way. Since the Constitutional Court of the Russian Federation traditionally avoids giving a proper assessment of the unconstitutional Federal Law "On Freedom of Conscience and Religious

Associations," this largely predetermines the glaring ineffectiveness of the entire Russian judicial system in protecting freedom of conscience.

It may also be concluded that court decisions based on the application of "anti-extremist" and "anti-terrorist" legislation demonstrate the transformation of the judicial system into an instrument for violating freedom of conscience and for unlawful persecution.

Among the important results, it should be noted that after the adoption of the "Yarovaya amendments" in the summer of 2016, hundreds of cases had been initiated by the end of 2018 against individuals and legal entities under Article 5.26 of the Code of Administrative Offences for improper "missionary" activity. Administrative liability was imposed mainly on Pentecostals, Baptists, Adventists, Muslims, Catholics, Krishna devotees, Buddhists and others. In the first six months of 2018 alone, Russian courts considered 234 cases of "illegal missionary activity" and held 147 persons liable, in most cases imposing fines, according to a published report of the Judicial Department. The total amount of fines imposed by courts during those six months amounted to 2,471,000 roubles.

By the end of 2018, the number of cases concerning "illegal missionary activity" had risen to 419, and the number of persons held liable reached 250 (172 individuals and 78 legal entities). The total amount of fines reached 3,700,000 roubles. Written warnings were issued five times, compulsory work was imposed twice, confiscation was applied five times, and expulsion from the country was ordered twice.

It should be recalled that prosecutors and law-enforcement bodies began identifying "illegal missionaries" after the State Duma adopted the "Yarovaya amendments" in July 2016. Representatives of religious groups not connected with the Russian Orthodox Church of the Moscow Patriarchate are treated as "illegal missionaries." They are punished for children's holidays, for giving away books and even for gathering fellow believers in someone's home. The legal machine, once set in motion, rarely pauses to ask whether it is protecting public order or simply chasing people for prayer, tea and paperbacks.

In recent years, the situation concerning freedom of conscience in Russia has been characterized by a transition from isolated restrictions to the systematic suppression of any religious groups not integrated into state ideology. The main blow has fallen on Muslims outside official structures, Protestants, pagans and alternative Orthodoxy, which have faced mass liquidation of legal entities and refusals of registration. The authorities systematically restrict communities in their rights to land and premises, block religious websites and seize literature, thereby effectively pushing "non-systemic" faith underground. The activity of organizations whose leaders are foreigners or have foreign links is subjected to severe filtering, and their work is often qualified as "interference in internal affairs." Increased control by the security services has led to a rise in criminal cases under "extremist" articles for peaceful religious practice, prayers or the reading of texts. The erosion of the constitutional principle of secularity, combined with state support for "official" confessions and media stigmatization of minorities, has provoked an increase in vandalism and physical violence motivated by intolerance.

The activity of the Supreme Court of Russia in 2020–2026 became the final stage in the formation of a rigid law-enforcement practice aimed at the liquidation of "undesirable" religious organizations. The highest judicial instance effectively legitimized the use of vague formulations of "anti-extremist" legislation, routinely rejecting complaints by Muslim and Protestant communities against decisions of lower courts. This position of the Supreme Court made it impossible for hundreds of believers to seek justice inside the country, creating a legal dead end. It led to a sharp increase in appeals to international structures, within the remaining available UN mechanisms and intergovernmental committees, which have become the only instruments for recording human rights violations. As a result, the Supreme Court turned from a body of legal protection into an instrument of state control, finally consolidating the priority of ideological expediency over freedom of conscience.

6.3. Problems in the Realization of the Right to Freedom of Conscience in Relation to Supporters of the Worldview Neutrality of the State

In recent years, numerous protests have continued in various regions of Russia in connection with violations of the principle of state secularity and the resulting violations of the lawful rights and interests of citizens. In particular, protests continued against the strengthening of links between religion and the state; state financing of the activity of religious associations; the teaching of religion within the state education system; the cancellation of cultural events under the pretext of protecting the "feelings of believers"; the establishment of clerical censorship in the media; the transfer to the Russian Orthodox Church of the Moscow Patriarchate of dozens of green-zone plots that, by law, could not be removed from public use; the unlawful construction of churches of the Russian Orthodox Church of the Moscow Patriarchate on natural complexes; and the unlawful transfer of cultural objects to that Church.

For example, on 11 January 2019, an appeal was prepared and sent through the online reception office to the Administration of the President of the Russian Federation. It concerned the inadmissibility, unconstitutionality and illegality of imposing a religious worldview and religious consciousness on servicemen of the Armed Forces of the Russian Federation, as well as the illegality and inappropriateness of constructing a cathedral of the Armed Forces of the Russian Federation in Patriot Park precisely as an Orthodox church dedicated to the Armed Forces. In addition to legal arguments based on the Constitution of the Russian Federation and the Federal Law "On Freedom of Conscience and Religious Associations," concerning the separation of state activity and religious organizations, the appeal emphasized that the Armed Forces of the Russian Federation, like the Russian people as a whole, include not only Orthodox Christian believers but also members of other confessions and people who do not consider themselves to belong to any religious practice. Moreover, the construction of a "military" cathedral was financed not only by donations from private persons who identify themselves as Orthodox believers, but also from accounts of legal entities operating at the expense of the state budget. In such a case, the act may fall under Article 285.1, "Misuse of Budgetary Funds." Part 2 of that article establishes liability in cases of collusion and amounts exceeding one and a half million roubles. This was presumably known to the military prosecutor's office, from whose accounts, in several districts, funds were also transferred for the construction of the cathedral. The addressees and participants of the appeal were also asked to note that the matter involved not only signs of a criminal offence and manipulation of budget funds, but also coercion into "donations" of private individuals, Armed Forces servicemen and civilian employees dependent on the Armed Forces leadership, who, according to some reports, did not receive bonuses because they had been written off for the construction of the military cathedral.

On 17 February 2019, a protest rally against the construction of a church of the Russian Orthodox Church of the Moscow Patriarchate in Svyatoslav Fyodorov Park was held in the Beskudnikovo district of Moscow. The park, opened in October 2018, had 175 larches and oaks and about six thousand shrubs planted in it. Yet in January 2019, the district administration website published a notice about the construction of a church of the Russian Orthodox Church of the Moscow Patriarchate at that site, already the seventh in the district. Opponents of construction feared that the laying of utilities for the church complex would damage the landscaped park area. However, the Moscow Patriarchate managed to mobilize its supporters, who came to the rally in greater numbers than the opponents of park development.

In April 2019, residents of the Moscow districts of Perovo and Novogireevo protested in Afghan Park against the construction of an Orthodox church on park territory. A small green area on the border of the two districts, one of the few places transferred for a "church complex" of the Russian Orthodox Church. The plans included a 21-metre-high church building for 500 people under the "200 Churches" programme, a Sunday school, a bell tower and a conference hall, as well as the felling of more than 30 trees and 160 shrubs. Sanitary felling was also envisaged over an area of more than four hectares. A cable had already begun to be laid for the construction of the church, although project documentation had not yet been obtained. The project had not passed expert review, but a construction permit had somehow been issued. Despite these legal violations, the alienation of part of the park territory took place through a simplified procedure, since the park, like many parks in Russia, was not registered in the cadastral records.

At the beginning of 2019, it became known that about one hundred and fifty new Orthodox churches would be opened in the Russian capital in the foreseeable future, according to a report by Vladimir Resin, State Duma deputy and curator of the "Programme 200." According to him, construction was being carried out on plots selected and approved by the Russian Orthodox Church: 232 complexes in total, including 192 church complexes in Old Moscow, 18 in New Moscow and 22 additional church complexes beyond the programme. Since 2010, when the programme for the construction of new churches in Moscow began, 85 churches had been built in the capital: 67 under the programme and 18 beyond it. In total, 61 churches had been put into operation, 12 complexes were being prepared for commissioning and construction and installation work had been completed on 12 sites. In 2019, 29 churches were to move to various stages of construction, 25 under the programme and 4 outside it. Sixteen programme churches were being prepared for construction, and another 32 were under design according to the approved schedule. Beyond the approved plan, at the initiative of parish rectors, work was also being conducted on designing another 21 church complexes. Resin noted that the programme had set a very high bar in terms of the pace of implementation, the system of control, the formation of communities and the selection of plots. "Programme 200" was one of Patriarch Kirill's first initiatives after his election as Primate of the Russian Church. The stated reason for building new churches in the capital was that Moscow allegedly had the smallest number of churches per density of Orthodox population among Russian regions. Resin explained that the construction of new churches in Moscow was planned on the basis of one church per 20,000 residents, with a maximum walking distance of about one kilometre.

Protests against the development of park zones by churches were noted not only in Moscow but also in various constituent entities of the Russian Federation.

Thus, on 4 January 2018, a protest against the development of park zones was held on Nevsky Prospekt in Saint Petersburg. About 40 activists took part, including opponents of the construction of a church of the Russian Orthodox Church of the Moscow Patriarchate in Malinovka Park. The protesters made public a "Green Manifesto," an appeal to citizens for the active defence of parks and squares. Residents of Saint Petersburg had protested for many years against the construction of a church in Malinovka Park, and the city authorities eventually cancelled the decision to build it. However, in June 2017, the Legislative Assembly of Saint Petersburg amended the general plan to permit the construction of the church.

At the beginning of 2018, public activists and environmental defenders in Rostov protested against the construction of a church of the Russian Orthodox Church of the Moscow Patriarchate in Elektroapparat Park. Activists of the Green Union stated that the construction of the church would damage the ecology of the square. Meanwhile, workers had already fenced off the territory where the construction was to take place. Environmentalists argued that Rostov is only 26 per cent greened, and that a church built in the square would also violate the constitutional rights of citizens who wish to live in a favourable environment. They also considered the construction illegal from a legal point of view and appealed to the city prosecutor's office.

In the winter of 2018, representatives of the Izhevsk public organization GradSovet prepared two appeals on 14 February protesting against the construction of a church of the Russian Orthodox Church of the Moscow Patriarchate in Cosmonauts Park. One letter, demanding that an alternative site be found for the church, was to be sent to the local diocese of the Russian Orthodox Church; the second, requesting that the opinion of protesters be taken into account, was to be sent to the city administration. In their letters, GradSovet activists recalled that any construction in recreational zones is prohibited.

In November 2018, residents of Kurgan who participated in an online vote did not support the idea of building a church of the Russian Orthodox Church of the Moscow Patriarchate opposite the former club building. Most participants in the poll in the VKontakte community "Kurgan LIVE" voted against construction. It had been proposed to move the Church of the Nativity of Christ, occupying the building of the former club of the Longitudinal Water Pipeline section of the South Ural Railway, to a plot located across the road. Ninety-five per cent of those who participated in the vote considered that the site could instead be used for a square, a school, a kindergarten or even an industrial facility. Acting Governor Vadim Shumkov, at a meeting with citizens, also did not support the idea of building a church on that plot.

In the autumn of 2018, residents of Nizhnevartovsk opposed the construction of a church of the Russian Orthodox Church of the Moscow Patriarchate at the intersection of Lenin and Chapaev streets, in the eighth microdistrict of the city. Residents wanted to see a park there, not a church, but representatives of the city administration plainly stated that "the church will be built with private funds, whereas the park would have to be developed with budgetary, municipal funds."

On 4 February 2018, a rally entitled "A Kingdom of Churches Is a Kingdom of Chains!" was held in Saint Petersburg against the clericalization of the Russian Federation and the fusion of the state with the Russian Orthodox Church of the Moscow Patriarchate. According to the page of the association that organized the action despite obstacles from the city authorities, the rally was directed against the clericalization of Russia. Participants stated that the position of the then Governor of Saint Petersburg fitted into this dark trend: statements and actions in support of transferring state museums to the Russian Orthodox Church, renaming streets in favour of clerical figures, and building churches on the territories of squares and parks. This, in their view, only emphasized the need for such an action. Speakers also noted that the situation with freedom of conscience in Russia was deteriorating, as reflected, in particular, in the report of the International Humanist and Ethical Union. Russia was marked by serious violations in three of four categories: education, public interaction and freedom of expression. The participants emphasized that the Russian Orthodox Church of the Moscow Patriarchate was aggressively invading all spheres of public life, which was extremely dangerous for the preservation of public and state security.

In September 2018, a picket took place in Saint Petersburg against the construction of a church of the Russian Orthodox Church of the Moscow Patriarchate in a square on Kanonersky Island. One protester held a poster reading: "The square to the residents, construction to the wasteland." This was not the only protest against the construction of a church in that square. Opponents of construction argued that the square, laid out one year earlier, should have been included in the list of public green spaces. Nevertheless, the Saint Petersburg Commission on Land Use and Development had already approved church construction on that territory.

In the autumn of 2018, opponents of the construction of the church on Kanonersky Island also went to court. Saint Petersburg deputy Sergey Vinnichenko and local residents challenged the decision to allocate the plot for the construction of a church of the Russian Orthodox Church of the Moscow Patriarchate. On 29 October, the Kuibyshevsky District Court of Saint Petersburg accepted their claim against the city Committee for Urban Planning, seeking cancellation of the permit for construction of the church on Kanonersky Island. The claimants argued that the Committee's order of 13 September, transferring the land plot on Kanonersky Island for construction of a church, chapel, bell tower, parish house and parking lot of the Russian Orthodox Church of the Moscow Patriarchate, violated their rights to a favourable environment and living conditions. According to local residents, neither the land-surveying project nor the layout of the plot on the cadastral plan had been approved, meaning that the Committee's order had been adopted in violation of the law.

In December 2018, another protest was held against the transfer to the local diocese of the Russian Orthodox Church of the Moscow Patriarchate of Saint Isaac's Cathedral in Saint Petersburg. The action was timed to the day of consecration of the Chapel of Saint Catherine. Fifty people took part. Participants visited the museum-monument and began collecting signatures "against the transfer of the Chapel of Saint Catherine to the Church." Earlier, residents of Saint Petersburg had repeatedly and persistently opposed the transfer of Saint Isaac's Cathedral to the Church of the Moscow Patriarchate, hoping to preserve its status as a museum-monument.

In May 2019, residents of Saint Petersburg opposed the development by the Russian Orthodox Church of the Moscow Patriarchate of South Primorsky Park. In a petition addressed to the Acting Governor of Saint Petersburg, A. D. Beglov, with copies to the Legislative Assembly of Saint Petersburg, residents recalled that South Primorsky Park had been laid out in 1960 as a continuation of the city's "emerald necklace," a kind of "Soviet-era Peterhof" with a ceremonial exit to the Gulf of Finland. Of the park area planned in the previous century, only about 10 per cent remained. The park's access to the sea was being built over with residential blocks in which no new public green spaces were provided, and any further reduction of existing park territories was painfully perceived by residents. More than ten years earlier, without public discussion, a plot of 16,672 square metres had been removed from the territory of South Primorsky Park and its zoning changed for church activity, although such activity had already been possible under the previous zoning. According to the petitioners, later events showed that the ultimate purpose of taking the park plot had been commercial management of the land. In 2016, the parish of the Church of the Intercession of the Mother of God in South Primorsky Park requested that the zoning of the plot on Doblesti Street be changed in order to place a reli ermitted use, and that the height regulation be increased from 15 metres to 40 metres. Under the requested main use, trade facilities and residential construction became possible instead of the existing wooden church of tree height. In the eyes of the petitioners, a socially significant public recreational green area had been taken from the city in order to develop commercial activity on it.

The petitioners asked the authorities to order the return to South Primorsky Park of the undeveloped part of the church plot that had been removed from it without public discussion, and to accelerate the capital repair of the park so that it could fully perform its recreational functions. The appeal ended with the words: "With the hope of being heard, taxpayers and residents of Saint Petersburg," noting that the collection of signatures was continuing.

In December 2018, residents of Yekaterinburg collected more than one thousand signatures against the construction of a church of the Russian Orthodox Church of the Moscow Patriarchate in a square. A group of activists opposing the construction of the Church of Saint Catherine near the Drama Theatre submitted the list of signatures on 14 December to the representative of the President of the Russian Federation in the Ural Federal District. Protests against the construction of this church had been taking place in the city for several years. However, the protests did not stop the arbitrariness of the Yekaterinburg authorities, which led in May 2019 to a spontaneous protest action to preserve the square at the site of the planned church. Protesters demolished a temporary fence erected on the construction site during the night of 12 to 13 May and occupied the square, from which they were later driven out by athletic young men. The number of protesters, according to different estimates, ranged from several hundred to two thousand people.

It is indicative that three quarters of Yekaterinburg residents voted against development of the central square by the Russian Orthodox Church of the Moscow Patriarchate in a poll conducted after President Putin's intervention. The Russian Public Opinion Research Center (VTsIOM) officially presented data on 22 May showing what residents of Yekaterinburg thought about the construction of the Church of the Holy Great Martyr Catherine in the city's central square. Only 18 per cent of residents opposed the construction of the church in principle. Only 7 per cent supported construction in the square near the Drama Theatre under the original project, while another 14 per cent considered that the church could be built there if the project were adjusted with due regard for citizens' opinions. At the same time, 58 per cent of residents believed that the church should be built in another part of the city. When asked directly whether the square near the Drama Theatre was a suitable place for the church, only 10 per cent answered positively, while an absolute majority, 74 per cent, described the site as unsuitable. Only 49 per cent of residents said that they had enough information about the plans for construction. On 16 May, at a forum of the All-Russia People's Front in Sochi, President Vladimir Putin commented on the events in the Ural capital: "There is a simple way: hold a survey, and the minority must submit to the majority. That is the principle of democracy. But the opinion of that minority must also be taken into account." At the same time, most civil activists who had for several years opposed the development of the city's central square by objects of the Russian Orthodox Church called for an official city referendum rather than a sociological poll.

In May 2019, residents of Omsk also protested against the construction of a church of the Russian Orthodox Church of the Moscow Patriarchate in the green zone on Dianova Street, near the Saint Elijah Cathedral. Citizens instead proposed organizing a park there. The initiative had both opponents and supporters. According to local residents, functioning churches already existed in the area: the Church of the Holy Great Martyr and Healer Panteleimon, the Cathedral of the Nativity of Christ and the Church of the Hieromartyr Sylvester. Recreational places for residents, by contrast, were almost absent; the area was densely built up, a "stone jungle" around them. In a public appeal published in the Omsk VK community, residents suggested pedestrian and bicycle paths, lighting, benches, bins, a dog-walking area with barriers and the city's first water play zone, similar to the Kazan embankment. Such a gift to citizens, they argued, would be remembered and appreciated far more strongly.

In May 2019, residents of Ulyanovsk opposed the construction of a stone church of the Russian Orthodox Church of the Moscow Patriarchate in a park instead of the existing wooden one, fearing the felling of trees. The prosecutor's office began checking the legality of the construction permit, and the Public Chamber prepared a poll of citizens. Residents of the Zasviyazhsky district protested against construction planned in Family Park, near the UAZ House of Culture and the 50th Anniversary of the Komsomol Avenue. According to Vasily Gvozdev, First Deputy Chairman of the Legislative Assembly of the Ulyanovsk Region, public hearings had been held on 27 March, with a majority of participants voting for construction, and on 26 April the city administration issued a construction permit. However, some residents were dissatisfied with the decision. Opponents objected to the felling of trees. Vitaly Kuzin, a Communist Party deputy who supported the protesters, noted that the administration's website contained no information either about the results of the public hearings or about the transfer of land to the Russian Orthodox Church. The decision to build had been made on the basis of a meeting attended by only 60 people, all supporters of construction. On 17 May, a gathering took place in the rs and opponents.

Activist Timur Sharkaev, speaking in defence of the square, explained that residents were not objecting to a church as such, but to the destruction of trees. "There are already churches everywhere. How many more do we need? Build a hundred more churches if you want, but we will not let you destroy the trees." He added that the priest's promise to transplant 40-year-old trees was implausible. Opponents also drew attention to the nearby entertainment centre, whose proximity would likely interfere with the church. Father Ioann, however, stated that the church would become part of a single complex with the reconstructed UAZ House of Culture, the future Patriot Centre.

Overall, in five years protests against the construction of churches of the Russian Orthodox Church of the Moscow Patriarchate took place in 28 Russian cities. The protests in Yekaterinburg became the largest and received the greatest press coverage, according to monitoring by M13 cited by Vedomosti on 22 May. During that period, anti-church-construction actions took place six times in Moscow and three times in Saint Petersburg; rallies were held twice in Blagoveshchensk, Krasnoyarsk, Nizhny Novgorod and Tolyatti. The Yekaterinburg protests that began in May became the leader in media attention: they accounted for 58 per cent of all publications on actions against church construction. By comparison, all six Moscow actions together accounted for only 17 per cent of the total number of materials. The second most covered action also took place in Moscow, where in 2015-2016 residents protested against the construction of a church in Torfyanka Park.

In May 2019, residents of a district on the outskirts of Kazan protested against the construction of a mosque. It was planned near residential buildings and a school, on Godovikov Street. Protesters noted that the construction site was located 30 metres from residential houses, with a school and kindergarten nearby. Local resident Svetlana Leontyeva, who created a Change.org petition calling for construction to be stopped, stated that the fence had been installed at night, that the majority were against the project, including Muslims, and that this did not seem to matter to the initiators. According to information boards, construction had been initiated by the Rakhmatulla mosque parish. The permit was issued on 23 January 2019 by the executive committee of Kazan. The works were to begin in the second quarter of 2019 and end in the third quarter of 2022. A white-stone mosque with two minarets was planned. Residents stressed that they were not seeking cancellation of construction as such, but wanted the mosque to be built in another, more convenient place. Leontyeva noted that children had physical education lessons there in winter and that no one could answer where they would do so after construction began. Initially, the project had not provided for children's or sports grounds; they were added only after residen

The strengthening of protests against the fusion of the state with religious associations coincided with continued threats, persecution and violence against supporters of freedom of conscience, state secularity and critics of clerical ideologization of the state. For example, at the end of 2017 it became known that in Bryansk the prosecutor's office of the Fokinsky district had sent to the Investigative Committee a decision to initiate a criminal case under Article 148 of the Criminal Code, concerning violation of the right to freedom of conscience and religion, against a blogger who had opposed a religious procession and the installation of a worship cross in the city. The prosecutor's office reached this conclusion after analysing an article by N. Vitkevich published in the media on 31 August 2016. The blogger accused the mayor of "demonstrative violations of the Constitution of the Russian Federation connected with depriving citizens of the city of the right freely to move and violating their freedom of conscience through the forced imposition of the faith demonstratively adhered to by city officials." He cited two examples: blocking traffic because of a religious procession and replacing a monument dedicated to the thousandth anniversary of Bryansk with a worship cross.

In the winter of 2018, the Saint Petersburg city authorities attempted to obstruct the Anticlericalism 2018 rally, which nevertheless took place on 4 February despite their pressure. The organizers' statement described serious difficulties faced by the campaign against the growing influence of the Church on the state. The Anticlericalism action traditionally takes place in Saint Petersburg in February, and that year it was to be dedicated to the centenary of the Decree on the Separation of Church from State and School from Church. However, under various pretexts, city authorities consistently refused to approve the action in all places designated for such rallies: Pionerskaya Square, Lenin Square, the area near the Planetarium and the area near Kazan Cathedral. Particularly outrageous, almost comic, was the refusal to approve a rally on Chernyshevsky Square. The Moskovsky District Administration refused permission because the organizers had not specified in their notification the "route of movement of rally participants," even though under existing legislation participants in a rally, unlike a march or demonstration, clearly do not move along a route. The organizers stated that the authorities' opposition was politically motivated and connected it with the loss of the secular character of the state and society: religion, they argued, was being imposed at all levels, while freedom to express alternative points of view would soon cease to exist even on paper. The organizers appealed to the Moskovsky District Court of Saint Petersburg, which found unlawful the authorities' refusal to approve the annual rally dedicated to the centenary of the Decree separating church from state and school from church.

In 2018, unlawful pressure by law-enforcement bodies and the judicial system against communities of people, including online communities, and against individuals continued. For example, in the winter of 2018, a court found a resident of Barnaul guilty of insulting the "feelings of believers" and sentenced her to a suspended term of imprisonment with a probation period of one year and six months, according to a 12 February report by the press service of the Altai Territory prosecutor's office. According to the prosecutor's office, from November 2014 to September 2016 the convicted woman posted on an open page and in a social-network group images insulting the feelings of believers. Some of the images were accompanied by texts degrading a group of persons by origin and attitude to religion. The court decision entered into force.

In the summer of 2018, it became known that a criminal case had been initiated against 35-year-old Andrey Bezborodov, accused of inciting hatred against a social group because of several captioned images on the social network VKontakte, under Part 1 of Article 282 of the Criminal Code. His lawyer V. Vasin reported that, in particular, the case concerned a photograph of President Vladimir Putin with Patriarch Kirill accompanied by a joke about the Trinity. After the criminal case over VKontakte collages was opened, a psychiatric examination declared Bezborodov, a security guard with a weapons permit, legally insane. Bezborodov himself stated that he was completely healthy and that the controversial images now imputed to him by investigators had merely been saved in his "Saved" folder on VKontakte.

On 15 April 2019, the Severodvinsk City Court of the Arkhangelsk Region upheld the decision of a magistrate's court that had punished local resident Anatoly Kazikhanov for publishing atheist memes. The activist was fined 15,000 roubles under Part 2 of Article 5.26 of the Code of Administrative Offences for intentional public desecration of religious attributes.

In May 2019, administrative cases were initiated against Ilnur Kamaldinov, a resident of Novocheboksarsk in the Chuvash Republic and a member of the punk band Barrikady Protesta, for desecration of religious symbols under Part 2 of Article 5.26 of the Code of Administrative Offences and for displaying Nazi symbols under Part 1 of Article 20.3 of the same Code. The cases concerned memes posted on VKontakte in 2014 and 2016. Kamaldinov explained that they were "just funny memes," funny at the time, and that they mainly expressed his attitude toward churches rather than toward believers.

In May 2019, residents of Yekaterinburg protesting against the unlawful construction of a church of the Russian Orthodox Church of the Moscow Patriarchate in a square were subjected to violence by certain "athletes" while the police failed to intervene, and then by law-enforcement officers themselves.

Police and the National Guard detained more than 100 protesters, who were later subjected to fines and administrative arrests.

On 3 June 2019, the Leninsky District Court of Yekaterinburg fined Anna Baltina, one of the creators of the movement Parks and Squares of Yekaterinburg, 20,000 roubles for "organizing an unauthorized mass event" on the first day of protests in the square near the Drama Theatre under Part 2 of Article 20.2 of the Code of Administrative Offences. The administrative offence report was based on posts on her personal Facebook page and in the VKontakte group Parks and Squares of Yekaterinburg.

On 5 June 2019, the Leninsky District Court of Yekaterinburg issued decisions concerning three participants in unauthorized protests against the construction of the Church of Saint Catherine of the Russian Orthodox Church of the Moscow Patriarchate in the square near the Drama Theatre. The court considered materials against four participants in unauthorized actions, found guilty under Part 5 of Article 20.2 of the Code of Administrative Offences for violating the established procedure for holding meetings, rallies, demonstrations, marches or pickets. Three defendants received fines, while the hearing of one case was postponed. Two participants were fined 10,000 roubles each, and one was fined 15,000 roubles.

Violations of freedom of conscience were also identified in the functioning of cultural institutions, which in recent years have become objects of clerical xenophobia and persecution. For example, in the autumn of 2018, in Rostov, the authors of the opera Khovanshchina were forced to coordinate the production in order to avoid possible difficulties connected with the law on protecting the "feelings of believers." This was reported by Andrey Ivanov, chief conductor of the Rostov Musical Theatre, at a press conference. One of the issues raised at the meeting with the press was the impact of the law on protecting the "feelings of believers" on culture and art, using the staging of Khovanshchina as an example. The work tells of the events of 1682, when Prince Ivan Khovansky briefly came to power in Moscow during the Streltsy uprising. The theatre's general director discussed the staging of the opera with the metropolitan and received his approval. The director added that this was connected with the increased attention paid to religion in society and noted that art is a sphere that does not affect the "feelings of believers," but that where a work contains references to historical events, for example, delicacy should be shown.

In this context, it should be recalled that in 2017 the film Matilda was subjected to large-scale persecution. In addition to numerous threats, on the night of 31 August 2017 the building in Saint Petersburg housing the Lendoc studio and the office of Alexei Uchitel's Rock studio was attacked with Molotov cocktails. The Kosmos cinema and concert hall in Yekaterinburg was also set on fire, as were two cars near the office of the law firm representing the interests of the film's director, Alexei Uchitel. Leaflets reading "Burn for Matilda" were scattered at the scene of the arson. In addition, allegedly at the suggestion of figures of the Russian Orthodox Church of the Moscow Patriarchate, the Bolshoi Theatre cancelled Kirill Serebrennikov's ballet Nureyev three days before its premiere.

In the spring of 2019, the Ministry of Culture of Russia recommended postponing the start of screenings of French director Francois Ozon's film By the Grace of God because of the celebration of Orthodox Easter. This was reported by Meduza, citing the distributor A-One Films. The film was scheduled for release on 25 April and had already been shown in several cinemas in the morning; evening screenings, according to the distributor, were most likely to be cancelled. A-One Films added that the distribution certificate had not been signed, although the company had been confident there would be no problem. As a result, representatives of the Ministry of Culture "orally recommended" that distributors postpone the premiere until the 29th, "after Holy Week." The ministry did not officially comment on the situation. Francois Ozon's film By the Grace of God tells the story of victims of abuse by a Catholic priest and is based on the real story of priest Bernard Preynat, who in 2017 was accused of molesting children at a scout camp in Lyon.

Thus, unlawful judicial prosecutions for "insulting the feelings of believers" and "desecrating objects of religious veneration" continued in Russia.

It is indicative that in 2018 the Supreme Court of Russia refused to review the sentence in the case of blogger Ruslan Sokolovsky, who in May 2017 had been sentenced by the Verkh-Isetsky District Court of Yekaterinburg for catching Pokemon in a church of the Russian Orthodox Church of the Moscow Patriarchate. He received a suspended sentence of three and a half years of imprisonment with a probation period of three years, as well as compulsory work. The judge found Sokolovsky guilty of inciting hatred under Part 1 of Article 282 of the Criminal Code, insulting the religious "feelings of believers" under Part 1 of Article 148, and illegal trafficking in technical means for covert receipt of information under Article 138.1. In particular, the expert opinion underlying the sentence stated that Sokolovsky had committed a crime by "denying the existence of God."

In 2020–2026, protest activity in Russia connected with violations of the principle of neutrality evolved from mass street actions into local conflicts and legal resistance against the background of tightened legislation. Resonant protests against the construction of churches in parks and green zones continued in different regions. One of the most significant conflicts took place in Krasnodar, where residents of the Yubileiny microdistrict protested en masse against the development of the embankment with a church, despite approval by the authorities. Such protests often arose from resistance to infill development and the transfer of public-use land plots to the Russian Orthodox Church of the Moscow Patriarchate, gradually becoming a struggle to preserve familiar living environments and constitutional rights to participation in local self-government.

Dissatisfaction also persisted with the introduction of a religious component into state schools, including the courses ORKSE and ODNKNR, and with the cancellation of secular cultural events under pressure from religious activists. However, because of the risk of persecution, such protests increasingly moved online. Criticism continued regarding budget expenditures for the needs of religious organizations and the transfer, considered illegal by activists, of cultural objects and natural complexes to the jurisdiction of the Russian Orthodox Church of the Moscow Patriarchate. By 2026, opportunities for lawful protest had narrowed, as participants increasingly faced administrative pressure and criminal cases, while expressions of dissatisfaction with clericalization were often equated with "insulting the feelings of believers" or "extremism."

6.4. The Activity of State and Non-Governmental Institutions in Protecting the Right to Freedom of Conscience in the Russian Federation

Against the background of the negative evolution of the principles and norms governing freedom of conscience in the Russian Federation, judicial practice, as well as the activity of law-enforcement bodies and the prosecutor's office, has not contributed to the realization of this right.

In particular, the activity of the Supreme Court of the Russian Federation reflects the crisis of the Russian judicial system and contributes to the growing number of appeals to international mechanisms for the protection of rights and freedoms. The Constitutional Court of the Russian Federation has failed to give a proper assessment to legislation in the field of freedom of conscience that is far from fully constitutional. This, in turn, has predetermined the striking ineffectiveness of the entire Russian judicial system in protecting freedom of conscience.

The protection of the right to freedom of conscience by the Office of the Commissioner for Human Rights in the Russian Federation is also insufficient. The same criticism applies to regional ombudsmen, to human rights commissions under the heads of executive authorities in the constituent entities of the Russian Federation, and to the Civic Chamber of the Russian Federation.

Finally, there has been a further decline in the level of state human-rights protection and a continued contraction of the non-governmental human-rights movement in the field of freedom of conscience. For example, the overwhelming majority of current problems related to the realization of freedom of conscience are not adequately reflected in the annual Report on the Activity of the Commissioner for Human Rights in the Russian Federation.

The report also largely ignores the fundamental constitutional principles concerning freedom of conscience and the secularity of the state. It states, in particular, that "the basic principles and guarantees for the realization of the right to freedom of conscience and religion are established by the Federal Law 'On Freedom of Conscience and Religious Associations'." In reality, the provisions of this Federal Law grossly contradict the Constitution and Russia's international legal obligations. This has been stated in human-rights reports and monographs, as well as in the Opinion of the Commissioner for Human Rights in the Russian Federation, O. O. Mironov, "On Verifying the Compliance of the Federal Law 'On Freedom of Conscience and Religious Associations' with the International Legal Obligations of the Russian Federation."

Equally indicative are the repeated formal replies to complaints from a number of religious organizations, stating that "on the basis of the Commissioner's appeal, the prosecutor's offices of the constituent entities of the Russian Federation carried out inspections, as a result of which no violations of federal legislation were identified." This is the familiar bureaucratic magic trick: the violation disappears because the institution responsible for seeing it has chosen not to look.

Among the positive aspects of the Report on the Activity of the Commissioner for Human Rights in the Russian Federation, one should note the recognition of "vague criteria for classifying religious materials as 'extremist'," which predetermine judicial arbitrariness, as well as the recognition of the need "to clarify the signs of 'extremist activity' specified in Article 282.2 of the Criminal Code of the Russian Federation." The report states: "Undoubtedly, any extremist activity is inadmissible; however, vague criteria for classifying religious materials as extremist are also inadmissible, when virtually any federal judge, at their own discretion, may prohibit any book, image, video or audio recording."

The activity of religious associations and confessionally oriented structures in the protection of freedom of conscience cannot withstand serious criticism at all. The priority of individual rights is alien to their institutional nature, since they are primarily oriented toward doctrinal attitudes and corporate interests. Quite often, some of them act as direct persecutors or initiators of persecution against those of other faiths. It must be stated that certain religious associations and confessionally oriented structures, which present themselves as actors in the protection of freedom of conscience, have finally turned into ideological appendages of state power, and sometimes into forces that initiate and legitimize persecution.

Although some confessionally oriented structures position themselves as scientific and human-rights organizations, they are in principle incapable of generating an objective position. Moreover, an international system has been identified in which such structures manipulate the academic community and create obstacles to scholarly exchange.

Nor can the protection of freedom of conscience be expected from other worldview organizations. As a rule, they oppose any strengthening of the influence of religion not only in the state, but also in society.

With rare exceptions, human-rights defenders have traditionally paid little attention to the issue of state neutrality. Moreover, some of them have for many years legitimized legally incorrect concepts that predetermine arbitrariness and persecution in their application, and have even received state grants to combat "extremism."

The activity of the Russian Association for the Protection of Religious Freedom (RAPRF), registered by the Ministry of Justice of Russia on 28 November 2014, appears highly ambiguous. In practice, it functions as a hybrid successor to the Russian branch of the International Association for Religious Freedom, but under the direct control of the Russian Orthodox Church of the Moscow Patriarchate and the Russian authorities.

The declared priority of RAPRF is "the preservation and support of traditional spiritual values." It is also stated that "the focus of RAPRF will include issues of protecting the personal freedom of conscience of citizens and the rights of religious associations regardless of their confessional affiliation," as well as the conduct of relevant monitoring and the presentation of its results to the Russian and international public. At the same time, representatives of RAPRF support highly controversial legislative innovations and make statements implying that there are supposedly no significant problems in the field of freedom of conscience at the federal level in Russia, but only certain abuses by regional officials.

By 2026, RAPRF had effectively transformed into an instrument of "state human-rights protection," transmitting on international platforms the thesis that freedom of conscience in Russia is in a satisfactory condition. The activity of the association is focused almost exclusively on supporting "traditional values" and loyal confessions under the patronage of the Russian Orthodox Church of the Moscow Patriarchate, while systemic persecution of minorities is ignored.

At the same time, independent monitoring has been almost entirely pushed out of the legal field, since objective documentation of violations is now interpreted as a threat to "spiritual security." The entire burden of real protection of believers' rights has fallen on scattered civic groups working under unprecedented administrative pressure. As a result, an imitative model of protection of freedom of conscience has taken shape in the country, where official RAPRF reports substitute for the real picture of law-enforcement practice.

The leading role in protecting freedom of conscience in contemporary Russia belongs to civic organizations. Despite numerous difficulties and a certain underestimation of this field, they remain capable, in principle, of providing objective monitoring of violations of freedom of conscience despite pressure from the state apparatus. Even under the status of "foreign agents" or the threat of liquidation, these structures retain the potential for constructive dialogue, pointing to legal gaps and excesses in legislation.

Their activity is critically important for recording cases of discrimination against "non-systemic" believers, which are systematically ignored by official structures such as RAPRF. Cooperation between such organizations and international legal mechanisms of the United Nations remains the last channel for verifying data on the real state of religious freedoms in Russia. Thus, in contemporary conditions, it is civil society that acts as the guarantor of the preservation of neutrality and the protection of citizens' constitutional rights.

REFERENCES TO CHAPTER 6

List of Sources Used

1. Federal Law No. 184-FZ of October 6, 1999 "On the General Principles of Organization of Legislative (Representative) and Executive Bodies of State Power of the Subjects of the Russian Federation" (as amended and supplemented) // Garant. Electronic resource. URL: http://base.garant.ru/12117177/3/ (accessed: 21.12.2019).
2. Decree of the President of the Russian Federation No. 849 of May 13, 2000 " On the Plenipotentiary Representative of the President of the Russian Federation in a Federal District" (as amended and supplemented) // Garant. Electronic resource. URL: http://base.garant.ru (accessed: 7.02.2019).
3. Linkin V.N. The Principle of the Secular State as One of the Foundations of the Constitutional Order of the Russian Federation: Abstract of Candidate of Legal Sciences Dissertation. Moscow, 2015. pp. 13-14.
4. Passport of the Federal Target Program "Strengthening the Unity of the Russian Nation and the Ethnocultural Development of the Peoples of Russia (2014-2020)" // URL: http://government.ru/media/files/41d4862001ad2a4e53 59.pdf.
5. Bocharova S., Rustamova F. The Russian Orthodox Church Was Allocated Two Billion Rubles for Spiritual Centers. To "Strengthen the Unity of the Russian Nation" // RBC. November 28, 2014.
6. An Agreement on Cooperation between the Russian Orthodox Church and the Ministry of Health of the Russian Federation Was Signed // Press Service of the Patriarch of Moscow and All Rus'. June 18, 2015.
7. Agreement on Cooperation between the Russian Orthodox Church and the Ministry of Health of the Russian Federation // Patriarchia.ru. June 18, 2015. URL: http://www.patriarchia.ru/db/text/4124569.html.
8. Under a Special Program, 22 Churches and 79 Chapels Were Built in Moscow // Interfax. December 29, 2014.
9. The Head of the Moscow Region Will Engage in the Restoration of the Joseph-Volokolamsk Monastery // Interfax. July 16, 2015.
10. A Mosque Was Opened in Dagestan That Will Become a Source for the Dissemination of "Traditional" Islam, Says Abdulatipov // Interfax. September 28, 2015.
11. Minnikhanov Warns of Harsh Measures against Those Intending to Spread Radical Ideas in Tatarstan // Interfax. September 28, 2015.
12. Methodological Recommendations on Issues Related to the Resocialization of Adolescents Subjected to Destructive Psychological Influence by Supporters of Religious-Extremist and Terrorist Ideology (appendix to Letter No. PZ-1608/09 of the Ministry of Education and Science of the Russian Federation dated November 23, 2017) // Garant.ru. January 25, 2018. URL: http://www.garant.ru/products/ipo/prime/doc/71738738/?prime.
13. Recommendations of the Synodal Department for Educational and Methodological Support for Teaching Academic Subjects, Courses, and Modules of Orthodox Culture // Orthodox Education. May 24, 2019. URL: https://pravobraz.ru/rekomendacii-sinodalnogo-otdela-po-uchebno-metodicheskomu-obespecheniyu-prepodavaniya-uchebnyx-predmetov-kursov-modulej-pravoslavnoj-kultury/.
14. In Smolensk, Diplomas Were Awarded to Graduates of the Professional Retraining Courses "Cultural Studies. Orthodox Culture" // Orthodox Education. November 6, 2018. URL: https://pravobraz.ru/v-smolenske-vrucheny-diplomy-vypusknikam-kursov-professionalnoj-perepodgotovki-kulturologiya-pravoslavnaya-kultura/.
15. Advanced Training Courses for Teachers of the Fundamentals of Orthodox Culture Were Completed in Moscow // Orthodox Education. May 24, 2019. URL: https://pravobraz.ru/v-moskve-zavershilis-kursy-povysheniya-kvalifikacii-uchitelej-osnov-pravoslavnoj-kultury/.
16. A Series of Lectures for Teachers of General Education Schools Was Completed in the [illegible] May 21,

2019. URL: https://pravobraz.ru/v-ekaterinburgskoj-eparxii-zavershilsya-cikl-lekcij-dlya-uchitelej-obshheobrazovatelnyx-shkol/.

17. An Agreement on Cooperation between the Moscow Theological Academy and the All-Russian State Institute of Cinematography Was Signed // Patriarchia.ru. September 27, 2018. URL: http://www.patriarchia.ru/db/text/5274535.html.

18. Theology Is Taught in More Than 50 Higher Education Institutions in the Russian Federation, Says Metropolitan Hilarion // Interfax. April 27, 2018. URL: http://www.interfax-religion.ru/?act=news&div=69740.

19. It is directly regulated by several articles of the Federal Laws "On the Status of Military Personnel" of 1998 (Art. 8), "On Freedom of Conscience and Religious Associations" of 1997 (Art. 6), as well as by the Internal Service Regulations of the Armed Forces of the Russian Federation and Orders of the Minister of Defense of the Russian Federation No. 70 of March 11, 2004 " On Bodies of Educational Work of the Armed Forces of the Russian Federation" and No. 79 of February 25, 2005 "Fundamentals of the Organization of Education in the Armed Forces" (previously, Orders of the Minister of Defense of the Russian Federation No. 235 of July 3, 1995 "On Improving the System of Educational Work in the Armed Forces of the Russian Federation" and No. 226 of July 6, 1995 "On Bodies of Educational Work of the Armed Forces of the Russian Federation" were in force).

20. The Consecration of a House Military Church in the Engineering Troops Took Place // Pobeda.ru. February 20, 2018. URL: http://pobeda.ru/sostoyalos-osvyashhenie-domovogo-voinskogo-hrama-v-inzhenernyih-voyskah.html.

21. Icons with Particles of Relics of Orthodox Saints Were Transferred to the Church at the General Staff of the Armed Forces of the Russian Federation // Pobeda.ru. September 25, 2018. URL: http://pobeda.ru/ikonyi-s-chastitsami-moshhey-pravoslavnyih-svyatyih-peredanyi-v-hram-pri-genshtabe-vs-rf.html.

22. The Patronal Feast Was Celebrated in the House Church of the Russian Ministry of Defense // Pobeda.ru. July 18, 2018. URL: http://pobeda.ru/v-domovom-hrame-minoboronyi-rossii-otmetili-prestolnyiy-prazdnik.html; On the Day of Commemoration of the Heavenly Patron of Rosgvardia, the Personnel of the "Peresvet" Detachment Took Part in a Church Service // Pobeda.ru. August 1, 2018. URL: http://pobeda.ru/v-den-pamyati-nebesnogo-pokrovitelya-rosgvardii-lichnyiy-sostav-otryada-peresvet-prinyal-uchastie-v-bogosluzhenii.html.

23. The Head of the Rosgvardia Sector Performed the Baptism of Servicemen of a Special Forces Unit // Pobeda.ru. August 31, 2018. URL: http://pobeda.ru/zaveduyushhiy-sektorom-rogvardii-sovershil-kreshhenie-voennosluzhashhih-podrazdeleniya-spetsnaza.html.

24. The Chairman of the Synodal Department Consecrated the Standard of the Head of the Main Political Directorate of the Armed Forces // Pobeda.ru. August 31, 2018. URL: http://pobeda.ru/predsedatel-sinodalnogo-otdela-osvyatil-shtandart-nachalnika-glavnogo-politupravleniya-vooruzhennyih-sil.html.

25. Icons with Particles of Relics of Orthodox Saints Were Transferred to the Church at the General Staff of the Armed Forces of the Russian Federation // Pobeda.ru. September 25, 2018. URL: http://pobeda.ru/ikonyi-s-chastitsami-moshhey-pravoslavnyih-svyatyih-peredanyi-v-hram-pri-genshtabe-vs-rf.html.

26. Awards of the Russian Orthodox Church Were Presented to Employees of the Russian Ministry of Defense // Pobeda.ru. March 9, 2017. URL: http://pobeda.ru/nagradyi-russkoy-pravoslavnoy-tserkvi-vruchenyi-sotrudnikam-minoboronyi-rossii.html.

27. Resolution of the "Military" Section of the XXVI International Christmas Educational Readings // Pobeda.ru. January 25, 2018. URL: http://pobeda.ru/rezolyutsiya-voennoy-sektsii-xhvi-mezhdunarodnyih-rozhdestvenskih-obrazovatelnyih-chteniy-duhovnyie-osnovyi-voinskogo-sluzheniya-otechestvu-nravstvennyie-tsennosti-i-patrioticheskoe-vospitani.html.

28. Regulation on the Organization of Work with Believers in the Armed Forces of the Russian Federation // Bulletin of Military and Naval Clergy. March izacii-raboty-s-ver

uyushhimi-v-vooruzhennyx-silax-rf/.
29.Penal Enforcement Code of the Russian Federation No. 1-FZ of January 8, 1997 (as amended on July 1, 2010) // Collection of Legislation of the Russian Federation, January 13, 1997, No. 2, Art. 198.
30.Brief Characterization of the Penal Enforcement System // Federal Penitentiary Service. URL: https://fsin.gov.ru/structure/inspector/iao/statistika/Kratkaya%20har-ka%20UIS/.
31.Agreement on Cooperation between the Federal Penitentiary Service of Russia and the Russian Orthodox Church // Official Website of the Russian Orthodox Church of the Moscow Patriarchate. February 22, 2011. URL: http://www.patriarchia.ru/db/text/1414730.html.
32.Opinion of the Commissioner for Human Rights in the Russian Federation " On Reviewing the Compliance of the Federal Law 'On Freedom of Conscience and Religious Associations' with the International-Legal Obligations of the Russian Federation." Rossiyskaya Gazeta. April 22, 1999.
33.Report on the Activities of the Commissioner for Human Rights in the Russian Federation for 2018. URL: http://ombudsmanrf.org/upload/files/docs/lib/doclad_2018.pdf.
34.Freedom of Conscience: Problems of Theory and Practice. Monograph (edited by F.M. Rudinsky, S.A. Buryanov). Moscow: ZAO TF MIR, 2012. 1120 p.
35.Buryanov S.A. International Recognition of the Right to Freedom of Conscience and Problems of Its Implementation in the Russian Federation under Conditions of Contemporary Global Processes: Monograph. Moscow: Poligraf Servis, 2020. 624 p.
36.Civil Human Rights in Russia: Modern Problems of Theory and Practice / compiled by F.M. Rudinsky. Transaction Publishers, New Brunswick, New Jersey (U.S.A.), 2008. 458 p.
37.Buryanov S.A. State Worldview Neutrality in the Context of Deteriorating Imbalances in Globalization: The Case and Current State of Affairs in the Russian Federation from 2016 to the Beginning of 2017. Amsterdam, 2018. 228 p.

CONCEPT BOX

PART VI / CHAPTER 7

EXIT FROM THE CONCEPTUAL DEADLOCK

A legal paradigm for inner freedom

Chapter 7 moves from diagnosis to legal reconstruction.

The problem is not solved by replacing one dominant worldview with another. It is solved by limiting the authority of public power in the sphere of meaning.

The chapter proposes liberty of worldview choice and public-law worldview neutrality as the basis for a new legal paradigm.

This paradigm protects the person's right to form, preserve, revise, and express a worldview without coercion. It also creates a bridge between classical freedom of conscience and new challenges arising from digital platforms, artificial intelligence, and neurotechnologies.

PART VI. A CHANGE OF PARADIGM

CIVILIZATIONAL THESIS

"A free society begins with limits on the power to define truth."

The deadlock can be overcome through a new legal paradigm: liberty of worldview choice guaranteed by public-law worldview neutrality in analog and digital environments. This is the turn from fragmentary protection to a systematic legal architecture of inner freedom.

CHAPTER VII

EXIT-FROM THE CONCEPTUAL DEADLOCK

A legal paradigm for inner freedom

Inner freedom requires a paradigm.
Law transforms it into reality.

Chapter 7. How to Find a Way Out of the Conceptual and Legal Deadlock

7.1. Where to Look for a Strategy? Directions for Reforming International and Domestic Law on Freedom of Conscience

On the basis of the study conducted, it is possible to conclude that the realization of freedom of conscience in contemporary Russia and in the world is in a deep systemic crisis.

This means that the state of freedom of conscience in Russia and globally is unsatisfactory, while the existing instruments designed to ensure its realization are inadequate. As a result, numerous problems arise.

According to research by the Pew Research Center's Forum on Religion and Public Life in Washington, conducted from July 2006 to June 2008 and covering 198 states, in 64 countries, whose residents make up almost 70 percent of the world's population, violations of the right to freedom of religion were characterized as high or very high. The researchers noted that in 178 countries, or 90 percent of the total number studied, religious organizations are required to undergo registration. In 117 countries, religious communities face serious difficulties in this process. The authorities of 75 countries restrict the missionary activity of religious groups. In almost half of the countries of the world, the activity of foreign missionaries is restricted or prohibited. In 86 percent of countries, the authorities provide financial or other support to religious organizations; however, in 76 percent of countries, certain religious groups enjoy an advantage in receiving such support.

V. Yelensky's article provides a more detailed analysis of the results of the above-mentioned study. It reports, in particular, that the constitutions and fundamental laws of 96 countries undertake to protect the freedom only of a specific religion or religions. In 45 percent of all countries, constitutions or equivalent laws recognize the special status of one or more religions, while in 38 countries religious communities are prohibited. Moreover, in 23 countries states discriminate against certain religions by declaring them dangerous sects or cults, while in 175 countries such practice is absent. In 42 countries, proselytism is prohibited with respect to certain religious groups, and in 33 countries proselytism is prohibited altogether.

More than a decade later, the situation has not improved. According to the Pew-Templeton Global Religious Futures project, which analyzes religious change and its impact on societies around the world, the global picture remains deeply troubling.

A 2019 Pew Research Center publication noted that over the decade from 2007 to 2017, government restrictions on religion, including laws, policies and actions by state officials that restrict religious beliefs and practices, had increased significantly worldwide.

It was further noted that government restrictions had grown in several different ways. Laws and policies restricting freedom of religion, such as requirements that religious groups register in order to operate, rose globally from 2007 to 2017, and government favouritism toward religious groups also increased. In other words, state power did not simply regulate religion more often; it increasingly began to rank religious communities according to their usefulness, loyalty or political convenience.

According to the Pew Research Center report "Government Restrictions on Religion Stayed at Peak Levels Globally in 2022," published on 18 December 2024, the global level of government restrictions on religion remained at 3.0 points on a ten-point scale, which is the highest level since the beginning of the research in 2007. Government harassment, including physical and verbal pressure on religious groups, was recorded in 186 of the 198 states studied. In addition, interference by authorities in worship and religious rituals reached peak levels, affecting 170 countries.

The global median score on the Social Hostilities Index remained at 1.6, while 45 countries displayed a high degree of aggression by private individuals and groups. The study emphasizes a direct connection between pressure by the authorities and hostility in society: in 75 percent of cases, high government restrictions are accompanied by a high level of social intolerance. Authoritarian regimes most often demonstrate a combination of strict control over religion and suppression of civil liberties. Countries with "very high" scores on both indices, governmental and social, include India, Pakistan, Egypt and Afghanistan.

At the same time, it should be noted that the relevant Pew Research Center studies are concerned with violations of religious freedom, not with violations of liberty of worldview choice. This means that their findings actually soften the real picture.

The reports of the International Ethical and Humanist Union examine violations of the rights of atheists in various states, including Brunei Darussalam, Sudan, Malaysia, Mauritania, the United Arab Emirates, Pakistan, the Maldives, Afghanistan, Iran and Saudi Arabia. Within the United Nations and other international institutions, the International Ethical and Humanist Union continues to highlight cases of persons accused of "apostasy" and calls on states to abolish these medieval laws.

According to the 2023 Freedom of Thought Report prepared by Humanists International [illegible] s and non-religious people was recorded in 186 countries. In 11 countries, state officials or state bodies openly marginalize, persecute, or incite hatred or violence against non-religious people. In 15 countries, identifying oneself as an atheist or non-religious person is illegal or not recognized. In 23 countries, non-religious people are prohibited from holding certain public offices. In 34 countries, state legislation is partly or wholly based on religious norms.

Blasphemy remains a criminal offence in at least 87 countries. In 48 countries, religious courts are used to resolve family or moral matters. In 85 countries, discriminatory financing of religion is observed. In 32 countries, it is difficult or illegal to create openly humanist organizations. The report also notes that in 33 countries public schools provide mandatory religious education without a secular or humanist alternative. The document emphasizes that discrimination against non-religious people is often connected with the establishment of a state religion, prohibitions on apostasy, the use of "blasphemy" laws and social stigmatization.

The international Christian human-rights organization Open Doors publishes an annual list of countries where Christians are most often persecuted and their rights violated because of their faith. In 2018, for the first time in seven years, Russia appeared on the Open Doors list. It ranked 41st, between Kenya in 40th place and Malaysia in 42nd place. The authors of the ranking noted that the authorities give preference only to the Russian Orthodox Church of the Moscow Patriarchate to the detriment of other confessions. In addition, the report pointed to persecution of Christians in the North Caucasus, a Muslim region of Russia.

According to Open Doors reports from recent years, the level of persecution of Christians in the world has reached a critical threshold of more than 388 million people. North Korea remains at the top of the list of the most dangerous countries, while Nigeria remains the global epicentre of physical violence and killings of Christians. The main trends of the period include the growth of radicalism in Africa and the introduction of digital surveillance systems targeting believers in authoritarian states. A sharp deterioration has also been recorded in Latin America and Central Asia because of pressure from organized crime and the tightening of religious legislation.

Since 2020, Russia has not appeared among the countries with an "extreme" or "very high" level of persecution. In 2026, Russia ranks 56th in the extended Open Doors list, with a "high" level of pressure on Christians. The main source of persecution is state regulation, as communities face administrative fines for missionary activity, seizure of literature and restrictions on the use of premises. In the North Caucasus, converts from Islam to Christianity continue to face risks of physical violence from society and relatives.

Churches with foreign connections are increasingly branded as channels of

"Western influence," which leads to intensified surveillance by security services. Overall, the situation is characterized not by mass violence, but by systemic bureaucratic pressure and the shrinking of the legal space for the activity of "non-traditional" denominations.

As for the annual reports of the United States Commission on International Religious Freedom, they contain incomplete information on the situation in the field of freedom of conscience and highly debatable recommendations for improving it. The Commission was established under the International Religious Freedom Act of 1998. It monitors violations of the right to freedom of religion around the world and provides recommendations to the President, the Secretary of State and the United States Congress. The Commission consists of ten members and is not part of the Department of State. Three members are appointed by the President and six by congressional leadership. The Ambassador-at-Large for International Religious Freedom, a position within the Department of State established by the same Act, is a member of the Commission without voting rights.

The annual report contains information on the state of freedom of religion in countries of concern to the Commission and provides recommendations for strengthening freedom of religion as an integral component of United States foreign policy. It includes chapters on countries that the Commission recommends to the Secretary of State for designation as countries of particular concern on the basis of serious violations of freedom of religion under the International Religious Freedom Act; on countries placed by the Commission on the Special Watch List because of violations in the field of religious freedom that do not place them among countries of particular concern but nevertheless require close attention; and on countries whose situations the Commission continues to monitor closely.

For example, the 2026 report lists the following states as the most serious violators of freedom of religion: Afghanistan, Burma (Myanmar), China, Cuba, Eritrea, India, Algeria, Azerbaijan, Egypt, Indonesia, Iran, Libya, Nicaragua, Nigeria, North Korea, Pakistan, Russia, Saudi Arabia, Syria, Tajikistan, Turkmenistan and Vietnam. The Special Watch List includes Iraq, Kazakhstan, Kyrgyzstan, Malaysia, Qatar, Turkey and Uzbekistan.

In addition to the Commission, the United States Department of State has an Office of International Religious Freedom and the position of Ambassador-at-Large for International Religious Freedom, also created to implement the 1998 International Religious Freedom Act. Each year, the Secretary of State and the Ambassador, on the basis of the Commission's recommendations, submit to Congress a report on international religious freedom. For example, another Annual Report of the United States Department of State on the state of religious freedom in the world was presented on 28 June 2024 in Washington. Among the states violating the rights of religious minorities, the report named Myanmar, China, Cuba, Eritrea, Iran, North Korea, Nicaragua, Pakistan, Saudi Arabia, Tajikistan, Turkmenistan and Russia.

Undoubtedly, the idea of concern for the state of religious freedom throughout the world is beyond question and deserves every approval. However, the "underwater rocks" of implementing this idea, both conceptually and in particular cases, require close and, as far as possible, impartial consideration. Above all, attention is drawn to the fact that the issue is only the protection of religious freedom, which is sometimes, for the sake of variety, called freedom of conscience.

It follows that the object of special concern is not the rights of every person, but only the rights of believers, and even then primarily in their institutional expression. In other words, the protection in question is effectively the protection of the rights of religious organizations. In practice, it increasingly turns out that not even all such organizations are protected, but mainly the influential ones. As a result, the systemic crisis in the realization of freedom of conscience is not adequately reflected in various studies and reports.

I believe that the systemic nature of the crisis of freedom of conscience in the Russian Federation and in the world is connected with the fact that it has affected all the conditional levels at which its realization is formed:

the scientific-theoretical and educational level;

the legislative level;

the activity of state authorities, law enforcement and judicial practice;

the human-rights protection level;

the level of mass media.

At the same time, as already noted, the scientific-theoretical and educational level is system-forming in relation to legislation, to the activity of state authorities, to law enforcement and to judicial practice.

At the same time, while the study confirms the enormous significance of freedom of conscience and the system-forming role of the scientific-theoretical and educational level in its realization, it also reveals extremely serious defects at this level. This gives grounds to speak of a crisis in the scientific understanding of freedom of conscience.

7.2. Substitution of Concepts as a Source of Legal Powerlessness

Understanding that the system-forming causes of the crisis in the realization of freedom of conscience lie at the scientific and theoretical level makes it possible to begin searching for its fundamental roots.

As has already been noted more than once, the weak scholarly development of this issue and the underdevelopment of the above-mentioned principles have led international legal norms and national legislation, including Russian legislation, in this field to focus almost exclusively on freedom of religion, rather than on freedom of conscience as a whole.

The right of everyone to freedom of conscience is mentioned everywhere, but largely as a declaration. In practice it remains outside the effective field of law. As a result, legislation which, by its very logic, should be aimed at realizing liberty of worldview choice is replaced by "special religious" legislation intended to regulate the activity of religious associations.

Religious associations have traditionally been objects of political interest and "special" control by public authorities, both in Russia and in many other states. The application of this "special" legislation therefore leads not only to violations of the rights of believers and religious minorities, but also to the erosion of a number of democratic constitutional principles that form the foundation of the constitutional order.

An analysis of the real situation also shows that "official" scholarship and the law-making process, with the interested or silent consent of the leaders of a number of confessions, especially those that regard themselves as "traditional", are often controlled by the authorities. In such circumstances, the role of scholarship is reduced to providing a quasi-scientific justification for the actual policy of the state.

It was probably these numerous contradictions that the Kazakh researcher G. P. Luparev had in mind when he proposed a radical replacement of the "archaic and steadily degrading institution of freedom of conscience" with a new institution of freedom of worldview convictions, broader in content and more complete in form, while still preserving "detailed legal regulation" of freedom of religion.

The researcher rightly notes that "the constitutional institution of freedom of conscience does not meet the most important requirement of modern legal technique, which obliges legislators and law-enforcement bodies to use only such terms and expressions in normative acts and practice as are perceived by everyone and do not have a double meaning."

While recognizing the adequacy of the problem raised by G. P. Luparev, the author does not consider the proposed approaches sufficient to solve the problem of realizing freedom of conscience.

A more promising direction is to understand and fill with modern content the historically established and normatively consolidated concept of "freedom of conscience."

The problem of improving the legal framework for the realization of freedom of conscience in various states, including Russia, is closely connected with the international level.

The reason is that the ideal of religious freedom has gained much wider recognition throughout the world, and its principles have therefore acquired clear normative significance.

Freedom of conscience, by contrast, is merely declared alongside freedom of religion in international law and in the constitutions of many states. Thus, the right of everyone to freedom of conscience is mentioned everywhere, but in practical legal terms remains largely outside the field of law.

Historically, in the principal international legal instruments, and in the Constitution of the Russian Federation following their example, standards of freedom of conscience have been considered only in connection with its religious aspects, more precisely, with religious-confessional aspects. Accordingly, the legal concept of freedom of conscience has acquired a narrower meaning: the right of an individual to decide independently whether to be guided in the evaluation of their actions and thoughts by the teachings of religion or to reject them.

It is impossible to deny the enormous role played by the Universal Declaration of Human Rights, the International Covenant on Civil and Political Rights, the International Covenant on Economic, Social and Cultural Rights, the Declaration on the Elimination of All Forms of Intolerance and of Discrimination Based on Religion or Belief, and other international legal instruments in the development of modern legal and political philosophy.

Yet because of the above-mentioned inconsistency, the broad and substantial concept of “freedom of conscience”, which concerns literally every individual, was reduced exclusively to freedom of religion, that is, to a freedom concerning only those individuals who consider themselves believers. Even more narrowly, it was reduced to the activity of religious associations, meaning the collective form of realizing the right to freedom of religion.

Since the criteria and boundaries of freedom of conscience in international legal documents and national legislation are defined without regard to the essence, nature and significance of this right, its realization becomes subject to unlawful “special” restrictions in accordance with state religious policy.

Thus, legal regulation in the field of freedom of conscience is built on principles that are legally incorrect, lack clear legal criteria and rely on a conceptual apparatus partly borrowed from theology. Freedom of conscience is replaced by freedom of religion; human rights are replaced by the rights of associations; religion is replaced by ideology; and, as a result, the priority of law is replaced by the pri l “elites.” A neat trick, if one's goal is to turn constitutional law into a decorative curtain.

One part of the problem is that the principle of religious freedom was in fact built on the division of society into "believers" and "non-believers", a division largely shaped by the opposition between knowledge and faith at the level of the individual, and between science and religion at the level of society.

Another part of the problem is that in legal science and law-making freedom of conscience is usually treated as a synonym for religious freedom. In other words, the broader concept is replaced by the narrower one, reflecting the absence of an understanding of freedom of conscience as a universal human phenomenon.

Accordingly, legislation that should be aimed at realizing liberty of worldview choice is substituted by "special religious" regulation of the activity of religious associations.

This "special" legislation obstructs the realization by every person of their constitutional rights because it allows the state to intervene in the worldview sphere and in the activity of religious associations. And the state usually does intervene. Religious associations, owing to their integrative capacities, have traditionally been objects of political interest and "special" control by public authorities, both in Russia and in many other countries.

As a result of the application of this "special" legislation, not only are the rights of religious associations violated, and not only is a whole range of democratic principles forming the foundation of the constitutional order blurred. Freedom of worldview choice is suppressed, and individual states, as well as the world community, probably lose the opportunity to make full use of the possibilities of globalization and to find an adequate response to its challenges.

The legally incorrect foundational principles and conceptual apparatus used today in the sphere of freedom of conscience do not allow for the creation of a legal framework capable of fully realizing this basic, system-forming right. The definition and essential content of freedom of conscience as a legal category are in fact distorted from the outset, creating nearly insurmountable obstacles to the realization of the declared principles.

It is therefore clear that an effective legal mechanism for realizing the right to freedom of conscience requires the broadest possible legal definition of this freedom, one free from narrowly confessional as well as atheistic limitations.

7.3. Can Religion Be Legally Defined? The Main Dispute over the Boundaries of Freedom of Conscience

Because the definition and essential content of freedom of conscience as a legal category are in fact distorted and determined through the undefined concept of "fr [illegible] lefining "religion" comes to the foreground.

According to the American scholar Cole Durham, "the most obvious issue is the significance of the term 'religion' itself in determining the validity of claims to the right to freedom of religion. Closely related to these problems are a number of other terms frequently encountered in questions of religious freedom, such as 'church', 'religious community', 'sect', and so on. The difficulty is that formulations almost inevitably create the risk of discrimination against religious groups that fall on the boundary of the definition."

Analyzing the case law of the European Court of Human Rights on freedom of conscience, particularly Kimlya and Others v. Russia, J. Murdoch notes that for a certain time there was no need to define the term "religion" precisely. The Court itself specifically emphasized that "it is of course not for the Court to decide in abstracto whether a particular body of beliefs and related practices may be regarded as a religion." However, X v. the United Kingdom of 1977 indicates that, where doubts arise, "an applicant may be required to prove that a particular 'religion' actually exists." At the same time, it is emphasized that Article 9 is essentially aimed at preventing ideological indoctrination by the state, allowing a person to hold, develop, refine and even completely change personal thoughts, convictions and religion. All of this concerns what is usually called the forum internum, as in Van den Dungen v. the Netherlands.

The problem is aggravated by the fact that the term "religion" is now used very widely in many sciences and in journalism, but often in a very narrow sense borrowed from theology: a confessional sense which, as V. S. Polosin notes, "is not always unambiguous even for believers of the same confession." At the same time, it is obvious that any discussion of religion, especially legal discussion, faces serious terminological difficulties. As Erich Fromm observed, the concept of religion is commonly associated with a system centred on God and supernatural forces. It therefore becomes questionable whether religions without a god, such as Buddhism, Taoism and Confucianism, can be called religions. Secular systems, including modern authoritarianism, are not usually called religions at all, although psychologically they may deserve that name. We simply do not have, and probably cannot have, a word for religion as a phenomenon that is not tied to a specific type of religion and does not bear its imprint.

Yu. E. Fedotova examined the problems of legally defining "religion" and related concepts. She concludes that there is no definition of the fundamental concepts in Russian legislation regulating the external manifestations of the religious sphere of human life. This, in turn, complicates law enforcement, permits expanded subjective interpretation of norms, or, conversely, makes their application impossible. The same researcher points to the inadmissibility of using unclear, polysemous and vague terms, and stresses the need to improve the principles and norms of freedom of conscience on the basis of the laws of logic and the rules of legal technique.

The problem of legally defining religion and religious organizations has fundamental theoretical, methodological and practical significance and has been thoroughly studied by R. A. Podoprigora. In his doctoral dissertation, he gives numerous examples showing the difficulty, and sometimes the impossibility, of legally defining religion both in national legislation and in international legal acts. On the basis of this research, he quite rightly concludes that "it is impossible and inexpedient to formulate the concept of religion in legislation."

Closely linked to this conclusion is another important thesis: the state must recognize religious organizations as existing even outside any organizational and legal form established by the state itself. It is necessary to distinguish between characteristics that describe a religious formation as a social structure and those that describe it as a legal structure. For recognition of the latter, the state may, and often does, establish certain criteria. Yet this does not mean that those criteria apply to all religious organizations, or that if they do not fit a particular social community, that community cannot be recognized as religious.

E. N. Markova states that the term "religion" is not defined in international law and that the constitutions of most states proclaim and guarantee freedom of religion as a fundamental human right while leaving the term "religion" itself undefined. Considering the risks of arbitrary law enforcement, she argues that if one must choose "between two evils", the lesser evil is still an expansive approach to defining religion. In a later work, she notes that religion may be interpreted only within a specific historical and geographical context and in light of accumulated comparative experience in defining religion.

Markova identifies several principal approaches to the legal definition of religion: first, a subjective-functional approach focused on the role that certain beliefs and convictions play in a person's life; second, an essential or substantive approach that seeks a set of characteristics common to all religious beliefs and attempts to identify their essence and features; third, an approach that bypasses the problem of defining religion, more typical of states of the Romano-Germanic legal family, where this issue is not considered to fall within the competence of courts. She further summarizes that in practice, interpretation of the term "belief" offers more opportunities for qualifying conduct that falls under the protection of freedom of conscience and religion. If a person's beliefs concern important and fundamental questions of human life and behaviour, it is not of great importance whether they are classified as religious or not, since both religion and beliefs receive equal protection, including protection of the right to refuse to perform legal duties on grounds of conscience and religious conviction.

In this context, it should be especially emphasized that the use in jurisprudence of terms that are not entirely correct and that lack clear legal criteria provokes deliberately false situations, creating initially insurmountable obstacles to the realization of declared human rights. In the context of freedom of conscience, legal terms must be distinguished from non-legal terms. It is perfectly obvious that, for example, a sociological or religious-studies definition and a legal definition are not always the same thing. Humanity, naturally, noticed this only after building entire regulatory systems on the confusion.

Since no legal definition of "religion" exists and since creating one appears impossible, this term should be replaced in jurisprudence by a broader term, such as worldview, and applied uniformly in accordance with the requirements of modern legal technique. Reasonable proposals by some researchers, including the author's earlier articles, that the term "religion" should be interpreted as broadly as possible and without any discrimination are extremely difficult to implement in practice, if they can be implemented at all.

At the same time, taking into account the trend toward the pluralization and even personalization of religiosity, and the inseparability of religiosity from the personal experience of each human being, it would not be wrong to say that there may be as many religions, in the legal and broadest sense, as there are people on Earth. Thus, in addition to the dominance of a narrow understanding, the crisis of the legal understanding of freedom of conscience and the chain of contradictions in the Russian and international legal framework are rooted in the problem of legally defining religion. Closely related to this is the problem of using knowledge from religious studies in resolving legal questions.

Because of the insufficient legal development of freedom of conscience, tolerance and the fight against intolerance, and because legally incorrect divisive principles are used, religious studies has gained considerable influence in science and education as a compensating scholarly and educational instrument. This sphere should, in principle, be approached with great caution when its conclusions are used for legal purposes.

It should be stressed that the application of religious-studies expertise in law enforcement, especially in the form of state religious expert examinations, has become one of the most controversial elements of Russian practice. Where the law lacks clear legal criteria, expert interpretations may begin to replace legal norms themselves. Then a scholar, expert or commissioned specialist effectively receives the power to decide which worldview is "normal", which is "traditional", and which is "dangerous." That is already a legal problem, not merely an academic one.

In this regard, the Institute for Freedom of Conscience has repeatedly drawn attention to the fact that the existence of religious studies expert examinations, bodies dealing with religion, "anti-extremist" legislation, lists of "extremist" literature and specialized "anti-extremist" security structures raises serious constitutional questions.

The Institute for Freedom of Conscience considers that the very existence of the institution of religious-studies expert examination and other bodies dealing with religion, "anti-extremist" legislation, lists of "extremist" literature and specialized "anti-extremist" security structures contradicts the constitutional principles of freedom of conscience and state secularity, and creates threats to human rights, social stability and state security.

It also emphasizes that constitutional freedom of conscience and the secularity of the state are incompatible with non-legal xenophobic labels and stereotypes such as "sect", "Wahhabism", "religious extremism", "Islamic terrorism", and similar constructions.

The concept of "extremism", despite its legislative consolidation, remains legally defective and highly controversial from a general scientific perspective.

Legal technique does not permit the use in normative legal acts of concepts that have multiple and subjective interpretations and that from the outset predetermine corruption risks and abuses in law enforcement.

Clear legal, and perhaps even scientific, criteria for the worldview sphere, including a legal concept of "religion", are absent, and their creation is impossible.

Especially given the current condition of religious-studies scholarship and the religious-studies community, the use of religious studies, not to mention theology, in the field of freedom of conscience and relations between the state and religious associations inevitably leads to unlawful interference by authorities in the religious sphere, to corruption, and even to a modern version of the Inquisition: a "witch hunt."

Proposals to create "forensic religious studies" or "religious-studies jurisprudence" are therefore untenable from a legal point of view.

The Institute for Freedom of Conscience states that the Expert Council on Religious Studies Expert Examination under the Ministry of Justice of the Russian Federation should be dissolved, and the institution of religious-studies expert examination should be abolished.

It further states that "anti-extremist" legislation and lists of literature should be repealed, and the relevant security structures disbanded.

Public officials should bear responsibility for using in their work non-legal concepts, especially those that incite hostility and intolerance on grounds of religion or belief, such as "sect", "Wahhabism", "religious extremism" and "Islamic terr

Officials, religious-studies scholars, theologians, lawyers and others who are personally interested in benefiting from this sphere should leave liberty of worldview choice alone. Its realization should be based on the same constitutional and legal grounds as those applicable to other public non-profit associations.

It must be acknowledged that contemporary Russia lacks independent and objective religious studies and a corresponding independent scholarly community. Moreover, there are many facts showing the transformation of religious-studies scholars into service personnel for power groups, religious corporations and professional corporations. There are also facts of direct and indirect xenophobia by some religious-studies scholars, including through the linguistic construction of inequality. The influence of religious studies on the state of freedom of conscience and tolerance, and on observance of the constitutional principle of state secularity, cannot be regarded as positive.

It appears that the use of religious-studies knowledge in jurisprudence, particularly in the formation of "legal religious studies", is not always correct, since the very prospects of religious studies are connected with increasing pluralism of methodological approaches. An equally important condition for the development of religious studies should be its distancing not only from theology, from the interests of religious corporations and from private financial interests, but also from political commissions by the state.

A similar attitude is warranted toward a number of disciplines adjacent to religious studies, above all the sociology of religion. The use of sociological tools and methods to study religion as a social subsystem has undoubtedly proved effective and justified. The same cannot be said of the use of sociological research results in jurisprudence. Such data are often contradictory because different research centres produce results with a very wide spread.

For this reason, the proposal by A. V. Pchelintsev and other authors to form "legal religious studies" or "religious jurisprudence" is doubtful. Pchelintsev speaks seriously of forensic religious studies as part of legal religious studies. In his view, the subject of this science could include state-confessional relations in the broadest sense, the theory of freedom of conscience and religion, problems of legal guarantees of freedom of conscience and religion, the relationship between religious and legal norms and conflicts between them, state religious-studies expert examinations, forensic religious studies itself, and other issues.

Thus, the problems of Russian religious studies and the religious-studies community are a significant factor in the crisis of scholarly understanding and in the systemic crisis of realizing freedom of conscience. Moreover, the crisis of religious studies as a science predetermines problems in the teaching of religious-studies disciplines within the education system.

Ultimately, it may be concluded that teaching religious-studies disciplines within the education system is highly controversial and does not fully contribute to the realization of human rights, the achievement of tolerance or the fight against intolerance. It therefore does not correspond to contemporary global processes.

No less important for finding ways to improve the legal framework for realizing freedom of conscience is the relationship between "freedom of conscience" and "relations between the state and religious associations", "confessional policy" and "state secularity". As already noted, in a modern secular state there should be no "special" state-confessional relations and no state religious or confessional policy. Relations between a democratic, law-governed, secular state and religious associations should be built on the same legal grounds as relations with other public non-profit associations.

In conclusion, the term "religion" is legally undefined, which predetermines arbitrariness and human rights violations in its application. Since it is impossible to create a legal definition of "religion", the term should, in the author's view, be abandoned in the legal system. As for the possibility, proposed by some authors, of using the term "beliefs", this term is also narrower than "worldview" and is only one of its elements, giving an individual or social group confidence in their views, knowledge and assessments of reality. A worldview is a system of views about the world and the place of the human being in it, about a person's relation to surrounding reality and to themselves, as well as the main life positions, ideals, beliefs, principles and value orientations conditioned by those views. In the context of improving the principles and norms of international and domestic law on freedom of conscience, the use of terms derived from "worldview" appears more correct.

7.4. Beyond the Division into "Believers" and "Non-Believers": Toward a New Paradigm of Worldview Choice

Perhaps the most fundamental problem directly underlying the crisis of the legal understanding and realization of freedom of conscience is the principle of division into believers and non-believers. This division, in turn, rests on the opposition between science and religion, knowledge and faith.

This division has no clear legal criteria. It implies not only distinction but opposition, and therefore predetermines conflict from the very beginning. An additional argument against the correctness of the division into "believers" and "non-believers" is provided by sociological research showing that about 30 percent of Russians fluctuate between belief and non-belief and are characterized by worldview uncertainty.

The principle of dividing people into "believers" and "non-believers" underlies the ntly, between their followers and those who do not belong to any confession. The same division also underlies the division of peoples and nations through the identification of religious-confessional and national identity. The coincidence of national and confessional self-identification, religious and national consciousness, manifests itself in the fact that not only believers but also non-believers associate themselves with one or another religious tradition: Russians with Orthodoxy, Tatars with Islam, Buryats and Kalmyks with Buddhism, and so on.

In the author's view, the legally incorrect principle of dividing society into "believers" and "non-believers", which lies at the foundation of legislation on freedom of conscience, is one of the main factors allowing certain power structures to define and control the worldview sphere at their discretion and in their own interests, contrary to the interests of society.

As already noted, the principle of dividing society into "believers" and "non-believers" is rooted in the relationship between science and religion, which even in the twenty-first century is described by many researchers as "contradiction", "confrontation", "conflict" and even "information war." Indirectly, including through relations between politics and religion, this conflict permeates all spheres of social relations and ultimately obstructs positive political integration and the solution of global problems.

The author agrees with A. N. Whitehead, who argued that when we consider the significance of religion for humanity and the significance of science, we may say without exaggeration that the future course of history depends on the decision of the present generation regarding the relations between them.

At the level of each individual, the conflict between science and religion manifests itself in the opposition between knowledge and faith, which leads to the division of society into "believers" and "non-believers." Once indirectly fixed in legal norms, this conflict paradigm created almost insurmountable obstacles to the realization of the system-forming right to free worldview choice, which lies at the basis of personal self-realization.

Because confessions have traditionally been within the sphere of political and state interests, the believer/non-believer paradigm allows the realization of the need for self-consciousness, the formation of a picture of the world and the self-determination of each individual to become, to a very significant extent, controllable by power groups.

It should be noted that overcoming the confrontation between science and religion through awareness of the objective change in their relationship during the process of evolution would contribute to the separation of politics and religion, or of the state and religious associations. As a result, it would reduce, and perhaps overcome, the conflict potential of the ethno-confessional factor through the realizatio [illegible] iere.

In turn, the formation of an effective legal mechanism for realizing the right to liberty of worldview choice, understood as the free formation of one's picture of the world, and recognizing diversity as well as the changing relationship between knowledge and faith in each individual, in accordance with the realities of the twenty-first century and the dynamics of global processes, would contribute to global thinking, political integration and the resolution of global problems.

The analysis of the factors underlying the crisis in the realization of freedom of conscience reveals that the crisis rests on an old paradigm based on the division "believer/non-believer." This paradigm is incorrect when applied in the legal system and implies the following:

1. The opposition between science and religion in society, and between knowledge and faith in each person;

2. The opposition between the individual and the collective;

3. The merging of the worldview sphere and politics, and the use of religious associations for political purposes;

4. The dependence of the realization of constitutional rights on relations between the state and religious organizations and on a religious or confessional policy formed at the discretion of public authorities;

5. The opposition between national interests and global trends, and between nation-states and the international community;

6. The use of legally incorrect principles and a conceptual apparatus that lack clear legal criteria.

As a consequence, a dead-end yet self-reproducing quasi-scientific system has formed in this sphere. It is based on the dominance of "official scholarship", which fulfils an ideological state order, and on certain scientific and educational structures belonging to particular religious confessions and focused on purely corporate interests. Confessionally oriented structures are usually incapable of maintaining a principled scientific position aimed at realizing liberty of worldview choice for everyone, while state scientific institutions often perform the task of justifying unconstitutional state-confessional relations and confessional policy.

Positive change requires a transition to a new paradigm of legal regulation of freedom of conscience, one that does not use legally incorrect divisive principles and that implies the following:

1. Rejection of the opposition between knowledge and faith, and transition to the concept of recognizing diversity and respecting the different relationships between knowledge and faith among different people in society;

2. Rejection of the opposition between the individual and the collective. Collective interests are derivative of individual interests;

3. Formation of a legal framework for separating worldview, including religious, associations from politics within the legal field;

4. Limitation of "special", in essence non-legal, regulation of the worldview sphere, and independence of the realization of constitutional rights from relations between the state and religious organizations and from religious or confessional policy;

5. Rejection of opposition between national interests and global trends, and their harmonious combination;

6. Rejection of legally incorrect principles and conceptual apparatus that lack clear legal criteria.

The division of society into believers and non-believers, whether explicitly or implicitly fixed in the legal system, obstructs the realization of the principles of liberty of worldview choice. Removing such incorrect principles from the legal foundations of society is a necessary condition for the realization of human rights, the overcoming of global security problems and the prevention of conflicts.

7.5. Globalization 4.0: How to Avoid Digital Dictatorship while Preserving Freedom and the Neutrality of the State

Today humanity lives amid intensifying global processes aimed at the formation of a planetary socio-techno-natural system. The growing integrity of the modern world is inseparable from the globalization of social relations, understood as the integration of social interactions into a single planetary system. These relations are changing dynamically, yet unevenly. This unevenness is accompanied by instability in international relations and by the aggravation of global risks that threaten the very existence of human civilization. In such conditions, humanity needs a transition toward a new system for managing global processes, capable of overcoming global risks and achieving sustainable development.

Ethno-confessional barriers remain one of the serious obstacles to international cooperation for sustainable development. For this reason, the improvement of international and domestic legal principles and norms in the field of the right to freedom of conscience becomes especially relevant. Strengthening the effectiveness of these principles and norms may help weaken such barriers and address problems of global security.

As noted earlier, globalization should be understood as a set of objective, yet contradictory and uneven, processes aimed at forming a unified planetary socio-techno-natural system. [illegible] new state of globality. Globality, in turn, should be understood as the integration, interpenetration, openness and interdependence of all social subsystems, as well as the technological and natural subsystems connected with them, on a planetary scale.

The present stage of human development is inseparable from the introduction into the economy and other spheres of life of artificial intelligence, robotics, neurotechnologies and many other technologies. The concept of a qualitatively new stage of digital globalization 4.0, based on the technologies of the Fourth Industrial Revolution, was proposed by Klaus Schwab and later interpreted by other researchers. Digital globalization 4.0 is a set of macro-scale, multidimensional and internally contradictory integration processes aimed at forming a single planetary system in which the innovative technologies of the Fourth Industrial Revolution play a decisive role.

Digital globalization 4.0 has enormous potential to improve human living conditions. At the same time, it carries serious challenges, including existential ones. The most recent technological leap is connected with the launch of large mathematical models with a very wide range of capabilities. Humanity has acquired tools of extraordinary power, and, with its usual genius for timing, has done so before fully agreeing on what they should be used for.

The study of the development of social subsystems intended to form the global system mentioned above reveals their uneven development. This predetermines imbalance and weak governability within the social system itself, as well as problems in its interaction with technological and natural spheres.

Most foreign and Russian researchers reasonably argue that global risks affect the interests of all humanity and that the survival of civilization depends on their resolution. Accordingly, unresolved global risks constitute a threat to the existence of human civilization.

In the context of this study, among the challenges of the new stage of globalization caused by the Fourth Industrial Revolution, the following should be mentioned: digital militarization; indiscriminate surveillance; hatred and hostility intensified through digital channels; cybersecurity threats; loss of control over artificial intelligence; and inequality caused by the absence of access to digital technologies and to the benefits they generate.

According to the 2025 Technology and Innovation Report of the United Nations Conference on Trade and Development (UNCTAD), the introduction of artificial intelligence may affect up to 60 percent of jobs in developed countries and around 40 percent in developing countries. This may intensify social inequality and income gaps if governments do not adapt education and social protection systems. On the one hand, AI may raise labour productivity. On the other, many [illegible] neans of subsistence.

The core problem is that the main financial gains from automation are usually captured by company owners, while ordinary workers and entire countries dependent on low-cost labour may become poorer. AI may further enrich owners of capital and highly qualified specialists, leaving everyone else behind.

In January 2026, on the eve of the World Economic Forum in Davos, Oxfam published the report Takers Not Makers: The Unjust Poverty and Unearned Wealth of Colonialism, or, in another public framing, Resistance to the Rule of the Rich: Protecting Freedom from Billionaire Power. The report states that against the background of unprecedented global digital development, the wealth of billionaires had grown by more than 16 percent by 2025, three times faster than the average over the previous five years, reaching 18.3 trillion US dollars, the highest level in history. This occurred while one in four people regularly experienced food insecurity and almost half of the world's population lived in poverty. In other words, the super-rich strengthen their political power in order to shape the rules of our economy and society in their own interests and to the detriment of the rights and freedoms of people across the world.

It may be assumed that the fundamental cause of global risks threatening the existence of human civilization, and at the same time one of the factors preventing their effective resolution, is the imbalance in the development of social subsystems and in their interaction with technological and natural spheres. Preventing global risks therefore requires a transition to the sustainable development of the planetary techno-socio-natural system. This transition must be based on levelling the development of its constituent subsystems and achieving a state of balance and governability.

The main conditional models for levelling development and achieving balance in the planetary techno-socio-natural system are: first, limiting the development of advanced subsystems, that is, slowing growth; and second, overcoming the lag of insufficiently developed subsystems, that is, managed growth. In the author's view, the most realistic, correct and promising approach is the formation and implementation of the second model: overcoming the lag of insufficiently developed subsystems in order to achieve balanced, manageable and sustainable development and to overcome global risks in the planetary social system and, ultimately, in the techno-socio-natural system as a whole.

In scholarly literature and in the documents of international organizations, sustainable development is understood as a managed, integrated and balanced process of realizing the vital interests of humanity, including future generations, in harmony with nature. The transition to sustainable development is inseparably linked with overcoming global risks and ensuring the survival of human civilization.

Thus, the need for global governance is understood here as an instrument for solving global problems and moving civilization toward sustainable development. Global governance is a planetary system of principles, norms, institutions, techniques and methods aimed at overcoming global risks and achieving sustainable development. International law, in interaction with domestic legal systems, as well as international institutions, may be regarded as elements of the emerging system of global governance. The survival and flourishing of civilization through sustainable development and the resolution of global risks may be identified as the main purpose and tasks of such a system.

In conditions of acceleration, complexity and globalization of social relations, hierarchical systems objectively lose effectiveness and gradually give way to new non-hierarchical models. Attempts to create an institutional model of global cooperation and governance adequate to modern realities, while relying on outdated scientific approaches, outdated and ineffective legal and regulatory systems, and personnel who have not been prepared for contemporary conditions, are hardly correct or effective.

Systemic continuity and evolutionary transition are essential conditions for the effective formation of a system for managing global processes in order to overcome global risks and achieve sustainable development. A smooth evolutionary transition from a hierarchical to a polyhierarchical and, ultimately, non-hierarchical system of global governance requires systematic work at all main levels: scientific and educational, normative, law-enforcement, media and mass communications, and civil society. This work should be based on global cooperation, primarily in the field of law and legal education.

As the primary foundation of adequate global governance, it is necessary to form a global normative system, or global law, understood as a system of knowledge, principles, norms, other social regulators, institutions and academic disciplines aimed at forming and reproducing a conceptual, normative and professional base for overcoming global risks and transitioning to sustainable development. In order to regulate global processes, overcome global risks and ensure the sustainable development of human civilization, innovative development of the normative system, as science, as a system of norms and as an academic discipline, is needed. In particular, it is necessary to define the modern concept and content of global law and education, including global legal education. The possibility of creating a system for managing global processes and achieving sustainable managed development in the interests of every inhabitant of planet Earth depends on this.

It should also be emphasized that the creation of a system of global process management adequate to modern realities and oriented toward the future requires the formation of a global normative system and global legal education supported by digital technologies. These may include decentralized, distributed, transparent network platforms and registries such as blockchain and derivative technologies. Global digital law and education, including global digital legal education, appear as an image of the future and as the basis of digital global governance for solving global problems and moving toward sustainable development.

The basic characteristics of blockchain, including planetary scale, publicity, openness, transparency, continuity, speed, decentralization, security, confidentiality, data protection, independence, stability, memory, equality of participants, consensuality, real-time operation, absence of intermediaries, low transaction costs and a wide range of applications, help solve the problem of institutionalizing trust and minimizing operational costs. In many respects, these characteristics correspond to the tasks of forming an adequate system of global governance. In particular, they may help solve the problems of demonopolizing power and property, transitioning to direct democracy, ensuring self-execution of legal norms, realizing human rights, distributing social benefits more evenly, strengthening social justice, overcoming corruption and improving the effectiveness of governance. It is precisely the characteristics and possibilities of innovative blockchain technologies, especially in combination with other technologies, that create grounds for the real embodiment of advanced theoretical developments in the field of globalization of law and governance, ideas that only yesterday seemed almost fantastic. It is necessary to move toward the creation of a global legal platform based on digital technologies.

In this context, highly promising directions requiring further development include global digital law and education, including global digital legal education, and global digital governance based on them.

International law, as a science, as a system of principles and norms, and as an academic discipline, in interaction with domestic legal systems and other regulatory systems, should serve as the initial material for the formation of global law. In this context, one of the first tasks is to develop directions for reforming the principles and norms of international law and its interaction with the legal systems of states. Only after reforming principles and norms can we speak of adequate reform of the institutions of international law in the context of global governance. In some cases, parallel reform of the international normative and institutional systems may also be possible.

A contemporary understanding of international law, as a system of principles and norms governing interactions between states and other subjects, must be connected with the need to realize human rights. This connection will help overcome global risks and move human civilization toward sustainable development. Since modern legal understanding is inseparably linked with human rights, and contemporary social relations are inseparably linked with global processes and problems, this must be reflected in the definition of law itself.

The principal promising directions for the progressive development of international law, intended to move humanity from spontaneous instability in international relations toward governability of modern processes and sustainable development, and requiring the development of theoretical-legal, normative and institutional foundations, include the following: human rights; freedom of conscience as the possibility of worldview choice; tolerance as equality and mutual respect, together with resistance to intolerance; worldview neutrality of subjects of international law; reform and development of new principles of international law as its foundational beginnings; the supremacy of international law based on the principle of legal certainty; universalization of principles and norms of international law; a non-polar system of international relations, meaning a rejection of the "right of force" and a transition to the "force of law"; strengthening the role of the United Nations and other international organizations; and the development of global digital law and legal education.

Among the main priorities for the development of international legal institutions, aimed at achieving balance and governability and overcoming the lag of the global political-social subsystem, the following should be identified: modernization of universal international organizations, primarily the United Nations and its specialized intergovernmental institutions and non-governmental agencies; modernization of regional international organizations, both intergovernmental and non-governmental; modernization of interaction between universal and regional international organizations; modernization of interaction between intergovernmental and non-governmental international organizations; and modernization of interaction between international organizations and other subjects of international law, including states, peoples and nations struggling for independence, quasi-states, individuals, legal entities and others. In light of the understanding of globalization described above, the modernization of interactions among global political institutions should be directed toward their integration, interpenetration and openness.

It is vitally important to develop theoretically and implement practically an adequate model of international relations. In conditions of intensifying global risks, unipolar, bipolar and even multipolar models cannot solve problems of peace and security on the planet because they all allow the use of force. It is urgently necessary to create, legally consolidate and apply an innovative non-force, non-polar model of international relations based on a radical reform of international law and institutions, excluding the use or threat of force by states. It appears advisable to form an international scientific and educational programme, ıs and International

Organizations in the Context of the Formation of a Global Normative System (Law) and Governance System."

In the author's view, human rights should become the flagship branch of the global normative system, or global law, for the purpose of forming global governance to overcome global risks and achieve sustainable development. The increasing complexity and globalization of social relations require greater freedom for the individual, understood as the ability to act in one's own interests without harming others. This freedom is embodied in human rights. For this reason, human rights serve as the core of modern international and domestic law and as the basis for the formation of future global law.

The system-forming right to freedom of conscience, understood as worldview choice, is a necessary condition for the formation of a global normative system, or global law, and for global governance aimed at overcoming global risks and achieving sustainable development. In conditions of intensifying global risks, an effective legal mechanism for realizing the right to freedom of conscience, together with the secularity of the state as its guarantee, is a necessary condition for separating religious and political institutions, overcoming xenophobic ethno-confessional dividing principles exploited by domestic political structures, and removing obstacles to the formation of an adequate system of global governance.

The realization of liberty of worldview choice is also a necessary condition for transforming value orientations, forming global thinking and moving toward the sustainable development of civilization. In addition, the realization of liberty of worldview choice is a key condition for reducing xenophobia, intolerance and ethno-confessional violence, and for ensuring global security.

Thus, liberty of worldview choice, as a system-forming right within the system of human rights, must occupy an important place in the global normative system, or global law, for the formation of global governance aimed at overcoming global risks and achieving sustainable development.

7.6. From Declarations to Guarantees: How to Turn Liberty of Worldview Choice into a Working Legal Instrument

The identified main trends in the development of global processes, the causes of the crisis in the realization of freedom of conscience, and the strategic directions for overcoming it through a new paradigm make it possible to formulate several prospects for improving the legal framework for realizing the right to freedom of conscience. A fundamental change in the existing order of things is necessary not only in Russia but also at the international level, since international legal principles and norms in the field of freedom of conscience are based on the same divisive principles and fail to meet the requirements of modern legal technique.

Creating an effective system for the realization of the right to freedom of conscience at the global and domestic levels would limit the possibility of using religion as an ideology for political purposes. It would undermine the very foundation of totalitarianism: the possibility of "special" restrictions, control and suppression of the worldview sphere. It would also help overcome ethno-confessional barriers and threats in the field of global security.

A comprehensive scientific and theoretical elaboration of a broad concept of freedom of conscience is required. It should include the following blocks:

1. development of a modern theoretical-legal model of freedom of conscience based on a new paradigm, eliminating contradictions in foundational principles and conceptual apparatus;

2. reform of universal international legal principles and norms concerning the right to freedom of conscience on the basis of this modern theoretical-legal model;

3. reform of regional international legal principles and norms concerning the right to freedom of conscience on the basis of the modern theoretical-legal model and universal international legal principles and norms;

4. reform of national domestic legal institutions of freedom of conscience on the basis of the modern theoretical-legal model and international legal principles and norms;

5. formation and improvement of international legal and national domestic systems of guarantees for realizing the right to freedom of conscience on the basis of the principle of worldview neutrality of states and other subjects of international law.

In the search for strategic approaches to the realization of the right to freedom of conscience, the issue of legal regulation of religious associations inevitably arises. If we follow the logic of the broadest possible understanding of freedom of conscience, the rights of religious communities should be considered derivative of the right of every individual to freedom of conscience. This means that the activity of religious associations should be regulated on general grounds, as the activity of public and non-profit associations. In other words, the existence of a special legal status for religious organizations appears not entirely correct, including because there is no legal definition of religion. Incidentally, during the discussion of the Russian draft law on freedom of religion in 1990, a similar point of view was voiced, but rejected for a number of reasons, one of which was the weak scholarly development of the issue.

A special "religious" law should not exist at all in the legal system of a democratic state. This assumption is, of course, far from indisputable, but there are serious grounds for it. The main one is the need to limit special regulation by the state of the spiritual and moral sphere, to which freedom of conscience belongs. A democratic state, in order to observe neutrality in the worldview sphere, should not have the power to determine at its own discretion what is religion and what is not. There are no clear legal criteria for this and there cannot be. There is not even a single definition of religion in religious studies. According to the Russian scholar of religion E. S. Elbakyan, there are more than two hundred such definitions. According to the Ukrainian scholar L. Filippovich, there are around three hundred. Humanity, naturally, looked at this conceptual fog and decided to build legislation inside it.

Administrative legal regulation of religious organizations, like that of other public non-profit associations, should be carried out only within the limits and scope related to their activity as legal entities and to their compliance with non-profit status. Abuses in the sphere of freedom of conscience should be addressed exclusively by criminal justice. Criminal justice should be made more effective so that there is no need to duplicate it through special legislation.

In order to realize the principles of freedom of conscience and reduce the gap between power and society, power groups must be deprived of legal mechanisms for controlling and using the worldview sphere, mechanisms usually based on legally incorrect terms borrowed from theology.

Society needs state and legal guarantees for the protection of freedom of conscience, first of all from special administrative legal restrictions. Preventive powers enshrined in legislation in an area that lacks clear legal criteria, combined with the ancient desire of power to control and restrict the worldview sphere, create from the outset the preconditions for arbitrariness and for depriving society of declared rights and freedoms. In this context, administrative restrictions introduced to prevent unlawful activity by religious associations in the field of freedom of conscience may inflict incomparably greater damage on civil society than the supposed violations themselves.

Another important aspect of state and legal protection of freedom of conscience is criminal law protection of relations arising in the realization of the right to freedom of conscience. Criminal acts must be suppressed regardless of what they are disguised as. The course toward building a democratic law-governed state and civil society is incompatible with clerical ideologization of state institutions, the legislative consolidation of rigid control, and inequality toward other confessions.

In the author's view, the legal regulation of religious associations in accordance with the modern theoretical-legal model of freedom of conscience as liberty of worldview choice should be based on the following principles:

1. the rights of associations must be regarded as derivative of human rights and freedoms, in order to avoid replacing the latter with the former;

2. the registration of religious associations, as with other public and non-profit associations, should be declaratory in nature;

3. special legal status and special regulation of religious associations should be abandoned; their activity should be regulated on general grounds together with other associations;

4. special benefits for religious associations should be abandoned; benefits may be granted on the same grounds as to other public non-profit associations;

5. special religious administrative legal and criminal legal restrictions introduced to prevent unlawful activity in the sphere of freedom of conscience by religious associations should be abandoned;

6. administrative legal regulation of religious organizations, like that of other public non-profit associations, should be carried out only within the limits and scope connected with their activity as legal entities and with compliance with non-profit status;

7. criminal prosecution is permissible only for unlawful acts, not for worldview.

Accordingly, the prospects for improving the legal framework are inseparable from the need for systemic measures aimed at realizing freedom of conscience as liberty of worldview choice. These measures include:

1. overcoming the politicization of religion and the clericalization of politics;

2. ending the religious policy of the state;

3. overcoming the dominance of special interests of political power and confessional bureaucracy;

4. acceptance of the ideals of freedom of conscience by religious confessions;

5. acceptance of the ideals of freedom of conscience by political leaders and parties;

6. improvement of law-enforcement practice;

7. improvement of the institution of constitutional responsibility in the field of freedom of conscience;

8. creation of a system of public oversight over compliance with the principles of worldview neutrality of the state, equality of religious associations and ideological diversity;

9. development and implementation of an anti-corruption strategy in relations between the state and religious associations;

10. monitoring studies in the field of freedom of conscience and in relations between the state and religious associations, including long-term conceptual trends and legislative transformation;

11. transparency, publicity and public awareness concerning relations between the state and religious associations;

12. adequate media coverage of problems of freedom of conscience;

13. reform of the activity of state human-rights bodies in the protection of freedom of conscience;

14. systematic work on freedom of conscience within the state and municipal education systems;

15. elimination of the thematic gap in the protection of freedom of conscience by human-rights NGOs;

16. systematic educational work on freedom of conscience in society.

As for the "American" model of religious freedom, often presented as the most advanced, it is indeed the best, but only among the worst. Its main difference from the models of most other countries is that, among the main components of religious policy, the "stick and carrot" of persecution and privilege, the United States mainly uses the carrot. Yet in the essential point these models remain similar: they assume the dominance of political interests, which always prove stronger than the ideals of freedom.

This means that there are no religious persecutions in the usual sense, but "quiet" discrimination appears at the level of benefits. A religious organization may be registered and receive official status through a declaratory procedure. However, to obtain tax exemption, it must prove to the Internal Revenue Service of the US Department of the Treasury the sincerity of religious beliefs, the connection between those beliefs and rituals or practice, and compliance with national policy. Obviously, not everyone can prove "truth" before a tax authority, which is a sentence one would prefer not to have to write in a serious book, but here we are.

It is impossible to deny the enormous importance of the First Amendment, and of its interpretations by the Supreme Court of the United States, which declares that Congress shall make no law respecting an establishment of religion or prohibiting the free exercise thereof. Yet it must be recognized that it does not guarantee liberty of worldview choice to everyone. One reason is that the Amendment cannot protect freedom from corruption in relations between the state and religious associations, corruption conditioned by the alliance of religion and politics.

The "American model" presupposes separation between the state and religious associatior nd political institutions. Characteristically, the scholarly and legal framework does not prevent this. Consequently, state-confessional separation is not irreversible. Undoubtedly, religious freedom formed one of the foundations of the stability of the federal structure of the American state and of the prosperity of its multi-ethnic and multi-confessional society. The Founding Fathers of the United States had good reason to believe that small groups, when treated with respect, are more loyal to the political regime than they would be under the dominance of one confession.

For more than two hundred years, a global system of legal guarantees of religious freedom has been developing. Yet it cannot be said that this process has been completed. Even after the adoption of the US Constitution in 1787 and the Bill of Rights in 1791, which directly contains the First Amendment, some states continued officially to preserve confessional preferences. In reality, the United States is a country not only of religious freedom, but also of its contrasts. Students in public schools may not pray publicly, but sessions of Congress are opened with prayer, as are conventions of both major parties. A teacher in the same school may be prohibited from displaying a cross, yet the President takes the oath on the Bible, and chaplains are an integral part of the US military. And this is not even to mention the inscription "In God We Trust" on US currency.

If the "American" model is copied blindly at the international legal level, as well as in Russia and other countries, then even if it takes root, it will not solve most of the problems. Even in the United States it does not solve many problems and often malfunctions. The American scholar J. Wood once stated that the rejection of the principle of religious freedom in relation to any one group of people represents a simultaneous threat to the religious freedom of all; that the infringement of the religious rights of a follower of any one religion is at the same time a denial of the religious rights of followers of all other religions; and, finally, an attack on the very foundation of human rights in general. These words reflect the American evolution from religious tolerance to religious freedom.

Today the next step is needed: a transition toward freedom of conscience as liberty of worldview choice. Accordingly, religious freedom for believers is possible only within the broader boundaries of freedom of conscience for everyone and without division. This means that today liberty of worldview choice should take the place of religious freedom, not by denying it, but by expanding its boundaries.

In this context, it should be noted that rejection of the principle of liberty of worldview choice in relation to any one group of people simultaneously threatens the religious freedom of all; infringement of the rights of the follower of any one worldview simultaneously denies the rights of followers of all other worldviews; and, finally, undermines the very foundation of human rights in general. Overall, it is necessary to create an independent international cooperation programme within which modern and forward-looking scholarly models should be formed, aimed at improving the system of legal guarantees of liberty of worldview choice at the international, global and national, domestic levels.

It appears advisable to study the phenomenon of freedom of conscience as a complex of interactions among elements, categories, concepts and definitions, belonging to the field of many sciences: global studies, history, political science, psychology, religious studies, sociology, jurisprudence and others. The priority direction of research, its red thread, should be the search for fundamentally new approaches taking into account the global perspective of world development. In this context, a model should be formed that contributes to solving global security problems facing humanity and to ensuring its survival.

In general, improving the situation requires systematic work by international and domestic expert and human-rights organizations at all levels: scientific-theoretical and educational; legislative; the activity of public authorities, law enforcement and judicial practice; human-rights activity; and mass media. The relevance of this work increases in conditions of intensifying global challenges and threats to global security.

7.7. Digital Technologies, Cognitive Freedom and Digital Human Rights by Design. Can Consciousness Be Protected from External Intrusion?

Violations of the worldview neutrality of the state and of liberty of worldview choice, and the resulting growth of intolerance and discrimination, multiply the threats to the security of the individual, society and the state. They are also connected with the risk of disintegration of the federal system of Russia. In current conditions, a critical rethinking and fundamental reform of the principles and norms governing the right to freedom of conscience is a necessary condition for preventing large-scale bloody conflicts and preserving a multi-confessional and multi-ethnic country.

In the context of the unprecedented development of global digital technologies, the risks of using them to violate human rights, including the right to freedom of conscience, are becoming especially acute.

According to Klaus Schwab, the nature of the ongoing changes is so fundamental that world history has never known an era of this kind: a time of both great opportunities and potential dangers. He points to emerging and unprecedented possibilities across a wide range of fields, including artificial intelligence, robotics, the Internet of Things, autonomous vehicles, three-dimensional printing, nanotechnology, biotechnology, materials science, energy storage and quantum computing. In another work based on the materials of the World Economic Forum [...] opment of digital technologies capable of completely changing established ways of perceiving the world, processing data, coordinating action, and producing goods and services. Over time, these technologies will change everything we currently take for granted: from the mechanisms of production to the tools of communication, work and perception of the surrounding world.

Technologies that can affect the functioning of the human brain against a person's will, as well as extract, store and process information from it, may pose a particular danger to liberty of worldview choice. These neurotechnologies, especially in combination with other technological capabilities such as the Internet of Things, artificial intelligence, facial recognition and related systems, may be used for control over the individual and may deprive a person of liberty of worldview choice.

Tony Roberts and Marjoke Oosterom warn that technology has ceased to be merely a tool of communication and has become one of the main methods by which authoritarian regimes retain power. They identify several key methods: digital surveillance through artificial intelligence, facial recognition and databases for monitoring citizens; censorship through the blocking of content and control over Internet providers; disinformation through bots and trolls used to manipulate public opinion; and shutdowns, meaning the full or partial blocking of the Internet during protests. Their work stresses that digital authoritarianism has become a global market, and that methods of digital control are increasingly penetrating democratic countries under the pretext of fighting terrorism or disinformation. This creates a threat to human rights worldwide. They also note that digital identification systems, including databases, biometrics and identity verification, were initially created to simplify public services, but in authoritarian contexts they can become instruments of surveillance, control and exclusion.

Overall, indiscriminate digital control through opaque collection, processing and use of personal, biometric, biological and other data by means of digital technologies 4.0, including video surveillance, facial recognition, biometrics and mobile applications, is already a serious factor in human-rights violations. Its global negative consequences remain far from clear. Using the fight against unlawful activity and the pandemic as a justification, some states have created digital systems for controlling their populations. One of the leading examples is China, which has built a system of total control in the Xinjiang Uyghur Autonomous Region, where almost every movement is monitored through facial-recognition cameras, geolocation sensors, drones, applications, big-data processing systems and other artificial-intelligence technologies. In recent years, this system has been continuously improved and expanded across the country and beyond.

To a significant extent, one must agree with the authors who warn that the accumulation of data and behavioural-prediction algorithms creates a new architecture of power that is not subject to democratic control. Political and legal resistance to this threat is highly relevant if constitutional human rights and the human future itself are to be preserved.

The volume Human Rights in the Digital Domain: Core Questions, edited by Triin Pajuste, shows that the rapid digitalization of society requires a deep adaptation of existing legal mechanisms for the protection of human rights. The authors emphasize the need for an international and interdisciplinary approach to the effective regulation of digital technologies and the prevention of new forms of discrimination and rights violations.

Susie Alegre examines threats to freedom of thought in the digital age, especially risks associated with algorithmic profiling, emotional monitoring and the collection of biometric data. She shows how emotion-recognition technologies and recommender algorithms limit intellectual autonomy by creating information bubbles and enabling manipulation of consciousness. As a response, Alegre proposes stronger protection of mental privacy, legal restrictions on the use of emotion-analysis technologies and an updated conception of human rights that reflects the challenges of the digital age.

Among the many technological innovations, emotion-recognition systems, or affective computing, deserve special attention. These systems allow computers to determine a person's emotional state by analysing facial expressions, voice, text and physiological indicators. The process usually includes three stages: data collection, feature extraction and classification through neural networks.

Facial analysis, for example, records the position of key facial points and compares them with supposedly universal expressions of joy, sadness, anger and other emotions. In voice analysis, the system evaluates timbre, pitch, speech rate and intonation. Text analytics analyses the meaning of words and the context of a message in order to determine mood. In some cases, pulse, blood pressure, temperature and breathing rate are read through wearable devices or contactless sensors.

The fields of application for emotion-recognition systems are very broad: marketing and retail, where they are used to assess consumer reactions to advertising or new products; customer service, where they monitor the mood of clients in call centres; medicine, where they are proposed for diagnosing depression and monitoring patients in real time; and security, where they are used to identify suspicious behaviour or excessive stress in public spaces. The education system deserves separate mention, since such systems may be used to monitor student engagement

For example, at Moscow City University, cameras with artificial intelligence that record video for neural-network analysis of students' emotions were controversially installed in at least 24 classrooms. As a result, the data of hundreds of people were tracked, collected and processed through a neural network. The initiators of the emotion-analysis system argued that this was very valuable for students preparing to become teachers, since while practising teaching they would receive not merely the subjective opinion of a methodologist, but supposedly objective data helping them build their individual teaching style.

It must be emphasized, however, that artificial-intelligence data are not always objective. AI relies on the existing body of human experience, and that experience includes prejudices embedded in public consciousness. Consequently, AI also learns from human bias. A well-known example is Amazon's attempt to create an artificial-intelligence algorithm to automate recruitment. The system began to downgrade applications from women. The programme was later edited so that the word 'women' and related terms would not be treated as negative, but the problem was not solved, and the project was eventually closed.

The problems created by emotion-recognition systems include serious violations of human rights, privacy and freedom of expression; inadequate consent and transparency; risks to confidentiality and personal-data protection; distorted interpretation and bias; psychological pressure on students; demotivation; and a reduction in the role of the teacher. Taken together, these problems require rejection of this deeply troubling technology. At the same time, proposals for a complete ban on emotion recognition are still largely absent in the academic literature. Existing proposals usually focus on technological accuracy, privacy risks, ethical dilemmas and potential discrimination.

S. S. Kulov, discussing the need to improve legal norms, notes that the digital transformation of society has generated de facto new forms of interference with the internal freedom of belief that were unknown to classical human-rights law. Algorithmic influence on a person's information environment, behavioural data analysis, emotion-recognition systems and emerging neurotechnologies create risks to freedom of thought, conscience and belief that are not always obvious, yet are large in scale.

At present, most states do not have a direct prohibition on this technology. However, several international organizations indicate that its use violates fundamental rights to privacy and personal-data protection. In the European Union, the EU AI Act, Regulation (EU) 2024/1689, prohibits the use of emotion-recognition systems in educational institutions and workplaces, for example for assessing the condition of employees or students. Regulation (EU) 2024/1689 of the European Parliament and of the Council of 13 June 2024 establishes harmonized rules on artificial intelligence and amends a number of EU regulations and directives. Under Article 5 of the Regulation, the use of emotion-recognition systems in workplaces and educational institutions is prohibited, except for medical or safety purposes.

In the Russian Federation there is no special law prohibiting emotion recognition. Nevertheless, such systems must comply with the Constitution of Russia and with current legislation. In particular, Article 23 provides that everyone has the right to privacy, personal and family secrecy, and the protection of honour and good name. Article 24 establishes that the collection, storage, use and dissemination of information about a person's private life without consent is not permitted.

Federal Law No. 152-FZ of 27 July 2006 On Personal Data and Federal Law No. 572-FZ of 29 December 2022 On Identification and/or Authentication of Individuals Using Biometric Personal Data regulate and restrict the use of technologies involving biometric data. Article 11 of Federal Law No. 152-FZ, as amended on 24 June 2025, permits the processing of biometric personal data only with the written consent of the data subject, except in cases related to defence, security, counter-terrorism, justice, mandatory fingerprinting or genomic registration, readmission agreements, operational-search activity, public service, criminal procedure, penitentiary law, migration, citizenship and other grounds specified by Russian legislation.

Federal Law No. 572-FZ strictly regulates the collection and storage of biometrics through the Unified Biometric System. Accordingly, the use of private facial-analysis systems without integration into this system may be recognized as a violation.

Federal Law No. 273-FZ of 29 December 2012 On Education in the Russian Federation, as amended on 29 December 2025, provides in Article 34 that students have the right to respect for human dignity, protection from all forms of physical and mental violence, protection from insult to personality, and protection of life and health. It also guarantees freedom of conscience, freedom of information and free expression of views and beliefs. Article 47 guarantees educators the right to respect for human dignity, protection from all forms of physical and mental violence, protection of professional honour and dignity, and a fair and objective investigation of violations of professional ethics.

From the perspective of international law, emotional monitoring may be regarded as interference with cognitive freedom and private life. Among the relevant norms are Article 12 of the Universal Declaration of Human Rights, Articles 17 and 19 of the International Covenant on Civil and Political Rights, Article 16 of the Convention on the Rights of the Child and other provisions.

Article 12 of the Universal Declaration of Human Rights provides that no one shall be subjected to arbitrary interference with privacy, family, home or correspondence, or to attacks upon honour and reputation, and that everyone has the right to the protection of the law against such interference or attacks. Article 17 of the International Covenant on Civil and Political Rights establishes similar protection against arbitrary or unlawful interference with privacy, family, home or correspondence and against unlawful attacks on honour and reputation. Article 19 guarantees the right to hold opinions without interference and the right to freedom of expression, including the freedom to seek, receive and impart information and ideas of all kinds. Article 16 of the Convention on the Rights of the Child establishes that no child shall be subjected to arbitrary or unlawful interference with privacy, family, home or correspondence, or to unlawful attacks on honour and reputation, and that the child has the right to legal protection against such interference or attacks.

When considering the relationship between the terms emotion recognition, affective computing and biometric psychoanalysis, the following should be noted. Emotion recognition is one component of affective computing, while biometric psychoanalysis may be one of the methods used in affective computing to obtain data. Emotion recognition often uses machine learning, neural networks, computer vision, natural-language processing and other technologies. Systems may analyse facial landmarks such as eyebrows, eyes and mouth, voice parameters such as volume, intonation and speech rate, or textual features. Affective computing seeks not only to recognize emotions, but also to react to them, for example by adapting content, changing the style of interaction or making decisions based on the user's emotional state. Biometric psychoanalysis may include analysis of dermatoglyphics, vibraimage methods assessing a current state from bodily movement, electrical activity of the skin, brain and facial muscles, and other parameters. All three concepts are connected with the study and processing of emotional information, but they differ in scale, purpose and method of application.

A ban on the development, deployment and operation of emotion-recognition systems, affective computing and biometric psychoanalysis in the specified spheres is necessary in order to protect the constitutional rights of citizens, prevent ethical and social risks, and maintain trust in institutions of power, education and labour. Such a ban would help prevent the creation of systems of total control, discrimination, psychological pressure and other negative consequences incompatible with the principles of the rule-of-law state and human dignity.

The relationship between neurotechnologies, virtual reality and cognitive protection also requires attention. Neurotechnologies may assist in the diagnosis and correctic n more vulnerable to manipulation. Virtual reality may be used to create training scenarios in which a person learns to recognize disinformation or to develop critical thinking in a controlled environment. In this sense, neurotechnologies work with the physiological basis of thinking, virtual reality creates environments for its training or study, and cognitive protection defines the goals and tasks of such training or correction.

In the context of neurotechnologies, virtual reality and cognitive protection, a prohibition on monitoring, forming, changing or suppressing beliefs may be implemented through restrictions on the use of neurotechnologies for unethical interference in cognitive processes, including manipulation of consciousness or forced alteration of beliefs. Another method is the regulation of virtual-reality applications so as to exclude their use for ideological programming or psychological pressure. A third approach is the development of cognitive protection, including educational and technological measures that help individuals critically evaluate information and resist external manipulative influence.

In general, a ban on the technological monitoring, formation, alteration or suppression of beliefs is a complex system of legal and ethical norms aimed at protecting personal autonomy. It includes guarantees of liberty of worldview choice and public-law worldview neutrality, as well as regulation of technological instruments that may be used to influence a person's beliefs.

Thus, the introduction of the above-mentioned emotion-recognition technologies and related systems violates the right to harmonious personal development in an environment free from total control. Emotions, thoughts and feelings must remain a person's private space unless that person freely decides to share them. Surveillance of emotions may be regarded as psychological pressure on the individual, and therefore as a violation of the right to respect for human dignity.

The education system deserves separate emphasis. The introduction of emotion-recognition systems into educational spaces may be presented under various benevolent masks, including the promise of 'absolute security'. In reality, however, it marks a shift from the protection of public order to preventive management of consciousness. The argument of threat prevention, including threats such as school shootings, is used to justify total transparency of the inner world of the person. Here lies the main trap of the post-digital age: security is replaced by algorithmic profiling, where a 'suspicious' emotion is treated almost as an offence. This deprives a person of the right to the natural complexity of feeling, forces them into 'facial loyalty', and turns a university from a territory of the search for truth into a training ground for predictable behaviour. This is precisely the geopolitical deadlock in which the state, for the illusion of absolute security, is ready to sacrifice the worldview freedom of people, thereby generating global risks to human autonomy.

In this context, it appears advisable to introduce a full and unconditional legislative ban on the development, deployment and operation of emotion-recognition systems, affective computing and biometric psychoanalysis in the activities of public authorities, educational organizations at all levels and labour relations. It is necessary to ensure the complete and irreversible destruction of all existing software systems and databases containing the results of emotional profiling and affective monitoring of the individual accumulated by public or private structures. Strict administrative and criminal liability should be established for heads and officials of educational organizations who authorize the use of such technologies under any pretext, including the alleged improvement of the educational process. Violation of the ban should entail immediate withdrawal of the state licence or accreditation of the institution. The development and dissemination of such technologies for use in the specified spheres should be directly prohibited.

It is extremely important to prevent the use of neurotechnologies, virtual, mixed and extended reality, biometric systems, behavioural profiling and other technologies for unlawful influence on the worldview sphere of the individual. Any use of such technologies in the public sphere must meet the requirements of legality, necessity, proportionality, respect for human dignity and worldview neutrality. The use of technologies to monitor, form, change or suppress a person's beliefs must be prohibited.

The report by Janna Anderson and Lee Rainie, Building a Human Resilience Infrastructure for the AI Age (2026), based on a detailed survey of more than 500 experts, calls for a shift from personal adaptation to institutional protection that ensures human resilience in the face of total automation. The authors propose moving the emphasis from individual adaptation to AI toward the creation of a multi-level resilience infrastructure: international agreements, legal boundaries for AI, independent safety audits, contestability of algorithmic decisions and labelling of machine-generated content. They also propose developing existential literacy, critical thinking, reflection and skills for navigating an algorithmized environment. The final part of the report emphasizes that mechanisms of restraint and adaptation under deep AI integration can make it possible to reach a new level of development while avoiding negative consequences.

In general, overcoming global challenges, including those arising from the introduction of new technologies, requires an evolutionary transition from a state-centred world order to a human-centred and just world order ensured through the realization of digital human rights.

Therefore, as an antidote both to the use of digital technologies for total control over people, the 'Big Brother' scenario, and to the potential threats arising from super-powerful artificial intelligence escaping control, the 'machine uprising' scenario, the development of digital human rights appears especially promising.

It seems that digital human rights are opportunities belonging to everyone from birth and enshrined in forms of international and domestic law, and in the future in forms of global law. These opportunities include: access to the Internet, means of digital individualization and storage of digital assets, and other digital technologies; protection of life and personal, biometric, biological and other data; access to participation in public affairs, and in the future global governance, through digital technological platforms and complexes; rights concerning the possession, use and disposal of digital property and assets, the exercise of digital economic activity, transactions and digital deals, and access to digital-security systems; access to social services based on digital technological platforms; and digital access to cultural values and education. It is understood that the realization of these opportunities must be based on equality and freedom from discrimination, and must serve the preservation and development of the human being rather than the convenience of technological systems.

Maksim Buryanov advances the concept of digital human rights by design, or Human Rights by Design, as a fundamental response to the existential challenges of the technological age. His position is based on the idea that human rights must not be an external legal addition to technology, but an embedded technical standard at the stage of creating any technology. He argues, with scientific justification, that technologies are not neutral. If the values of human rights and subjectivity are not embedded in their code from the beginning, they automatically become instruments of control and dehumanization. To prevent the transformation of the human being into an object of algorithmic governance, the architecture of AI and neurotechnologies must technically guarantee cognitive freedom and mental autonomy.

M. Buryanov emphasizes that the absence of worldview neutrality in the state leads to the creation of biased software that discriminates against the individual. At the centre of his approach stands the realization of digital rights and the protection of the 'silence of consciousness' from technological intrusion. This, in his view, is the only way to preserve human subjectivity under non-human-centred technological deployment. He calls for a transition from declarative rights to 'programmable law', where violations of human rights become technically impossible due to the very architecture of systems.

The fundamental proposals of M. Buryanov concerning the development and legal consolidation of digital human rights in the system of international and domestic law appear extremely relevant for realizing liberty of worldview choice in the context of digital technologies.

On a planetary scale, it is worldview freedom, the free formation of a picture of the world and, consequently, the determination of one's place within it, that constitutes a necessary condition for an adequate understanding of global processes, interactions and problems. The modern world has indeed changed, and it continues to change rapidly. The present and future of human civilization depend on the adequacy of our responses to these changes. Accordingly, worldview neutrality and liberty of worldview choice are necessary conditions for transforming value orientations, forming global thinking, overcoming the problems of global security and moving toward the sustainable development of an integrated world.

–

REFERENCES TO CHAPTER 7

List of Sources Used

1. Elensky V. Religious Freedom: Global Dimensions // Electronic resource. URL: https://religion.in.ua/main/analitica/5607-religoznaya-svoboda-globalnye-izmereniya.html (accessed: 8.12.2019).
2. Many Countries Favor Specific Religions, Officially or Unofficially // Pew Research Center. Electronic resource. URL: http://www.pewforum.org/2017/10/03/many-countries-favor-specific-religions-officially-or-unofficially/ (accessed: 4.12.2019).
3. A Closer Look at How Religious Restrictions Have Risen Around the World. Tenth Annual Report Dives Deeper into the Ways Government Restrictions on Religion and Social Hostilities Involving Religion Have Changed, from 2007 to 2017 // Pew Research Center. Electronic resource. URL: https://www.pewforum.org/2019/07/15/a-closer-look-at-how-religious-restrictions-have-risen-around-the-world/ (accessed: 4.12.2019).
4. Pew Research Center. Government Restrictions on Religion Stayed at Peak Levels Globally in 2022 (2024). URL: https://www.pewresearch.org/religion/2024/12/18/government-restrictions-on-religion-stayed-at-peak-levels-globally-in-2022/ (accessed: 21.04.2026).
5. Freedom of Thought Report. Electronic resource. URL: https://fot.humanists.international/editorial-introduction/ (accessed: 8.12.2019).
6. Freedom of Thought Report 2023. Humanists International. 2023. URL: https://fot.humanists.international/ (accessed: 21.04.2026).
7. Russia Was Included in the List of Countries Persecuting Christians for the First Time in Seven Years // RBC. January 16, 2019. Electronic resource. URL: https://www.rbc.ru/rbcfreenews/5c3f228d9a794778ad909e4e (accessed: 8.12.2019).
8. Luparev G.P. Freedom of Conscience: A "Sacred Cow" or a Constitutional-Legal Anachronism? // Religion and Law. 2002. No. 2. p. 6.
9. Ibid. p. 4.
10. Durham W.C. Freedom of Religion or Belief: Laws Affecting the Structuring of Religious Communities. ODIHR Background Paper 1999/4. p. 22.
11. Murdoch J. Protecting the Right to Freedom of Thought, Conscience and Religion under the European Convention on Human Rights. Council of Europe Human Rights Handbooks Series. Voronezh: Elit Firm LLC, 2014. 163 p. p. 27.
12. Cited work. p. 28.
13. Cited work. p. 29.
14. Polosin V.S. Myth, Religion, State. Moscow, 1998. p. 17.
15. Fromm E. Psychoanalysis and Religion // To Have or To Be. Moscow: Progress, 1990. p. 236.
16. Fedotova Yu.E. Problems of Defining the Concept of "Religion" and Related Concepts in Russian Legislation // Russian Juridical Journal. 2017. No. 4. pp. 20-23.
17. Cited work. p. 23.
18. Podoprigora R.A. The State and Religious Organizations: Administrative-Legal Issues. Almaty: Arkaim, 2002. pp. 45-80.
19. Podoprigora R.A. Administrative-Legal Regulation of the Activities of Religious Organizations in Foreign Countries: Abstract of Doctor of Legal Sciences Dissertation. Almaty, 2002. p. 20.
20. Cited work. p. 22.
21. Markova E.N. The Problem of the Legal Definition of Religion // Constitutional and Municipal Law. 2018. No. 11. pp. 34-45.
22. Cited work. p. 43.
23. Markova E.N. Freedom of Conscience and Religion: The Negative Dimension and Protection of Personal Convictions: Candidate of Legal Sciences Dissertation. Moscow, 2019. 277 p. p. 18.
24. Markova E.N. Freedom of Conscience and Religion: The Negative Dimension and Protection of Personal Convictions: Candidate of Legal Sciences Dissertation. Moscow, 2019. 277 p. pp. 18-19.

25. Freedom of Conscience: Problems of Theory and Practice. Monograph edited by F.M. Rudinsky and S.A. Buryanov. Moscow: ZAO TF MIR, 2012. 1120 p.
26. Krasnikov A.N. Trends in Contemporary Religious Studies // Freedom of Conscience in a Rule-of-Law State: Legal and Information Aspects. Seminar Materials. Moscow, 2000. p. 97.
27. Fromm E. Psychoanalysis and Religion // To Have or To Be? Moscow, 1990. p. 236.
28. Filatov S. Contemporary Russia and Sects // Foreign Literature. 1996. No. 8.
29. Filatov S., Shchipkov A. Non-Traditional Religiosity in Post-Communist Russia // Truth and Life. 1997. Nos. 5-7.
30. Religion and Society: Essays on the Religious Life of Contemporary Russia / ed. and comp. S.B. Filatov. Moscow; Saint Petersburg: Letny Sad, 2002. pp. 401-449.
31. Contemporary Religious Life in Russia: An Experience of Systematic Description. Vol. IV / ed. M. Bourdeaux and S.B. Filatov. Moscow: Universitetskaya Kniga; Lotos, 2006.
32. Tikhonravov Yu.V. Forensic Religious Studies. Moscow, 1998. p. 204.
33. Statement of the Institute for Freedom of Conscience "The Institute of Religious-Studies Expert Examination Must Be Abolished, and 'Anti-Extremist' Legislation, Lists of Religious Literature, and Structures Must Cease to Exist" // Buryanov S.A. Freedom of Conscience and Secularity of the State in the Russian Federation. Theoretical and Applied Study for 2009 - Early 2010. Moscow, 2010. p. 36.
34. Garadzha V.I. Sociology of Religion. Moscow, 1996. p. 226.
35. Fundamentals of Religious-Studies Expert Examination (ed. A.V. Pchelintsev). Moscow, 2002.
36. Pchelintsev A.V. Is There a Future for Legal Religious Studies? Reflections on Yu.V. Tikhonravov's Book "Forensic Religious Studies" // Fundamentals of Religious-Studies Expert Examination. Moscow, 2002. p. 114.
37. Aleinikova S.M., Buryanov S.A. The Secular State in Questions and Answers: Briefly, Accessibly, and Relevantly / S.M. Aleinikova, S.A. Buryanov. Moscow: Zdravomyslie Foundation, 2015. 59 p.
38. Buryanov S.A. International Legal Documents in the Field of Freedom of Conscience and the Practice of Their Implementation in the Russian Federation: Theoretical and Applied Study for 2011. Moscow: Russian Helsinki Group, 2012. 244 p.
39. Buryanov S.A. Xenophobia, Intolerance and Discrimination on Grounds of Religion or Belief in the Subjects of the Russian Federation. Specialized Information and Analytical Report for 2006 - First Half of 2007. Moscow: Russian Helsinki Group, 2007. 240 p.
40. Buryanov S.A. Religion in Elections in Russia: The Factor of Relations between the State and Religious Associations in the Federal Electoral Cycle of 2003-2004. Moscow: Institute for Freedom of Conscience, 2005. 200 p.
41. Buryanov S.A. Freedom of Conscience and Secularity of the State in the Russian Federation. Theoretical and Applied Study for 2009 - Early 2010. Moscow, 2010. 173 p.
42. Buryanov S.A. Secularity of the State and Internationally Recognized Freedom of Conscience. Theoretical and Applied Study for 2015 - Early 2016. Moscow: Poligraf Servis, 2016. 258 p.
43. Mchedlov M.P. The Basis of Civil Concord // Freedom of Conscience in the Spiritual Revival of the Fatherland. Moscow, 1994. p. 11.
44. Mozgovoy S.A. Ethnoreligious Relations and Problems of Civil Peace in Russia // Religion and Nationalism. Moscow, 2000. p. 157.
45. Whitehead A.N. Selected Works on Philosophy. Moscow, 1990. p. 242.
46. Buryanov S.A. A New Paradigm of Freedom of Conscience // Contemporary Law. No. 5. 2003. pp. 11-15.
47. Buryanov S.A. A New Paradigm of Freedom of Conscience: In Addition to the Published Article // Contemporary Law. No. 7. 2003. pp. 39-41.
48. Schwab K. Shaping the Fourth Industrial Revolution. Cologny; Geneva: Portfolio Penguin, 2018. 287 p.
49. Baldwin R. The Globotics Upheaval: Globalization, Robotics, and the Future of Work.

50. Burianov M. Global Digital Human Rights in the Implementation of the Global Digital Compact: The Toolkit for Human-Centered GovTech, AI and Global Governance. United Humans Press, 2025. 380 p.
51. Steger M.B. Globalization: A Very Short Introduction. 5th ed. Oxford UP, 2024. 192 p.
52. Buryanov M.S. Digital Human Rights in the Context of Global Processes: Theory and Practice of Implementation. Monograph. Scientific editor S.A. Buryanov. Moscow: Rusains, 2024. 148 p.
53. Buryanov S.A., Buryanov M.S. In Search of an Adequate World Order: The Future of the Rule of Law in the Era of Digital Globalization 4.0 // Age of Globalization. 2024. No. 2 (50). pp. 127-140. DOI: 10.30884/vglob/2024.02.10.
54. AI's $4.8 Trillion Future: UN Trade and Development Alerts on Widening Divides. URL: https://unctad.org/press-material/ais-48-trillion-future-un-trade-and-development-alerts-divides-urges-action (accessed: 21.02.2026).
55. Billionaire Wealth Jumps Three Times Faster in 2025 to Highest Peak Ever, Sparking Dangerous Political Inequality. URL: https://www.oxfam.org/en/press-releases/billionaire-wealth-jumps-three-times-faster-2025-highest-peak-ever (accessed: 21.01.2026).
56. Buryanov S.A. The Nature and Limits of Regulation in the Sphere of Freedom of Conscience in a Democratic State // Law and Politics. No. 5. 2000. p. 67.
57. Schwab K. The Fourth Industrial Revolution = Die Vierte Industrielle Revolution. Eksmo, 2016. 208 p. p. 9.
58. Cited work. p. 8.
59. Schwab K. Technologies of the Fourth Industrial Revolution = Shaping the Fourth Industrial Revolution. Eksmo, 2018. 320 p.
60. Roberts T., Oosterom M. Digital Authoritarianism: A Systematic Literature Review // Information Technology for Development. 2024. Vol. 30, No. 5. pp. 860-884.
61. Zuboff S. The Age of Surveillance Capitalism: The Fight for a Human Future at the New Frontier of Power / Shoshana Zuboff. New York: Public Affairs, 2019. 704 p.
62. Human Rights in the Digital Domain: Core Questions / ed. by T. Pajuste. Cambridge: Cambridge University Press, 2025. 492 p.
63. Alegre S. Freedom to Think: The Long Struggle to Liberate Our Minds / S. Alegre. London: Atlantic Books, 2023. 400 p.
64. Gritsenko D. Anxious Universities: AI Will Assess the Probability of Student Expulsion. How Enhanced Control Through New Technologies Will Affect the Emotional State of University Students. URL: https://iz.ru/1546434/denis-gritcenko/trevozhnyi-vuzov-ii-otcenit-veroiatnost-otchisleniia-studentov (accessed: 21.03.2026).
65. How AI Technologies Are Changing the Approach to Analyzing Pedagogical Practices. URL: https://www.mgpu.ru/na-ai-journey-rasskazali-kak-tehnologii-ii-menyayut-podhod-k-analizu-pedagogicheskih-praktik/ (accessed: 21.03.2026).
66. What Risks Are Posed by Artificial Intelligence Bias. URL: https://cloud.vk.com/blog/kak-ii-mozhet-navredit-biznesu/ (accessed: 21.03.2026).
67. White Paper on Ethics in Artificial Intelligence / ed. by A.V. Neznamov. Moscow: Nova Creative Group, 2024. 200 p. p. 127.
68. AI Tools for Analyzing Student Emotions. URL: https://www.restack.io/p/ai-for-personalized-medicine-answer-ai-emotion-analysis-cat-ai (accessed: 7.03.2026).
69. Microsoft Abandoned a Creepy Emotion-Recognition Technology. URL: https://tech.onliner.by/2022/06/22/microsoft-otkazalas-ot-zhutkovatoj-texnologii-raspoznavaniya-emocij (accessed: 7.03.2026).
70. Kulov S.S. New Forms of Interference with Freedom of Belief: International-Law Doctrine under Conditions of Digital Transformation // International Law and International Organizations. 2025. No. 3. pp. 78-92. DOI: 10.7256/2454-0633.2025.3.74730.
71. Anderson J. Building a Human Resilience Infrastructure for the AI Age [Electronic resource] / J. Anderson, L. Rainie // Pew Research Center. 2024. 147 p.

72. Buryanov S.A., Buryanov M.S., Nikitaev D.M. The Right to Freedom of Conscience in the Context of Global Processes: Theory and Practice of Implementation in the Russian Federation. (Master's Program). Monograph. Scientific editor S.A. Buryanov. Moscow: Rusains, 2020. 236 p.
73. Buryanov S.A., Buryanov M.S. In Search of an Adequate World Order: The Future of the Rule of Law in the Era of Digital Globalization 4.0 // Age of Globalization. 2024. No. 2 (50). pp. 127-140. DOI: 10.30884/vglob/2024.02.10.
74. Buryanov M.S. The Gates to Global Law: Global Digital Human Rights // Scientific Works of Moscow University for the Humanities. 2020. No. 2. pp. 63-66. DOI: 10.17805/trudy.2020.2.11.
75. Buryanov M.S. Prospects for the Development of Russian Statehood under Conditions of Contemporary Global Processes and Challenges // Current Problems of Global Studies: Russia in a Globalizing World. Collection of Scientific Papers of Participants of the VI All-Russian Scientific and Practical Conference with International Participation. Edited by I.V. Ilyin. Moscow, 2019. pp. 67-73.
76. Buryanov M.S. Digital Human Rights in the Context of Global Processes: Theory and Practice of Implementation. Monograph. Scientific editor S.A. Buryanov. Moscow: Rusains, 2024. 148 p.
77. Burianov M. Global Digital Human Rights in the Implementation of the Global Digital Compact: The Toolkit for Human-Centered Tech, AI and Global Governance. United Humans Press, 2025. 380 p.
78. Buryanov M.S. Digital Rights of the Child: Monograph. Moscow: Rusains, 2023. 294 p.
79. Buryanov M.S. Draft Declaration of Global Digital Human Rights // Eurasian Law Journal. 2025. No. 12 (211). pp. 522-526.
80. Buryanov M.S. Problems and Prospects of the Development of Law in the Context of the Digital Transformation of the State // Eurasian Law Journal. 2025. No. 10 (209). pp. 502-504.
81. Buryanov S.A. The Right to Freedom of Conscience under Conditions of Global Processes and Challenges 4.0 // Eurasian Law Journal. 2023. No. 6 (181). pp. 18-21.
82. Freedom of Conscience: Problems of Theory and Practice. Monograph (edited by F.M. Rudinsky and S.A. Buryanov). Moscow: ZAO TF MIR, 2012. 1120 p.
83. Buryanov S.A. International Recognition of the Right to Freedom of Conscience and Problems of Its Implementation in the Russian Federation under Conditions of Contemporary Global Processes: Monograph. Moscow: Poligraf Servis, 2020. 624 p.
84. Civil Human Rights in Russia: Modern Problems of Theory and Practice / compiled by F.M. Rudinsky. Transaction Publishers, New Brunswick, New Jersey (U.S.A.), 2008. 458 p.
85. Buryanov S.A., Buryanov M.S., Nikitaev D.M. The Right to Freedom of Conscience in the Context of Global Processes: Theory and Practice of Implementation in the Russian Federation. (Master's Program). Monograph. Edited by S.A. Buryanov. Moscow: LitRes, 2022. 234 p.
86. Digital Human Rights in the Context of Global Processes: Theory and Practice of Implementation: Monograph / M.S. Buryanov; scientific editor S.A. Buryanov. Moscow: Rusains, 2024. 148 p.

CONCLUDING PART. THE FUTURE BEYOND THE DEADLOCK. LEGAL CONDITIONS FOR THE SURVIVAL OF HUMANITY

CIVILIZATIONAL THESIS

"Civilization survives where the human mind remains legally inviolable."

The future depends on the protection of the inner human world. Worldview liberty, cognitive freedom, mental sovereignty, and public-law neutrality must become working principles of global law, education, technology, and governance - principles capable of protecting the human being before power reaches the source of thought itself.

CONCLUSIONS. LIBERTY OF WORLDVIEW CHOICE, GUARANTEED BY PUBLIC-LAW WORLDVIEW NEUTRALITY, AS A NEW LEGAL PARADIGM OF THE GLOBAL DIGITAL AGE 4.0

The evolution of the concept of internationally recognized freedom of conscience has moved from the protection of the rights of religious minorities through interstate agreements in the sixteenth century to the emergence of international protection of religious freedom in the eighteenth century and the first half of the twentieth century, within the framework of the League of Nations, and then to the recognition and development of freedom of thought, conscience and religion as part of the international system of human rights protection from the mid-twentieth century to the end of the twentieth century, within the United Nations and regional and subregional organizations. At the beginning of the twenty-first century, the doctrinal development of a new paradigm of the internationally recognized right to freedom of conscience began: freedom of conscience as the opportunity for every person to make a free worldview choice without discrimination and on the basis of the worldview neutrality of states and other subjects of international law.

This study has revealed the system-forming influence of the degree of scientific development of theoretical and legal problems in the field of human rights related to freedom of conscience on the effectiveness of the principles and norms of international and domestic law. These deficiencies have predetermined systemic problems at the level of law enforcement and, ultimately, an insufficient level of realization of this right in contemporary states and in Russia. Among the main theoretical and legal problems in the field of the internationally recognized right to freedom of conscience are the relationship between the concepts of 'freedom of thought, conscience and religion' and 'freedom of conscience', 'freedom of religion', 'state secularity', 'state religious policy', 'relations between the state and religious associations', and related categories. No less controversial is the use in international law and in Russian domestic law of contested terms borrowed from religious studies, such as 'traditional religions' or 'traditional religious organizations', 'non-traditional religions' or 'non-traditional religious organizations', 'sect', as well as the concepts of 'feelings of believers', 'missionary activity', 'spiritual security', 'religious extremism', 'Islamic extremism', and similar terms. The extremely weak scientific elaboration of key and related terms in the field of freedom of conscience has predetermined the low effectiveness of international legal principles and norms and their interaction with the domestic legislation of the Russian Federation.

Another block of theoretical and legal problems in the sphere of freedom of conscience concerns the relationship between the sources and forms of this right. From the standpoint of distinguishing between the sources and forms of the right to freedom of conscience, the Russian Federation has a single system of international and domestic forms of law relating to freedom of conscience, understood as the opportunity for every person to make a free worldview choice without discrimination and on the basis of the worldview neutrality of states and other subjects of international law. In addition, the system of forms of the right to freedom of conscience in the Russian Federation exists in unity and interdependence with the system of forms of human rights law in the Russian Federation, which consists of international human rights law and Russian human rights law.

The concepts of the supremacy of international law and of the law-governed state provide an important basis for reforming the principles and norms of international and domestic law on freedom of conscience. The concepts of the rule of law and of the law-governed state are interconnected with each other and with the legal certainty of legal principles and norms as a stable and predictable foundation for the realization of human rights in the sphere of freedom of conscience and non-discrimination. Among the many factors of legal certainty, legal terminology is fundamental, above all key legal concepts through which the content of normative prescriptions is expressed, because legal constructions and legal acts are built from them. When key legal terminology does not meet the requirements of legal certainty, uniform understanding and interpretation of the principles and norms of international and domestic law on freedom of conscience become impossible, which predetermines their low effectiveness. It appears advisable, on the basis of the 2012 Declaration of the High-Level Meeting of the General Assembly on the Rule of Law at the National and International Levels, to adopt a legally binding instrument establishing the legal certainty of the key legal concepts underlying the principles and norms of international and domestic law.

The concept of public-law worldview neutrality, interconnected with the concepts of the supremacy of international law and the law-governed state and oriented toward the realization of human rights, is one of the most important foundations for reforming the forms of international and domestic law on freedom of conscience. An analysis of global experience as a whole allows us to conclude that, under the influence of global processes, interaction between international and domestic legal systems is intensifying, different legal families are moving closer together, and the objective tendency in relations between the state and religious associations is a gradual departure from the privileged status of religious assoc   eutrality of states, including equality of religious associations. This is reflected in constitutions and legislation. In Russia, the concept of worldview neutrality is reflected in Article 14 of the Constitution of the Russian Federation, which establishes the secular nature of the state and the equality of religious associations as one of the foundations of the constitutional order.

In contemporary conditions, it is especially important to understand the significance of realizing the right to freedom of conscience for achieving tolerance and overcoming discrimination, intolerance and xenophobia on grounds of worldview affiliation. The problem of the influence of the ethno-confessional factor on security has become sharply aggravated at both the international and domestic levels. These problems require study in light of the contemporary character of social relations and the requirements of legal technique.

The internationally recognized right to freedom of conscience and the concept of the worldview neutrality of states are closely interconnected with tolerance and with countering xenophobia, intolerance and discrimination, as enshrined in the principles and norms of international law and in Russian domestic law. It is necessary to emphasize the importance of systematic work to develop tolerance as mutual respect based on equal rights regardless of differences, and to counter intolerance through international organizations and state education systems. It appears advisable, on the basis of the 1995 Declaration of Principles on Tolerance, to adopt a legally binding instrument establishing tolerance, understood as mutual respect based on equality regardless of differences, as an important condition for realizing the internationally recognized right to freedom of conscience.

The study of global experience in the legal regulation of relations between states and religious associations has revealed an objective tendency toward a gradual departure from political control and inequality through the constitutional and legislative recognition of the right to freedom of conscience. At the same time, the persistence of outdated and legally controversial models of interaction between the state and religious associations remains an objective reality.

A global problem lies in the contradiction between universal norms of freedom of conscience and the intensification of what is, in essence, a corrupt religious policy of states pursued in the narrow interests of elites under the pretexts of supporting a 'religious majority', 'spiritual heritage' or whatever else happens to be labeled 'traditional' by those in power. While European countries are moving toward a model of soft partnership with 'national churches', Islamic and Buddhist regions show a tendency toward a rigid rejection of neutrality in favor of the state establishment of an official faith.

This leads not only to institutional discrimination, where personal freedom of belief is sacrificed to political interests, but also to a significant intensification of geopolitical confrontation, making the sustainable development of the world community impossible and bringing humanity closer to a point of no return. The conflict is most acute and obvious in states based on Islamic law, where religious sovereignty directly blocks international norms concerning the right to change religion or to adopt atheism. Even in developed democracies, religious law continues to control the personal status of citizens, creating legal enclaves within domestic legal systems. As a result, the modern world demonstrates a trend toward the politicization of religion, transforming it from a private matter of citizens into a rigid instrument of state administration and restriction of freedoms.

Before it is too late, global experience in the legal regulation of relations between states and religious associations must be summarized and rethought through the prism of the right to liberty of worldview choice and the worldview neutrality of states, within the expert structures of international organizations, with a corresponding international programme developed on this basis.

The study of the relationship between the right to freedom of conscience, state secularity and relations between states and religious associations has revealed the close interconnection of these concepts. Under contemporary legal conditions, relations between states and religious associations must be subordinated to the principle of public-law worldview neutrality, which should be recognized in international law as one of the most important guarantees for the realization of human rights, including the right to freedom of conscience.

Among the many debated approaches to defining the concept and content of the secularity of states and other subjects of international law, the most legally correct understanding is secularity as public-law worldview neutrality. This implies the indifference of states and other authorized subjects of international law toward the worldview sphere in all its lawful manifestations. Its features should therefore include: non-identification with the worldview sphere; indifference in the worldview sphere; a prohibition on discrimination on grounds of worldview affiliation; legal equality of public non-profit associations; refusal of special relations with worldview associations; refusal of special privileges and restrictions for worldview associations; refusal to give legal force to the compulsory nature of any worldview or ideology; provision of a system of guarantees for the realization of liberty of worldview choice for every individual; refusal to create international intergovernmental organizations on worldview grounds; and observance by international organizations of the principles of worldview neutrality recognized by the international community.

Public-law worldview neutrality is one of the most important guarantees for the realization of internationally recognized human rights, including the right to freedom of conscience. It appears advisable to adopt a Global Declaration on the Right to Liberty of Worldview Choice and the Principle of Worldview Neutrality of Subjects of Public International Law in Analog and Digital Environments. On its basis, a legally binding convention could then be adopted as part of the codification and progressive development of international law at the universal level of the United Nations. In the longer term, this would make it possible to reform domestic legal systems.

Globalization should be understood as a set of mostly objective, yet contradictory and uneven processes directed toward the formation of a single planetary techno-socio-natural system. Completion of these processes may lead to a qualitatively new state of globality in social and related interactions. The state of globality should be understood as the integration, interpenetration and, consequently, openness and interdependence of all social subsystems, as well as related technological and natural subsystems, on a planetary scale. Thus, globality covers the totality of parameters and characteristics in different spheres of social relations, or subsystems, from which the planetary system is formed. The study of the dynamics of development of the social subsystems intended to form the global system mentioned above has revealed the unevenness of their development. This predetermines imbalance and lack of governability both within the social system itself and in its interaction with the technogenic and natural spheres. Since the level of development of social subsystems is uneven, this ultimately predetermines instability in international relations and global challenges and threats that hinder the sustainable development of the world community and individual states. Global processes therefore have a significant impact on international relations and international law. The formation of integrated, open and interdependent social subsystems requires a radical rethinking of international relations and a search for innovative directions in the reform of international law.

Global challenges have affected many spheres of social relations, as well as the interactions between society, the technosphere and the forces of nature. This results from the uneven development of global processes. Among the consequences of uneven development of global subsystems and imbalance in global processes, special attention should be paid to global challenges and threats in the international-legal sphere, including security. These challenges bring to the forefront the problem of moving toward sustainable, manageable development of the international community and its participants. In particular, awareness of the importance of sustainable development has been reflected in documents of the United Nations and of regional, interregional and subregional organizations.

The complication and globalization of social relations require the realization of greater individual freedom, embodied in internationally recognized human rights, including the right to freedom of conscience. This, in turn, predetermines the need to improve thc norms and mechanisms of international cooperation.

Important conditions for overcoming global challenges and moving toward sustainable, manageable development of civilization include overcoming barriers to the integration of states; eliminating imbalance in the development of global processes; and achieving balance and governability within global subsystems. One of the most serious obstacles to the positive integration of states and the formation of an integrated socio-techno-natural system is the historically formed set of divisive ethno-confessional contradictions based on the union of religion and politics and embedded in the foundation of nation-states. The realization of the internationally recognized right to freedom of conscience and the principle of worldview neutrality of states and other subjects of international law are intended to overcome, softly and evolutionarily, the ethno-confessional contradictions that obstruct global security and the positive integration of subjects of international law for the purpose of sustainable development. Since, under conditions of globalization of social relations, modern international law is called upon to serve as an instrument for maintaining international legal order and ensuring equal cooperation among states and peoples in solving the global problems of the contemporary world, the importance of realizing human rights, including the right to freedom of conscience, increases substantially. In the context of contemporary global processes, a global integrated system for the realization of human rights, including in the sphere of freedom, must be formed on the basis of the reformatting and integration of international and domestic subsystems.

The contemporary system of protection of the human right to freedom of conscience includes international and domestic subsystems interconnected through the mechanism of implementation. The contemporary subsystem of international protection of the human right to freedom of conscience includes universal international legal principles and norms; universal international legal mechanisms; regional, interregional and subregional international legal principles and norms; and regional, interregional and subregional international legal mechanisms. The contemporary domestic subsystem, within a specific state, includes domestic principles and norms and domestic mechanisms. This means that, in a modern state, there operates a system for protecting the human right to freedom of conscience consisting of international and domestic subsystems interconnected through implementation mechanisms and including universal international legal principles, norms and mechanisms; relevant regional, interregional and subregional international legal principles, norms and mechanisms; and relevant domestic principles, norms and mechanisms.

Within the progressive development of international law, it is necessary to move from the concept of protection, as well as respect, promotion and observance, to the concept of realization of the human right to freedom of conscience. This ultimately means the ability to benefit from the relevant social goods in the interests of the development of every person, regardless of differences. A modern and effective system for realizing human rights in the sphere of freedom of conscience must be formed, consisting of international and domestic subsystems and based on the worldview neutrality of subjects of international law, as well as tolerance regardless of differences and resistance to intolerance.

Universal international legal principles and norms for the protection of human rights in the sphere of freedom of conscience are contained in the UN Charter, the Universal Declaration of Human Rights, the 1966 Covenants and other foundational acts. It should be noted that these universal international legal principles and norms are not fully effective and require substantial reform on the basis of a new paradigm of freedom of conscience as the opportunity for worldview choice; the principle of the supremacy of international law, including the legal certainty of key terms; the principle of worldview neutrality; and the principle of mutual respect regardless of differences and of overcoming xenophobia, intolerance and discrimination.

Regional, interregional and subregional international legal principles and norms in the sphere of freedom of conscience are contained in the documents of the relevant organizations. It should also be noted that these regional, interregional and subregional principles and norms likewise do not fully correspond to contemporary conditions and require universalization and substantial reform on the same foundations proposed for the universal norms.

Universal mechanisms for the protection of the right to freedom of conscience include the subsystem of protection by the main, charter-based organs of the United Nations, including the Third Committee of the General Assembly on social, humanitarian and cultural issues, the Security Council, the Economic and Social Council and the International Court of Justice; the subsystem of protection by auxiliary, non-charter bodies, among which the principal ones include the Human Rights Council, including the Universal Periodic Review and the Special Rapporteur on freedom of religion or belief, the Working Group on Arbitrary Detention and the Office of the United Nations High Commissioner for Human Rights; the subsystem of protection by specialized agencies of the United Nations, including UNESCO and its confidential procedure 104 EX/3.3 for examining complaints of human rights violations in the fields of education, science, culture and communication; and the subsystem of treaty bodies protecting the right to freedom of conscience, including the UN Human Rights Committee, its concluding observations on reports, decisions on complaints and General Comments on the 1966 International Covenant on Civil and Political Rights.

The activity of the principal, charter-based organs of the United Nations in the field of the human right to freedom of conscience is based on the relevant principles and norms contained in foundational documents, which require substantial reform. It appears advisable for the General Assembly to develop and adopt a conceptual document on the development of international cooperation in the sphere of realizing the right to liberty of worldview choice through the formation of a system of legal guarantees, including the principle of the supremacy of international law, including the legal certainty of key terms; the principle of worldview neutrality; the principle of mutual respect regardless of differences; and the overcoming of xenophobia, intolerance, discrimination and other destructive phenomena. These approaches could be used by the General Assembly in its norm-setting activity and in the exercise of its supervisory powers when interacting with the system of treaty bodies.

Given the exceptional importance of realizing the right to freedom of conscience for maintaining international peace and security, and especially for overcoming ethno-confessional conflicts, it appears advisable to establish an international tribunal for the prosecution of persons responsible for large-scale violations of the right to freedom of conscience that threaten international peace and security.

It must be acknowledged that the powers of ECOSOC to promote the realization of the right to freedom of conscience have not been fully used. This is especially true of its norm-setting powers to develop and adopt relevant principles and norms. In particular, ECOSOC could assist in the development, discussion and adoption of a Convention on the Realization of the Human Right to Freedom of Conscience, a Declaration, and in the longer term a convention, on the worldview neutrality of states and other subjects of international law, as well as a number of other recommendatory documents. ECOSOC's supervisory powers could also be more fully implemented and improved, for example through the establishment of a Subcommission on the Realization of the Human Right to Freedom of Conscience and other auxiliary bodies in this field.

The insufficient effectiveness of the principles and norms in the field of the human right to freedom of conscience has largely predetermined the less than effective activity of the International Court of Justice in this area. Improvement of the relevant principles and norms, together with procedural reform, especially expansion of the range of subjects entitled to bring cases, could increase the effectiveness of the International Court of Justice in protecting the human right to freedom of conscience.

Analysis of the organizational structure of the Secretariat and its auxiliary bodies suggests that the right to freedom of conscience is not a priority for them, nor for the UN Secretary-General. Establishing the posts of Special Representative and Special Adviser on the human right to freedom of conscience could help activate the powers of the Secretariat and the Secretary-General, provided that the relevant principles and norms are also improved.

Among the auxiliary, non-charter bodies of the United Nations, the Human Rights Council should be singled out. It uses the following instruments: the Universal Periodic Review, prepared by a Working Group mainly on the basis of reports by states parties, which often provide very limited information on violations of the human right to freedom of conscience; and the Special Rapporteur on freedom of religion or belief, who, within the mandate of the Special Procedures, conducts research, monitoring and consultations and prepares annual reports for the Human Rights Council, the Commission on Human Rights and the United Nations General Assembly. Since the very broad activity of the Special Rapporteur on freedom of religion or belief is based on principles and a conceptual apparatus that are not fully correct from the standpoint of legal certainty and modern legal technique, this activity does not fully contribute to the realization of the right to freedom of conscience. As a result, the Special Rapporteur's stated main objective, to identify existing and emerging obstacles to the enjoyment of the right to freedom of religion or belief and to present recommendations on ways and means to overcome such obstacles, remains unrealized.

The UN Working Group on Arbitrary Detention, also acting within the mandate of the Special Procedures, investigates allegations of detention connected with religious affiliation and contrary to international law. However, the investigation and international publicity given to the detention of Jehovah's Witnesses in contemporary Russia did not halt the intensifying wave of repression.

Although the Office of the United Nations High Commissioner for Human Rights supports the activity of the Special Rapporteur on freedom of religion or belief, the right to freedom of conscience is not a priority for it.

The International Law Commission, an auxiliary body of the UN General Assembly, prepares annual reports to the General Assembly, and its potential could assist in improving the principles and norms concerning the human right to freedom of conscience and the worldview neutrality of states. In particular, the Commission could assist in drafting a Convention on the Realization of the Human Right to Freedom of Conscience, as well as a Declaration, and in the longer term a convention, on the worldview neutrality of states and other subjects of international law.

Among the specialized agencies of the United Nations, UNESCO should be singled out. Its activity includes the confidential and relatively simple 104 EX/3.3 procedure for the soft examination of complaints concerning human rights violations in the fields of education, science, culture and communication. The recommendatory nature of decisions on complaints, combined with the insufficient effectiveness of principles and norms in the field of the right to freedom of conscience, has predetermined the absence of a significant influence by UNESCO on the situation in this area. At the same time, UNESCO's positive potential influence on the realization of human rights in the sphere of freedom of conscience could be significantly strengthened through the drafting and adoption of a Convention on the Realization of the Principles of Tolerance.

Within the subsystem of treaty bodies protecting the right to freedom of conscience, the UN Human Rights Committee stands out as the monitoring mechanism of the 1966 International Covenant on Civil and Political Rights. The problems in the Committee's activity in the field of the human right to freedom of conscience are determined, as has already been noted repeatedly, by the imperfection of the relevant principles and norms. Moreover, analysis of General Comment No. 22 (48) of the UN Human Rights Committee, 'Freedom of thought, conscience and religion' under Article 18, reveals an inability to overcome the numerous contradictions underlying the far from encouraging situation in the sphere of freedom of conscience in the contemporary world.

Overall, the extremely low effectiveness of universal mechanisms for the protection of the right to freedom of conscience is largely predetermined by the quality of the principles and norms, which, as noted above, require substantial reform. In addition, their effectiveness depends on overcoming problems of interaction with domestic principles, norms and mechanisms, as well as with regional, interregional and subregional principles, norms and mechanisms.

The subsystem of regional, interregional and subregional mechanisms for the protection of the right to freedom of conscience includes European mechanisms, Inter-American mechanisms, African mechanisms, mechanisms of the Organization for Security and Co-operation in Europe, mechanisms of the Commonwealth of Independent States, mechanisms of the Association of Southeast Asian Nations, mechanisms of the Organization of Islamic Cooperation and others. The relevant regional mechanisms for the protection of the human right to freedom of conscience should be understood as those formed in the corresponding regions of the world.

The Russian Federation no longer participates in European mechanisms and has to a significant extent suspended interaction within the Organization for Security and Co-operation in Europe. At the same time, the central role in protecting the right to freedom of conscience belongs to the European Court of Human Rights, established on the basis of the 1950 European Convention on Human Rights. It should be noted that the mechanism for protecting the right to freedom of conscience through the European Court of Human Rights is the most effective of the existing human rights mechanisms, both at the regional and universal levels. However, it must be acknowledged that the insufficient effectiveness of the principles and norms of international law, as well as of domestic legal systems, led to a situation in which the ECtHR was unable to exert a significant influence on the realization of the right to freedom of conscience in Russia.

The European system for the protection of human rights in the sphere of freedom of conscience includes the principles, norms and mechanisms of the Council of Europe, including the European Convention for the Protection of Human Rights and Fundamental Freedoms, the former European Commission, the European Court of Human Rights, the Committee of Ministers, the Parliamentary Assembly, the Secretary General of the organization, the Secretariat, the Steering Committee for Human Rights, the European Commission against Racism and Intolerance, the Council of Europe Commissioner for Human Rights, the European Committee on Social Cohesion, Human Dignity and Equality, the European Commission for Democracy through Law, known as the Venice Commission, the Congress of Local and Regional Authorities of Europe and others; the mechanisms of the European Union, including the Charter of Fundamental Rights of the European Union and the European Union Agency for Fundamental Rights; and the mechanisms of the OSCE, including the Final Act of the Conference on Security and Co-operation in Europe, the Helsinki Accords, the Charter of Paris for a New Europe, the Organization for Security and Co-operation in Europe, the Office for Democratic Institutions and Human Rights and the Advisory Council of Experts on Freedom of Religion or Belief.

Regional, interregional and subregional human rights systems also include the Inter-American system, consisting of the American Convention on Human Rights, the Organization of American States, the Inter-American Commission on Human Rights and the Inter-American Court of Human Rights; the African system, including the African Charter on Human and Peoples' Rights, the African Union, the African Commission on Human and Peoples' Rights, the African Court on Human and Peoples' Rights and the African Committee of Experts on the Rights and Welfare of the Child; the system of the Commonwealth of Independent States, including the Declaration of the Heads of State of the CIS on International Obligations in the Field of Human Rights and Fundamental Freedoms, the CIS Convention on Human Rights and Fundamental Freedoms and the Human Rights Commission, provided for by the Charter but not created; the system of Southeast Asian states, including the ASEAN Human Rights Declaration, the Association of Southeast Asian Nations and the ASEAN Intergovernmental Commission on Human Rights; the Islamic system, including the Declaration on Human Rights in Islam, the Cairo Declaration, the Covenant on the Rights of the Child in Islam, the Plan of Action for the Advancement of Women, the Organization of Islamic Cooperation and the Human Rights Commission; and the Arab system, including the Arab Charter on Human Rights, the League of Arab States, the Permanent Arab Commission on Human Rights and the Arab Court of Human Rights.

Thus, in addition to universal mechanisms for the protection of human rights, regional mechanisms also exist, including some that do not fully correspond to universal standards. As for the right to freedom of conscience, even European and American mechanisms remain far from perfect, to say nothing of African, Asian and especially Islamic mechanisms.

Overall, the effectiveness of regional, interregional and subregional mechanisms for the protection of the right to freedom of conscience depends to a significant extent not only on the quality of principles and norms, which, as noted above, require substantial reform, but also on their interaction with universal and domestic principles, norms and mechanisms. Under conditions of intensified global integration processes, the universalization of principles and norms in the sphere of protecting the human right to freedom of conscience is objective in character.

International mechanisms for the protection of the right to freedom of conscience, based on principles and norms that are not fully effective and require reform, must be systematically reformatted. Point reforms, especially given the unprecedented transformation of social relations, cannot improve the situation.

Effective implementation of international legal principles and norms on the right to freedom of conscience in domestic legislation is an important stage in its realization. In the context of global integration processes, not only convergence but also mutual penetration of international and domestic principles and norms is objective in character. Formation of effective mechanisms for implementing the right to freedom of conscience will help overcome ethno-confessional contradictions and support the positive integration of subjects of international law for overcoming global challenges and transitioning to sustainable development.

The study has revealed problems in the implementation of international legal principles and norms on the right to freedom of conscience in the domestic legislation of the Russian Federation. In terms of recognizing the right to freedom of conscience, the Constitution of the Russian Federation does not contradict international principles and norms, but it also requires improvement. The Basic Law of Russia contains declarations of the principle of the secular state as one of the foundations of the constitutional order, but does not define it, which substantially reduces the effectiveness of the norms and mechanisms for realizing the right to freedom of conscience. Russian legislation, and above all the basic 1997 Federal Law 'On Freedom of Conscience and Religious Associations', contradicts the Constitution of the Russian Federation and the letter and spirit of Russia's international obligations.

In recent years, the evolution of Russian legislation has relied on legally undefined concepts and has contradicted the constitutional principle of the secular state. This predetermines arbitrariness and violations of the human right to freedom of conscience. The evolution of Russian legislation in the sphere of the right to freedom of conscience largely contradicts the Constitution of the Russian Federation and international obligations.

The study of the problems of realizing the right to freedom of conscience and the principle of the worldview neutrality of the state in various spheres of social relations has revealed the systemic nature of the crisis in these areas. The crisis has affected all main levels, from the theoretical, legal and educational levels to the normative and law-enforcement levels. Problems in international legal and domestic principles and norms in the sphere of the right to freedom of conscience have predetermined mass systemic violations of this right, as well as the growth of xenophobia, intolerance and discrimination on grounds of worldview affiliation.

The study has revealed systemic problems in the realization of the right to freedom of conscience and the principle of the worldview neutrality of the state in various spheres of social relations. In particular, violations of the principle of worldview neutrality in the state and municipal education system have led to violations of the rights of parents and children to upbringing and education in accordance with their own convictions. Violations of the principle of worldview neutrality in the Armed Forces, other troops and military formations, as well as in places of detention and deprivation of liberty, have led to violations of the right to freedom of conscience. Violations of the principle of worldview neutrality by all branches of state power have significantly intensified the problems of realizing the right to freedom of conscience.

The development of global digital technologies, especially neurotechnologies combined with the Internet of Things, artificial intelligence, emotion recognition and related systems, carries potential threats of total control over the worldview sphere of the human person.

The geopolitical deadlock of the present age also lies in the paradoxical desire for absolute security through total digital transparency. Yet in pursuing this transparency, the state destroys the very worldview freedom that remains one of the sources of the human dimension in the life of civilization. By sacrificing the human right to inner secrecy for the illusion of security, power creates a new existential threat: dehumanization through the transformation of people into a data mass managed by algorithms and deprived of a future.

In these realities, the realization of the right to liberty of worldview choice requires the development and legal consolidation of digital human rights. It also requires a full and unconditional prohibition, at the level of international law, on the development, deployment and operation of emotion-recognition systems, affective computing and biometric psychoanalysis in the activities of public authorities, educational institutions at all levels and the sphere of labour relations.

Overall, a prohibition on the monitoring, formation, alteration or suppression of beliefs through technology should be understood as a comprehensive system of legal and ethical norms aimed at protecting personal autonomy. It includes guarantees of liberty of worldview choice, secured by public-law worldview neutrality, as well as the potential regulation of technological instruments that may be used to influence human beliefs.

RECOMMENDATIONS TO THE WORLD COMMUNITY. FROM FRAGMENTARY PROTECTION TO A GLOBAL ARCHITECTURE OF GUARANTEES FOR THE REALIZATION OF LIBERTY OF WORLDVIEW CHOICE, SECURED BY PUBLIC-LAW WORLDVIEW NEUTRALITY

It is recommended to develop and adopt, on the basis of the 2012 Declaration of the High-Level Meeting of the General Assembly on the Rule of Law at the National and International Levels, a legally binding instrument establishing the legal certainty of the key legal concepts underlying the principles and norms of international and domestic law, including the right to freedom of conscience.

It is recommended to legally establish the principle of worldview neutrality of states and other subjects of international law as one of the most important guarantees for the realization of the internationally recognized right to freedom of conscience. In particular, it appears advisable to discuss this issue within the expert structures of universal international organizations and to form a corresponding international programme. As a first stage, it is recommended to develop and adopt a Global Declaration on the Right to Liberty of Worldview Choice and the Principle of Worldview Neutrality of Subjects of Public International Law in Analog and Digital Environments. Thereafter, on the basis of the Declaration, it is recommended to adopt a legally binding Global Convention on the Right to Liberty of Worldview Choice and the Principle of Worldview Neutrality of Subjects of Public International Law in Analog and Digital Environments.

Taking into account the problem of reconciling different approaches to the understanding of secularity, I propose, in the context of international law, to use the concept of 'international-law worldview neutrality of subjects of international law'. This should be understood as the indifference of states and other competent subjects of international law toward the worldview sphere in all its lawful manifestations.

It is recommended to develop and adopt, on the basis of the 1995 Declaration of Principles on Tolerance, a legally binding instrument, for example a Convention on the Principles of Tolerance understood as mutual respect based on equality regardless of differences. Such an instrument would become an important condition for the realization of the internationally recognized right to freedom of conscience.

To achieve these objectives within the framework of a United Nations international programme, the following tasks are proposed:

- developing and agreeing on contemporary theoretical and legal models based on a new paradigm that removes contradictions between fundamental principles and conceptual apparatus: first, the rule of law based on the principle of legal certainty, in order to limit subjective, multiple and arbitrary interpretations of the principles and norms of international and domestic law and to prevent human rights violations, including violations of freedom of conscience; second, freedom of conscience as liberty of worldview choice; third, secularity as worldview neutrality of states and other subjects of international law; and fourth, tolerance as mutual respect based on equality regardless of differences;

- developing and agreeing on contemporary theoretical and legal models for reforming universal, regional, interregional and subregional international-law principles and norms on the basis of modern theoretical and legal models of the rule of law, legal certainty, liberty of worldview choice, worldview neutrality of states and other subjects of international law, and mutual respect based on equality regardless of differences;

- developing and agreeing on contemporary theoretical and legal models for reforming domestic legal systems on the basis of modern theoretical and legal models and international-law principles and norms;

- developing and agreeing on contemporary theoretical and legal models for creating and improving international-law and domestic systems of guarantees for the realization of internationally recognized human rights, including freedom of conscience, on the basis of the principle of worldview neutrality of states and other subjects of international law;

- developing and agreeing on contemporary theoretical and legal models for reforming the system of international protection of freedom of conscience on the basis of a new paradigm that removes contradictions in fundamental principles and concepts by moving from the concept of religious freedom for believers, without denying it but expanding it, toward freedom of conscience as liberty of worldview choice for every person, without division into believers and non-believers;

- developing and agreeing on contemporary theoretical and legal models for reforming do on the basis of modern theoretical and legal models and international-law principles and norms.

It is recommended to form a programme to support the scientific development and legal consolidation of digital human rights for the purpose of protecting the right to freedom of conscience from violations committed through the use of digital technologies.

RECOMMENDATIONS TO THE RUSSIAN FEDERATION. HOW TO RETURN THE FUTURE, FREEDOM, NEUTRALITY, LEGAL CERTAINTY AND TRUST

The Russian Federation is recommended to carry out a fundamental reform of the norms and institutions governing freedom of conscience, state secularity and the prevention of intolerance:

- to establish a developed interpretation of the constitutional principle of state secularity as worldview neutrality in federal law and, in the future, in the Constitution itself;

- to reform the domestic legal institution of freedom of conscience on the basis of a contemporary theoretical and legal model, international-law principles and norms, the principle of legal certainty and other requirements of modern legal technique;

- to repeal the Federal Law 'On Freedom of Conscience and Religious Associations', while restoring democratic mechanisms of legal regulation common to all public non-profit associations;

- to repeal 'anti-extremist' legislation, including the Federal Law 'On Countering Extremist Activity', amendments to legislation at all levels adopted in connection with it, corresponding legislation of the constituent entities of the Russian Federation, and to abolish 'anti-extremist' structures and lists of 'extremist' organizations and literature;

- to abolish the institution of state religious-studies expertise, including the Expert Council for State Religious-Studies Expertise under the Ministry of Justice of the Russian Federation, the Scientific Advisory Council for the Study of Religious Information Materials for Signs of 'Extremism' under the Ministry of Justice, and expert councils under executive authorities in the constituent entities of the Russian Federation;

- to abolish agreements and contracts between public authorities at different levels and religious organizations;

- to repeal legal norms aimed at introducing confessionally oriented subjects into the state and municipal education system of the Russian

Federation, including 'Fundamentals of Religious Cultures and Secular Ethics', 'Fundamentals of the Spiritual and Moral Culture of the Peoples of Russia' and similar courses;

- to abolish the state educational standard in the speciality or field of 'Theology' and to remove the speciality 26.00.01 'Theology' from the list of academic disciplines of the Higher Attestation Commission;

- to cancel the introduction of military clergy in the Armed Forces of Russia and to abolish the position of assistant unit commander for work with believing servicemen;

- to ensure equal opportunities for the realization of freedom of conscience in security structures and in the penitentiary system on the basis of worldview neutrality and the prevention of intolerance;

- to prohibit public officials from any public demonstration of worldview preferences, including symbols, participation in worship services and similar forms;

- to repeal the Federal Law 'On the Transfer to Religious Organizations of Religious Property in State or Municipal Ownership' and to stop the unlawful transfer of state property and cultural assets to religious associations;

- to repeal the Federal Law 'On Amendments to the Criminal Code of the Russian Federation and Certain Legislative Acts of the Russian Federation for the Purpose of Countering Insults to Religious Beliefs and Feelings of Citizens, Desecration of Objects and Items of Religious Veneration, Pilgrimage, and Places of Religious Rites and Ceremonies';

- to repeal Federal Law No. 374-FZ of 6 July 2016 'On Amendments to the Federal Law on Countering Terrorism and Certain Legislative Acts of the Russian Federation in Terms of Establishing Additional Measures to Counter Terrorism and Ensure Public Safety', including amendments to the Federal Law of 26 September 1997 No. 125-FZ 'On Freedom of Conscience and Religious Associations' aimed at regulating 'missionary activity';

- to adopt a Federal Law 'On the Right to Liberty of Worldview Choice and the Principle of Worldview Neutrality of the State in Analog and Digital Environments';

- to stop other trends in legislative transformation aimed at tightening unlawful control over the worldview sphere and contributing to intolerance;
- to repeal norms establishing non-working public holidays in connection with religious holidays, while introducing provisions into the Labour Code of Russia under which every employee may take one or two days off at any convenient time;
- to adopt a set of measures aimed at improving law-enforcement practice in the sphere of preventing intolerance;
- to stop the persecution of civil activists defending the constitutional secularity of the state and internationally recognized freedom of conscience;
- to annul unjust court decisions and rehabilitate victims of unlawful persecution on grounds of worldview affiliation;
- to bring state policy at all levels in the sphere of freedom of conscience and the prevention of intolerance into conformity with the Constitution of Russia, including by ending the unconstitutional religious policy of the state;
- to develop and implement a comprehensive programme aimed at observing the principles of worldview neutrality of the state, equality of religious associations, ideological diversity and the prevention of intolerance;
- to stop the unlawful construction of places of worship in protected natural zones;
- to stop the unlawful transfer of cultural and natural heritage sites to religious associations;
- to develop and implement a comprehensive programme aimed at implementing an anti-corruption strategy in relations between the state and religious associations;
- to ensure transparency, openness and public awareness in the sphere of relations between the state and religious associations;
- to bring state information policy in the sphere of freedom of conscience and the prevention of intolerance into conformity with the

constitutional principles of worldview neutrality and ideological diversity, and to stop the dissemination of xenophobic stereotypes through the media and mass communications;

- to adopt a general education standard on 'Human Rights' for school pupils and university students, including issues of freedom of conscience, worldview neutrality of the state, tolerance, and the diagnosis and prevention of intolerance;

- to create within school subjects, special courses and electives a system for reproducing knowledge about human rights, including liberty of worldview choice, worldview neutrality of the state, tolerance, and the diagnosis and prevention of intolerance;

- to introduce university courses on 'Human Rights', 'Tolerance', and 'Diagnosis and Prevention of Intolerance';

- to introduce, at law faculties and law universities, special courses on the right to liberty of worldview choice and the principle of worldview neutrality of subjects of public international law in analog and digital environments;

- to conduct and support systematic educational work on liberty of worldview choice and the principle of worldview neutrality of subjects of public international law in analog and digital environments;

- to reform the activities of state human rights structures in the sphere of realizing liberty of worldview choice, worldview neutrality and the prevention of intolerance.

Epilogue: Beyond Armageddon

When Cognitive Freedom Becomes the Foundation of a New Global Order and a Question of Civilization's Future

The contemporary systemic crisis of global security forces a fundamental reconsideration of the nature of international conflicts since systemic control over the worldview sphere and the manipulation of the mass consciousness of the majority of the global population constitute a basic and indispensable condition for unleashing wars in the interests of narrow elite groups because without total informational and cognitive dictate a forceful geopolitical confrontation for the benefit of a small fraction of humanity would be a priori impossible. The recognition of this fundamental dependency compels the author to depart from classical political science paradigms and to declare that the root of the civilizational deadlock lies not in the dimension of material resource distribution or the balance of armaments but in the sphere of the total alienation of individual freedom of thought under which critical circumstances we conclude this study at the height of one of the greatest civilizational fractures of our time which manifests as a systemic crisis of liberty of worldview choice and public-law worldview neutrality.

The global situation in the worldview sphere directly accelerates the threat of losing human subjectivity. According to Pew Research Center, government restrictions have reached a historical maximum, affecting 186 countries and more than 75 per cent of the world's population. At the same time, Open Doors records an extreme level of persecution of Christians, affecting 388 million people, while Humanists International points to criminal punishment for blasphemy in 87 states and records discrimination against humanists, atheists and non-religious people in 186 countries around the world.

In a number of states, religious prescriptions are treated as superior to obligations under international treaties, which makes interaction with international law difficult. In particular, Islamic doctrine often interprets human rights as obligations before faith rather than as guarantees of individual freedoms, while the vague wording of archaic laws allows them to be applied arbitrarily and to discriminate on worldview grounds. Reforms required for convergence with international norms are blocked by resistance from religious and political elites and are further complicated by deep socio-cultural factors.

In recent decades, representatives of centres that call themselves scholarly while belonging to wealthy religious organizations have extended their influence across international organizations and much of the world, reproducing corrupted approaches to relations between the state and religious associations. They participate in the development of international norms and in drafting laws on religious associations in dozens of countries, promoting a model that often grants religious organizations special collective rights and privileges contrary to the principle of neutrality. For example, one of many such centres, acting as a powerful instrument of global legal influence for its church, has over decades covered more than 50 countries through direct consultations on legislative reform. Through annual symposia, the centre has trained more than a thousand state officials from two thirds of the world's countries, including judges and ministers, who then introduce the approaches promoted by the centre into national legal systems. Each year, experts from the centre participate in dozens of legal projects, steering the development of world legislation toward the priority of collective rights of religious organizations over public-law worldview neutrality.

The growing international influence of confessional-political lobbying, supported by opaque financial instruments, creates a risk of deforming international law. In the absence of clear frameworks of worldview neutrality, the sacred status of faith institutions becomes an instrument for obtaining unjustified privileges. This requires the adoption of norms that equalize the rights and obligations of all actors in the public sphere and exclude the possibility of political manipulation through identity.

The case of Brigham Young University, recognized as undesirable in Russia, clearly confirms that the era in which state institutions could serve as a screen for private dogmas has exhausted itself and leads toward global collapse. We propose not a struggle against faith, but a radical disinfection of public administration from prophecies that turn science into a target for security services, and science and education into instruments of corrupt expansion. The only path to saving civilization from geopolitical Armageddon lies through the implementation of worldview neutrality as a basic technical standard guaranteeing humanity's right to rational survival and sustainable development.

As a consequence, in conditions where most of the planet's population lives under severe state control over beliefs, technologies such as AI, biometrics and social scoring are transformed from instruments of development into tools of worldview training.

The absence of neutrality leads to a situation in which digital algorithms, especially in authoritarian and clerical regimes, are programmed to detect 'wrong' thoughts, making the preservation of personal autonomy impossible. As a result, technological progress deprived of the legal foundation of worldview neutrality leads to the creation of a global system in which the human person ceases to be a subject of choice and becomes an object of algorithmic control. This creates an existential threat: the transformation of the person from a free subject into an algorithmically governed biological object.

Without a neutral legal space, any innovation is directed toward standardizing the inner world, leading to an irreversible atrophy of free will. The global crisis of worldview neutrality turns technology into an instrument of direct suppression of human subjectivity, transforming free choice into an algorithmically managed process.

Under such conditions of worldview training, biometric and neuromarketing technologies deprive the person of the right to the privacy of thought, turning the human being from a subject into a predictable data object. On the scale of humanity, this means the risk of a complete loss of subjectivity and uniqueness, where the person is replaced by a predictable digital avatar deprived of the right to worldview error and spiritual search. Thus, the struggle for the neutrality of the state today is a struggle to preserve Homo sapiens as a biological species capable of independent awareness of itself and the world.

The digital revolution 4.0, with its pervasive artificial intelligence, metaverses and biometric surveillance, has ceased to be merely a technological background. It has become a new habitat, where freedom of conscience, as the fundamental human right to independent worldview choice, becomes the last line of defence for the human essence itself.

For a long time, freedom of conscience was perceived as a special topic. As one section of human rights law. As an area where religion, the state, minorities, discrimination, sometimes courts, sometimes international treaties and sometimes academic debates on secularity intersect. All of this is true. But it is no longer enough.

One should agree with Nazila Ghanea that freedom of religion is not a 'privilege of believers', but a fundamental right of every person to personal conscience. Indeed, today freedom of conscience is not only a question of whether a person may profess a religion or reject it. It is a question of something much greater. Does a person have the possibility to form their own picture of the world and find themselves within it?

This right was born in history slowly and painfully. At first, it concerned the fragile and fragmentary protection of religious minorities. Then it concerned partial religious tolerance. Later, international law began to recognize religious freedom as an important part of civilized order. After the Second World War, freedom of thought, conscience and religion entered the universal language of human rights. Yet even this proved insufficient. The world changed faster than its legal formulas.

The crises in the Middle East are an alarming signal for all humanity. If the world does not learn to overcome ethno-confessional barriers and move toward multilateral cooperation, scenarios of civilizational Armageddon will become reality. Salvation should be sought in economic interdependence, joint solutions to global challenges and the human-centred implementation of technologies. Only by uniting around common goals can we build a sustainable planetary system.

Today it is already clear that the old framework is insufficient. The human person lives not only among religions and states. The person lives in a world of global networks, conflicting identities, political myths, digital platforms, neurotechnologies, algorithmic systems, automated surveillance and increasingly subtle mechanisms of influence on attention, memory, fear and beliefs. In these conditions, freedom of conscience can no longer be understood only as the right to religious choice or refusal of religion. It must be understood and legally consolidated more broadly as liberty of worldview choice.

This is where a new legal era begins. The essence of this era is not that the human person needs beliefs less. On the contrary. In a world of acceleration, uncertainty and global fractures, the need for a picture of the world has only intensified. At the same time, the forces seeking to control that picture of the world have also grown stronger. Some do so through ideology. Others through political loyalty. Others through the privileged position of certain religious structures. Others through the language of law, saturated with vague and conflict-generating concepts. Others through technology.

For this reason, freedom of conscience ceases to be a narrow freedom of religion and becomes a broad possibility of worldview formation, one of the central themes of humanity's future.

This problem also has another dimension, less visible but no less important. Law weakens not only when it is violated. It weakens when it is poorly understood. When key concepts are vague. When legal categories are replaced by ideological ones. When clear and predictable norms give way to constructions suited more to arbitrary application than to the protection of human rights. That is why legal certainty is not a technical detail here, but a question of the very quality of freedom. If the key words of law are unclear, freedom itself becomes dependent on the will of the interpreter.

From this follows one of the most important conclusions of this book: in the field of freedom of conscience, the problem does not begin at the final stage, not in court and not at the police station. It begins at the level of definitions. At the level of theory. At the level of the language through which international and domestic law attempts to describe human freedom in the worldview sphere.

When the broad right to freedom of conscience is replaced by a narrow freedom of religion, the person loses part of their legal visibility. When the state does not limit itself by the principle of worldview neutrality, it quietly begins to transform from a guarantor of freedom into a participant in the struggle for meaning in the interests of a narrow part of society. When undefined and politically loaded categories are fixed in law, privileges, discrimination, arbitrariness and destabilization arise. When there is no systematic work to cultivate tolerance, society begins to reproduce xenophobia, intolerance and fear of difference.

All this is especially dangerous in the age of globalization. The world no longer consists of isolated spaces where common problems can be postponed endlessly. Today humanity lives within a single, although unevenly developing, planetary system, where social, technological and natural processes are increasingly intertwined. In such a system, no stable order is possible without more mature forms of international cooperation. Yet cooperation itself is impossible if the world remains split by ethno-confessional barriers, historical fears and the claim of politics to a monopoly on truth.

Therefore, freedom of conscience in its new understanding, as liberty of worldview choice, becomes not simply one right among others. It becomes a condition for the transition to a more mature civilization. To a civilization in which the human person has the right not only to live, but also to understand what they believe, how they think, what they trust and how they build their picture of the world.

This is also where the idea of worldview neutrality grows from. It does not mean indifference to the person or the abandonment of values. It means that the state and other holders of public power must not appropriate the right to final truth in the worldview sphere. Their task is not to appoint the correct picture of the world, but to guarantee a space in which each person can form it freely, without fear, without privileges for some and without stigma for others.

This is especially important today, when a new threat of total technological control over the worldview sphere of the human person appears on the horizon. If in the past power fought primarily for behaviour, it now increasingly gains tools for fighting over attention, emotions, reactions, habits and beliefs. Neurotechnologies, artificial intelligence, the Internet of Things, facial recognition, platform algorithms and digital profiling create conditions in which interference in the inner world of the person becomes not an exception, but an infrastructural possibility.

Today traditional threats to freedom of conscience from states are being replaced by a more subtle and more dangerous mechanism of algorithmic control over consciousness. We have already entered a zone of unprecedented risk. Neural networks analyse our beliefs through digital traces, turning the intimate sphere of the spirit into a transparent object for manipulation. Filter bubbles and predictive analytics narrow the horizon of thought, replacing the free search for truth with an algorithmically assigned vector. The risk of automated deprivation of rights and social benefits based on analysis of a person's views, extracted by AI black boxes, is growing.

In effect, the digital age presents humanity with an ultimatum. Either algorithms will dictate meanings to us, or we will manage to build human-centred global law 4.0. In this new architecture, technologies must become not an instrument of suppression, but a digital armour guaranteeing the worldview sovereignty of every person.

Susie Alegre rightly calls for a rethinking of the concept of human rights for the twenty-first century, in order to preserve people's ability to think independently in a world where algorithms increasingly dictate what and how we should think.

Today the battle for the human right to one's own picture of the world ceases to be a private matter of the individual. In the age of the post-digital fracture, cognitive freedom becomes the only constant capable of holding civilization back from disintegration. We face a choice. Either we turn the world into a transparent matrix of algorithmic control. Or we recognize the sovereignty of human consciousness as the foundation of a new global order. The future will arrive not when technologies become perfect, but when the right to an uncomputable truth becomes absolute.

Civilizational Armageddon is not a flash in the sky, but silence in the human soul after it has replaced its meanings with algorithmic code. Our search for a way out of the geopolitical deadlock has led to the only correct answer. Peace will come not only when the guns fall silent, but when states recognize worldview sovereignty as the highest value. In the age of global existential threats, cognitive freedom and strict neutrality of power become the only foundation of a new world order. To prevent catastrophe, it is necessary to protect every person's right to their own picture of the world as the last territory of freedom that cannot be digitized, subordinated or destroyed.

The new era requires a new legal paradigm. Not a paradigm of control over difference, but a paradigm of respect for free worldview choice. Not a paradigm of special privileges, but a paradigm of equality of rights. Not a paradigm of religious policy, but a paradigm of worldview neutrality. Not a paradigm of formal protection, but a paradigm of real implementation of human rights.

This book grows precisely out of such an understanding of the problem. Its task is to show that freedom to form one's picture of the world must be rethought as one of the most important conditions of human self-realization, social resilience, international solidarity and the survival of civilization in the face of global challenges.

When a person loses the right to their own picture of the world, they lose not merely one freedom. They lose the inner point of support from which freedom itself begins. That is why the conversation about freedom of one's picture of the world today is not a conversation about the past. It is a conversation about the future.

In these conditions, law cannot remain in the language of the past. Recognizing the scale of these challenges, we appeal to the United Nations, UNESCO and global technology regulators. It is time to acknowledge that declarations of the previous century are powerless before the code of the future. We call for the development and adoption of a Global Declaration and a Global Convention on the Right to Liberty of Worldview Choice and the Principle of Worldview Neutrality of Subjects of Public International Law in Analog and Digital Environments, securing the following: the right to autonomy of consciousness and a prohibition on the use of AI for hidden manipulation of beliefs and social scoring on worldview grounds; the protection of neurodata and recognition of personal views and psychophysical reactions as a zone of the highest privacy, removed from commercial and state circulation; algorithmic transparency and mandatory code audit, including open-source requirements, for all systems that influence the formation of the information agenda and access to rights; the right to analog life and the guaranteed possibility of full participation in society without mandatory digital identification and biometric surveillance.

The realization of liberty of worldview choice in analog and digital environments is the program code of the future and the core of human-centred development. Our choice today will determine whether the human spirit remains sovereign or becomes a statistical error in a global database. The realization of the right to beliefs in the digital age is not merely a legal task. It is a battle to preserve the human being as such. The time has come to inscribe freedom to form one's picture of the world into the foundation of world order, transforming it from a declaration into an unshakeable algorithm of our common freedom.

Overcoming the rule of force in the digital age is possible only through recognition that the true power of civilization is measured not by the efficiency of its control systems, but by the degree of freedom of its citizens. The transition to a model of universal basic income and global education is not an act of charity, but the material embodiment of worldview neutrality. Only a person freed from fear for survival and from the burden of imposed meanings can become a full subject, rather than an object, in interaction with artificial intelligence. The systemic work begun today lays the foundation for a world in which technology serves people rather than absorbs them.

Tomorrow is not predetermined. It is variable and depends on our readiness to embed the values of human rights into the foundation of the global network. If the Declaration and the Convention become the genetic code of unified AI, we will see not the decline of humanity, but its true Renaissance. Worldview neutrality will become the space in which the diversity of pictures of the world ensures the resilience of the entire system. We owe future generations not a digital prison, but a free planet where every person's right to an authentic human experience is protected by the force of global consensus. This path should have begun yesterday, but today we still have a chance to choose life.

The deadlocked system of international relations that has taken shape rested on the confidence of world elites that, as before, they would be able to maintain control by using ideological division, censorship, surveillance and local conflicts. Yet in the age of globalization 4.0, this strategy leads to the strengthening of existential threats and to civilizational Armageddon, where everyone loses.

Today, understanding is gradually emerging that the digital militarization of the new century will sooner or later deprive elites of the illusory possibility of hiding in bunkers, while global social chaos, civil wars and ecological collapse will finally devalue money and power. All this is taking place against the background of a total inability to reach agreement because of irrational ethno-confessional and ideological barriers.

In these realities, understanding of our concept and of the need for change will most likely come to global political elites not through reading academic monographs, but through the basic instinct of self-preservation. Accordingly, the model of a worldview-blind, human-centred service state will be accepted by them as the last lifebuoy under conditions threatening their own biological survival. Provided it is not too late. This is why the transition to worldview neutrality of institutions of power is no longer a matter of philosophical choice, but an unavoidable condition for the survival of human civilization.

Draft Global Declaration on the Right to Liberty of Worldview Choice and the Principle of Worldview Neutrality of Subjects of Public International Law in Analog and Digital Environments

(Developed jointly by Sergey A. Buryanov and Maxim S. Buryanov)

Global Declaration

on the Right to Liberty of Worldview Choice and the Principle of Worldview Neutrality of Subjects of Public International Law in Analog and Digital Environments

Preamble

States and other subjects of public international law,

reaffirming their commitment to the purposes and principles of the Charter of the United Nations, the Universal Declaration of Human Rights, the International Covenant on Civil and Political Rights and other universal and regional international legal instruments in the field of human rights,

proceeding from the understanding that human dignity is inseparable from the ability of a person independently to form a picture of the world, determine their place in the world, and relate knowledge and faith, conviction and doubt, tradition and critical thinking,

recognizing that the historical evolution of the internationally recognized right to freedom of conscience has passed from the protection of individual religious minorities to the affirmation of freedom of thought, conscience and belief as a vital part of the universal system of human rights,

noting that in the twenty-first century the need has matured for the further development of this system on the basis of a new paradigm that understands freedom of conscience as liberty of worldview choice for every person, without discrimination or coercion,

aware that uncertainty in key legal concepts, the substitution of the broad right to freedom of conscience by a narrow understanding of freedom of religion, and the use of vague and conflict-generating categories undermine the effectiveness of international and domestic law,

taking into account that the worldview sphere of the human person cannot be an object of privileged control by the state, political, religious, corporate or other centres of power,

proceeding from the view that the worldview neutrality of states and other subjects of international law is one of the most important guarantees for the realization of human rights, including the right to freedom of conscience, recognizing that under conditions of globalization, increasing interdependence, the growth of ethno-confessional contradictions, the spread of intolerance, xenophobia and discrimination, and the rapid development of digital technologies, artificial intelligence, neurotechnologies, facial recognition systems and other means of mass influence on consciousness, the protection of the inner freedom of the human person acquires planetary significance,

considering that the sustainable, manageable and peaceful development of civilization requires a transition from a limited approach to the protection of freedom of religion to a broader and more contemporary concept of realizing the liberty of worldview choice of every person,

proclaim this Declaration.

Article 1. The Right to Liberty of Worldview Choice

Every person has the right to liberty of worldview choice. This right includes the freedom to form, hold, preserve, express, change or reject beliefs, values, views of the world, religious, non-religious, philosophical, ethical and other worldview positions.

Article 2. Inadmissibility of Coercion

No one may be subjected to coercion capable of limiting the freedom of their worldview choice.

All forms of pressure, violence, threats, discrimination, abuse of dependence, and administrative, political, economic, informational, technological or other influence aimed at imposing or suppressing beliefs shall be prohibited.

Article 3. Equality and Non-Discrimination

All people are equal in dignity and rights regardless of their worldview affiliation, the absence of such affiliation, a change of beliefs or a refusal to express them publicly.

Discrimination on the grounds of religion, beliefs, worldview, philosophical position, criticism of beliefs, doubt, unbelief or membership in relevant associations shall be prohibited.

Article 4. Freedom of Worldview Self-Determination

Liberty of worldview choice belongs to every person from birth and does not depend on recognition by the state, religious organizations, political structures, corporations or other actors.

No one shall be required to identify themselves in the worldview sphere against their own will.

Article 5. Individual and Collective Dimensions of Freedom

The right to liberty of worldview choice includes both an individual and a collective dimension.

Persons have the right to associate on the basis of shared beliefs, to establish associations acting in accordance with law and human rights, and to carry out their activities on equal grounds with other public and non-profit associations.

Article 6. Public-Law Worldview Neutrality

States and other subjects of public international law shall observe the principle of public-law worldview neutrality.

This principle means that no worldview, ideology or religion shall be established as compulsory, privileged or officially preferred for all.

Article 7. The Content of Worldview Neutrality

Public-law worldview neutrality includes:

• non-identification of public authority with any worldview;

• a prohibition on discrimination on the grounds of worldview affiliation;

• legal equality of worldview associations and other public associations;

• refusal of special privileges and restrictions on worldview grounds;

• refusal of special relations between public authority and worldview associations where such relations violate the principle of equality;

• refusal to legally establish the universal obligatory character of any worldview or ideology;

• the provision of guarantees for free worldview choice for every person.

Article 8. Neutrality of Public Institutions

The principle of worldview neutrality shall be observed in all public-law spheres, including the legislative, executive and judicial branches, local self-government, the state and municipal education system, the armed forces, law-enforcement bodies, the penitentiary system, state and municipal mass media, and other institutions exercising public functions.

Article 9. Legal Certainty

Guarantees of liberty of worldview choice shall be based on the principle of legal certainty.

Key legal concepts in this sphere shall be clear, unambiguous, predictable and consistent with the requirements of modern legal technique.

The use of uncertain, stigmatizing and conflict-generating categories that create conditions for arbitrary restrictions on human rights shall be inadmissible.

Article 10. Rejection of Coercive Regulation of the Worldview Sphere

Legal regulation may concern only unlawful acts, not worldview as such.

Criminal, administrative and other forms of prosecution shall be permitted only for specific unlawful acts, and not for belonging to beliefs or to religious, philosophical or other worldview systems.

Article 11. Tolerance and Mutual Respect

States and other subjects of international law undertake to develop tolerance as mutual respect on the basis of equality, regardless of differences.

They shall carry out systematic work to prevent xenophobia, intolerance and discrimination on grounds of worldview affiliation through education, culture, public communication and international cooperation.

Article 12. Education and the Rights of Parents and Children

Every person has the right to education that does not violate their liberty of worldview choice.

Parents and other legal representatives have the right to ensure the upbringing of children in accordance with their convictions, within the limits of human rights and the dignity of the child.

The state and municipal education system shall be built on the principles of worldview neutrality, human rights and respect for diversity.

Article 13. Mass Media and Public Communication

States undertake to prevent the use of public media and other channels of mass communication for the dissemination of xenophobic stereotypes, stigmatization, discrimination and the hidden imposition of a worldview hierarchy.

Public information policy shall comply with the principles of neutrality, equality and mutual respect.

Article 14. Digital Technologies, Algorithms and Artificial Intelligence

The development, deployment and use of digital technologies, artificial intelligence systems, neurotechnologies, facial recognition systems, platform and algorithmic tools shall not violate the liberty of worldview choice of the human person.

States undertake to prevent the use of such technologies for total control over the worldview sphere, hidden manipulation of beliefs, mass profiling and discrimination.

Article 15. The Principle of Algorithmic Neutrality of Subjects of Public International Law

In the digital environment, the worldview neutrality of the state and other subjects of public international law includes an unconditional prohibition on the use of state algorithms, artificial intelligence systems, automated recommendation systems, profiling and forecasting for the manipulation of consciousness, the imposition of exclusive ideologies or hidden influence on the political, religious, philosophical and other beliefs of the individual.

It is recognized that the use of digital technologies to suppress worldview pluralism or to create artificial information filters at the state level constitutes a violation of the right to the free development of the person and a threat to the civilizational foundations of society.

Subjects of public international law confirm that state digital systems and algorithmic infrastructures shall be designed, applied and subjected to regular independent audit for compliance with the principles of worldview neutrality, non-discrimination, transparency and accountability to society.

The duty of public authorities to ensure the algorithmic impartiality of public services is proclaimed, guaranteeing that automated systems do not become instruments of ideological segregation or of restricting access to the diversity of worldview positions.

Article 16. The Responsibility of States to Ensure Worldview Neutrality in the Private Digital Sector

States recognize their primary role and responsibility in ensuring that the activities of private actors, including transnational digital corporations, do not violate the right of the individual to free and independent worldview choice.

States confirm their duty to create conditions under which private digital platforms function on the basis of algorithmic impartiality, transparency and accountability to society.

The need for state assistance in eliminating any forms of hidden worldview manipulation carried out through artificial intelligence technologies and personal profiling systems is recognized.

States shall strive to establish standards of content moderation that exclude discrimination on worldview grounds and prevent the creation of isolated information spaces.

States recognize the right of every person to effective protection and access to fair appeal procedures in cases where decisions of private algorithmic systems violate their worldview freedom.

The importance of international cooperation among states is emphasized in order to prevent the use of private technological resources as instruments of ideological domination that undermine the foundations of global stability and civilizational diversity.

Article 17. Prohibition of the Technological Exploitation of Human Emotions

A complete and unconditional prohibition is proclaimed on the development, deployment and use of systems for recognizing, analysing and classifying human emotional states in public space, education, labour relations and public administration.

Any attempt at automated penetration into the inner emotional world of the person shall be recognized as a gross violation of human dignity, the right to mental integrity and the fundamental liberty of worldview choice.

Article 18. International Organizations

International organizations exercising public functions shall observe the principle of worldview neutrality and promote the realization of liberty of worldview choice.

The creation of international intergovernmental structures on the basis of compulsory worldview affiliation shall be inadmissible where it contradicts the principles of equality and the universality of human rights.

Article 19. Realization Instead of Formal Protection

The international community shall move from a limited approach oriented only toward the formal protection of, or respect for, freedom of conscience to a broader concept of its full realization, ensuring that every person can enjoy the benefits of liberty of worldview choice in the interests of their own development, self-realization and dignified life.

Article 20. International Cooperation and Legal Reform

States and other subjects of international law undertake to promote:

- reform of international and domestic legal institutions in this sphere;
- coordination of contemporary theoretical and legal models of liberty of worldview choice;
- formation of international and national guarantees for the realization of the relevant right;
- development of international mechanisms for monitoring and cooperation;
- preparation of a legally binding convention on the basis of this Declaration.

Article 21. Final Provision

This Declaration is intended to become a basis for the progressive development of international law, the reform of national legal systems and the formation of a planetary culture of mutual respect, in which the human right to one's own picture of the world is recognized as one of the most important conditions for peace, sustainable development and the survival of civilization.

Draft Global Convention on the Right to Liberty of Worldview Choice and the Principle of Worldview Neutrality of Subjects of Public International Law in Analog and Digital Environments

(Developed jointly by Sergey A. Buryanov and Maxim S. Buryanov)

Global Convention

on the Right to Liberty of Worldview Choice and the Principle of Worldview Neutrality of Subjects of Public International Law in Analog and Digital Environments

Preamble

The States Parties to the present Convention,

reaffirming their commitment to the purposes and principles of the Charter of the United Nations, the Universal Declaration of Human Rights, the International Covenant on Civil and Political Rights, the International Covenant on Economic, Social and Cultural Rights and other universal and regional international legal instruments in the field of human rights,

proceeding from the understanding that human dignity is inseparable from the ability of a person independently to form a picture of the world, determine their place in the world, and relate knowledge and faith, conviction and doubt, tradition and critical thinking,

recognizing that the historical evolution of the internationally recognized right to freedom of thought, conscience, religion or belief requires further development toward a broader and more contemporary legal paradigm that understands freedom of conscience as the liberty of worldview choice of every person, without discrimination or coercion,

aware that uncertainty in key legal concepts, the substitution of the broad right to freedom of conscience by a narrow understanding of freedom of religion, and the use of vague, stigmatizing and conflict-generating categories undermine the effectiveness of international and domestic law,

taking into account that the worldview sphere of the human person cannot be an object of privileged control by the state, political, religious, corporate or other centres of power,

confirming that the worldview neutrality of states and other holders of public authority is one of the most important guarantees for the realization of internationally recognized human rights, including the right to liberty of worldview choice,

recognizing that, under conditions of globalization, increasing interdependence, the growth of ethno-confessional contradictions, the spread of intolerance, xenophobia and discrimination, and the rapid development of digital technologies, artificial intelligence, neurotechnologies, virtual reality, facial recognition systems and other means of influence on consciousness, the protection of the inner freedom of the human person acquires planetary significance,

considering that the sustainable, manageable and peaceful development of civilization requires a transition from a limited approach to the protection of freedom of religion to a broader and more contemporary concept of realizing the liberty of worldview choice of every person in analog and digital environments,

have agreed as follows:

PART I

GENERAL PROVISIONS

Article 1

Terms and Scope of Application

For the purposes of the present Convention:

"Liberty of worldview choice" means the right belonging to every person freely to form, have, preserve, express, reconsider, change or reject beliefs, values, views of the world, and religious, non-religious, philosophical, ethical and other worldview positions.

"Worldview neutrality" means the obligation of the state and other holders of public authority to refrain from establishing, imposing, privileging or discriminatorily restricting any worldview, ideology or religion, and to ensure equal legal conditions for the free worldview choice of every person.

"Analog and digital environments" means all forms of the creation, storage, transmission, search, processing, dissemination and perception of information, including printed and audiovisual media, telecommunication networks, the Internet, platform services, algorithmic systems, artificial intelligence systems, neurotechnologies, biometric systems, extended, mixed and virtual reality technologies, and other existing or emerging technological forms of communication and information storage.

"Algorithmic system" means any automated or semi-automated system, including an artificial intelligence system, that affects the sorting, ranking, recommendation, restriction, removal, amplification, suppression, prediction or personalization of information content, access to such content, or user behaviour.

"Discriminatory moderation" means the removal, blocking, restriction of visibility, hidden downgrading of reach, demonetization, labelling or other limitation of the dissemination of content or an account solely or predominantly on the grounds of the worldview expressed, where such content does not contain direct and intentional incitement to the commission of a specific violent crime against a person.

"Holders of public authority" means states, their bodies and institutions, international organizations, and other entities to which public functions are delegated under international or domestic law.

The provisions of the present Convention apply to all forms of communication and information storage, including analog and digital environments.

Article 2

Subject Matter and Purpose of the Convention

The purpose of the present Convention is to ensure the effective realization of the right of every person to liberty of worldview choice and to enshrine the worldview neutrality of states and other holders of public authority in analog and digital environments.

The present Convention is aimed at:

a) protecting the inner freedom of the human person and its external manifestations from coercion, discrimination and manipulation;

b) ensuring equality regardless of worldview affiliation;

c) preventing abuses of public and digital power in the worldview sphere;

d) establishing international and domestic guarantees for the realization of liberty of worldview choice.

Article 3

Fundamental Principles

States Parties shall be guided by the following principles:

a) human dignity;

b) liberty of worldview choice;

c) equality and non-discrimination;

d) legal certainty;

e) worldview neutrality of public authority;

f) tolerance as mutual respect on the basis of equality, regardless of differences;

g) pluralism of opinions and free access to a diversity of ideas;

h) inadmissibility of coercion in the worldview sphere;

i) transparency and accountability of algorithmic systems affecting liberty of worldview choice.

PART II

THE RIGHT TO LIBERTY OF WORLDVIEW CHOICE

Article 4

Content of the Right

Every person has the right to liberty of worldview choice.

This right includes the freedom to:

a) form, have and change beliefs;

b) refrain from disclosing one's beliefs;

c) express or refrain from expressing one's beliefs, individually or jointly with others;

d) seek, receive and impart information and ideas necessary for the formation of one's own picture of the world;

e) refuse worldview identification imposed by the state, society, a platform or any other actor.

Article 5

Inadmissibility of Coercion

No one may be subjected to coercion capable of limiting the freedom of their worldview choice.

All forms of pressure, violence, threats, discrimination, abuse of dependence, administrative, political, economic, informational, technological, algorithmic or other influence aimed at imposing, suppressing or manipulating beliefs shall be prohibited.

Article 6

Equality and Non-Discrimination

All people are equal in dignity and rights regardless of their worldview affiliation, the absence of such affiliation, a change of beliefs or a refusal to express them publicly.

Discrimination on the grounds of religion, beliefs, worldview, philosophical position, criticism of beliefs, doubt, unbelief or belonging to relevant associations shall be prohibited.

Article 7

Individual and Collective Dimensions

The right to liberty of worldview choice has both an individual and a collective dimension.

Persons have the right to associate on the basis of shared beliefs and to carry out the corresponding activities on equal grounds with other public and non-profit associations.

The rights of associations shall be regarded as derivative from human rights and freedoms and may not be used to substitute for, or restrict, the rights of the individual.

PART III

PERMISSIBLE LIMITATIONS AND LEGAL CERTAINTY

Article 8

General Standard of Permissible Limitations

Limitations on the freedom to express a worldview are inadmissible except where such expression constitutes direct, intentional and immediate incitement to the commission of a specific violent crime against a person.

Any permissible limitation must be:

a) provided by law;

b) strictly necessary;

c) proportionate;

d) individualized;

e) subject to independent review.

The expression of a worldview position shall not, in itself, constitute an unlawful act.

Article 9

Prohibition of Expansive Interpretation of Undefined Categories

The absence of international consensus on the definitions of "terrorism", "extremism" or analogous categories may not serve as a basis for an expansive interpretation of such concepts for the purpose of introducing censorship, discriminatory moderation or restrictions on liberty of worldview choice.

States Parties undertake not to use uncertain, stigmatizing and politically loaded categories to restrict the rights guaranteed by the present Convention.

Article 10

Legal Certainty

Guarantees of liberty of worldview choice shall be based on the principle of legal certainty.

Key legal concepts in this sphere shall be clear, unambiguous, predictable and consistent with the requirements of contemporary legal technique.

States Parties undertake to review legislation and practice based on legally uncertain and discriminatory concepts.

PART IV

WORLDVIEW NEUTRALITY OF THE STATE AND PUBLIC AUTHORITY

Article 11

General Obligation of Neutrality

States Parties undertake to observe the principle of public-law worldview neutrality.

No worldview, ideology or religion shall be established as compulsory, privileged or officially preferred for all.

Article 12

Content of Worldview Neutrality

Worldview neutrality includes:

a) non-identification of public authority with any worldview;

b) a prohibition on discrimination on grounds of worldview affiliation;

c) legal equality of associations;

d) refusal of special worldview privileges and restrictions;

e) refusal of special relations between public authority and worldview associations where such relations violate the principle of equality;

f) refusal to legally establish the universal obligatory character of any worldview or ideology;

g) provision of real guarantees of free worldview choice for every person.

Article 13

Neutrality of Public Institutions

The principle of worldview neutrality shall be binding on all branches of power, local self-government, the education system, the armed forces, law-enforcement bodies, the penitentiary system, state and municipal media, and other institutions exercising public functions.

Civil servants, when exercising public functions, shall refrain from using their status to impose or privilege worldview preferences.

Article 14

Education, Children and Parents

Every person has the right to education that does not violate their liberty of worldview choice.

The state and municipal education system shall be built on the principles of worldview r diversity.

Parents and other legal representatives have the right to ensure the upbringing of children in accordance with their convictions within the limits of human rights, the dignity of the child and the child's evolving autonomy.

Article 15

Public Communication and Mass Media

States Parties undertake to prevent the use of public media and other channels of mass communication for stigmatization, xenophobia, discrimination and the hidden imposition of a worldview hierarchy.

Public information policy shall comply with the principles of neutrality, equality, reliability and mutual respect.

PART V

ANALOG AND DIGITAL ENVIRONMENTS

Article 16

Algorithmic Neutrality

In the digital environment, the worldview neutrality of the state includes a prohibition on the use of state algorithms, artificial intelligence systems, automated recommendation systems, profiling and forecasting for the manipulation of consciousness, the imposition of a single ideology, the suppression of worldview pluralism or hidden influence on political, religious, philosophical or other beliefs.

States Parties undertake to ensure that state digital systems and algorithmic infrastructures are designed, applied and reviewed for compliance with the principles of neutrality, non-discrimination, accountability and legal certainty.

Article 17

Digital Sovereignty of the Individual

Every person has the right to digital sovereignty over their worldview sphere.

This right includes:

a) the right independently to determine the degree of openness and personalization of one's information environment;

b) the right to protection from hidden algorithmic confinement within a limited information circuit;

c) the right of access to a diversity of ideas and sources without forced filtering imposed by the state, a platform or any other actor;

d) the right to know about significant mechanisms of algorithmic influence on access to information.

Nothing in the present article may be interpreted as justifying the manipulative restriction of a person's access to a diversity of ideas under the pretext of their "protection".

Article 18

The Right to Digital Search and Access to a Diversity of Ideas

Every person has the right to digital search and access to a diversity of ideas, views and worldview positions in global data networks.

States Parties shall not apply forced filtering, state control and filtering of information flows, hidden restriction of search results, blocking or other forms of systemic restriction of worldview search, except in the cases provided for in article 8 of the present Convention.

Article 19

Prohibition of Discriminatory Moderation

States Parties undertake to prevent discriminatory moderation in the digital information space.

State bodies, as well as platforms and digital intermediaries regulated by them, shall not restrict content solely on the ground that it expresses a different worldview, unless such content falls within the exception provided for in article 8 of the present Convention.

Pluralism of opinions in global data networks shall be protected as an important condition for liberty of worldview choice.

Article 20

Algorithmic Transparency

States Parties undertake to ensure disclosure of the basic principles of operation of algorithmic systems that may have a significant impact on access to information, the visibility of worldview content, ranking, recommendations, removal, hidden downgrading of reach or other forms of influence on liberty of worldview choice.

Persons shall have the right to:

a) be notified of significant algorithmic influence on access to information;

b) receive an explanation of the reasons for restriction, removal, downgrading of reach or other interference;

c) appeal such decisions before an independent body.

Practices of hidden reach restriction without proper notification and a review procedure shall not be permitted.

Article 21

Obligations of States in Relation to Private Platforms and Digital Corporations

States Parties shall take the necessary legislative, administrative, judicial and other measures to ensure that private digital platforms, including transnational corporations operating within their jurisdiction or providing services on their territory, realize the rights enshrined in the present Convention.

Such measures include:

a) a prohibition on discriminatory moderation on worldview grounds;

b) requirements for transparency and accountability of algorithmic systems;

c) the existence of effective procedures for notification, review and appeal of decisions of automated systems;

d) obligations to prevent hidden worldview manipulation of users.

States Parties shall ensure that systemically significant digital platforms and owners of algorithmic systems observe the principle of worldview neutrality in their public activities.

States Parties shall guarantee that recommendation, search and content-moderation algorithms are not used for hidden manipulation of users' worldview choice or for the artificial promotion of certain doctrines to the detriment of others.

States Parties shall establish requirements for algorithmic transparency, obliging operators to disclose the principles by which information flows affecting a person's worldview self-determination are formed.

States Parties undertake to introduce mechanisms of state supervision and independent audit of algorithms in order to prevent worldview discrimination.

States Parties shall promote international cooperation in developing universal norms of responsibility for violations of the worldview sovereignty of the individual.

Article 22

Complete Prohibition and Elimination of Emotion-Recognition Systems and Legal Responsibility

States Parties shall establish a complete and unconditional legislative prohibition on the development, deployment and operation of systems for emotion recognition, affective computing and biometric psychoanalysis in the activities of public authorities, educational organizations at all levels and labour relations.

States Parties undertake, within [specify time period] from the entry into force of the present Convention, to ensure the complete and irreversible destruction of all existing software complexes and databases containing the results of emotional profiling and affective monitoring of the individual accumulated by public and private structures.

States Parties shall establish strict administrative and criminal liability for heads and officials of educational organizations who authorize the use of such technologies under any pretext, including the "improvement of the educational process". A violation of this prohibition shall entail the immediate withdrawal of the state licence or accreditation of the institution.

The development and distribution of emotion-recognition systems for use in public and educational spaces shall be equated with unlawful interference with the mental integrity of the person and shall entail criminal prosecution of the responsible persons and the compulsory liquidation of the relevant technology companies.

Any data obtained in violation of the present prohibition shall be legally void and may not be used for the adoption of any administrative, academic or other decisions concerning the individual.

Article 23

Neurotechnologies, Virtual Reality and Cognitive Protection

States Parties undertake to prevent the use of neurotechnologies, virtual, mixed and extended reality, biometric systems, behavioural profiling and other technologies for unlawful influence on the worldview sphere of the human person.

Any use of such technologies in the public sphere shall meet the requirements of:

a) legality;

b) necessity;

c) proportionality;

d) respect for human dignity;

e) worldview neutrality.

The use of technologies for monitoring, forming, changing or suppressing a person's beliefs shall be prohibited.

PART VI

IMPLEMENTATION, CONTROL AND COOPERATION

Article 24

Domestic Guarantees and Remedies

States Parties shall ensure effective judicial, administrative and other remedies for persons whose rights under the present Convention have been violated.

Such remedies shall be accessible, effective, independent and capable of ensuring the prompt cessation of violations, restoration of violated rights and appropriate compensation.

Article 25

National Implementation Mechanisms

Each State Party shall establish or designate one or more independent national mechanisms to promote the implementation of the present Convention.

Such mechanisms may include ombudspersons, national human rights institutions, specialized commissions, independent digital-environment regulators and other bodies meeting the principles of independence and competence.

Article 26

Committee on Liberty of Worldview Choice and Worldview Neutrality

For the purpose of monitoring the implementation of the present Convention, a Committee on Liberty of Worldview Choice and Worldview Neutrality shall be established.

The Committee shall consist of independent experts of high moral standing and recognized competence in international law, human rights, digital regulation and related disciplines.

The Committee shall consider reports of States Parties, formulate concluding observations and adopt general recommendations.

Article 27

Reports of States Parties

States Parties shall submit to the Committee reports on the legislative, judicial, administrative and other measures they have adopted to implement the present Convention.

The initial report shall be submitted within two years after the entry into force of the Convention for the State concerned, and subsequent reports every four years o

Article 28

Inter-State Communications and Individual Petitions

A State Party may recognize the competence of the Committee to consider communications concerning a violation of the present Convention by another State Party.

A State Party may also recognize the competence of the Committee to receive and consider individual communications from persons or groups of persons subject to its jurisdiction who claim to be victims of a violation of the rights enshrined in the present Convention.

The procedure for implementing the present article may be further regulated by an optional protocol.

Article 29

International Cooperation

States Parties undertake to cooperate with one another and with competent international organizations for the purposes of:

a) harmonizing standards;

b) exchanging best practices;

c) developing common approaches to digital challenges;

d) promoting education, awareness-raising and tolerance;

e) preventing transboundary violations of liberty of worldview choice.

PART VII

FINAL PROVISIONS

Article 30

Relationship with Other International Treaties

Nothing in the present Convention shall be interpreted as diminishing a higher standard of protection established by international law or domestic legislation.

Article 31

Signature, Ratification and Accession

The present Convention shall be open for signature by all States. It shall be subject to ratification. It shall be open for accession by any State.

Article 32

Entry into Force

The present Convention shall enter into force on the ninetieth day after the date of deposit of the twentieth instrument of ratification or accession.

Article 33

Amendments

Any State Party may propose an amendment to the present Convention. Amendments shall be adopted in accordance with the procedure established by the General Assembly of the United Nations or by a conference of States Parties.

Article 34

Denunciation

Any State Party may denounce the present Convention by written notification to the depositary. Denunciation shall take effect one year after receipt of such notification.

Article 35

Depositary

The Secretary-General of the United Nations shall be the depositary of the present Convention.

Article 36

Authentic Texts

The Arabic, Chinese, English, French, Russian and Spanish texts of the present Convention shall be equally authentic.

Explanatory Note and Article-by-Article Commentary to the Draft Global Convention

Explanatory Note

to the Draft Global Convention on the Right to Liberty of Worldview Choice and the Principle of Worldview Neutrality of Subjects of Public International Law in Analog and Digital Environments

(Developed jointly by Sergey A. Buryanov and Maxim S. Buryanov)

1. General Idea of the Convention

The draft Convention proceeds from the understanding that the internationally recognized right to freedom of thought, conscience, religion or belief requires further development in the conditions of the twenty-first century. The existing international legal architecture has laid the foundation for protecting the human person against direct coercion in the sphere of beliefs. Yet it no longer fully reflects the contemporary reality in which the worldview sphere is influenced not only by the state and traditional institutions, but also by digital platforms, algorithms, artificial intelligence systems, neurotechnologies and other means of hidden influence on consciousness.

The draft proposes to develop existing norms not by denying the classical right to freedom of religion or belief, but by expanding and systematizing it within the broader category of liberty of worldview choice. This approach makes it possible to cover religious, non-religious, philosophical, ethical and other forms of human self-determination, thereby bringing international law into closer alignment with the principle of non-discrimination and the real diversity of human experience.

2. Why a Declaration Alone Is Not Enough

A declaration plays an important conceptual, political and normative role, but it does not create legally binding obligations for states. A convention, by contrast, transfers principles into the sphere of an international treaty by providing for:

- the duty of states to bring legislation and practice into conformity with the treaty;
- the creation of a monitoring system;
- the formation of common standards of interpretation;
- the gradual harmonization of domestic and international guarantees.

Thus, the draft Convention is aimed not merely at proclaiming values, but at realizing the right to liberty of worldview choice.

3. Main Innovations of the Draft

The draft Convention contains several key innovations.

First, it enshrines liberty of worldview choice as an independent and broader legal category that cannot be reduced solely to freedom of religion.

Second, it formulates worldview neutrality as a duty of the state and other holders of public authority to refrain from establishing, imposing, privileging or discriminatorily restricting any worldview.

Third, it transfers this issue into analog and digital environments, recognizing that in the contemporary world control over access to information, content ranking, algorithmic filtering and behavioural profiling can affect freedom of belief no less significantly than traditional mechanisms of coercion.

Fourth, the draft introduces contemporary categories, including:

- algorithmic neutrality;
- digital sovereignty of the individual;
- the right to digital search;
- the prohibition of discriminatory moderation;
- algorithmic transparency.

4. The Balance between Universality and Political Feasibility

The draft is deliberately constructed in a way that preserves the possibility of international consensus. Therefore, it:

- does not abolish existing norms on freedom of religion, but expands them;
- does not require the unification of all models of statehood, but introduces a minimum universal standard;
- does not substitute national sovereignty, but limits the abuse of public and quasi-public power in the worldview sphere;
- permits limitations only in the extremely narrow case of direct incitement to a specific violent crime against a person.

It is this balance that makes the draft suitable not only for theory, but also for international discussion.

5. Practical Purpose of the Draft Convention

The draft Convention may serve as a basis for:

- the preparation of a universal United Nations treaty;
- the development of an optional protocol on the digital environment;
- the formulation of general comments by a treaty body;
- the reform of national legislation;
- the formation of standards of responsibility for digital platforms;

• the legal recognition of new digital human rights in the sphere of protecting worldview autonomy.

Article-by-Article Commentary

to the Draft Global Convention on the Right to Liberty of Worldview Choice and the Principle of Worldview Neutrality of Subjects of Public International Law in Analog and Digital Environments

(Developed jointly by Sergey A. Buryanov and Maxim S. Buryanov)

Article 1. Terms and Scope of Application

This article establishes the conceptual framework. It is especially important because it rightly emphasizes the crisis of legal certainty. The Convention is significant from the very beginning because it applies not only to traditional forms of communication, but also to algorithms, artificial intelligence, neurotechnologies and virtual reality.

Article 2. Subject Matter and Purpose of the Convention

This article makes clear that the Convention is not concerned with the private protection of religion alone, but with the broader realization of the right to liberty of worldview choice and the establishment of neutrality in analog and digital environments.

Article 3. Fundamental Principles

This is the normative core of the Convention. Human dignity, freedom, equality, legal certainty, neutrality, tolerance and algorithmic accountability are brought together here. For international law, this is important as a unified interpretive framework.

Article 4. Content of the Right

This article makes the right to worldview choice living and operational. What matters is not only the right to have beliefs, but also the right to change them, to refrain from disclosing them and to seek information necessary for their formation.

Article 5. Inadmissibility of Coercion

This article expands the classical understanding of coercion. It takes into account not only physical and administrative pressure, but also technological, informational and algorithmic forms of interference.

Article 6. Equality and Non-Discrimination

This is a universal anti-discrimination norm that takes protection beyond the division between "believers" and "non-believers".

Article 7. Individual and Collective Dimensions

This article prevents an imbalance. On the one hand, the individual is protected. On the other hand, the right of association is recognized. At the same time, collective ri

Article 8. General Standard of Permissible Limitations

This is one of the most important articles. It deliberately makes the standard of permissible limitations very narrow. Only direct, intentional and immediate incitement to the commission of a specific violent crime against a person may be limited.

Article 9. Prohibition of Expansive Interpretation of Undefined Categories

This article responds to the abuse of terms such as "extremism", "terrorism" and other elastic legal batons. It is needed to prevent expansive interpretation for the sake of censorship and other abuses.

Article 10. Legal Certainty

This is the legal axis of the entire construction. Without it, the Convention would become a noble set of words that everyone later interprets as they please, which usually means as the stronger party finds convenient.

Article 11. General Obligation of Neutrality

This article establishes that the state cannot be the bearer of compulsory truth. It marks a principled break with confessional and ideological models of power.

Article 12. Content of Worldview Neutrality

This article unfolds neutrality into its elements: rejection of privileges, rejection of compulsory ideology, equality of associations and guarantees of choice.

Article 13. Neutrality of Public Institutions

This is a very important practical article. It translates the principle into the language of institutions: courts, schools, the army, the police, the media and local self-government.

Article 14. Education, Children and Parents

This article establishes a balance between the rights of parents, the rights of the child and the duty of the state to preserve neutrality in education.

Article 15. Public Communication and the Media

This article protects the public sphere from being transformed into a machine of xenophobia, stigma and hidden worldview normalization.

Article 16. Algorithmic Neutrality

This is a new central article. It affirms that the digital power of the state must also be neutral. State artificial intelligence cannot become a machine for ideologically designing the citizen.

Article 17. Digital Sovereignty of the Individual

This article enshrines the right of the person not to dissolve in another actor's algorithmic environment. It is one of the strongest and most contemporary ideas of the document.

Article 18. The Right to Digital Search

This is a very important norm for the twenty-first century. Freedom of belief without freedom of worldview search becomes a paper decoration.

Article 19. Prohibition of Discriminatory Moderation

This article addresses digital censorship and the hidden suppression of worldview-inconvenient content.

Article 20. Algorithmic Transparency

If a system affects the visibility of ideas, a person must know the rules by which this is done. Otherwise, the appearance of freedom remains, while real access to it disappears.

Article 21. Duties of States in Relation to Private Platforms

This is a very important compromise bridge. The Convention does not turn Big Tech into subjects of international law, but it imposes on states the duty to regulate their conduct.

Article 22. Full Prohibition, Elimination of Emotion Recognition Systems and Legal Responsibility

This article establishes a legal prohibition on interference with the inner world of the human person, protecting emotions from control by the state and/or artificial intelligence in public administration, education and labour relations. It prevents digital slavery and preserves the person as a free subject by prohibiting sanctions based on feelings rather than actions.

Article 23. Neurotechnologies, Virtual Reality and Cognitive Protection

This is an article for the future. It introduces the right not to become the object of hidden intrusion into the very structure of attention, perception and belief.

Article 24. Domestic Guarantees

Without remedies, any treaty has no future.

Article 25. National Implementation Mechanisms

This article makes it possible to institutionalize the Convention within states.

Article 26. Committee

This article creates an international body for interpretation and monitoring.

Article 27. Reports of States

This is the classic United Nations treaty mechanism.

Article 28. Inter-State Communications and Individual Communications

This article opens the prospect of individual complaints and inter-state communications, possibly through an optional protocol.

Article 29. International Cooperation

Without this article, the treaty would be purely defensive. Here, a constructive agenda is established.

Articles 30–36. Final Provisions

These articles contain the standard treaty block: the relationship with other treaties, ratification, entry into force, amendments, denunciation, depositary and authentic languages.

Draft Model Law on the Right to Liberty of Worldview Choice and the Principle of Worldview Neutrality of Subjects of Public Law in Analog and Digital Environments

(Developed jointly by Sergey A. Buryanov and Maxim S. Buryanov)

MODEL LAW

on the Right to Liberty of Worldview Choice and the Principle of Worldview Neutrality of Subjects of Public Law in Analog and Digital Environments

Preamble

This Model Law is adopted for the purpose of ensuring human dignity, protecting inner freedom and creating legal guarantees for the free formation, preservation, change, expression or rejection of convictions, values, religious, non-religious, philosophical, ethical, scientific and other worldview orientations in analog and digital environments.

This Law proceeds from the premise that, in contemporary legal development, freedom of conscience must be disclosed more fully and precisely through the category of liberty of worldview choice, which covers both religious and non-religious forms of human self-determination.

This Law is based on the principles of human dignity, equality, non-discrimination, legal certainty, worldview neutrality of public authority, pluralism of opinions, inadmissibility of coercion in the worldview sphere, and the protection of the inner cognitive freedom of the person in the context of technological development.

Chapter 1. General Provisions

Article 1. Subject Matter of This Law

1. This Law establishes:

1. the content and guarantees of the right to liberty of worldview choice;

2. the foundations of the worldview neutrality of the state and other holders of public authority;

3. requirements applicable to analog and digital environments regarding the inadmissibility of worldview coercion, discrimination and manipulation;

4. the duties of state bodies, organizations, officials, digital platforms and other information intermediaries within the limits provided by legislation;

5. mechanisms for the protection of violated rights, oversight and responsibility.

2. This Law applies to the regulation of social relations connected with the formation, expression, dissemination, search, receipt, storage and processing of information affecting the worldview sphere of the person, as well as relations connected with the exercise of public authority in analog and digital environments.

Article 2. Purposes of This Law

The purposes of this Law are:

1. to ensure the effective realization of every person's right to liberty of worldview choice;

2. to enshrine the worldview neutrality of the state and other holders of public authority;

3. to prevent discrimination, coercion and manipulation in the worldview sphere;

4. to protect worldview pluralism in analog and digital environments;

5. to create legal mechanisms for countering abuses of public, informational, algorithmic and technological power.

Article 3. Basic Concepts

For the purposes of this Law, the following basic concepts shall apply:

1. Liberty of worldview choice means the right of every person freely to form, hold, preserve, express, revise, change or reject convictions, values, ideas about the world, religious, non-religious, philosophical, ethical, scientific and other worldview orientations.

2. Worldview neutrality means the duty of the state and other holders of public authority to refrain from establishing, imposing, privileging or discriminatorily restricting any worldview, ideology, religion, philosophical doctrine, system of convictions or form of worldview self-determination.

3. Analog and digital environment means all forms of creation, storage, transmission, search, processing, dissemination and perception of information, including print and audiovisual media, telecommunications networks, the Internet, platforms, artificial intelligence systems, virtual and augmented environments, neurotechnologies and other technological systems.

4. Algorithmic system means an automated or semi-automated system, including an artificial intelligence system, that influences the sorting, ranking, recommendation, limitation, deletion, amplification or distribution of information, content, services or access to them.

5. Discriminatory moderation means the deletion, blocking, limitation of visibility, hidden downranking, demonetization, labelling or other restriction of content or an account exclusively or predominantly on worldview grounds, where such expression does not fall within a lawfully established and proportionate restriction.

6. Holders of public authority means state bodies, local self-government bodies, state and municipal institutions, state and municipal organizations, as well as other subjects endowed by law with public powers or performing public functions.

7. Algorithmic neutrality means the prohibition on the use of algorithmic systems, including state systems, for the hidden or direct imposition, privileging, suppression or manipulative formation of religious, non-religious, philosophical, political or other convictions.

8. Digital sovereignty of the person in the worldview sphere means the right of a person independently to determine the degree of openness, personalization and diversity of their information environment, and to be protected from hidden algorithmic influence on the formation, change or suppression of convictions.

9. Worldview impact assessment means a mandatory procedure for identifying, analyzing and minimizing risks connected with the possible impact of a law, state programme, digital system, platform or algorithmic technology on the liberty of worldview choice, worldview neutrality and worldview pluralism.

Article 4. Basic Principles of This Law

Legal regulation in the sphere covered by this Law shall be based on the following principles:

1. human dignity;
2. liberty of worldview choice;
3. equality and non-discrimination;
4. legal certainty;
5. worldview neutrality of public authority;
6. pluralism of opinions and free access to a diversity of ideas;
7. inadmissibility of coercion in the worldview sphere;
8. transparency, explainability and accountability of algorithmic systems;
9. the priority of protecting the inner freedom of the person;
10. judicial and administrative guarantees of rights and freedoms.

Chapter 2. The Right to Liberty of Worldview Choice

Article 5. Content of the Right to Liberty of Worldview Choice

1. Every person has the right to liberty of worldview choice.

2. This right includes the freedom:

1. to form, hold, preserve and change convictions;

2. to express or not to express one's convictions;

3. not to disclose one's convictions, worldview affiliation or the absence thereof;

4. to seek, receive, create and disseminate information and ideas necessary for the formation of one's own picture of the world;

5. to refuse imposed worldview identification;

6. not to be subjected to manipulative influence aimed at forming, suppressing or changing convictions without free and informed consent;

7. to use analog and digital means of communication without discrimination on worldview grounds.

3. The inner freedom of a person to form and hold convictions shall not be subject to limitation.

Article 6. Inadmissibility of Coercion

1. No one may be subjected to coercion capable of restricting their liberty of worldview choice.

2. Any form of pressure, violence, threats, abuse of dependency, administrative, political, economic, informational, technological, algorithmic or other influence is prohibited where it is aimed at:

1. the imposition of convictions;

2. the suppression of convictions;

3. the alteration of convictions against a person's will;

4. compelling worldview self-disclosure;

5. denial of access to rights, services, education, work or participation in public life on worldview grounds.

3. The state shall not create or maintain mandatory registers of citizens' worldview, religious, philosophical or other conviction-based affiliation, except in cases where such disclosure is made by the person voluntarily, freely, specifically and for a lawful purpose.

Article 7. Equality and Non-Discrimination

1. All persons are equal in dignity and rights regardless of their worldview affiliation, the absence of such affiliation, a change of convictions or a refusal publicly to express them.

2. Any direct or indirect discrimination is prohibited on the grounds of:

1. religion;
2. convictions;
3. worldview;
4. philosophical position;
5. criticism of convictions;
6. doubt;
7. non-belief;
8. belonging or not belonging to relevant associations.

3. Discrimination on the ground of a person's presumed worldview is prohibited.

Article 8. Individual and Collective Dimensions of the Right

1. The right to liberty of worldview choice has both an individual and a collective dimension.

2. Persons have the right to associate on the basis of shared convictions and to carry out corresponding activities on equal terms with other public and non-profit associations.

3. The rights of associations are derivative of human rights and freedoms and may not be used to substitute, restrict or suppress the rights of the individual.

Chapter 3. Permissible Limitations and Legal Certainty

Article 9. General Standard of Permissible Limitations

1. Limitation of the external expression of a worldview position shall be permissible only where such expression constitutes direct, intentional and immediate incitement to the commission of a specific violent crime against the person or directly threatens the rights and freedoms of others in the manner prescribed by law.

2. Any limitation shall be permissible only where all of the following conditions are met simultaneously:

1. it is expressly provided by law;
2. it is strictly necessary in a democratic society;
3. it is proportionate;
4. it is individualized;
5. it is subject to independent administrative or judicial review;
6. it does not affect the essence of the right and does not turn into an instrument of worldview censorship.

3. The expression of a worldview position, by itself, shall not constitute an unlawful act.

Article 10. Prohibition of Expansive Interpretation of Undefined Categories

1. Undefined, stigmatizing or politically charged categories shall not be used to restrict the rights guaranteed by this Law.

2. Concepts used in legislation to limit rights and freedoms in the worldview sphere must be clear, unambiguous, foreseeable and subject to narrow interpretation.

3. The absence of a precise and verifiable legal criterion may not serve as grounds for censorship, discriminatory moderation, blocking of content or restriction of liberty of worldview choice.

Article 11. Legal Certainty and Review of Legislation

1. The state shall ensure the review of legislation, administrative practice and digital regulation based on legally undefined or discriminatory concepts capable of leading to violations of liberty of worldview choice.

2. Normative legal acts affecting the worldview sphere shall be subject to mandatory assessment for compliance with the principles of:

1. legal certainty;
2. neutrality;

3. equality;
4. non-discrimination;
5. proportionality.

Chapter 4. Worldview Neutrality of the State and Public Authority

Article 12. General Obligation of Worldview Neutrality

1. The state and other holders of public authority shall observe the principle of worldview neutrality.

2. No worldview, ideology or religion may be established as mandatory, officially preferred or privileged for all.

3. Public authority shall not use its institutions, resources, symbolic power, budgetary mechanisms, information channels or digital systems to impose, encourage or suppress a particular worldview.

Article 13. Content of Worldview Neutrality

Worldview neutrality includes:

1. non-identification of public authority with any worldview;

2. prohibition of discrimination on worldview grounds;

3. legal equality of associations;

4. renunciation of special worldview privileges and restrictions, unless otherwise required for the protection of human rights on the basis of law and the principle of proportionality;

5. renunciation of the legal establishment of the universal obligatoriness of any worldview or ideology;

6. ensuring real guarantees of free worldview choice for every person;

7. neutrality of state algorithmic and informational influence.

Article 14. Neutrality of Public Institutions and Officials

1. The principle of worldview neutrality shall be binding on:

1. legislative, executive and judicial authorities;

2. local self-government bodies;

3. the education system;

4. the armed forces;

5. law enforcement bodies;

6. the penitentiary system;

7. state and municipal mass media;

8. other bodies and organizations performing public functions.

2. State and municipal officials, in the exercise of public powers, shall refrain from using their status for:

1. imposing worldview preferences;

2. exerting public pressure on dependent persons;

3. discrimination on the grounds of convictions;

4. hidden or direct ideological influence through official communication channels.

Article 15. Education, Children and Parents

1. The state and municipal education system shall be built on the principles of worldview neutrality, human rights, pluralism, equality, non-discrimination, peace, respect for the dignity of the child and the diversity of convictions.

2. The educational process shall not be used for the hidden or direct imposition of a mandatory worldview.

3. Parents and other legal representatives have the right to ensure the upbringing of children in accordance with their convictions, within the limits of human rights, the dignity of the child and the child's evolving autonomy.

4. The state shall ensure children's access to education that develops critical thinking, the capacity for independent worldview choice and respect for the rights of others.

Article 16. Public Communication and Public Media

1. State and municipal mass media, official information resources and other public communication channels shall observe the principles of neutrality, equality, accuracy and mutual respect.

2. Public media and official communication channels shall not be used for:

1. stigmatizing persons or groups on worldview grounds;

2. xenophobia and discrimination;

3. the formation of an official worldview hierarchy;

4. hidden ideological influence on the population.

Chapter 5. Liberty of Worldview Choice in Analog and Digital Environments

Article 17. Algorithmic Neutrality

1. The use by state bodies of algorithmic systems, artificial intelligence systems, automated recommendation systems, profiling and forecasting systems shall not lead to:

1. manipulation of consciousness;

2. imposition of a single ideology;

3. suppression of worldview pluralism;

4. hidden influence on a person's religious, philosophical, political or other convictions.

2. State digital systems and algorithmic infrastructures shall be designed, implemented and operated on the basis of the principles of:

1. neutrality;

2. non-discrimination;

3. accountability;

4. explainability;

5. legal certainty;

6. independent verifiability.

3. State digital systems shall not be used to form mandatory ideological loyalty, worldview scoring, hidden behavioural nudging or restriction of information search on worldview grounds.

Article 18. Digital Sovereignty of the Person

1. Every person has the right to digital sovereignty over their worldview sphere.

2. This right includes:

1. the right independently to determine the degree of openness and personalization of one's information environment;

2. the right to protection against hidden algorithmic enclosure within a limited information circuit;

3. the right of access to a diversity of ideas and sources without forced filtering;

4. the right to know about significant mechanisms of algorithmic influence on access to information;

5. the right to rofiling;

6. the right to the inviolability of one's inner cognitive sphere.

3. Manipulative restriction of a person's access to a diversity of ideas under the pretext of protection, upbringing, prevention or convenience of use of a digital service shall not be permitted.

Article 19. Right to Digital Search and Access to a Diversity of Ideas

1. Every person has the right to digital search and access to a diversity of ideas, views and worldview positions in global data networks.

2. State bodies shall not apply systematic filtering, hidden limitation of search results, blocking or other forms of restricting worldview search, except in cases expressly provided by law and consistent with Article 9 of this Law.

3. Search, recommendation and other similar digital systems that substantially affect access to information shall provide user-understandable means for controlling the degree of personalization and diversity of results.

Article 20. Prohibition of Discriminatory Moderation

1. Discriminatory moderation in the digital information space shall not be permitted.

2. State bodies, as well as digital platforms and information intermediaries in cases provided by legislation, shall not restrict content solely on the ground that it expresses a different worldview, unless such content falls within a lawfully established, necessary and proportionate limitation.

3. Hidden downranking, limitation of visibility, demonetization, labelling or blocking of content on worldview grounds without proper notification, explanation of reasons and a review procedure shall be prohibited.

Article 21. Algorithmic Transparency and the Right to Explanation

1. A person has the right to be notified of significant algorithmic influence on access to information, content visibility, ranking, recommendations, deletion, limitation of reach or other interference capable of affecting liberty of worldview choice.

2. A person has the right:

1. to receive an understandable explanation of the reasons for the decision;

2. to know whether automated or semi-automated means were used;

3. to request human review of the decision;

4. to challenge the decision through administrative and judicial procedures.

3. The explanation shall be sufficient to understand the essence of the interference, the legal basis, the criteria applied and the means of appeal.

Article 22. Duties of Digital Platforms and Information Intermediaries

1. Digital platforms, social networks, video-hosting services, search engines, content marketplaces, messengers and other information intermediaries substantially affecting access to information shall, within the limits provided by legislation:

1. take measures to prevent discriminatory moderation;

2. ensure transparency of the main moderation rules and principles of operation of algorithmic systems;

3. notify users of significant restrictions on content and accounts;

4. ensure effective procedures for review and appeal;

5. prevent hidden worldview manipulation of users;

6. ensure the possibility of configuring the level of personalization and diversity of content;

7. retain information on significant moderation decisions and restrictions to the extent sufficient for verifying the legality of such decisions.

2. The duties provided for in this Article shall be specified by national legislation, taking into account the scale of the platform, the nature of its functions, the number of users and the degree of influence on the public information environment.

Article 23. Neurotechnologies, Virtual Reality and Cognitive Protection

1. The use of neurotechnologies, virtual, mixed and augmented reality, biometric systems, behavioural profiling and other technologies for unlawful influence on the worldview sphere of the person shall be prohibited.

2. Any use of such technologies in the public sphere shall be permissible only subject to the requirements of:

1. lawfulness;

2. necessity;

3. proportionality;

4. respect for human dignity;

5. worldview neutrality;

6. free and informed consent, unless otherwise expressly provided by law and strictly necessary for the protection of the rights and lawful interests of the person.

3. The hidden use of technologies to form, change or suppress a person's convictions without their free and informed consent shall be prohibited.

Article 24. Worldview Impact Assessment

1. Before state bodies introduce algorithmic systems, information platforms, digital profiles, recommendation systems, systems for monitoring information behaviour or other technologies capable of affecting the worldview sphere of the person, a worldview impact assessment shall be conducted.

2. The worldview impact assessment shall include:

1. a description of the purposes and functions of the system;

2. an analysis of risks of discrimination, censorship, manipulation, algorithmic enclosure and hidden ideological influence;

3. verification of the legal basis;

4. assessment of necessity and proportionality;

5. measures to mitigate risks;

6. the procedure for independent audit and review;

7. a public part of the conclusion, except for information with restricted access under law.

3. Where a disproportionate risk of violating liberty of worldview choice is identified, the implementation of such a system shall not be permitted until the identified violations are remedied.

Chapter 6. Mechanisms of Implementation and Protection

Article 25. National System for Implementing This Law

1. The state shall create or designate independent mechanisms for the implementation of this Law.

2. Such mechanisms may include:

1. the human rights commissioner or ombudsperson;

2. a national human rights institution;

3. an independent regulator of the digital environment;

4. a specialized commission on liberty of worldview choice and worldview neutrality;

5. other bodies meeting the requirements of independence, competence and public accountability.

3. The bodies referred to above shall cooperate with one another and shall ensure:

1. the receipt and consideration of applications;

2. the preparation of recommendations;

3. the conduct of inspections and audits;

4. monitoring of law-enforcement practice;

5. education and the preparation of methodological materials.

Article 26. Right to Petition and Procedure for Reviewing Decisions

1. Any person who believes that their right to liberty of worldview choice has been violated has the right to apply to:

1. the relevant state body or organization;

2. the owner of a digital platform or an information intermediary;

3. an independent regulator;

4. the human rights commissioner or ombudsperson;

5. a court.

2. In applications connected with the removal of content, limitation of reach, blocking of an account, algorithmic interference or another restriction of digital access, the applicant shall be provided with:

1. notification of the decision taken;

2. an explanation of the reasons;

3. information on the review procedure;

4. the possibility of prompt human consideration of the complaint.

3. Complaints against actions or omissions of state bodies and other holders of public authority shall be considered as a priority where the violation is connected with the risk of an ongoing restriction of the right.

Article 27. Judicial Protection and Remedies

1. Everyone shall be guaranteed judicial protection of the rights provided for in this Law.

2. A court may:

1. declare actions, decisions or omissions unlawful;

2. order the violation to be terminated;

3. order the restoration of access to information, content, an account or a service;

4. declare normative and individual acts violating the principle of worldview neutrality invalid;

5. award compensation for pecuniary and non-pecuniary damage;

6. order a review of an algorithmic system or suspend its use.

3. In considering cases involving automated decisions, a court may require the provision of information sufficient to verify lawfulness, necessity, proportionality and the absence of discrimination.

Article 28. Public Reporting and Monitoring

1. State bodies that introduce digital systems capable of affecting liberty of worldview choice shall ensure regular public reporting on the measures applied, the risks identified and the ways of minimizing them.

2. The independent bodies referred to in Article 25 of this Law shall prepare a periodic public report on the state of liberty of worldview choice and compliance with the principle of worldview neutrality.

3. The report shall reflect:

1. the principal violations and trends;

2. the practice of using digital systems;

3. problems of legal certainty;

4. proposals for improving legislation and practice.

Chapter 7. Responsibility

Article 29. Responsibility for Violation of This Law

1. Persons guilty of violating this Law shall bear disciplinary, administrative, civil-law or other responsibility in accordance with legislation.

2. Grounds for responsibility shall include, in particular:

1. discrimination on worldview grounds;

2. unlawful coercion to express, change or conceal convictions;

3. abuse of public powers for the purpose of worldview imposition;

4. implementation or use of algorithmic systems in violation of the requirements of this Law;

5. hidden worldview manipulation;

6. unlawful discriminatory moderation;

7. failure to provide notification, explanation or a review procedure in cases where this is mandatory.

3. Where a violation is committed through the use of an automated system, this shall not release the operator, owner, administrator, customer, authorized body or other responsible person from the obligation to prove the lawfulness of the relevant decision or action.

Article 30. Inadmissibility of Justifying Violations by Automation

1. Reference to the autonomy of an algorithmic system, artificial intelligence, machine learning, a technical error or trade secret may not, by itself, constitute sufficient grounds for refusing protection of the right provided for in this Law.

2. Trade secrets and other regimes restricting access to information shall not exclude the duty to provide a supervisory body or court with information necessary to verify compliance with the requirements of this Law.

Chapter 8. International Cooperation and Final Provisions

Article 31. International Cooperation

The state shall promote international cooperation for the purposes of:

1. harmonizing standards for the protection of liberty of worldview choice;

2. exchanging best practices;

3. developing common approaches to digital and cross-border violations;

4. education, development of tolerance and strengthening of a culture of human rights;

5. forming international standards of responsibility for platforms and other digital corporations.

Article 32. Relationship with Other Normative Legal Acts

1. Where an international treaty or another normative legal act establishes a higher standard of protection of liberty of worldview choice, the higher standard of protection shall apply.

2. Legislative norms shall be interpreted and applied in accordance with this Law, proceeding from the priority of the dignity of the person, the inner freedom of the person, the worldview neutrality of public authority and legal certainty.

Article 33. Bringing Legislation into Conformity with This Law

1. State bodies shall, within the period established by national legislation, ensure that normative legal acts, administrative procedures, rules of digital moderation and state algorithmic systems are brought into conformity with this Law.

2. Until legislation is brought into conformity with this Law, normative and individual acts shall apply only insofar as they do not contradict its provisions.

Article 34. Entry into Force

This Law shall enter into force in accordance with the procedure established by national legislation.

REFERENCES TO THE CONCLUDING PART

List of Sources Used

1. Buryanov S.A. Current State and Main Directions of Reforming the Internationally Recognized Freedom of Thought, Conscience and Religion // Eurasian Law Journal. No. 1 (92). 2016. pp. 126-130.
2. Buryanov S.A. Formation and Development of Freedom of Thought, Conscience and Religion in International Law // Eurasian Law Journal. No. 2 (105). 2017. pp. 183-189.
3. Buryanov S.A. Current State and Main Directions of Reforming the Internationally Recognized Freedom of Thought, Conscience and Religion // Eurasian Law Journal. No. 1 (92). 2016. pp. 126-130.
4. Buryanov S.A. Violations of the Right to Freedom of Conscience and Religion // Human Rights in the Russian Federation: Collection of Reports on Events of 2017 / [responsible editor and compiler N. Kostenko]. Moscow, 2018. 218 p. pp. 7-30.
5. Buryanov S.A. Violations of the Right to Freedom of Conscience and Religion // Human Rights in the Russian Federation: Collection of Reports on Events of 2018 / [responsible editor and compiler N. Kostenko]. Moscow, 2019. 338 p. pp.26-55.
6. Buryanov S.A. Current State and Main Directions of Reforming the Internationally Recognized Freedom of Thought, Conscience and Religion // Eurasian Law Journal. No. 1 (92). 2016. pp. 126-130.
7. Buryanov S.A. Formation and Development of Freedom of Thought, Conscience and Religion in International Law // Eurasian Law Journal. No. 2 (105). 2017. pp. 183-189.
8. Ghanea N. Freedom of Religion or Belief and the Right to Life: Report of the Special Rapporteur: A/HRC/52/38 / Nazila Ghanea; UN Human Rights Council. Geneva: UN, 2023. 25 p.
9. Alegre S. Freedom to Think: The Long Struggle to Liberate Our Minds / S. Alegre. London: Atlantic Books, 2023. 400 p.
10. Freedom of Conscience: Problems of Theory and Practice. Monograph (edited by F.M. Rudinsky and S.A. Buryanov). Moscow: ZAO TF MIR, 2012. 1120 p.

11. Buryanov S.A. International Recognition of the Right to Freedom of Conscience and Problems of Its Implementation in the Russian Federation under Conditions of Contemporary Global Processes: Monograph. Moscow: Poligraf Servis, 2020. 624 p.

12. Civil Human Rights in Russia: Modern Problems of Theory and Practice / compiled by F.M. Rudinsky. Transaction Publishers, New Brunswick, New Jersey (U.S.A.), 2008. 458 p.

13. Buryanov S.A. State Worldview Neutrality in the Context of Deteriorating Imbalances in Globalization: The Case and Current State of Affairs in the Russian Federation from 2016 to the Beginning of 2017. Amsterdam, 2018. 228 p.

14. Buryanov S.A., Buryanov M.S., Nikitaev D.M. The Right to Freedom of Conscience in the Context of Global Processes: Theory and Practice of Implementation in the Russian Federation. (Master's Program). Monograph. Edited by S.A. Buryanov. Moscow: LitRes, 2022. 234 p.

15. Digital Human Rights in the Context of Global Processes: Theory and Practice of Implementation: Monograph / M.S. Buryanov; scientific editor S.A. Buryanov. Moscow: Rusains, 2024. 148 p.

BIBLIOGRAPHIC LIST

1. Abashidze A.Kh. Principles of International Law: Problems of Conceptual and Substantive Content // Moscow Journal of International Law. No. 4. pp. 19-30.

2. Abashidze A.Kh. Problems of the International Legal Protection of Minorities: Doctor of Legal Sciences Dissertation. Moscow, 1997. 58 p.

3. Abashidze A.Kh., Klishas A.A. Protection of Human Rights and Freedoms in Accordance with International Legal Standards under the New Realities of a Globalizing World // Bulletin of Peoples' Friendship University of Russia. Series: Legal Sciences. 2013. No. 2. pp. 251-255.

4. Abashidze A.Kh., Melshina K.Yu. The Rule of Law as Understood by the United Nations // Bulletin of Volgograd State University. Series 5: Jurisprudence. 2016. Vol. 15. No. 4 (33). pp. 10-18.

5. Abashidze A.Kh., Solntsev A.M. Anniversary of the African Charter on Human and Peoples' Rights // Eurasian Law Journal. 2012. No. 2 (45). pp. 22-25.

6. Abashidze A.Kh., Solntsev A.M., Vorobyev D.V. The Competence of the Organisation of Islamic Cooperation in the Field of Human Rights Protection // Bulletin of Volgograd State University. Series 5: Jurisprudence. 2013. No. 3. pp. 213-219.

7. Avakyan S.A. Constitutional Law of Russia: Textbook in 2 Volumes. Vol. 1. Moscow: Yurist, 2005.

8. Avakyan S.A. Freedom of Religion as a Constitutional-Legal Institution // Bulletin of Moscow State University. Series 11. No. 1. 1999.

9. Azarov A., Reuter V., Hüfner K. Protection of Human Rights: International and Russian Mechanisms. Moscow: Moscow School of Human Rights, 2000.

10. Akchurin T.F. The Right to Freedom of Conscience: An International Legal Aspect // Russian Law Online. 2018. No. 3. pp. 19-28.

11. Aleinikova S.M. On the Question of a "Vulgar Understanding of Secularity": Political Myth-Making or a New Path of Development? // Bulletin of Brest University. Series 1. No. 2. 2013.

12. Aleinikova S.M., Buryanov S.A. The Secular State in Questions and Answers: Briefly, Accessibly, and Relevantly / ed. by I. Kondratiev. Moscow, 2015. 62 p.

13. Alekseev S.S. The Holiest Thing God Has on Earth: Immanuel Kant and Problems of Law in the Contemporary Era. Moscow: NORMA Publishing House, 1998.

14. Alekseeva L.M. The History of Dissent in the USSR: The Latest Period. 3rd ed., stereotyped. Moscow: Russian Helsinki Group, 2012. 384 p. pp. 150-205.

15. Alibekova S.Ya. National-Ethnic and Religious Aspects of Public Security // Theory and Practice of Social Development. 2015. No. 19.

16. Al-Ali Naser Abdel Rahim. Universalism and the Islamic Concept of Human Rights and Freedoms: Candidate of Legal Sciences Dissertation. Moscow, 1999. 188 p.

17. Alting von Geusau Frans A.M. How Can the Transition to a Law-Governed Society Be Achieved? Moscow, 1997.
18. Amini A.R. The Political System of Afghanistan and Prospects for the Secularity of the State // PolitBook. 2018. No. 2. pp. 93-103.
19. Anti -Globalism and Global Governance: Reports, Discussions, Reference Materials. Moscow: MGIMO University, Ministry of Foreign Affairs of Russia, 2006. Electronic resource. URL: http://www.mgimo.ru/files/6195/global-anti[1].pdf (accessed: 23.09.2019).
20. Antonov I.Yu. The Rule-of-Law State and Fundamental Human Rights and Freedoms: Abstract of Candidate of Legal Sciences Dissertation. Moscow, 2005.
21. Apanasyuk L.A. Theoretical and Methodological Foundations for Preventing Xenophobia in the Student Environment by Means of Socio-Cultural Activity: Abstract of Doctor of Pedagogical Sciences Dissertation. Tambov, 2014.
22. Babiy M.Yu. Freedom of Conscience in the Context of the Humanization of Ukrainian Society: New Challenges and Prospects. Electronic resource. URL: http://archive.nbuv.gov.ua/portal/soc_gum/ukrr/2012_62-63/Babiy.htm (accessed: 8.12.2018).
23. Baboshina E.V., Kosov G.V., Stankevich G.V. Theoretical and Legal Analysis of Universal and Regional International Documents on Freedom of Conscience, Belief and Religion // Caspian Region: Politics, Economy, Culture. 2018. No. 4 (57). pp. 76-85.
24. Baglai M.V. Constitutional Law of the Russian Federation. Moscow, 2001.
25. Baglai M.V. Constitutional Law of the Russian Federation: Textbook for Law Universities and Faculties. Moscow: NORMA-INFRA Publishing Group, 1998.
26. Bazikov R.V. Social Stereotypes: A Conceptual Aspect: Candidate of Philosophical Sciences Dissertation. Rostov-on-Don, 1999.
27. Barabanov O.N. Global Governance as a Topic for Scientific Analysis. Electronic resource. URL: http://globalanti.risa.ru/reports.php?cat_id=31&doc_id=13

(accessed: 23.09.2019).

28. Barenboim P.D., Kravchenko D.V. The Doctrine of the Russian Rule-of-Law State: Components, Principles, Features // International Legal Doctrines: Seeking Mutual Understanding / Collection of Articles edited by E.G. Tarlo and D.V. Kravchenko. Moscow: LUM, 2013.
29. Bakhvalov S.V. Legislative Technology: Selected Problems of Theory and Practice: Abstract of Candidate of Legal Sciences Dissertation. Nizhny Novgorod, 2006.
30. Bezrukova L. The Constitutional Court of the Russian Federation Permitted the Reconsideration of Civil Cases When the European Court of Human Rights Has Issued a Judgment on Them // Rossiyskaya Gazeta. February 26, 2010.
31. Bezuglov A.A., Soldatov S.A. Constitutional Law of Russia. In 3 Volumes. Moscow, 2001. p. 340.
32. Bekyashev K.A. Globalization and International Law // Eurasian Law Journal. 2014. No. 1. pp. 25-29.
33. Biryukov P.N., Buryanov S.A. Insur Farkhutdinov: Ensuring Peace and Security in Eurasia: An International Legal Assessment of the Events in Syria. Interview with Insur Zabirovich Farkhutdinov, Doctor of Legal Sciences and Editor-in-Chief of the Eurasian Law Journal // Eurasian Law Journal. 2015. No. 10. pp. 8-15.
34. Great Legal Dictionary. Akademik.ru. Moscow, 2010. Electronic resource. URL: http://jurisprudence.academic.ru (accessed: 3.01.2018).
35. Bryantseva O.V. Implementation of the Constitutional Right to Freedom of Conscience in the Russian Federation: The Case of Volgograd Region: Abstract of Candidate of Legal Sciences Dissertation. Volgograd, 2005.
36. Buryanov M.S. Global Digital Human Rights in the Context of Implementing Sustainable Development Goal No. 16 // Current Problems of Science and Practice: Gatchina Readings 2020: in 2 Volumes. Vol. 2: Tribune of the Young Scholar: Collection of Scientific Papers Based on the Materials of the VII International Scientific and Practical Conference (Gatchina, May 22, 2020) / ed. by V.R. Kovalev and T.O. Boziev. Gatchina: Publishing House of GIEFPT, 2020. Vol. 2.

487 p.

37. Buryanov M.S. Current Aspects of the Development of Information Law in the Context of the Risks of the Fourth Industrial Revolution and the Digital Divide // Current Problems of Administrative, Financial and Information Law in Russia and Abroad: Materials of an Interuniversity Scientific Conference at the Department of Administrative and Financial Law of the Law Institute of Peoples' Friendship University of Russia. Moscow, March 20, 2020. Moscow: RUDN University, 2020. 155 p. pp. 67-72.
38. Buryanov M.S. The Path to Global Law: Global Digital Human Rights // Scientific Works of Moscow University for the Humanities. 2020. No. 2. pp. 63-66. DOI: 10.17805/trudy.2020.2.11.
39. Buryanov M.S. Global Digital Human Rights in the Context of Digitalization Risks // Age of Globalization. 2020. No. 3. pp. 54-70. DOI: 10.30884/vglob/2020.03.05.
40. Buryanov M.S. Global Digital Human Rights as a Condition for Overcoming Social Differentiation in Russia and the World // Legal Support for Social Justice under Conditions of Digitalization: Collection of Materials of the All-Russian Scientific and Practical Conference with International Participation / ed. by T.A. Soshnikova. Moscow: Publishing House of Moscow University for the Humanities, 2020. 402 p. pp. 317-321.
41. Buryanov M.S. Deconstruction of Law against the Background of Globalization and Planetary Risks // Intellectual Culture of Belarus: Cognitive and Predictive Potential of Socio-Philosophical Knowledge: Materials of the Fourth International Scientific Conference (Minsk, November 14-15, 2019). In 2 Volumes. Vol. 2 / Institute of Philosophy of the National Academy of Sciences of Belarus; editorial board: A.A. Lazarevich (chair) [et al.]. Minsk: Four Quarters, 2019. 384 p.
42. Buryanov M.S. The Significance and Prospects of the Development of Law for Sustainable Development // Sustainable Development Strategy: Environmental Rights and Other Components / Materials of a Scientific and Practical Conference; ed. by T.A. Soshnikova, N.V. Kolotova and E.E.

Pirogova. Moscow: Publishing House of Moscow Universityfor the Humanities, 2018. 267 p. pp. 235-240.

43. Buryanov M.S. The Significance of Law in the Context of Contemporary Global Processes // Current Problems in the Formation and Development of the Legal System of the Russian Federation [Electronic resource]: Collection of Materials of the II All-Russian Scientific and Practical Conference of Students, Master's Students and Postgraduate Students (Syktyvkar, April 5-6, 2018). Syktyvkar: Publishing House of Pitirim Sorokin Syktyvkar State University, 2018.

44. Buryanov M.S. Prospects for the Development of Constitutional Law in Russia in the Context of Achieving Social Justice under Conditions of Globalization // Principles of Social Justice and Their Implementation in the Contemporary World / ed. by T.A. Soshnikova and E.E. Pirogova. Moscow: Publishing House of Moscow University for the Humanities, 2019. 379 p. pp. 284-287.

45. Buryanov M.S. Prospects for the Development of Russian Statehood in the Context of Contemporary Global Processes and Challenges // Current Problems of Global Studies: Russia in a Globalizing World. Collection of Materials of the VI All-Russian Scientific and Practical Conference, Lomonosov Moscow State University, June 4-6, 2019 / ed. by I.V. Ilyin. Moscow: International Public Organization for the Promotion of Scientific Research Named after N.D. Kondratiev, 2019. 466 p. pp. 67-73.

46. Buryanov M.S. Human Rights in the Networks of Technogenic Civilization // Law of Technogenic Civilization: Contemporary Transformations and Vectors of Development: Materials of the International Student Scientific and Practical Conference (Moscow, October 24, 2019) / Yu.N. Kashevarova, I.A. Shulyatyev, E.K. Saifullin. Moscow: Institute of Legislation and Comparative Law under the Government of the Russian Federation, 2021. 222 p. pp. 119-125.

47. Buryanov M.S. The Law of Peace against the Law of War: At the Dawn of the Collapse of Old Approaches to Governing the World Order or of Human Oblivion // Current Problems of International Law: Public, Private and Integration Aspects. Collection of Articles and Dissertations. Edited by O.V.

Lutkova. Moscow, 2019. 104 p. pp. 63-66.

48. Buryanov M.S. Principles of Public-Law Regulation of Artificial Intelligence and Global Digital Human Rights // N.N. Moiseev on Russia in the XXI Century: Global Challenges, Risks and Solutions. Collection of Main Reports of the XXVIII Moiseev Readings, International Scientific and Practical Conference "N.N. Moiseev on Russia in the XXI Century: Global Challenges, Risks and Solutions" (March 2-6, 2020): in 2 Parts / general editor Academician M.Ch. Zalikhanov; ed. by Yu.G.; comp. by Prof. N.F. Vinokurova, Prof. S.A. Stepanov, Associate Prof. N.V. Martilova. Moscow; Nizhny Novgorod: Minin University, 2021. 352 p.
49. Buryanov M.S. Problems and Prospects of the Development of Law in the Context of the Digital Transformation of the State // Eurasian Law Journal. 2025. No. 10 (209). pp. 502-504.
50. Buryanov M.S. Draft Declaration of Global Digital Human Rights // Eurasian Law Journal. 2025. No. 12 (211). pp. 522-526.
51. Buryanov M.S. The Role of the UN in Shaping a New World Order under Conditions of Digital Globalization: From the Law of Force to the Force of Law // Collection of Articles by Finalists of the A.A. Gromyko Competition of Young International Scholars from CIS Countries, 2020 / ed. by V.V. Sutyrin and A.S. Peshenkov. Moscow: Institute of Europe of the Russian Academy of Sciences; A.A. Gromyko Association for Foreign Policy Studies, 2021. 625 p. pp. 71-83.
52. Buryanov M.S. Overarching Tasks of Global Governance for the UN in the Context of the Sustainable Development Goals // Programmatic Provisions and Practical Approaches of the UN and ILO in the Field of Achieving Social Justice and Improving Quality of Life: Collection of Materials of the International Scientific and Practical Conference / ed. by T.A. Soshnikova. Moscow: Publishing House of Moscow University for the Humanities, 2021. 462 p. pp. 415-420.
53. Buryanov M.S. The UN Sustainable Development Goals until 2030 and Their Implementation in the Context of Digital Human Rights // Global Dialogue on the Sustainable Development Goals: The Legal Dimension (on the 75th

Anniversary of the United Nations): Collection of Materials ofthe International Scientific and Practical Conference / ed. byT.A. Soshnikova. Moscow: Publishing House of MoscowUniversity for the Humanities, 2020. 364 p. pp. 263-267.

54. Buryanov M.S. Digital Human Rights under Conditions of Global Processes: Theory and Practice of Implementation: Monograph / scientific editor S.A. Buryanov. Moscow: Rusains, 2024. 148 p.
55. Buryanov M.S. Digital Human Rights as an Important Factor in Building European Labour Law in the Era of Industry 4.0 and Artificial Intelligence // Problems of Legal Regulation in the Works of Young Scholars: Materials of the XII Intra-University Scientific and Practical Conference of Master's Students (Nizhny Novgorod, January 10, 2020), the XVI All-Russian Scientific and Practical Student Conference " Current Problems of Contemporary Legal Science and Practice" (Nizhny Novgorod, May 15, 2020), the XIII Intra-University Scientific and Practical Conference of Master's Students (Nizhny Novgorod, June 8, 2020), the XIV Intra-University Scientific and Practical Conference of Master's Students (Nizhny Novgorod, June 25, 2020), and the II Student Research Paper Competition in European Law (Nizhny Novgorod, May 10-25, 2020) / ed. by V.I. Tsyganov and A.N. Fedyushkin. Novgorod, 2021. 1340 p. pp. 162-168.
56. Buryanov M.S. Digital Human Rights as a Response to the Threats of Globalization 4.0 // Globalistics: Global Problems and the Future of Humanity. Collection of Articles of the International Scientific Congress "Globalistics 2020", May 18-22 and October 20-24, 2020 / ed. by I.V. Ilyin. Moscow: International Public Organization for the Promotion of Scientific Research Named after N.D. Kondratiev, 2020. 969 p. pp. 395-399.
57. Buryanov M.S. Digital Human Rights as a Condition for the Effective Participation of Russia and Other Eurasian Economic Union Member States in Digitalization 4.0 // Technical and Technological Problems of Service. 2021. No. 1 (55). pp. 61-67.
58. Buryanov M.S. Digital Human Rights as a Condition for the Effective Participation of Russia and Other Eurasian

Economic Union Member States in Digitalization 4.0(Conclusion) // Technical and Technological Problems ofService. 2021. No. 2 (56). pp. 83-90.

59. Buryanov S.A. Secularity of the State and Internationally Recognized Freedom of Conscience. Theoretical and Applied Study for 2015 - Early 2016. Moscow: Poligraf Servis, 2016.

60. Buryanov S.A. IV International Scientific Congress " Globalistics 2015": Prospects for the Development of International Law under Conditions of the Globalization of Social Relations // Eurasian Law Journal. 2015. No. 10.

61. Buryanov S.A. Analysis of New Provisions of Russian Legislation in the Sphere of Freedom of Conscience from the Perspective of International Law // Human Rights in International and National Law. Collection of Scientific Articles Dedicated to the 10th Anniversary of the Department of International Law and Human Rights of the Law Institute of Moscow City University. Moscow: Human Rights, 2015. pp. 102-114.

62. Buryanov S.A. The Future of International Law under Conditions of the Globalization of Social Relations through the Prism of the Creative Legacy of Igor Ivanovich Lukashuk // Eurasian Law Journal. 2016. No. 7 (98). pp. 77-81.

63. Buryanov S.A. State Policy of the Russian Federation in the Sphere of Freedom of Conscience at the Turn of the Millennium // Law and Politics. 2000. No. 1.

64. Buryanov S.A. The Confessional Factor in the Context of the Problem of Ensuring Security: International Legal and Domestic Aspects // Religions and the Security of the State and Society: Materials of a Public Conference. Moscow: Poligraf Servis, 2015. pp. 34-43.

65. Buryanov S.A. Xenophobia, Intolerance and Discrimination on Grounds of Religion or Belief in the Subjects of the Russian Federation. Specialized Information and Analytical Report for 2006 - First Half of 2007. Moscow: Russian Helsinki Group, 2007. 240 p.

66. Buryanov S.A. International Recognition of the Right to Freedom of Conscience and Problems of Its Implementation in the Russian Federation under Conditions of Contemporary Global Processes. Moscow, 2020. 624 p.

67. Buryanov S.A. International Legal Documents in the Sphere of Freedom of Conscience and the Practice of Their Application in the Russian Federation. Theoretical and Applied Study for 2011. Moscow: Russian Helsinki Group, 2012. 244 p.
68. Buryanov S.A. A New Paradigm of Legal Regulation of Freedom of Conscience // Contemporary Law. 2003. No. 5. pp. 11-15.
69. Buryanov S.A. A New Paradigm of Freedom of Conscience: In Addition to the Published Article // Contemporary Law. 2003. No. 7. pp. 39-41.
70. Buryanov S.A. Legal Foundations, Essential Content and Guarantees of Freedom of Conscience // State and Law. 2001. No. 2.
71. Buryanov S.A. Problems of Implementing Freedom of Conscience and the Secularity of the State in the Russian Federation. Theoretical and Applied Study for 2012 / ed. by R.V. Bagdasarov. Moscow: Russia for All, 2013. 352 p.
72. Buryanov S.A. Problems of Improving the Legal and Regulatory Framework for Implementing the Right to Freedom of Conscience in the Context of Globalization // State and Law. 2002. No. 10. pp. 26-31.
73. Buryanov S.A. Manifestations of Systemic Corruption in Relations between the State and Religious Associations // Overcoming Corruption as the Main Condition for Establishing a Rule-of-Law State. Interdepartmental Scientific Collection / editor-in-chief A.I. Komarova. Moscow, 2009. Vol. 1 (39). pp. 399-419.
74. Buryanov S.A. Implementation of Constitutional Freedom of Conscience and Freedom of Religion in the Russian Federation: Monograph. Moscow: ZAO TF MIR, 2009. 288 p.
75. Buryanov S.A. Religion in Elections in Russia: The Factor of Relations between the State and Religious Associations in the Federal Electoral Cycle of 2003-2004. Moscow: Institute for Freedom of Conscience, 2005. 200 p.
76. Buryanov S.A. Secularity of the State in the Russian Federation. Theoretical and Applied Study for 2016 - Early 2017. Moscow, 2017. 196 p.

77. Buryanov S.A. Secularity of the State and Internationally Recognized Freedom of Conscience. Theoretical and Applied Study for 2015 - Early 2016. Moscow: Poligraf Servis, 2016. 258 p.
78. Buryanov S.A. Freedom of Conscience in Russian Scholarship, Law-Making and Law Enforcement // Law and Politics. 2001. No. 7.
79. Buryanov S.A. Freedom of Conscience and Secularity of the State in the Russian Federation. Theoretical and Applied Study for 2009 - Early 2010. Moscow: OOO TF MIR, 2010. 173 p.
80. Buryanov S.A. Freedom of Conscience as a Global Value: On the Path toward Political Unity and the Solution of Global Problems // Age of Globalization. 2009. No. 1. pp. 136-151.
81. Buryanov S.A. Current State and Main Directions of Reforming the Internationally Recognized Freedom of Thought, Conscience and Religion // Eurasian Law Journal. 2016. No. 1 (92). pp. 126-130.
82. Buryanov S.A. Current State and Main Directions of Reforming the Internationally Recognized Freedom of Thought, Conscience and Religion // Eurasian Law Journal. 2016. No. 1 (92). pp. 126-130.
83. Buryanov S.A. Formation and Development of Freedom of Thought, Conscience and Religion in International Law // Eurasian Law Journal. 2017. No. 2 (105). pp. 183-189.
84. Buryanov S.A. The Nature and Limits of Regulation in the Sphere of Freedom of Conscience in a Democratic State // Law and Politics. 2000. No. 5. pp. 60-67.
85. Buryanov S.A., Buryanov M.S. The Future of Human Rights under Conditions of Transition to the Digital Economy 4.0 // Interaction between the State and Civil Society When Introducing Special Measures in the Sphere of the Economy: Collection of Materials of the International Scientific and Practical Conference / ed. by T.A. Soshnikova. Moscow: Publishing House of Moscow University for the Humanities, 2022. 392 p.
86. Buryanov S.A., Buryanov M.S. In Search of an Adequate World Order: The Future of the Rule of Law in the Era of

Digital Globalization 4.0 // Age of Globalization. 2024. No. 2(50). pp. 127-140. DOI: 10.30884/vglob/2024.02.10.

87. Buryanov S.A., Buryanov M.S. Global Challenges of Digital Technologies // Philosophy and the Challenges of Modernity: On the 90th Anniversary of the Institute of Philosophy of the National Academy of Sciences of Belarus: Materials of the International Scientific Conference (Minsk, April 15-16, 2021). In 3 Volumes. Vol. 2 / Institute of Philosophy of the National Academy of Sciences of Belarus; editorial board: A.A. Lazarevich (chair) [et al.]. Minsk: Four Quarters, 2021. 378 p. pp. 89-92.
88. Buryanov S.A., Buryanov M.S. Global Challenges of Digital Technologies and Prospects for Their International Legal Regulation // Digital Technologies and Law: Collection of Scientific Papers of the First International Scientific and Practical Conference (Kazan, September 23, 2022) / ed. by I.R. Begishev, E.A. Gromova, M.V. Zaloilo, I.A. Filippova, A.A. Shutova. In 6 Volumes. Vol. 2. Kazan: Poznanie Publishing House of Kazan Innovative University, 2022. 556 p. pp. 266-273.
89. Buryanov S.A., Buryanov M.S. Global Challenges of the Digital Divide and Prospects for Their Legal Regulation // World Civilizations. 2022. Vol. 7. No. 2. pp. 7-13.
90. Buryanov S.A., Buryanov M.S. Global Challenges of Digitalization and Prospects for the Development of Law and Legal Education // Russia in the XXI Century: Education as an Important Civilizational Institution for Development and the Formation of Russian Cultural and Historical Identity. Collection of Reports and Materials of the XXX Moiseev Readings - Scientific and Practical Conference. Moscow, 2022. 477 p. pp. 279-285.
91. Buryanov S.A., Buryanov M.S. The Global Crisis of Information Civilization and Prospects for Transition to Sustainable, Governed, Human-Centered Development // Cultural and Civilizational Crisis under Conditions of the Information Society: Materials of the International Scientific and Practical Conference, Vitebsk, December 2, 2022 / Vitebsk State University; editorial board: A.A. Lazarevich (editor-in-chief), E.V. Davlyatova, E.I. Rudkovsky. Vitebsk: P.M. Masherov Vitebsk State University, 2022. 273 p. pp.

217-220.

92. Buryanov S.A., Buryanov M.S. The UN Global Digital Compact as an Important Step toward Overcoming Inequality and Achieving Social Justice // Social Justice and Law: Toward Strengthening Peace and Preventing Crises. Collection of Materials of the International Scientific and Practical Conference / ed. by T.A. Soshnikova. Moscow: Publishing House of Moscow University for the Humanities, 2023. 356 p. pp. 29-33.
93. Buryanov S.A., Buryanov M.S. Globalization of the World Order and the Law of Force: What Has Led to the Escalation of Geopolitical Confrontation by Force in the Contemporary World, and Is There a Prospect of Evolving toward Sustainable Human-Centered Development? Part 1 // Eurasian Law Journal. 2025. No. 5 (204). pp. 23-27.
94. Buryanov S.A., Buryanov M.S. Globalization of the World Order and the Law of Force: What Has Led to the Escalation of Geopolitical Confrontation by Force in the Contemporary World, and Is There a Prospect of Evolving toward Sustainable Human-Centered Development? Part 2 // Eurasian Law Journal. 2025. No. 7 (206). pp. 18-22.
95. Buryanov S.A., Buryanov M.S. On the Problems and Prospects of Adopting the UN Global Digital Compact // X Legal Readings: The Russian State and Law: Vectors of Development Based on Traditions: All-Russian Scientific and Practical Conference (Syktyvkar, April 7-8, 2023): Collection of Articles. Part 1 / ed. by V.D. Potapov and V.V. Vorobyev. Syktyvkar: Publishing House of Pitirim Sorokin Syktyvkar State University, 2023. 198 p. pp. 28-32.
96. Buryanov S.A., Buryanov M.S. The Concept of an Evolutionary Transition to Human-Centered Global Governance // Age of Globalization. 2021. No. 3. pp. 86-100. DOI: 10.30884/vglob/2021.03.07.
97. Buryanov S.A., Buryanov M.S. Selected Problems of the Development of International Law under Conditions of Digital Globalization // Materials of the Republican Scientific and Practical Conference "The Constitution as a Factor of State Stability" (November 5, 2022). Dushanbe, 2022. 430 p. pp. 86-94.

98. Buryanov S.A., Buryanov M.S. New Threats to Global Security and Prospects for the Development of International Law // Eurasian Law Journal. 2020. No. 11 (150). pp. 35-40.
99. Buryanov S.A., Buryanov M.S. On the Prospects for the International Legal Consolidation of Digital Human Rights // Law. Economy. Social Partnership [Electronic resource]: Collection of Scientific Papers of the International University " MITSO"; editorial board: V.F. Ermolovich (editor-in-chief) [et al.]. Minsk: International University "MITSO", 2023. 1044 p. pp. 394-397.
100. Buryanov S.A., Buryanov M.S. Prospects for Human-Centered Development of Law in the Context of Contemporary Digital Global Processes and Challenges // Law in Contemporary Belarusian Society: Collection of Scientific Papers. Issue 16. National Center of Legislation and Legal Research of the Republic of Belarus. Minsk: Colorgrad, 2021. 832 p. pp. 16-25.
101. Buryanov S.A., Buryanov M.S. Human Rights in the Context of Digital Global Processes and Challenges // Ontology and Axiology of Law. Abstracts of Reports and Communications of the Tenth International Scientific Conference. Editorial board chaired by S.K. Buryakov. Omsk, 2021. 140 p. pp. 87-89.
102. Buryanov S.A., Buryanov M.S. Human Rights as a Key Factor in Achieving Sustainable Governed Development // Age of Globalization. 2022. No. 4 (44). pp. 97-110. DOI: 10.30884/vglob/2022.04.07.
103. Buryanov S.A., Buryanov M.S. Problems of Global IT Technologies and Selected Approaches to Their Legal Regulation // Science. Entrepreneurship. Innovation: Collection of Materials of the International Scientific and Practical Forum, Minsk, March 30-31, 2023 / editorial board: O.V. Bodakova [et al.]; general editor A.N. Shklyarevsky. Minsk: Belarusian State Economic University, 2023. 272 p. pp. 26-30.
104. Buryanov S.A., Buryanov M.S. Problems and Prospects for the Development of Global Studies under Conditions of a New Stage of Digital Globalization // Age of Globalization. 2026. No. 1. pp. 117-127. DOI: 10.30884/vglob/2026.01.11.

105. Buryanov S.A., Buryanov M.S. Formation of Global Education as a Key Factor in Overcoming Planetary Threats and Ensuring the Survival of Human Civilization // Intellectual Culture of Belarus: Problems of Interpreting Philosophical Heritage and Contemporary Tasks of the Humanities: Materials of the Sixth International Scientific Conference (Minsk, November 17-18, 2022). In 2 Volumes. Vol. 2 / Institute of Philosophy of the National Academy of Sciences of Belarus; editorial board: A.A. Lazarevich (chair) [et al.]. Minsk: Four Quarters, 2022. 372 p.

106. Buryanov S.A., Buryanov M.S., Nikitaev D.M. Protection of Human Rights in the Sphere of Freedom of Conscience through United Nations Mechanisms in the Context of Global Processes and Threats // Legal World. 2021. No. 1. pp. 45-49.

107. Buryanov S.A., Buryanov M.S., Nikitaev D.M. The Right to Freedom of Conscience under Conditions of Global Processes: Theory and Practice of Implementation in the Russian Federation (Master's Level). Monograph / scientific editor S.A. Buryanov. Moscow: Rusains, 2020. 236 p.

108. Buryanov S.A., Krivenky A.I. Strategic Prospects for the Development of Law and Legal Education under Conditions of Contemporary Global Processes // Main Trends and Prospects in the Development of Contemporary Law: Materials of the Annual International Scientific Conference in Memory of Professor Felix Mikhailovich Rudinsky, April 19, 2018 / general editor D.A. Pashentsev, Doctor of Legal Sciences, Professor. Moscow, 2018. pp. 247-251.

109. Buryanov S.A., Krivenky A.I., Pashentsev D.A., Romanova G.V. Issues of the Globalization of Culture and the Protection of Cultural Rights of the Human Being and Citizen: An Interdisciplinary Study. Monograph / general editors A.I. Krivenky and S.A. Buryanov. Moscow: Moscow City University, 2018. 200 p.

110. Buryanov S.A., Mozgovoy S.A. The Problem of Implementing Freedom of Conscience and Trends in Relations between the State and Religious Associations in Russia: Information and Analytical Report of the Institute for Freedom of Conscience (Second Half of 2001 - End of 2004) / S.A. Buryanov, S.A. Mozgovoy; Institute for Freedom of Conscience. Moscow, 2005. 2nd scientific ed., expanded and

revised. 526 p.

111. Buryanov S.A., Nikitaev D.M. The Problem of Freedom of Conscience as an Important Component of Global Studies // Global Processes and New Formats of Multilateral Cooperation: Collection of Scientific Papers of Conference Participants / ed. by I.V. Ilyin. Moscow: International Public Organization for the Promotion of Scientific Research Named after N.D. Kondratiev, 2016. 282 p. pp. 163-167.

112. Office for Democratic Institutions and Human Rights. Annual Report 2018. Electronic resource. URL: https://www.osce.org/ru/node/431876?download=true (accessed: 19.10.2019).

113. V.A. Kartashkin: On the 50th Anniversary of the Human Rights Covenants. Interview with Vladimir Alekseevich Kartashkin, Doctor of Legal Sciences, Professor, Honored Lawyer of the Russian Federation // Eurasian Law Journal. 2016. No. 2 (93). pp. 8-15.

114. Myasnikov V.A., Naydenova N.N., Tagunova I.A. Education in the Global Dimension: Monograph. Moscow: Institute for Theory and History of Pedagogy of the Russian Academy of Education, 2009. 512 p.

115. Valeev R.M. International Protection of Human Rights to Freedom of Conscience and Religion // Bulletin of V.N. Tatishchev Volga University. 2012. No. 2 (76).

116. Vashanova O.V. The Principle of Non-Discrimination of the Individual in International Law: Candidate of Legal Sciences Dissertation. Moscow, 2004. 232 p.

117. Weber A.B. The Contemporary World and the Problem of Global Governance // Age of Globalization. 2009. No. 1. pp. 3-16.

118. Confessional Policy of the Russian State: Textbook / ed. by M.O. Shakhov. Moscow: Publishing House of the Russian Academy of Public Administration, 2003.

119. Vitruk N.V. General Theory of the Legal Status of the Individual. Moscow: Norma, 2008.

120. Vlasenko N.A. The Categories of "Uncertainty" and " Certainty" in the Study of Contemporary Law // Legal Science and Practice: Bulletin of the Nizhny Novgorod Academy of

the Ministry of Internal Affairs of Russia. 2017. No. 1 (37).pp. 8-17.

121. Vlasenko N.A. Theory of State and Law: Textbook. 2nd ed., revised, expanded and corrected. Moscow: Prospekt, 2011.

122. The Influence of Hellenism on Contemporary Science, Culture and Education: Collective Monograph / scientific editor T.P. Minchenko. Tomsk: STT, 2016. 402 p.

123. Volkov V.V. Philosophy: Main Problems, Concepts and Terms. Textbook. St. Petersburg: Lenand, 2015. 136 p.

124. Voronkova M.L. Constitutional Foundations of the Secular State in the Russian Federation: Candidate of Legal Sciences Dissertation. Saratov, 2006.

125. Vyshkvartsev V.V. The Rule-of-Law State and Human Rights and Freedoms in the Concepts of Russian Scholars of the XX and Early XXI Centuries: Abstract of Candidate of Legal Sciences Dissertation. Moscow, 2009. p. 21.

126. Weisbrodt D., Fitzpatrick J., Newman F. International Human Rights: Law, Policy and Process. Third Edition. Human Rights Center, University of Minnesota, 2003. Electronic resource. URL: http://www1.umn.edu/humanrts/russian/hrtsbook/hrtsintroduction.html#note_3 (accessed: 27.01.2017).

127. Gavreev S.A. Socio-Psychological Characteristics of Ethnophobia: Candidate of Psychological Sciences Dissertation. Moscow, 2008.

128. Gavrilenkov A.F. The Policy of State Authority of the Russian Empire toward the Orthodox Church, Non-Orthodox Christian Denominations and Abrahamic Non-Christian Faiths in 1721-1917: Essence, Principles and Evolution: Abstract of Doctor of Historical Sciences Dissertation. Moscow, 2010.

129. Gadzhiev G.A. The Judicial Doctrine of the Rule of Law in Russia // International Legal Doctrines: Seeking Mutual Understanding / Collection of Articles edited by E.G. Tarlo and D.V. Kravchenko. Moscow: LUM, 2013.

130. Gakaev Z.Zh. Ethnic Stereotypes in the Press: Candidate of Historical Sciences Dissertation. Moscow, 2003.

131. Garadzha V.I. Sociology of Religion. Moscow, 1996.

132. Globalistics: Encyclopedia / editors-in-chief I.I. Mazur and A.N. Chumakov // Center for Scientific and Applied Programs " Dialogue". Moscow: Raduga Publishing House, 2003.

133. Global Governance in the XXI Century: Innovative Approaches / ed. by Al.A. Gromyko. Moscow: Institute of Europe of the Russian Academy of Sciences; Nestor-History, 2013.

134. Global Processes and the Formation of Global Education: An Interdisciplinary Study. Monograph in 2 Books / author team: S.A. Buryanov, E.S. Demidenko, E.A. Dergacheva et al.; general editors S.A. Buryanov and A.I. Krivenky. Book 1. Moscow: Moscow City University, 2019. 200 p.

135. Global Processes and the Formation of Global Education: An Interdisciplinary Study. Monograph in 2 Books / author team: S.A. Afanasyeva, S.A. Buryanov, A.V. Zvonarev et al.; general editors S.A. Buryanov and A.I. Krivenky. Book 2. Moscow: Moscow City University, 2019. 276 p.

136. Gontsovsky V.K. Ethnic Xenophobia among Young People in the Southern Federal District: The Case of Rostov Region: Abstract of Candidate of Sociological Sciences Dissertation. Rostov-on-Don, 2014.

137. Gorshkova S.A. Council of Europe Standards on Human Rights and Russian Legislation: Monograph. Moscow: NIMP, 2001. 348 p.

138. Goryan E.V. The Islamic Concept of Human Rights and the Progressive Development of International Law // International Law. 2015. No. 3. pp. 91-101.

139. Civil Human Rights in Russia: Contemporary Problems of Theory and Practice / ed. by F.M. Rudinsky. Moscow: ZAO TF MIR, 2006.

140. Civil Human Rights in Russia: Contemporary Problems of Theory and Practice / author team including Pavel Viktorovich Anisimov et al.; ed. by F.M. Rudinsky. 3rd ed. Moscow, 2010. 624 p.

141. Facets of Globalization: Difficult Questions of Contemporary Development. Moscow: Alpina Publisher, 2003.

142. Davydova M.L. Theoretical and Methodological Problems of the Concept and Composition of Legal Technique: Abstract

of Doctor of Legal Sciences Dissertation. Volgograd, 2010.

143. Davis D. The Evolution of Religious Freedom as a Universal Human Right: A Study of the Role of the 1981 UN Declaration on the Elimination of All Forms of Intolerance and of Discrimination Based on Religion or Belief // Protection of the Individual from Discrimination: Reader. In 3 Volumes. Vol. 3. Moscow: New Justice, 2006. 360 p.

144. Dorskaya A.A. Protection of Freedom of Conscience in International Humanitarian Law // International Public and Private Law. 2009. No. 4. pp. 8-11.

145. Dorskaya A.A. Church Law in the Legal System of the Russian Empire from the Late XVIII to the Early XX Century: Abstract of Doctor of Legal Sciences Dissertation. Moscow, 2008.

146. Drinova E.M. Religion and the Political Process: Religious-Political Expansion and the Politicization of Religion in the Contemporary World: Abstract of Doctor of Political Sciences Dissertation. Rostov-on-Don, 2012.

147. Durham W.C. Freedom of Religion or Belief: Laws Affecting the Structuring of Religious Communities. ODIHR Background Paper 1999/4.

148. Evstafyeva A.V. Hate Speech in the Mass Media: Linguistic and Extralinguistic Factors of Functioning: Abstract of Candidate of Philological Sciences Dissertation. Tolyatti, 2009.

149. Emelyanova N.N. International Legal Problems of Ensuring Global Security at the Contemporary Stage: Doctor of Legal Sciences Dissertation. Moscow, 2013. 404 p.

150. Ershov V.V. Current Problems of Expert Activity of the Civic Chamber of the Russian Federation in Light of Contemporary Legal Understanding, Law-Making and Law Enforcement // Russian Justice. 2009. No. 5.

151. Ershov V.V. International and Domestic Law from the Standpoints of Legalism and an Integrative Understanding of Law // Russian Justice. 2011. No. 8 (64).

152. Ershov V.V. The Relationship between International and Domestic Law under Conditions of Globalization // Russian Justice. 2011. No. 6 (62).

153. Zhuzha D.Yu. Political Aspects of Global Governance of Natural Resources: Abstract of Candidate of Political Sciences Dissertation. Moscow, 2012.

154. Zhuravleva O.O. The Principle of Legal Certainty and Acts of the Highest Judicial Bodies in Tax Law. Electronic resource. URL: http://www.justicemaker.ru/view-article.php?id=14&art=1821 (accessed: 11.01.2018).

155. Zanibekov M.M. Current Problems of the Globalization of International Trade // Eurasian Law Journal. 2015. No. 4 (83). pp. 54-58.

156. Protection of Human Rights: Collection of Documents 1998-2000. Moscow, 2001.

157. Zelentsov A.B. Legal Globalization and Public Law // Contemporary Problems of Public Law in Russia and Foreign Countries. Collection of Scientific Papers by Faculty Members, Postgraduate Students and Degree Candidates of the Department of Administrative and Financial Law, Law Faculty, Peoples' Friendship University of Russia. Moscow, 2012. pp. 6-25.

158. Zorkin V.D. The Limit of Compliance // Rossiyskaya Gazeta - Federal Issue No. 5325 (246). October 29, 2010.

159. Zubarev E. Pity and Irony against the Golden Censer: It Is No Longer Possible to Restore Russia's Former Secular Status without Revolutionary Restructuring; No One Voluntarily Gives Up a Sinecure // Rosbalt. Electronic resource. URL: http://www.rosbalt.ru/blogs/2013/05/01/1124348.html (accessed: 17.12.2019).

160. Zuikov R.S. International Relations at the Beginning of the XXI Century: Analysis of Criteria of Systemicity: Abstract of Candidate of Political Sciences Dissertation. Moscow, 2010.

161. Zuikov R.S. Problems of Forming a New System of International Relations from the Perspective of World-Systems and Civilizational Approaches // Space and Time in World Politics and International Relations. Materials of the 4th Convention of the Russian International Studies Association. In 10 Volumes. Vol. 6: New Trends in World Politics. Moscow: MGIMO University, 2007.

162. Ibrahim Ahmed Hadj Yusuf Abdalla. International and National Mechanisms for the Protection of Human Rights in the Middle East: Candidate of Legal Sciences Dissertation. Moscow, 1999. 192 p.

163. Ivaneev S.V. Global Problems and Selected Approaches to Their Solution // Rights and Freedoms of the Human Being and Citizen: Theoretical Aspects and Legal Practice. Materials of the Annual International Scientific Conference in Memory of Professor Felix Mikhailovich Rudinsky / general editor D.A. Pashentsev. 2016. pp. 117-121.

164. Ivaneev S.V. Humanization of Education as a Source for the Development of Human Capital // Human Capital. 2015. No. 5 (77). pp. 123-127.

165. Ivaneev S.V. On the Relationship between the Secular and the Religious in the System of Contemporary Education // Human Capital. 2014. No. 7 (67). pp. 44-48.

166. Ivaneev S.V. No Peace under Religious Olive Trees // History of State and Law. 2011. No. 1. pp. 23-24.

167. Ivaneev S.V. Rational and Ethical Guidelines in the Legal Education of Military Personnel // Military Law. 2017. No. 5 (45). pp. 138-142.

168. Ivanov D.V. Evolution of the Concept of Globalization // Telescope. 2002. No. 5. p. 1.

169. Ilyin I.V. Globalistics in the Context of Political Processes: Doctor of Political Sciences Dissertation. Moscow, 2011. 428 p.

170. Ilyin I.V., Kaverin M.A. Issues of Transforming International Organizations into Institutions of Global Governance // Age of Globalization. 2014. No. 2.

171. Integration Law in the Contemporary World: A Comparative Legal Study / ed. by S.Yu. Kashkin. Moscow, 2014. 112 p.

172. Isaev A.V. Socio-Political Partnership between Public Authorities and the Russian Orthodox Church in Contemporary Russia: Abstract of Candidate of Legal Sciences Dissertation. Orel, 2011.

173. Kalamkaryan R.A., Migachev Yu.I. International Law. Moscow: Eksmo, 2004.

174. Kanevsky K.G. Legal Regulation of State-Confessional Relations in the Russian Federation: Candidate of Legal Sciences Dissertation. Moscow, 2004.

175. Karpushkin A.V. Constitutional Foundations and Genesis of Relations between the State and Religious Associations in the Russian Federation: Abstract of Candidate of Legal Sciences Dissertation. Saratov, 2009.

176. Kartashkin V.A. The Supremacy of International Law in World Politics and at the National Level // Observer. 2015. No. 11 (310). pp. 112-122.

177. Kartashkin V.A. Human Rights in International and Domestic Law / responsible editor E.A. Lukasheva. Moscow: Publishing House of the Institute of State and Law of the Russian Academy of Sciences, 1995.

178. Kartashkin V.A. Human Rights and International Security // Proceedings of the Institute of State and Law of the Russian Academy of Sciences. 2013. No. 1. pp. 33-59.

179. Kartashkin V.A. Human Rights and Principles of International Law in the XXI Century: Monograph. Moscow: Norma; INFRA-M, 2018. 148 p.

180. Kartashkin V.A. Human Rights: The Correlation and Interaction of International and Domestic Law // Proceedings of the Institute of State and Law of the Russian Academy of Sciences. 2016. No. 4. pp. 14-34.

181. Kartashkin V.A. The UN Charter and the United Nations in a Globalizing World // International Lawyer. 2004. No. 1. pp. 4-8.

182. Kartashkin V.A., Lukasheva E.A. Rights and Freedoms of the Human Being and Citizen: Universalism, Regionalism, Realities // State and Law. No. 7. 2010. pp. 37-45.

183. Kartukhin V.Yu. Use of the Rules of Legal Technique in Law-Making by Constituent Entities of the Russian Federation // Contemporary Jurisprudence. Current Problems of Russian Law. Electronic resource. URL: http://www.ibil.ru/index.php?type=review&area=1&p=articles&id=358 (accessed: 09.01.2018).

184. Kenenov A.A., Chernobel G.T. Logical Foundations of the Legislative Process // Jurisprudence. 1991. No. 6.

185. Kenichi Ohmae. The Mind of the Strategist: The Art of Japanese Business. Moscow: Alpina Publisher, 2007. 224 p.

186. Kete K.Zh. Activities of the African Committee of Experts on the Rights and Welfare of the Child // Issues of Russian and International Law. 2016. No. 9. pp. 175-183.

187. King A., Schneider B. The First Global Revolution: A Report by the Club of Rome. Moscow: Progress-Pangea, 1991. 344 p.

188. Kirichenko M.G. Freedom of Conscience in the USSR. Moscow: Legal Literature, 1985.

189. Kiseleva E.V. The Concept of Discrimination in Relation to the Right to Freedom of Thought, Conscience and Religion: International Legal Norms and the Practice of International Universal and Regional Human Rights Bodies // Eurasian Law Journal. 2019. No. 1 (128). pp. 29-34.

190. Colson Ch. Conflict of Kingdoms. Moscow, 1996.

191. Koneva A.E. International Legal Aspects of Strengthening the System of Human Rights Treaty Bodies: Candidate of Legal Sciences Dissertation. Moscow, 2015. 235 p.

192. Konnevskaya A.Yu. Constitutional Freedom of Conscience: Content, Limits and Guarantees: Abstract of Candidate of Legal Sciences Dissertation. Chelyabinsk, 2010.

193. Constitutional Law of the Russian Federation. Moscow, 2002. Electronic resource. URL: http://slovari.yandex.ru (accessed: 03.01.2018).

194. Constitutional Rights and Freedoms of the Human Being and Citizen in the Russian Federation / ed. by O.I. Tiunov. Moscow: Norma, 2005.

195. Constitution of the Russian Federation. Moscow: Legal Literature, 1995.

196. Conflicts and Wars of the XXI Century: The Middle East and North Africa / Institute of Oriental Studies of the Russian Academy of Sciences. Moscow: Institute of Oriental Studies of the Russian Academy of Sciences, 2015. 504 p.

197. Konyukhova I.A. Sources of International and Constitutional Law in the Regulation of Human and Civil Rights and Freedoms: International, Foreign and Domestic Experience // International Legal Standards in Constitutional Law:

Collection of Scientific Papers. Part I / Russian Academy ofJustice; Institute of Scientific Information on Social Sciencesof the Russian Academy of Sciences; responsible editor I.A.Konyukhova. Moscow, 2006. pp. 108-138.

198. Korshunov A.N. The Idea of Global Law: Philosophical and Methodological Aspects: Abstract of Candidate of Philosophical Sciences Dissertation. Rostov-on-Don, 2010.

199. Kostenko A.N. Social Naturalism as a Methodological Principle of Legal Global Studies. Electronic resource. URL: http://lib.chdu.edu.ua/pdf/naukpraci/politics/2007/69-56-42.pdf (accessed: 14.04.2019).

200. Kochetov E.G. Main Characteristics of the Globalization Process and the Legal Dimension of the World // Journal of Russian Law. 2003. No. 3. pp. 88-97.

201. Krasnikov A.N. Trends in Contemporary Religious Studies // Freedom of Conscience in a Law-Governed State: Legal and Informational Aspects. Seminar Materials. Moscow, 2000.

202. Krivolapov P.S. New Trends in International Cooperation in the Field of Human Rights: Candidate of Legal Sciences Dissertation. Moscow, 2006. 213 p.

203. Kroz M.V., Ratinov N.A. Socio-Psychological and Legal Aspects of Xenophobia. Moscow, 2005.

204. Krygina I.A., Rybak S.V. The Legal Education Process in the Context of Developing the Idea of Tolerance under Contemporary Conditions: Monograph. Rostov-on-Don, 2012. 142 p.

205. Kulagin M.A. Real Rights of Religious Organizations: Abstract of Candidate of Legal Sciences Dissertation. Moscow, 2013. p. 19.

206. The Human Rights Course: Textbook for Staff Members of Ombudsman Offices and Human Rights Commissions in the Russian Federation. Moscow: Russian Helsinki Group, 2005.

207. Lazarev V.V., Saidov A.Kh. Patterns in the Development of Contemporary Law and Legal Global Studies // Bulletin of the International Institute of Economics and Law. 2011. No. 2 (2). pp. 85-91.

208. Lebedev S.A. Science in the Global World // Age of Globalization. 2012. No. 2 (10). pp. 145-151.

209. Lebedev S.D. Religious and Secular Cultures as Types of Systems of Social Knowledge. Belgorod: Publishing House of Belgorod State University, 2003.

210. Lebedeva M.M. World Politics. 2nd revised and expanded edition. Moscow: Aspect Press, 2007. 365 p.

211. Levina A.A. The Institution of Freedom of Conscience and Religion: Russia's Historical and Legal Experience: Abstract of Candidate of Legal Sciences Dissertation. Nizhny Novgorod, 2003.

212. Levitsky S.A. The Tragedy of Freedom. Posev Publishing House, 1984.

213. Linkin V.N. The Principle of the Secular State as One of the Foundations of the Constitutional Order of the Russian Federation: Abstract of Candidate of Legal Sciences Dissertation. Moscow, 2015.

214. Lovinyukov A.S. The 1990 USSR Law on Freedom of Conscience and Religious Organizations // Soviet State and Law. 1991. No. 4.

215. Lovinyukov A.S. Freedom of Conscience: Analysis, Practice, Conclusions // State and Law. 1995. No. 1. pp. 24-36.

216. Loginov A.V. Power and Faith: The State and Religious Institutions in History and the Present. Moscow: Great Russian Encyclopedia, 2005.

217. Lukashuk I.I. International Law. General Part: Textbook for Students of Law Faculties and Universities. 3rd revised and expanded edition. Moscow: Wolters Kluwer, 2005.

218. Lukashuk I.I. Interaction between International and Domestic Law under Conditions of Globalization // Journal of Russian Law. 2002. No. 3. pp. 115-128.

219. Lukashuk I.I. Globalization, the State and the XXI Century. Moscow, 2000.

220. Lukashuk I.I. International Law. General Part: Textbook for Students of Law Faculties and Universities. 3rd revised and expanded edition. Moscow: Wolters Kluwer, 2005.

221. Lukyantsev G.E. The Regional System for the Protection of Human Rights within the Council of Europe: The Example of the 1950 European Convention for the Protection of Human Rights and Fundamental Freedoms: Candidate of Legal

Sciences Dissertation. Moscow, 1998. 190 p.

222. Luparev G.P. Freedom of Conscience: A Sacred Cow or a Constitutional-Legal Anachronism? // Religion and Law. 2002. No. 2.

223. Luchin V.O. The Constitution of the Russian Federation: Problems of Implementation. Moscow: UNITI-DANA, 2002.

224. Lykov A.Yu. On the Problem of Integrating Legal Systems in the Formation of a World State // History of State and Law. 2013. No. 12.

225. Lykov A.Yu. The World State as the Future of the International Community: Monograph. Moscow: Prospekt, 2013. 248 p.

226. Magomedov A.M. Human Rights in Islamic and International Law: Candidate of Legal Sciences Dissertation. Moscow, 2011.

227. Malakhov V.P., Abashidze A.Kh. International Civil Society: Essence and Current State // Bulletin of Moscow University of the Ministry of Internal Affairs of Russia. 2014. No. 11. pp. 164-167.

228. Malayan R.E. The UN in the Emerging System of Global Governance: Abstract of Candidate of Political Sciences Dissertation. Moscow, 2009. 22 p.

229. Mamedov A.A. International Norms-Principles and Economic Globalization // Bulletin of Peoples' Friendship University of Russia. Series: Legal Sciences. 2014. No. 2. pp. 265-271.

230. Mamedov A.A. Contemporary Trends in International Legal Regulation of Banking under Conditions of Globalization // Eurasian Law Journal. 2013. No. 6 (61). pp. 37-40.

231. Markova E.N. The Problem of the Legal Definition of Religion // Constitutional and Municipal Law. 2018. No. 11. pp. 34-45.

232. Markova E.N. Freedom of Conscience and Religion: The Negative Dimension and Protection of Personal Beliefs: Candidate of Legal Sciences Dissertation. Moscow, 2019. 277 p.

233. Marochkin S.Yu. Legislative, Doctrinal and Practical Assimilation of the Constitutional Principle on Norms of

International Law in Russia's Legal System // Application ofInternational Human Rights Treaties in the Legal System ofthe Russian Federation. Materials of the All-Russian Scientificand Practical Conference. Yekaterinburg, 2003.

234. Marochkin S.Yu. The Problem of the Effectiveness of Norms of International Law / scientific editor G.V. Ignatenko. Irkutsk: Irkutsk University Press, 1988.

235. Marchenko M.N. Legal Systems of the Contemporary World: Textbook. Moscow: Zertsalo-M, 2001. pp. 292-293.

236. International Protection of Human Rights: Textbook / responsible editor A.Kh. Abashidze. Moscow: Peoples' Friendship University of Russia, 2017. 466 p.

237. International Law and Soviet Legislation: Collective Monograph / ed. by G.I. Kurdyukov. Kazan: Kazan University Press, 1991.

238. International Law: Textbook for Universities / responsible editors G.V. Ignatenko and O.I. Tiunov. Moscow: NORMA-INFRA, 1999. 584 p.

239. International Law. Special Part: Textbook for Students of Law Faculties and Universities / I.I. Lukashuk. Moscow: Wolters Kluwer, 2005.

240. International Law: Textbook / responsible editors Yu.M. Kolosov and E.S. Krivchikova. 2nd revised and expanded edition. Moscow: International Relations, 2005.

241. International Public Law: Collection of Documents / compilers and authors of the foreword K.A. Bekyashev and D.K. Bekyashev. Moscow: Prospekt, 2009. 1200 p.

242. International Public Law: Textbook / L.P. Anufrieva, D.K. Bekyashev, K.A. Bekyashev, V.V. Ustinov [et al.]; responsible editor K.A. Bekyashev. 4th revised and expanded edition. Moscow: TK Welby; Prospekt Publishing House, 2007.

243. Melnikova N.V. Formation of Ethnic Tolerance in a Metropolis: Sociological Analysis: Abstract of Candidate of Sociological Sciences Dissertation. Moscow, 2002.

244. Murdoch J. Protecting the Right to Freedom of Thought, Conscience and Religion under the European Convention on Human Rights. Council of Europe Handbooks Series.

Voronezh: Elist Firm LLC, 2014. 163 p.

245. Meshcheryakova A.F. The Secular State in Contemporary Russia: Constitutional and Legal Analysis: Abstract of Candidate of Legal Sciences Dissertation. Saratov, 2009.

246. Meshcheryakova O.M. Globalization of Law: Philosophical and Legal Reflection // Russian Journal of Legal Studies. 2015. No. 2 (3). pp. 28-32.

247. Mingazov L.Kh. The Effectiveness of Norms of International Law: Theoretical Problems: Abstract of Doctor of Legal Sciences Dissertation. Kazan, 2000.

248. Minniahmetov R.A. Shiite Concepts of Justice in Islamic Law // Eurasian Law Journal. 2013. No. 12.

249. Minchenko T.P. The Problem of Freedom of Conscience in the Age of Post-Secularity: Origins and Prospects: Doctor of Philosophical Sciences Dissertation. Tomsk, 2011.

250. Minchenko T.P. Development of the Right to Freedom of Conscience in the Post-Secular World // Rights and Freedoms of the Human Being and Citizen: Theoretical Aspects and Legal Practice. Materials of the Interuniversity Scientific Conference in Memory of Professor Felix Mikhailovich Rudinsky, April 26, 2012. Moscow: APPKiPRO, 2012. pp. 155-158.

251. Miroshnikova E.M. The Constitutional-Legal Mechanism of State-Church Relations in the Federal Republic of Germany // World Experience of State-Church Relations. Moscow, 1998. pp. 83-84.

252. Miroshnikova E.M. The Cooperation Model of State-Church Relations: Experience and Problems: Monograph. Moscow: Institute of Europe of the Russian Academy of Sciences; Tula: Publishing House of Tula State Pedagogical University named after L.N. Tolstoy, 2007.

253. Mozgovoy S.A. Confessional Clerical State // Religious Studies: Encyclopedic Dictionary. Moscow: Academic Project, 2006.

254. Mozgovoy S.A. Military-Religious Relations: Methodological Approaches to Defining the Concept // State, Religion and Church in Russia and Abroad. 2001. Vol. 19. No. 2. pp. 108-110.

255. Mozgovoy S.A. Selected Issues of Legal Regulation of Relations between the Armed Forces and the Russian Orthodox Church (1992 to the Present) // Military Law Journal. 2011. No. 3. pp. 27-32.

256. Mozgovoy S.A. The Secular State as Worldview-Neutral // Socionomicon. International Collection of Scientific and Practical Works. Issue 2. Rostov-on-Don: NB, 2006. pp. 70-73.

257. Mozgovoy S.A. Ethno-Religious Relations and Problems of Civil Peace in Russia // Religion and Nationalism. Moscow, 2000.

258. Moiseev A.A. Current Aspects of International Law under Conditions of Globalization // Bulletin of Peoples' Friendship University of Russia. Series: Legal Sciences. 2007. No. 2 (14). pp. 45-58.

259. Moiseev A.A. The Importance of International Law for Developing Dialogue among National Cultures under Conditions of Globalization // Dialogue of Cultures under Conditions of Globalization: Materials of the XII International Likhachev Scientific Readings (St. Petersburg, May 17-18, 2012). St. Petersburg: St. Petersburg University of the Humanities and Social Sciences, 2012. Vol. 1. pp. 484-486.

260. Moiseev A.A. The World at a Crossroads: International Law in an Era of Global Change // Eurasian Law Journal. 2026. No. 1 (212). pp. 13-21.

261. Moiseev A.A. Supranationality in Contemporary International Law. Moscow: Scientific Book, 2007. 288 p.

262. Moiseev A.A. The Phenomenon of Globalization and the International Community // Age of Globalization. 2015. No. 2 (16). pp. 59-72.

263. Moiseev N.N. Morality and the Phenomenon of Evolution: The Ecological Imperative and Ethics of the XXI Century // Social Sciences and Contemporary World. 1994. No. 6.

264. Moiseev N.N. The Ecological Background of Contemporary Politics // Social Sciences and Contemporary World. 1999. No. 4.

265. Moran G.M. What Is Religious Freedom? What Should the Law Guarantee? // Religion and Law. 1997. No. 1.

266. Mukhametzyanova R.M. Policy of the Contemporary Russian State in the Sphere of Freedom of Conscience: The Process of Forming a New Model: Abstract of Candidate of Political Sciences Dissertation. Ufa, 2004.

267. Nazaretyan A.P. The Clash of Civilizations and The End of History // Social Sciences and Contemporary World. 1994. No. 6.

268. Nazaretyan A.P. A Nonlinear Future: Megahistorical, Synergetic and Cultural-Psychological Prerequisites for Global Forecasting. Moscow: MBA Publishing House, 2013.

269. Nemchuk A.A. Global Governance in the Contemporary World: A Political-Science Analysis: Doctor of Political Sciences Dissertation. Moscow, 2004. 303 p.

270. Neshataeva T.N. International Organizations and Law: New Trends in International Legal Regulation. Moscow: Delo, 1998. 272 p.

271. Nizhnikov S.A. Philosophy: A Course of Lectures: Textbook for Universities. Moscow, 2006. 383 p.

272. Nikitaev D.M. On the Contemporary Definition of the Concept and Content of Freedom of Conscience // Contemporary Lawyer. 2018. No. 4 (25). pp. 66-75.

273. Nikitaev D.M. Problems of Implementing Freedom of Conscience in Contemporary Russia and the World / D.M. Nikitaev, S.A. Buryanov; scientific editor S.A. Buryanov. Moscow: RUSCIENCE, 2020. 146 p.

274. Nikitaev D.M., Buryanov S.A. On the Contemporary Definition of the Concept and Content of the Secularity of the State // Contemporary Lawyer. 2019. No. 3 (28), July-September. pp. 85-96.

275. Nikolaev A.M. The European Convention for the Protection of Human Rights and Fundamental Freedoms and Constitutional Human Rights and Freedoms // Science and Education. 2011. No. 11 (17). pp. 85-88.

276. Nikolaev A.M., Davtyan M.K. Enforcement of Judgments of the European Court of Human Rights and the Inter-American Court of Human Rights: A Comparative Analysis // Journal of Foreign Legislation and Comparative Law. 2018. No. 4 (71). pp. 40-46.

277. Nikolaev A.M., Folosyan M.E. Selected Problems in the Practice of the European Court of Human Rights in the Field of Freedom of Thought, Conscience and Religion // Issues of Russian and International Law. 2016. No. 9. pp. 184-195.

278. Novikov M.V. The Clerical State and the Church: Relations between Church and State in Israel and Neighboring Countries: Abstract of Candidate of Legal Sciences Dissertation. Moscow, 2013.

279. Orlova O.G. The Discursive Theory of the Stereotype: Abstract of Doctor of Philological Sciences Dissertation. Kemerovo, 2013.

280. Osipova L.V. The Legal Status of Religious Associations as an Institution of Civil Society in Contemporary Russia: Abstract of Candidate of Legal Sciences Dissertation. Moscow, 2009.

281. Osipova L.N. Relations between State Institutions and Religious Associations in Contemporary Russia: Candidate of Political Sciences Dissertation. Ufa, 2004.

282. Foundations of Religious Studies Expertise. Moscow, 2002.

283. Pavlenko V.B. Institutional Aspects of Global Governance of Political Processes: Doctor of Political Sciences Dissertation. Moscow, 2008.

284. Panteleev B. Legal Certainty or Unpredictability? Electronic resource. URL: http://www.center-bereg.ru/916.htm (accessed: 03.01.2018).

285. Pashentsev D.A. The Significance of Human Rights under Contemporary Conditions // Concepts for Developing the Institution of Human Rights under Conditions of Globalization of Contemporary Law and Politics: Collection of Scientific Papers Based on Materials of the International Scientific and Practical Round Table / responsible editor N.N. Kuleshova. Ryazan, 2016. pp. 195-197.

286. Pereplesnina E.M. Generally Recognized Principles and Norms of International Law and International Treaties of the Russian Federation in the Russian Legal System. Petrozavodsk: Verso, 2009.

287. Prospects for Geopolitical Dynamics and Interaction among Civilizations. Part 7 of the Global Forecast The Future of

Civilizations for the Period up to 2050 / ed. by Yu.V.Yakovets, A.I. Ageev and T.T. Timofeev. Moscow: MISK,2009. pp. 161-170.

288. Petyukova O.N. Legal Forms of Relations between the Soviet State and the Russian Orthodox Church in 1917-1945: Abstract of Doctor of Legal Sciences Dissertation. Moscow, 2011.

289. Pimenov A. The Past That Has Not Passed: Toward a Characterization of Nontheistic Religiosity // Religion. Magic. Myth. Moscow: URSS, 1997.

290. Podzigun I.M. Globalization and Global Problems: A Philosophical and Methodological Analysis: Doctor of Philosophical Sciences Dissertation. Moscow, 2003. 384 p.

291. Podoprigora R.A. Administrative-Legal Regulation of the Activities of Religious Organizations in Foreign Countries: Abstract of Doctor of Legal Sciences Dissertation. Almaty, 2002.

292. Podoprigora R.A. The State and Religious Organizations: Administrative-Legal Issues. Almaty: Arkaim, 2002.

293. Pokanina E.B. A Socio-Philosophical Analysis of the Transformation of State-Confessional Relations in the Republic of Kalmykia: Abstract of Doctor of Philosophical Sciences Dissertation. Moscow, 2012.

294. Polosin V.S. Myth, Religion, State. Moscow, 1998.

295. Understanding Human Rights: Manual on Human Rights Education / ed. by W. Benedek and M. Nikolova. Graz: European Training and Research Centre for Human Rights and Democracy; Russian version: Warsaw: Homework, 2005.

296. Ponkin I.V. Legal Foundations of the Secularity of the State and Education. Moscow, 2003.

297. Ponkin I.V. The Secularity of the State. Moscow: Educational and Scientific Centre for Pre-University Education, 2004.

298. Human Rights in the Russian Federation: Collection of Reports on the Events of 2016 / responsible editor and compiler N. Kostenko. Moscow, 2017.

299. Human Rights in the Russian Federation: Collection of Reports on the Events of 2017 / responsible editor and

compiler N. Kostenko. Moscow, 2018. 218 p.

300. Human Rights in the Russian Federation: Collection of Reports on the Events of 2018 / responsible editor and compiler N. Kostenko. Moscow, 2019. 338 p.

301. Human Rights in the Russian Federation: Collection of Reports on the Events of 2007. Moscow, 2008.

302. Human Rights in the Russian Federation: Collection of Reports on the Events of 2008. Moscow, 2009.

303. Human Rights in the Russian Federation: Report on the Events of 2010 / compiler N. Kostenko. Moscow, 2011.

304. Human Rights in the Russian Federation: Report on the Events of 2013 / compiler N. Kostenko. Moscow, 2014.

305. Human Rights in the Russian Federation: Collection of Reports on the Events of 2012 / compilers V. Karastelev, N. Kostenko and D. Meshcheryakov. Moscow, 2013.

306. Human Rights in the Russian Federation: Collection of Reports on the Events of 2014 / compiler N. Kostenko. Moscow, 2015.

307. Human Rights under Conditions of Globalization and Their Protection in Private International Law: An Interdisciplinary Study: Collective Monograph in Two Books / ed. by A.I. Krivenky. Moscow: Moscow City University, 2016.

308. Human Rights / responsible editor Corresponding Member of the Russian Academy of Sciences E.A. Lukasheva. Moscow: Norma, 2004.

309. Presnyakov M.V. The Principle of Legal Certainty in the System of Constitutional Principles // Human Rights: Theory, History, Practice. Collection of Scientific Papers Dedicated to the 65th Anniversary of the Universal Declaration of Human Rights / general editor E.M. Pavlenko. Volgograd: Publishing House of the Volgograd Branch of the Russian Presidential Academy of National Economy and Public Administration, 2014.

310. Pridvorov N.A., Tikhonova E.V. The Institution of Freedom of Conscience and Freedom of Religion in the Law of Contemporary Russia. Moscow: Jurisprudence, 2007. pp. 39-40.

311. Problems of Legal Understanding: Collective Monograph Based on the Results of the III Alekseev Readings / responsible editor V.D. Perevalov. Yekaterinburg: Publishing House of Ural State Law University, 2018. 276 p.

312. Pryakhina T.M. Constitutional Problems of Interaction between International and National Law: Teaching and Methodological Manual / ed. by V.O. Luchin. Moscow: Publishing House of the Russian State Social University, 2006.

313. Pryakhina T.M. Ensuring the Interaction of International and National Law by the Constitutional Court of the Russian Federation. Saratov: Nauka Information Centre, 2010.

314. Pchelintsev A.V. Freedom of Religion and Religious Associations in the Russian Federation: A Constitutional-Legal Study: Abstract of Doctor of Legal Sciences Dissertation. Moscow, 2012.

315. Guidelines for Review of Legislation Pertaining to Religion or Belief. OSCE/ODIHR. Warsaw, 2005. Electronic resource. URL: https://www.osce.org/ru/odihr/13994?download=true (accessed: 19.10.2019).

316. Religious Associations. Freedom of Conscience and Religion. Regulatory Acts. Judicial Practice. Moscow: Jurisprudence, 2001.

317. Religion and Society: Essays on Religious Life in Contemporary Russia / responsible editor and compiler S.B. Filatov. Moscow; St. Petersburg: Letny Sad, 2002.

318. Religion and the Secular State: The Principle of Laicite in the World and Eurasia. Moscow: Franco-Russian Centre for the Humanities and Social Sciences in Moscow, 2008. 288 p.

319. Rosenbaum Yu.A. The Soviet State and the Church / responsible editor B.M. Lazarev. Moscow, 1985. 174 p.

320. Romasevich D.I. Global Problems of the Contemporary World and Russia's Environmental Policy: Candidate of Political Sciences Dissertation. Moscow, 2006. 148 p.

321. Roan M. The Right to Freedom of Religion or Belief: Textbook. Minnesota: University of Minnesota Human Rights Center, 2003. Electronic resource. URL: http://www1.umn.edu/humanrts/russian/edumat/studyguides/Rreligion.html (accessed: 21.11.2019).

322. Rudinsky F.M. Law, Religion, Offences. Volgograd, 1971.

323. Rudinsky F.M. The Institution of Freedom of Conscience under Soviet State Law: Candidate of Legal Sciences Dissertation. Novosibirsk, 1963.

324. Rudinsky F.M. The Science of Human Rights and Problems of Constitutional Law. Moscow: MIR Trading Company CJSC, 2006.

325. Rudinsky F.M., Gavrilova Yu.V., Krikunova A.A., Soshnikova T.A. Economic and Social Rights of the Human Being and Citizen: Contemporary Problems of Theory and Practice / general editor F.M. Rudinsky. Moscow: Human Rights, 2009.

326. Guidelines on the Legal Personality of Religious or Belief Communities. OSCE/ODIHR. Warsaw, 2015. Electronic resource. URL: https://www.osce.org/ru/odihr/145431?download=true (accessed: 19.10.2019).

327. The Russian Orthodox Church in the XX Century: History and Relations with the State and Society. Moscow, 2002.

328. Savva S.S. Clericalization of Secular Education in Russia: A Constitutional-Legal Analysis // Religious Studies. 2006. No. 4. pp. 76-88.

329. Savva S.S. Constitutional-Legal Enshrinement of the Principle of Secularity in Education: Content and Implementation Practice // Politics and Society. 2006. No. 7-8. pp. 106-115.

330. Savelyev V.N. Freedom of Conscience: History and Theory. Moscow, 1991.

331. Saidov A.Kh. Generally Recognized Human Rights: Textbook / A.Kh. Saidov; ed. by I.I. Lukashuk; Academic Law University. Moscow: MZ Press, 2004.

332. Saidov A.Kh. Comparative Law: The Main Legal Systems of the Contemporary World: Textbook / ed. by V.A. Tumanov. Moscow: Yurist, 2003.

333. Samovich Yu.V. The International Mechanism for Protecting Human Rights and Freedoms: Universal and European Aspects: Candidate of Legal Sciences Dissertation. Yekaterinburg, 1999. 217 p.

334. Samovich Yu.V. Formation of the Institution for the Protection of Individual Rights and Freedoms in International Law // International Public and Private Law. 2006. No. 1. pp. 2-5.

335. Freedom of Conscience in the Spiritual Revival of the Fatherland. Moscow, 1994.

336. Freedom of Conscience in Contemporary Russia. Moscow, 2017. 200 p.

337. Freedom of Conscience, Religion and Belief: Problems of Theory and Practice. Russian Documents: Collection of Normative Legal Acts / compilers S.A. Buryanov, Yu.V. Gavrilova and S.S. Savva; ed. by S.A. Buryanov. Moscow: Moscow City University, 2011. 160 p.

338. Freedom of Conscience, Religion and Belief: Problems of Theory and Practice. International Documents: Collection of Normative Legal Acts / compilers S.A. Buryanov, Yu.V. Gavrilova and S.S. Savva; ed. by S.A. Buryanov. Moscow: Moscow City University, 2011. 122 p.

339. Freedom of Conscience, Religion and Belief: Problems of Theory and Practice. Documents of Foreign States: Collection of Normative Legal Acts / compilers S.A. Buryanov, Yu.V. Gavrilova and S.S. Savva; ed. by S.A. Buryanov. Moscow: Moscow City University, 2013. 196 p.

340. Freedom of Conscience, Religion and Belief: Problems of Theory and Practice. Collection of Judicial Practice Materials / compilers S.A. Buryanov, Yu.V. Gavrilova and S.S. Savva; ed. by S.A. Buryanov. Moscow: Moscow City University, 2013. 204 p.

341. Freedom of Conscience, Religion and Belief: Problems of Theory and Practice. Collection of Materials of State and Non-Governmental Human Rights Structures / compilers S.A. Buryanov, Yu.V. Gavrilova and S.S. Savva; ed. by S.A. Buryanov. Moscow: Moscow City University, 2014. 180 p.

342. Freedom of Conscience: Problems of Theory and Practice. Monograph / ed. by F.M. Rudinsky and S.A. Buryanov. Moscow: MIR Trading Company CJSC, 2012. 1120 p.

343. Us and Them: Tolerance, Stereotypes, Rights. Moscow: Russian Helsinki Group, 2016. 116 p.

344. Sebentsov A.E. The Principle of Freedom of Conscience and Its Implementation in the Russian Federation: State, Trends and Prospects: Abstract of Candidate of Philosophical Sciences Dissertation. St. Petersburg, 2009.

345. Freedom of Conscience: Problems of Theory and Practice. Monograph / ed. by F.M. Rudinsky and S.A. Buryanov. Moscow: TF MIR CJSC, 2012. 1120 p.

346. "Ours" and "Others": Tolerance, Stereotypes, Rights. Moscow: Russian Helsinki Group, 2016. 116 p.

347. Sebentsov A.E. The Principle of Freedom of Conscience and Its Implementation in the Russian Federation: State, Trends and Prospects: Abstract of Candidate of Philosophical Sciences Dissertation. St. Petersburg, 2009.

348. Sigalov K.E. The Sacralization of Power as an Immanent Feature of the Russian Political and Legal Tradition // Legal Traditions. Zhidkov Readings. Materials of the International Scientific Conference. Moscow, 2014. pp. 235-239.

349. Sigalov K.E., Prokuratov A.A. The Legal Environment and Features of Russian Legal Understanding in the Context of Historical Epochs // Law and Education. 2005. No. 4. pp. 246-259.

350. Silkenat D.R. The Rule of Law and Its Understanding // International Legal Doctrines: The Search for Mutual Understanding / Collection of Articles edited by E.G. Tarlo and D.V. Kravchenko. Moscow: LUM, 2013.

351. Sledzevsky I.V. The Desecularization of the World Community as a Source of Tension in International Relations and World Politics // Outlines of Global Transformations: Politics, Economics, Law. 2018. Vol. 11. No. 4. pp. 30-45. DOI: 10.23932/2542-0240-2018-11-4-30-45.

352. Sledzevsky I.V. The Civilizational Dimension of Contemporary World Development: Problems and Approaches // World Economy and International Relations. 2020. Vol. 64. No. 1. pp. 82-90. DOI: 10.20542/0131-2227-2020-64-1-82-90.

353. Smbatyan A.S. The Evolution of the International Legal Order: Universalization or Fragmentation? // Moscow University Bulletin. Series 11: Law. 2013. No. 6. pp. 48-58.

354. Smorgunov L.V. Human Prospects in a Globalizing World / ed. by V.V. Partsvaniya. St. Petersburg: St. Petersburg Philosophical Society, 2003.

355. So A.A. Constitutional and Legal Foundations of Freedom of Religion and the Activities of Religious Associations in Russia: Based on the Constituent Entities of the Russian Federation in the Northwestern Federal District: Abstract of Candidate of Legal Sciences Dissertation. St. Petersburg, 2010.

356. Contemporary Religious Life in Russia: An Attempt at Systematic Description. Vol. IV / executive editors M. Bourdeaux and S.B. Filatov. Moscow: University Book, Lotos, 2006.

357. Solntsev A.M. Selected Issues of Discrimination on Religious Grounds in the Contemporary World // Eurasian Law Journal. 2019. No. 1. pp. 25-28.

358. Stankevich G.V. The Religious Factor in the Contemporary Political Process: Abstract of Doctor of Political Sciences Dissertation. Pyatigorsk, 2012.

359. Stetskevich M.S. Freedom of Conscience: Textbook. St. Petersburg: St. Petersburg University Press, 2006.

360. Stetsko E.V. "Global Governance" and the Role of Non-Governmental Organizations in Its Formation // Society. Environment. Development (Terra Humana). 2012. No. 4.

361. Sultanov A.R. On Legal Certainty and Judicial Lawmaking. Electronic resource. URL: http://www.juristlib.ru/book_3659.html (accessed: 16.01.2018).

362. Theory of State and Law / executive editors V.M. Korelsky and V.D. Perevalov. Yekaterinburg, 1996.

363. Theory of State and Law: Textbook. 3rd expanded and supplemented edition / ed. by M.N. Marchenko. Moscow: Zertsalo Publishing House, 2000. 624 p.

364. Theory of State and Law: Textbook / ed. by N.M. Chepurnova and A.V. Seregin. Moscow: Eurasian Open Institute, 2007. 465 p.

365. Teteryatnikov N.Yu. Freedom of Conscience as a Subjective Right and a Legal Institution: Theory, History and Modernity. Krasnoyarsk, 2005.

366. Teteryatnikov N.Yu. Theoretical and Legal Aspects of Freedom of Conscience: Abstract of Candidate of Legal Sciences Dissertation. Yekaterinburg, 2002.

367. Tikhonova E.V. The Institution of Freedom of Conscience and Freedom of Religion in the Law of Contemporary Russia: Abstract of Candidate of Legal Sciences Dissertation. Tambov, 2006.

368. Tikhonravov Yu.V. Judicial Religious Studies. Moscow, 1998.

369. Trofimova A.Kh. Globalization: Processes of Formation and Development. Privatization and Globalization // Legal World. 2011. No. 10.

370. Tunkin G.I. Theory of International Law / general editor L.N. Shestakov. Moscow: Zertsalo, 2006.

371. Whitehead A.N. Selected Works on Philosophy. Moscow, 1990.

372. Ursul A.D. V.I. Vernadsky and the Noospheric Prospects of the Global Direction of Science // URL: http://gigabaza.ru/doc/41510.html (accessed: 17.07.2019).

373. Ursul A.D., Ursul T.A. Education for Sustainable Development: First Results, Problems and Prospects // Sociodynamics. 2015. No. 1. pp. 11-74.

374. Utyashev M.M. Human Rights and Freedoms in the System of Universal Human Values // Bulletin of Bashkir University. 2009. Vol. 14. No. 3-1. pp. 1244-1250.

375. Farkhutdinov I.Z. The American Doctrine of Preventive Military Strike: International Legal Aspects. Moscow, 2017. 338 p.

376. Farkhutdinov I.Z. The American Doctrine of Preventive Strike from Monroe to Trump: International Legal Aspects. Moscow, 2018. 419 p.

377. Farkhutdinov I.Z. The Israeli Doctrine of Preventive Self-Defense and International Law // Eurasian Law Journal. 2016. No. 8. pp. 15-32.

378. Farkhutdinov I.Z. The Iranian Doctrine of Preventive Self-Defense and International Law // Eurasian Law Journal. 2017. No. 1. pp. 15-26.

379. Farkhutdinov I.Z. The Iranian Doctrine of Preventive Self-Defense and International Law (Conclusion) // Eurasian Law Journal. 2017. No. 2. pp. 15-25.

380. Farkhutdinov I.Z. International or Global Law // International Lawyer. All-Russian Journal of International Law. 2004. No. 4. pp. 15-23.

381. Farkhutdinov I.Z. International Investment Law: Theory and Practice of Application / I.Z. Farkhutdinov. Moscow: Wolters Kluwer, 2005.

382. Farkhutdinov I.Z. International Law under Conditions of Globalization // Law and Politics. 2003. No. 8. pp. 141-149.

383. Farkhutdinov I.Z. International Law and the U.S. Doctrine of Preventive Self-Defense // Eurasian Law Journal. 2016. No. 2. pp. 23-31.

384. Farkhutdinov I.Z. International Law on the Use of Military Force by a State against Non-State Actors // Eurasian Law Journal. 2016. No. 7 (98). pp. 59-76.

385. Farkhutdinov I.Z. International Law on the Principle of Non-Use of Force or Threat of Force: History and Modernity // Eurasian Law Journal. 2015. No. 11 (90). pp. 34-38.

386. Farkhutdinov I.Z. International Law on the Self-Defense of States // Eurasian Law Journal. 2016. No. 1. pp. 91-100.

387. Farkhutdinov I.Z. International Legal Regulation of Investment Relations: Theory and Practice: Doctor of Legal Sciences Dissertation. Moscow, 2006. 390 p.

388. Farkhutdinov I.Z. The Evolution of International Law: From Westphalia to Versailles: Monograph. Moscow: INFRA-M, 2024. 446 p. (Scientific Thought). DOI: 10.12737/2135819.

389. Farkhutdinov I.Z. The Evolution of International Law from Versailles to Munich (1919-1938): Myths and Realities: Monograph. Moscow: INFRA-M, 2025. 496 p. (Scientific Thought). DOI: 10.12737/2188100.

390. Fedotova Yu.E. On Criminal Liability for Violating the Principle of State Secularity // Science and Religion in a Secular State. Moscow, 2017. pp. 62-64; Fedotova Yu.E. Freedom of Conscience and Religion as an Object of Criminal-Law Protection: Candidate of Legal Sciences Dissertation. Yekaterinburg. 2018. 229 p.

391. Fedotova Yu.E. Problems of Defining the Concept of "Religion" and Related Concepts in Russian Legislation // Russian Juridical Journal. 2017. No. 4. pp. 20-23.

392. Filatov S. Contemporary Russia and Sects // Foreign Literature. 1996. No. 8.

393. Filatov S., Shchipkov A. Non-Traditional Religiosity in Post-Communist Russia // Truth and Life. 1997. No. 5-7.

394. Philosophy: Encyclopedic Dictionary. Content / ed. by A.A. Ivin. Moscow: Gardariki, 2004. 1072 p.

395. Fromm E. Psychoanalysis and Religion // To Have or to Be? Moscow, 1990.

396. Khizhnyak V.S. The Constitutional and Legal Mechanism of Interaction between the Domestic Law of the Russian Federation and International Law: Doctor of Legal Sciences Dissertation. Saratov, 2008. 476 p.

397. Khlestov O.N. Globalization and the Fight against International Terrorism // International Lawyer. All-Russian Journal of International Law. 2003. No. 2. pp. 24-28.

398. Khlestov O.N. The USSR, Russia and the World Order: 70 Years of Service to the Fatherland / O.N. Khlestov. 4th ed. Moscow: Dashkov and Co. Publishing and Trading Corporation, 2022. 171 p.

399. Hordetsky B. On Problems with the Opacity of the Concept of "Freedom of Conscience and Religion" in United Nations Documents // Russian Political Science. 2016. No. 1. pp. 137-146.

400. Khorin I.S. The Globalization of Social Life: History and Modernity // Knowledge. Understanding. Skill. 2005. No. 2.

401. Tsivatyi V.G. Legal Education and Law under Transformational Conditions of Polycentricity: Globalization and Institutional Discourses // Russian and International Law: The Common and the Particular. Materials of the All-Russian Scientific and Practical Conference in Memory of Professor F.M. Rudinsky. 2019. pp. 370-375.

402. Chernikova N.V. Concepts of Global Education and Features of Their Implementation in the Practice of Russian Schools: Abstract of Candidate of Pedagogical Sciences Dissertation. Rostov-on-Don, 2005. 167 p.

403. Chernichenko S.V. Outlines of International Law: General Issues. Moscow: Scientific Book, 2014. 592 p.

404. Chernichenko S.V. The Individual and International Law. Moscow: International Relations, 1974. 166 p.

405. Chernichenko S.V. International Law: Current State and Prospects // Eurasian Law Journal. 2011. No. 1 (32). pp. 13-15.

406. Chernichenko S.V. Theory of International Law: In Two Volumes. Moscow: NIMP, 1999. Vol. 1: Contemporary Theoretical Problems. 336 p.; Vol. 2: Old and New Theoretical Problems. 526 p.

407. Chetvernin V.A. Introduction to the General Theory of Law and State. Moscow: Institute of State and Law, Russian Academy of Sciences, 2003. 204 p.

408. Cheshkov M.A. Global Studies as Scientific Knowledge: Essays on Theory and the Categorical Apparatus. Moscow: Scientific and Educational Forum on International Relations, 2005.

409. Cheshkov M.A. Global Studies: Subject, Problems and Prospects // Social Sciences and Contemporary World. 1998. No. 2.

410. Chumakov A.N. The Global World: The Problem of Governance // Age of Globalization. 2010. No. 1. pp. 3-15.

411. Chumakov A.N. Globalization: Outlines of an Integral World: Monograph. 3rd revised and supplemented ed. Moscow: Prospekt, 2017. 448 p.

412. Chumakov A.N. The Metaphysics of Globalization: A Cultural and Civilizational Context. Monograph. 2nd revised and supplemented ed. Moscow: Prospekt, 2017. 496 p.

413. Shakhov M.O. Russia Is Not Threatened by Clericalization: The Moscow Patriarchate Merely Creates the Appearance of Its Influence // NG-Religions. February 19, 2003.

414. Shakhov M.O. Conceptual Problems in Forming State Policy on Religion // Religion and Law. 2001. No. 1. pp. 4-6; On the Principles of State Support for the Activities of Religious Organizations // Religion and Law. 2001. No. 4. pp. 28-30.

415. Schwab K. Shaping the Fourth Industrial Revolution. Eksmo, 2018. 320 p.

416. Schwab K. The Fourth Industrial Revolution. Eksmo, 2016. 208 p.

417. Shmakova O.A. Institutions of Religious Legal Policy in a Contemporary Secular State: Abstract of Candidate of Legal Sciences Dissertation. Rostov-on-Don, 2008.

418. Shutova O.S. A Structural Model of State Policy in the Field of Relations with Religious Associations: Abstract of Candidate of Philosophical Sciences Dissertation. Moscow, 2004.

419. Ekimov A.I. The "Convergent Model" of Social Development and Law // Law and Contemporary States. 2015. No. 4. pp. 11-18.

420. Ekimov A.I. On the Contemporary Model of Development of Russian Society // Economy. Law. Society. 2016. No. 4 (8). pp. 23-29.

421. Expert Opinion on the Federal Law "On Freedom of Conscience and Religious Associations" // Religion and Law. 1997. No. 1. pp. 21-23.

422. Elementary Foundations of the General Theory of Law. Moscow, 2003.

423. Encyclopedia of the Lawyer. Moscow, 2005. Electronic resource. URL: http://slovari.yandex.ru (accessed: 30.01.2018).

424. Yusipova I.V. Law and Language in the Mechanism for Ensuring Human Rights: Abstract of Candidate of Legal Sciences Dissertation. Vladimir, 2008.

425. Yakunin G.P. On the Contemporary Situation of the Russian Orthodox Church and the Prospects for Religious Revival in Russia: Report of the Christian Committee for the Defense of the Rights of Believers in the USSR / priest Gleb Yakunin. Frankfurt am Main: Posev, 1979. 143 p.

426. Yashkova T.A. Political Modernization under Conditions of Global Transformational Challenges: Monograph. Moscow: MAKS Press, 2007. 255 p.

427. Alegre S. Freedom to Think: The Long Struggle to Liberate Our Minds. London: Atlantic Books, 2022. 320 p.

428. Alegre S. Freedom to Think: The Long Struggle to Liberate Our Minds. London: Atlantic Books, 2023. 400 p.

429. Alegre S. Protecting Freedom of Thought in the Digital Age // CIGI Policy Brief. 2021. No. 165. 7 p.

430. Alegre S. Rethinking Freedom of Thought for the 21st Century // European Human Rights Law Review. 2017. No. 3. pp. 221-233.

431. Alegre S. Toward an International Agenda to Protect the Forum Internum // CIGI Policy Brief. 2022. No. 176. 8 p.

432. Anderson J., Rainie L. Building a Human Resilience Infrastructure for the AI Age // Pew Research Center. 2024. 147 p.

433. Barfield W., Pagallo U. Advanced Introduction to Law and Artificial Intelligence // Elgar Advanced Introductions. Northampton: Edward Elgar Publishing, 2020. 208 p.

434. Beck U. What Is Globalization? Frankfurt am Main, 1998. pp. 28-44.

435. Berger M. Religion and Islam in Contemporary International Relations. The Hague: Netherlands Institute of International Relations, 2010.

436. Berger P. The Desecularization of the World: Resurgent Religion and World Politics. Washington: Ethics and Public Policy Center, 1999.

437. Burianov M. Global Digital Human Rights in the Implementation of the Global Digital Compact: The Toolkit for Human-Centered GovTech, AI and Global Governance. United Humans Press, 2025. 380 p.

438. Buryanov M. Global Digital Human Rights in the Implementation of the Global Digital Compact. Kindle Edition, 2024. 491 p.

439. Buryanov S.A. State Worldview Neutrality in the Context of Deteriorating Imbalances in Globalization: The Case and Current State of Affairs in the Russian Federation from 2016 to the Beginning of 2017. Amsterdam, 2018. 228 p.

440. Dawson S. The Religious Resurgence and International Relations Theory // Religious Studies Review. 2013. No. 39. pp. 201-221.

441. De Zayas A., Möller J.Th. The United Nations Human Rights Committee Case Law 1977-2008: A Handbook. Kehl am Rhein: N.P. Engel Publishers, 2009.

442. Drinan R.F. Can God and Caesar Coexist? Balancing Religious Freedom and International Law. New Haven: Yale University Press, 2004. 266 p.

443. Durand C. How Silicon Valley Unleashed Techno-Feudalism: The Making of the Digital Economy / C. Durand; translated by D. Broder. Brooklyn; London: Verso, 2024. 240 p.

444. Farahany N.A. Cognitive Liberty and the Need for New Human Rights Protections. Submission to the UN Human Rights Council Advisory Committee. Geneva: OHCHR, 2023.

445. Floridi L. Augmented Intelligence: A Guide to IT for Philosophers. Rome: Armando, 1996. 152 p.

446. Floridi L. Internet: An Epistemological Essay. Milan: Il Saggiatore, 1997. 188 p.

447. Floridi L. Open Problems in the Philosophy of Information // Metaphilosophy. 2004. Vol. 35. No. 4. pp. 554-582.

448. Floridi L. Philosophy and Computing: An Introduction. London; New York: Routledge, 1999. 256 p.

449. Floridi L. The Blackwell Guide to the Philosophy of Computing and Information. Oxford: Blackwell, 2003. 378 p.

450. Floridi L. The Philosophy of Information: Ten Years Later // Metaphilosophy. 2010. Vol. 41. No. 3. pp. 420-442.

451. Fox J. Religion as an Overlooked Element of International Relations // International Studies Review. 2001. Vol. 3. No. 3. pp. 53-73.

452. Gawdat M. Scary Smart: The Future of Artificial Intelligence and How You Can Save Our World / M. Gawdat. London: Bluebird, 2021. 338 p.

453. Ghanea N. Freedom of Religion or Belief and the Right to Life: Report of the Special Rapporteur: A/HRC/52/38. Geneva: UN, 2023. 25 p.

454. Ghanea N. Human Rights, the UN and the Bahá'ís in Iran. Liverpool: Liverpool University Press, 2021. 656 p.

455. Ghanea N., Bielefeldt H., Wiener M. Freedom of Religion or Belief: An International Law Commentary. Oxford: Oxford University Press, 2016. 640 p.

456. Giddens A. The Consequences of Modernity. Stanford: Stanford University Press, 1990. 186 p.

457. Gorski P.S. The Post-Secular in Question: Religion in Contemporary Society. New York: Social Science Research Council, 2012.

458. Haynes J. Religion, Politics, and International Relations: Change and Continuity // Global Policy. 2012. No. 3. pp. 251-252.

459. Harari Y.N. Nexus: A Brief History of Information Networks from the Stone Age to AI / Y.N. Harari. London: Fern Press, 2024. 528 p.

460. Hurd E.Sh. Beyond Religious Freedom: The New Global Politics of Religion. Princeton: Princeton University Press, 2015.

461. Hurd E.Sh. The Politics of Secularism in International Relations. Princeton: Princeton University Press, 2009.

462. Human Rights in the Digital Domain: Core Questions / ed. by T. Pajuste. Cambridge: Cambridge University Press, 2025. 492 p.

463. Levitt T. The Globalization of Markets // Harvard Business Review. May-June 1983. No. 3. pp. 2-11.

464. Muñoz J.M., Marinaro J.A. Neurorights as Reconceptualized Human Rights // Frontiers in Political Science. 2023. Vol. 5. Article 1322922.

465. Nikolaev A.M., Folosyan M.E. Freedom of Conscience and Religion in the System of Individual Rights and Freedoms of Human and Citizen in the Russian Federation // Contemporary Problems of Social Work. 2015. No. 1. pp. 107-111.

466. Ohmae K. The End of the Nation State: The Rise of Regional Economies. Free Press, 1995. 214 p.

467. Ohmae K. The Next Global Stage: Challenges and Opportunities in Our Borderless World. Wharton School Publishing, 2005. 312 p.

468. Pajuste T. (ed.). Human Rights in the Digital Domain: Core Questions. Cambridge: Cambridge University Press, 2025. 492 p.

469. Roberts T., Oosterom M. Digital Authoritarianism: A Systematic Literature Review // Information Technology for Development. 2024. Vol. 30. No. 5. pp. 860-884.

470. Robertson R., Lechner F. Modernization, Globalization and the Problem of Culture in World-Systems Theory // Theory, Culture & Society. 1985. No. 2. pp. 103-117.

471. Romero O.G. Cognitive Liberty and the Psychedelic Humanities // Frontiers in Psychology. 2023. Vol. 14. Article 1128996.

472. Rosenau J.N., Czempiel E.-O. Governance Without Government: Order and Change in World Politics. Cambridge: Cambridge University Press, 1992. 311 p.

473. Sandal N.A., Fox J. Religion in International Relations Theory: Interactions and Possibilities. New York: Routledge, 2013.

474. Sklair L. Sociology of the Global System. Hemel Hempstead, 1991. pp. 6-8.

475. Slaughter A.-M. A New World Order. Princeton; Oxford: Princeton University Press, 2004. 341 p.

476. Snyder J.L. Religion and International Relations Theory. New York: Columbia University Press, 2011.

477. Spichak S. The Controversial Push for New Brain and Neurorights // Journal of Medical Internet Research. 2025. Vol. 27. Article e72270.

478. Varoufakis Y. Technofeudalism: What Killed Capitalism / Y. Varoufakis. London: Bodley Head, 2023. 304 p.

479. Yakunin G., Regelson L. Letters from Moscow: Religion and Human Rights in the USSR. Keston College, 1978. 157 p.

480. Yuste R., Quadra-Salcedo T., García Fernández M. Neuro-Rights and New Charts of Digital Rights: A Dialogue Beyond the Limits of the Law // Indiana Journal of Global Legal Studies. 2023. Vol. 30. Issue 1. pp. 15-37.

481. Zuboff S. The Age of Surveillance Capitalism: The Fight for a Human Future at the New Frontier of Power. New York: PublicAffairs, 2019. 704 p.

FINAL PAGE OF THE BOOK

BURYANOVS GLOBAL LAW FORUM

Other Volumes in the Series:

ANTHROPOLOGY OF THE PLANETARY HUMAN-CENTERED CIVILIZATION 4.0

The monograph Averting Civilizational Armageddon: In Search of a Way Out of the Geopolitical Deadlock. Worldview Liberty and Public-Law Worldview Neutrality in an Age of Global Existential Threats is part of a long-term research project devoted to the protection of human identity in an age of global challenges.

We invite the academic community, experts from international organizations, and public servants to continue the discussion on the future of global law and the protection of human identity.

www.maxlaw.tilda.ws/globalization4ir

[QR CODE of website]

ALSO PUBLISHED BY UNITED HUMANS PRESS

Buryanov, M. Global Digital Human Rights in the Implementation of the Global Digital Compact: The Toolkit for Human-Centered GovTech, AI, and Global Governance. Geneva: United Humans Press, 2025. 350 p.

FORTHCOMING FROM UNITED HUMANS PRESS (2026-2030)

Buryanov, S., Buryanov, M. Global Reproduction of Knowledge as a Flagship for Preventing Civilizational Armageddon and Ensuring Human-Centric Sustainable Development: The Future of Education Amid the Latest Challenges of the Fourth Industrial Revolution.

Buryanov, S., Buryanov, M. Taming the Digital Leviathan: Global Human Rights as a Civilizational Security Code and the Foundation of a Sustainable, Human-Centered World Order.

Buryanov, S., Buryanov, M. Legal Globalization 4.0: A Human-Centered Turn in Legal Science as a Tool for Preventing Civilizational Armageddon.

Buryanov, S., Buryanov, M. The Human-Centered Future of International Law, Which Should Have Arrived Yesterday: From Geopolitical Deadlock to Sustainable Managed Development.

Buryanov, S., Buryanov, M. Global Governance 4.0 as an Imperative for Preventing Civilizational Collapse. Is It Too Late to Speak About a Human-Centered World Order?

Buryanov, M. Digital Transformation of the State 4.0. Designing a Human-Centered Strategy for Overcoming Technocratic Authoritarianism and Securing Individual Sovereignty.

www.ingramcontent.com/pod-product-compliance
Lightning Source LLC
LaVergne TN
LVHW040516210926
848480LV00007BA/1484